Secondary Metabolite Medicinal Plants

Medicinal plant-based synthesis of nanoparticles from various extracts is easy, safe, and eco-friendly. Medicinal and herbal plants are the natural source of medicines, mainly due to the presence of secondary metabolites, and have been used as medicine since ancient times. *Secondary Metabolites from Medicinal Plants: Nanoparticles Synthesis and their Applications* provides an overview on medicinal plant-based secondary metabolites and their use in the synthesis of different types of nanoparticles. It explores trends in the growth, characterization, properties, and applications of nanoparticles from secondary metabolites including terpenoids, alkaloids, flavonoids, and phenolic compounds. It also explains the opportunities and future challenges of secondary metabolites in nanoparticle synthesis.

Nanotechnology is a burgeoning research field, and due to its widespread application in almost every branch of science and technology, it creates many new opportunities. As part of the *Exploring Medicinal Plants* series, this book will be of huge benefit to plant scientists and researchers as well as graduates, postgraduates, researchers, and consultants working in the field of nanoparticles.

Exploring Medicinal Plants

Series Editor
Azamal Husen
Wolaita Sodo University, Ethiopia

Medicinal plants render a rich source of bioactive compounds used in drug formulation and development; they play a key role in traditional or indigenous health systems. As the demand for herbal medicines increases worldwide, supply is declining as most of the harvest is derived from naturally growing vegetation. Considering global interests and covering several important aspects associated with medicinal plants, the Exploring Medicinal Plants series comprises volumes valuable to academia, practitioners, and researchers interested in medicinal plants. Topics provide information on a range of subjects including diversity, conservation, propagation, cultivation, physiology, molecular biology, growth response under extreme environment, handling, storage, bioactive compounds, secondary metabolites, extraction, therapeutics, mode of action, and healthcare practices.

Led by Azamal Husen, PhD, this series is directed to a broad range of researchers and professionals consisting of topical books exploring information related to medicinal plants. It includes edited volumes, references, and textbooks available for individual print and electronic purchases.

Sustainable Uses of Medicinal Plants, *Learnmore Kambizi and Callistus Bvenura*

Omics Studies of Medicinal Plants, *Ahmad Altaf*

Traditional Herbal Therapy for the Human Immune System, *Azamal Husen*

Environmental Pollution and Medicinal Plants, *Azamal Husen*

Herbs, Shrubs and Trees of Potential Medicinal Benefits, *Azamal Husen*

Phytopharmaceuticals and Biotechnology of Herbal Plants, *Sachidanand Singh, Rahul Datta, Parul Johri, and Mala Trivedi*

Exploring Poisonous Plants: Medicinal Values, Toxicity Responses, and Therapeutic Uses, *Azamal Husen*

Plants as Medicine and Aromatics: Conservation, Ecology, and Pharmacognosy,*Mohd Kafeel, Ahmad Ansari, Bengu Turkyilmaz Unal, Munir Ozturk, and Gary Owens*

Medicinal Plant Responses to Stressful Conditions, *Arafat Abdel Hamed Abdel Latef*

Aromatic and Medicinal Plants of Drylands and Deserts: Ecology, Ethnobiology and Potential Uses, *David Ramiro Aguillón Gutiérrez, Cristian Torres León, and Jorge Alejandro Aguirre Joya*

Secondary Metabolites from Medicinal Plants: Nanoparticles Synthesis and their Applications, *Rakesh Kumar Bachheti* and *Archana Bachheti*

Secondary Metabolites from Medicinal Plants

Nanoparticles Synthesis and their Applications

Edited by

Rakesh Kumar Bachheti and Archana Bachheti

CRC Press is an imprint of the
Taylor & Francis Group, an informa business

First edition published 2023
by CRC Press
6000 Broken Sound Parkway NW, Suite 300, Boca Raton, FL 33487-2742

and by CRC Press
4 Park Square, Milton Park, Abingdon, Oxon, OX14 4RN

CRC Press is an imprint of Taylor & Francis Group, LLC

ISBN: 9781032075150 (hbk)
ISBN: 9781032100999 (pbk)
ISBN: 9781003213727 (ebk)

DOI: 10.1201/9781003213727

Typeset in Times
by Deanta Global Publishing Services, Chennai, India

Contents

Preface

Nanotechnology is a promising interdisciplinary research field. Because of its widespread application in almost every branch of science and technology, it creates new opportunities in a variety of fields. Medicinal plant-based syntheses of nanoparticles from various extracts are easy, safe, and eco-friendly. Medicinal and herbal plants are the natural source of medicines, mainly due to the presence of secondary metabolites, and have been used as medicine since ancient times. In addition to serving as competitive weapons against other bacteria, fungi, amoebas, plants, insects, and large animals, secondary metabolites also act as metal transporters, symbiotic agents between microbes and plants, nematodes, insects, and higher animals. Medicinal and herbal plants can produce a vast and diverse group of secondary metabolites. Most of the time, pure isolated bioactive compounds are more biologically active, hence researchers and scientists are focusing their research on the synthesis of nanoparticles using a particular class of secondary metabolites such as terpenoids, alkaloids, and phenolic compounds such as flavonoids.

This book consists of 16 chapters highlighting specifically medicinal and herbal plant-based secondary metabolites in synthesizing different types of nanoparticles and their potential applications. These chapters discuss the role of a variety of molecules, including proteins and different low-molecular-weight substances like terpenoids, alkaloids, amino acids, alcoholic substances, and polyphenols in nanoparticle synthesis.

We are grateful not only to those colleagues who kindly agreed to contribute chapters to this volume but also to those who assisted us in reviewing these contributions. We also thank the reviewer who provided specific assistance in vetting and finalizing the manuscripts. We are grateful to Prof. Azamal Husen, the Series Editor for his tireless overall help at every step in preparing this book.

Rakesh Kumar Bachheti
Archana Bachheti

Editors

Rakesh Kumar Bachheti graduated from the Hemwati Nandan Bahuguna Garhwal University (a Central university), India, in 1996. He completed his MSc in Organic Chemistry from Hemwati Nandan Bahuguna Garhwal University, India, in 1998. He completed a one-year Post Graduate Diploma in Pulp and Paper Technology from Forest Research Institute, India, in 2001. He obtained his Ph.D. in Organic Chemistry from Kumaun University, India, in 2007. He is presently working as an Associate Professor of Organic Chemistry in the Department of Industrial Chemistry at the Addis Ababa Science and Technology University (AASTU) of Ethiopia, where he teaches Ph.D., graduate, and undergraduate students. Before joining AASTU, Rakesh worked as Dean Project (Assistant) at Graphic Era University (A grade university by NACC), India. Rakesh also presented papers at international (Malaysia, Thailand, and India) and national conferences. He was also a member of important committees such as the Internal Quality Assurance Cell (IQAC), Anti-ragging Committee. His major research interests include natural products for industrial applications, biofuel and bioenergy, green synthesis of nanoparticles and their application, and pulp and paper technology. He retains a fundamental love for natural products, which permeates all of his research. He has also successfully advised 30 MSc and three Ph.D. students to completion, and countless undergraduates have researched in his laboratory. Dr Bachheti is actively involved in curriculum development for BSc/MSc/Ph.D. programmes. He has over 75 publications dealing with various aspects of natural product chemistry and nanotechnology and has 18 book chapters published by Springer, Elsevier, and Nova Publishing. Presently, he is supervising three Ph.D. students, three Master's students, and is also working on two research projects funded by AASTU.

Archana (Joshi) Bachheti earned a BSc in 1997 and MSc in 1999 from H.N.B. Garhwal University, India. She received her Ph.D. from Forest Research Institute, India, in 2006. She has carried out research projects and consultancy work in the areas of eco-restoration/development of wasteland, physico-chemical properties of *Jatropha curcus* seed oil and its relation to altitudinal variation and has been a Consultant Ecologist to a project funded by a government agency. Dr Joshi is currently a Professor at Graphic Era University, India. She has also served in many capacities in academia within India and provided expertise internationally for more than 15 years where she taught Ecology and Environment, Environmental Science, Freshwater Ecology, Disaster Management, and Bryophytes and Pteridophytes. Her major research interests encompass the broad, interdisciplinary field of plant ecology, with a focus on eco-restoration, green chemistry, especially the synthesis of nanomaterial, and medicinal properties of plants. The breadth of her research spans from the ecological amelioration of degraded land and the physical and chemical properties of plant oils, to plant-based nanomaterial. She has guided one PhD student and is presently supervising three scholars as well as guiding graduate and undergraduate students in their research projects. While it was the fascination with forest biodiversity that captured her interest, it has been her love for the exploration of values of biodiversity and social upliftment that has maintained that passion. Dr Joshi has published more than 50 research articles in international and national journals along with 10 book chapters. She has organized several national seminars/conferences at Graphic Era University, India.

1 Medicinal Plant-Based Metabolites in Nanoparticles Synthesis and Their Cutting-Edge Applications

An Overview

Yakob Godebo Godeto, Abate Ayele, Ibrahim Nasser Ahmed, Azamal Husen, and Rakesh Kumar Bachheti

CONTENTS

DOI: 10.1201/9781003213727-1

1.1 INTRODUCTION

As a rapidly emerging technology and new research area in applied sciences and engineering, nanotechnology lays a strong foundation for a technological revolution and transformation in various sectors. Nanotechnology is a popular area of contemporary research and provides significant contributions to the industrial, agricultural, information technology, energy, environmental, food, and health sectors by producing nanomaterials from nanosized particles (particles with diameters of less than 100 nm) with novel properties and functions (Bachheti et al. 2021; 2021a). These and other promising benefits of nanoparticles, from domestic to industrial sectors, are likely due to their unique and novel properties such as particle size, surface area, reactivity, surface charge, and shape compared to their bulk counterparts (Bundschuh et al. 2018). The size and surface properties determine the solubility, absorption, emission, conductivity, reactivity, and optical behaviour of synthesized nanoparticles. These properties can be manipulated by altering the initial size of the source material (Koul et al. 2018; Bachheti et al. 2020). Nanoparticles with various outstanding properties and functions have been synthesized physically, chemically, and biogenically following two major approaches (i.e., bottom-up and top-down approaches) (Rafique et al. 2017; Khan et al. 2019; Ijaz et al. 2020; Jadoun et al. 2021).

In the bottom-up synthesis approach (such as vapour deposition, sol-gel, aerosol, chemical vapour deposition (CVD), laser pyrolysis, and the biological synthesis process), materials are built up from atom to cluster and then to nanoparticles. In contrast, in the top-down synthesis approach, crushing, splitting, and milling are used as size reduction techniques to reduce suitable materials to nanosized units (Rafique et al. 2017; Khan et al. 2019; Jadoun et al. 2021). Nanoparticles' physical, chemical, and biological synthesis processes are categorized under these two approaches. Nanoparticles with different sizes, shapes, and surface properties have been synthesized physically through CVD, the plasma method, microwave irradiation, pulsed laser ablation, ball milling, spray pyrolysis, and gamma radiation; chemically through chemical reduction, photochemical, polyol, microemulsion, thermal decomposition, electrochemical, pyrolysis, solvothermal, and coprecipitation; and biologically by using plants, microorganisms, and biomolecules (Kuppusamy et al. 2016; Rafique et al. 2017; Sharma et al. 2019; Cele 2020).

Currently, the interest of many researchers is shifting from conventional techniques (chemical and physical pathways) to biogenic synthesis as a greener approach. This is mainly because the biological process is known to be energy efficient, economical, environmentally benign, and results in a safe product, unlike with the physical and chemical techniques using toxic synthetic chemicals, high energy, and highly concentrated reducing and stabilizing agents that are harmful to the environment and human health (Kuppusamy et al. 2016; Marslin et al. 2018; Ijaz et al. 2020). Nowadays, nanoparticles of different sizes, shapes, and surface properties are synthesized employing various biological components such as reducing, capping, and stabilizing agents to overcome the drawbacks of conventional methods.

The application of green chemistry to synthesize nanoparticles using extracts and/or components of living and dead plant biomass (Jadoun et al. 2021), bacteria (Mukherjee and Nethi 2019), fungi (Neethu et al. 2019), microalgae (Jacob et al. 2021), yeasts (Skalickova et al. 2017), proteins, and enzymes (Hong et al. 2020; Arib et al. 2021) is gaining popularity due to their sufficient availability, low cost, biocompatibility, biodegradability, and lower environmental and health toxicity. Of the aforementioned biological entities, plants are abundantly available natural resources with varieties of diversity. They are rich in natural compounds such as flavonoids, alkaloids, saponins, tannins, steroids, and other nutritional compounds. These natural metabolites are derived directly or indirectly from various parts of plants such as roots, stems, barks, leaves, shoots, flowers, seeds, and nuts. These parts of plants participate in nanoparticle synthesis intracellularly, extracellularly, or reduction comprising their individual phytoconstituent isolates (Dauthal and Mukhopadhyay 2016) as indicated in Figure 1.1

Unlike the microbial route, which involves a multi-step process including the isolation of potential microbes, maintenance of culture, specific culture preparation, and subculturing, biological synthesis of nanoparticles using a medicinal plant is easy to scale up for large-scale production

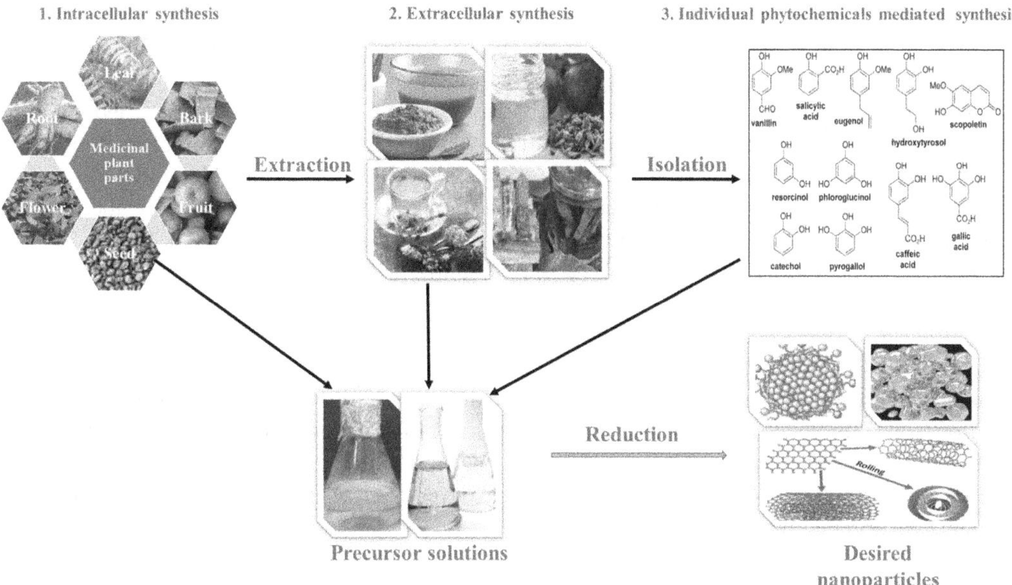

FIGURE 1.1 Medicinal plant-mediated synthesis of nanoparticles (intracellularly, extracellularly, or using individual phytoconstituents).

and does not require any complex methodologies (Husen 2019; Bachheti 2019a; Chandra et al. 2020). The major role of utilizing various parts of medicinal plants (roots, stems, barks, leaves, shoots, flowers, seeds, and nuts) in nanoparticles synthesis is the use of their primary and secondary metabolites as reducing as well as capping and stabilizing agents (Ovais et al. 2018; Aritonang et al. 2019; Bachheti et al. 2020a, 2020b; Chandra et al. 2020). Plants possess thousands of small to more complex organic molecules used in traditional, folk, and modern medicine for healing different ailments either directly or indirectly. They possess carbohydrates, proteins, and lipids as a primary metabolite (Zaynab et al. 2019) and phenolics, alkaloids, polyphenols, quinones, tannins, flavonoids, terpenes, coumarins, lectins, and polypeptides, saponins, etc. as secondary metabolites (Alamgir 2017; Anand et al.2019).

Currently, numerous nanomaterials and nanocomposites such as metal and metal oxide (Küünal et al. 2018; Yadi et al. 2018; Bachheti et al. 2020), carbon-based (Zare et al. 2020), polymeric (Roy et al. 2014; Tripathi et al. 2019), lipid-based (Fernandes et al. 2021), and other nanomaterials have been synthesized biologically using medicinal plant extracts for domestic, environmental, industrial, biomedicinal, and other applications. These metabolites have been extracted from different parts of the plant by applying various traditional and modern methodologies to synthesize nanoparticles, a major building block of nanomaterials. Hence, this review work provides an overview of medicinal plant metabolites, their role in synthesizing various nanoparticles, and their cutting-edge applications of resulting nanomaterials, especially in the case of human health and environmental protection.

1.2 MEDICINAL PLANTS AND THEIR METABOLITES

1.2.1 Medicinal Value of Plants

From the initial hunter-gather stage to several adaptation stages, plants have played incredible roles in human livelihoods. Almost all of the world relies on them, particularly for shelter, food, and medicines (Beshah et al. 2020; Godeto et al. 2021; Abate et al. 2022; Asfaw et al. 2022). Medicinal plants as a source of treatment for ailments can be dated back to the earliest civilizations in China, India, and the Near East. Plant-based ingredients provide a wide range of medicinal properties. Herbal medicine is used by 60–80% of people worldwide for their primary healthcare needs (Wanjohi et al. 2020). Long before pharmaceutical development, communities relied on their traditional knowledge, skills, and customary practices to prevent, diagnose, and cure health problems, utilizing various natural resources (Barata et al. 2016). Herbalism is a type of traditional or folk medicine that relies on plants and plant extracts (Pan et al. 2014). All plants that traditional and official medicine considered curative or were used for that purpose are called medicinal plants. Medicinal plants have traditionally been a valuable source of both curative and preventive medicinal therapeutic preparations for humans and a source of important bioactive components for extraction. Nearly 80% of the world's population, particularly in developing countries, depends on traditional medicine and products for its healthcare needs (Mbuni et al. 2020; Abate et al. 2021). Many individuals suffering from diseases in developing regions combine conventional medicine with traditional ones. Their application begins in the distant past because treatment with plants is as old as humanity itself. Humans needed food to survive and as they used plants in their diet they discovered their medicinal properties (Šantić et al. 2017). Many pharmaceutical medications are derived from plants that were originally used in traditional medicine systems; according to the World Health Organization (WHO), approximately 25% of these medicines are derived from plants used in traditional medicine systems (Barata et al. 2016).

1.2.2 Classes of Plant Metabolites and Their Medicinal Values

The intermediates and products of metabolism are known as metabolites (Tiwari and Rana 2015). Plants are naturally capable of producing a wide range of metabolites, and the number of metabolites in the plant kingdom is estimated to be around 1,000,000 (Yonekura-Sakakibara et al. 2019). The role of phytochemicals in plant metabolism determines whether they are categorized as primary or secondary constituents. They are naturally synthesized in all parts of the plant body, including the bark, leaves, stems, roots, flowers, fruits, seeds, and so on; in other words, active components can be found in any part of the plant (Jyothiprabha and Venkatachalam 2016). Carbohydrates, proteins, nucleic acids, purines and pyrimidines, chlorophyll, etc., are primary metabolites. While terpenes, flavonoids, alkaloids, lignans, curcumin, saponins, plant steroids, phenolic, flavonoids, glycosides, etc., are examples of secondary constituents (Saxena et al. 2013; Alzandi et al. 2021). Secondary metabolites are important sources of medicines and play a vital role in plant adaptation to their environment (Tumbde et al. 2021). Plant extracts have been used in phytomedicines since the dawn of humanity. This can be driven by bark, leaves, flowers, roots, fruits, and seeds (Yadav and Agarwala 2011).

1.2.2.1 Phenolics

Phenols are aromatic chemical compounds with weakly acidic properties that are distinguished by a hydroxyl (OH) group linked directly to an aromatic ring (Okigbo et al. 2009; Dai and Mumper 2010). It was discovered that, in addition to their primary antioxidant activity, this group of chemicals has a wide range of biological roles, the majority of which are associated with carcinogenesis modulation (Dai and Mumper 2010).

1.2.2.2 Alkaloids

Alkaloids, which are biosynthesized from amino acids like tyrosine, are poisonous chemicals that protect plants against diseases and are commercially utilized as medicine (Ebenezer et al. 2019). Initially, alkaloids are pharmacologically active, nitrogen-containing basic compounds derived from plants. They can also cause hallucinations, loss of coordination, seizures, vomiting, and death by blocking ion channels, inhibiting enzymes, or interfering with neurotransmission (Tiwari and Rana 2015). They have valuable pharmacological properties such as antimalarial (quinine), antiasthmatic (ephedrine), anticancer (homoharringtonine), vasodilatory (vincamine), antiarrhythmic (quinidine), analgesic (morphine), antibacterial (chelerythrine), and antihyperglycemic (e.g., piperine) (Umashankar 2020).

1.2.2.3 Saponins

Saponins are naturally occurring surface-active glycosides produced primarily by plants, some bacteria, and lower marine animals. The antidiabetic properties of saponins enabled physicians to discover new medications that are beneficial in treating diabetes mellitus (Marrelli et al. 2016). Saponins are amphiphilic compounds with carbohydrate, triterpenoid, or steroid aglycone moieties. Fungicidal, antibacterial, antiviral, anti-inflammatory, anticancer, antioxidant, and immunomodulatory properties are among their biological activities (Alzandi et al. 2021).

1.2.2.4 Terpenes

Terpenes, also known as isoprenoids, are the most copious and diverse group of naturally occurring compounds. They are chiefly found in plants, but larger classes of terpenes, such as sterols and squalene are also common in animals. The natural aroma, flavour, and pigment of plants are all due to this group of compounds. Among various therapeutic applications, they are known for antiplasmodial activity, which is remarkable because their mode of action is analogous to that of the widely used antimalarial medication chloroquine (Cox-Georgian et al. 2019).

1.2.2.5 Carotenoids

Plants, algae, fungi, and bacteria all contain carotenoids, which are a type of tetraterpenoid molecule. They are phytonutrients that promote health and contribute to the prevention of cardiovascular disease (such as heart disease, cancer, and diabetes), Alzheimer's disease, and other age-related diseases (Pott et al. 2019).

1.2.2.6 Tannins

Tannins are a type of polyphenol that can be divided into two categories: (1) condensed tannins (*syn.* proanthocyanidins), which are made up of flavan-3-of polymer subunits linked by 4–6 and 4–8 interflavan bonds, and (2) hydrolysable tannins, which are gallic acid esters with a central polyol, usually -D-glucopyranose (Pott et al. 2019). Tannins have a wide range of effects, from decreasing protein and other nutrient availability, such as amino acids and minerals, to protecting ruminants from bloat, improving rumen bypass protein, improving meat quality, and reducing helminth infestation (Makkar et al. 2009). Plant extracts containing tannins are employed as astringent, diuretic, anti-inflammatory, antiseptic, antioxidant, and haemostatic medicines and against stomach and duodenal tumours (Saxena et al. 2013).

1.2.2.7 Flavonoids

Flavonoids are a type of secondary metabolite found in plants made up of polyphenolic compounds. They have antioxidative, anti-inflammatory, anticarcinogenic, and antimutagenic properties. As a result, they are a key ingredient in the pharmaceutical, nutraceutical, and cosmetics industries (Karak 2019; Umashankar 2020).

1.2.2.8 Carbohydrates and Related Compounds

Carbohydrates and related chemicals derived from plants include fibres, cellulose and its derivatives, starch and its derivatives, mucilages (uronic acid-containing polymers), dextrins, fructans, pectins, and gums. They have been reported to have immune-modulatory, antitumour, hypoglycaemic, anticoagulant (e.g., heparin), and antiviral properties and their application as bulking agents in the pharmaceutical industry (Mustafa et al. 2017). Glycosides are secondary plant metabolites made up of two constituents (i.e., glycone (a carbohydrate component) and aglycone (a non-carbohydrate component)). The glycone component is typically composed of one or more glucose units, and the aglycone component may be any one of the plant's secondary metabolites from alkaloids, phenolics, or terpenoids (Mustafa et al. 2017).

1.2.2.9 Lipids

Lipids are large classes of macromolecules including, but not limited to, fixed oils, waxes, essential oils, sterols, fat-soluble vitamins (such as vitamins A, D, E, and K), phospholipids, and other naturally occurring compounds. They accomplish decisive biological tasks in the body as major structural components of all biological membranes, energy reservoirs, and fuel for cellular processes (Hussein and El-Anssary 2019).

Plants are naturally enriched with various biological compounds for self-protection, growth, and reproduction (Pang et al. 2021). Different parts of plants possess varied quantities and types of phytoconstituents. Naturally existing phytochemicals in various plant parts have participated in healing several ailments in humans and other animals directly or indirectly from the start of the human era (Asaduzzaman and Asao 2018, Godeto et al. 2021). Major medicinal properties as a promising biological function of some commonly used medicinal plants and their metabolites are given in Table 1.1. They are used in medicine individually or in combination with other compatible substances to this day.

1.3 ROLES OF PLANT METABOLITES IN NANOPARTICLES SYNTHESIS

Amongst repeatedly reported green synthesis pathways, plant-mediated synthesis of nanoparticles is the current research interest of many scientists due to their abundance, flexibility to choose any parts (leaves, stems, barks, roots, shoots, fruits, seeds, flowers, etc.); they are easy to scale up and do not require additional processes such as isolation, culturing and sub-culturing, commonly used steps in microbial-based synthesis. Extracts from different plant parts possess several compounds with hydroxides, carbonyls, methoxy, amines, amides, thiols, carboxyls, phosphates, and other functional groups that play a decisive role in reacting with precursor compounds to form nanoparticles (Küünal et al. 2018; Bachheti et al. 2019; Painuli et al. 2020). For instance, the –OH functional groups in eugenol of clove extract for the synthesis of aqueous colloidal solutions of silver nanoparticles (AgNPs) act as a reducing agent and other terpenoids that are believed to be the surface-active molecules stabilize the synthesized nanoparticle (Parlinska-Wojtan et al. 2018).

On the other hand, terpenoid compounds in the aqueous flower extract of chamomile medicinal plant (*Matricaria chamomilla*) are involved in reducing the metal ion (Ag^+) to its reduced form (Ag^0) and also act as a capping agent of synthesized AgNPs (Parlinska-Wojtan et al. 2016). Moreover, Alanazi et al. (2016) reported that protein molecules in the leaf extract of *Calendula officinalis*, popular herbal and cosmetic plants are crucial in reducing, capping, and stabilizing activities in the biosynthesis of AgNPs. This implies that the existence of phytochemicals such as flavonoids, alkaloids, phenolic acids, carotenoids, saponins, terpenoids, proteins, lipids, and carbohydrates in medicinal plants are the major driving force for exploiting plant extracts in nanoparticles synthesis. Thus, metabolites of medicinal plants, especially secondary metabolites, participate extensively in reducing the starting precursor compounds to the desired nanoparticles and preventing the particles from undesired agglomerations by capping and stabilizing.

1.3.1 Roles of Flavonoids in Nanoparticles Synthesis

Flavonoids, encompassing various polyphenolic functional groups have participated in reducing several nanoparticles. For instance, quercetin and other flavonoid compounds are involved reducing silver nitrate precursor to AgNPs in an alkaline medium and the resulting nanoparticle has been used as antimicrobial, antioxidant, and anti-inflammatory agents (Hussain et al. 2019). The key is that flavonoids, polyphenolic phytochemicals comprising flavanones, flavones, isoflavones, and poly-methoxylated flavones contribute to the synthesis of wide verities of nanoparticles (Mondal

TABLE 1.1
Some Commonly Used Medicinal Plants and Their Therapeutic Effects

S.N.	Medicinal plants	Parts used	Major metabolites (dominant compounds)	Therapeutic effects	References
1	Chamomile (*Matricaria chamomilla* L.)	Flowers, Leaves, Stems, Roots	Sesquiterpenes, flavonoids, coumarins, and polyacetylenes	Analgesic, anticancer, anti-inflammatory, antimicrobial, antioxidant, antiallergic, and activities	Singh et al. (2011); Roby et al. (2013)
2	*Bauhinia variegate* L.	Bark	Tannins, alkaloids, and saponins	Antibacterial activity	Parekh et al. (2006)
3	*Amaranthus spinosus* and *Boerhaavia erecta*	Stems	Tannins, alkaloids, saponins, polyuronides, betalains, amines, and amino acids	Antimalarial activities	Hilou et al. (2006)
4	*Allium sativum*	Bulbs	Cardiac glycosides, saponins, and terpenoids	Antibacterial activities	Pathmanathan et al. (2010)
5	*Eucalyptus citriodora*	Leaves	Tannins, alkaloids, cardiac glycosides, saponins, and terpenoid		
6	*Ocimum sanctum*	Leaves	Cardiac glycosides and tannins		
7	*Tribulus terrestris*	Leaves	Flavonoids, glycosides, alkaloids, saponin, and phenols	Antibacterial and antioxidant activities	Rehman et al. (2021)
8	*Vaccinium macrocarpon*	Leaves	Flavonoids, alkaloids, and phenols		
9	*Cuminum cyminum*	Leaves	Flavonoids, glycosides, alkaloids, saponin, phenols, and tannins		
10	*Rheum emodi*	Leaves	Flavonoids, alkaloids, phenols, and tannins		
11	*Piper cubeba*	Leaves	Flavonoids, glycosides, and phenols		
12	*Azadirachta indica*	Leaves	Saponin, phenols, flavonoids, and terpenoids	Antibacterial and antioxidant activities	Tiwari et al. (2021)
13	*Urtica parviflora*	Leaves	Phenols, flavonoids, and terpenoids		
14	*Cassia fistula*	Leaves	Saponin, phenols, flavonoids, and terpenoids		
15	*Crinum amoenum*	Leaves	Saponin and flavonoids		
16	*Aleuritopteris bicolor*	Leaves	Saponin, phenols, flavonoids, and terpenoids		
17	*Rosa indica* L.	Roots, Stems, Leaves	Flavonoids, tannin, alkaloids, and carbohydrates	Antifungal activities	Begum et al. (2021)
18	*Prunnus amygdalus L*		Flavonoids and tannin, oil, and fats		
19	*Prunus armeniaca* L.				
20	*Momordica charantia*	Fruits	Flavonoid, flavonols, and phenolic	Kidney function	Mardani et al. (2014)
21	*Azadirachta indica*	Leaves	Coumarins, alkaloids, glycosides, proteins, and saponin	Antimicrobial activities	Nurudeen and Falana (2021)
22	*Calotropis procera*	Leaves	Alkaloid, flavonoid, proteins, saponin, and terpenoid		

(Continued)

TABLE 1.1 (CONTINUED)
Some Commonly Used Medicinal Plants and Their Therapeutic Effects

S.N.	Medicinal plants	Parts used	Major metabolites (dominant compounds)	Therapeutic effects	References
23	*Carica papaya*	Leaves	Anthraquinone, alkaloids, flavonoid, glycosides, and saponin		
24	*Vernonia amygdalina*	Leaves	Anthraquinone, alkaloids, coumarin, saponin, tannin, and terpenoid		
25	*Myrtus communis* L..	Leaves	Flavonoid, alkaloids, and terpenoid	Antibacterial activities	Sharara et al. (2021)
26	*Artemisia annua*	Leaves	Polyphenols, catechins, and terpenes	Antimicrobial activities	Golbarg et al. (2021)
27	*Oxalis corniculata*	Leaves			
28	*Solenostemma argel*	Aerial parts	Proteins	Antibacterial and antioxidant activities	El-Zayat et al. (2021)
29	*Teucrium polium*	Aerial parts	Crude fats and total carbohydrates		
30	*Achillea fragrantissima*	Aerial parts	Proteins		
31	*Peganum harmala*	Aerial parts	Crude fats and total carbohydrates		
32	*Solanum virginianum*	Leaves	Phenols, flavonoids, saponins, terpenoids, steroids, and glycosides	Antibacterial activities	Gagare et al., 2021
33	*Physalis angulata*	Leaves			
34	*Mesembryanthemum crystallinum*, *Mesembryanthemum forsskaolii Hochst.* Ex. *Boiss*, and *Mesembryanthemum nodiflorum*	Aerial parts	Phenolics, flavonoids, tannins, alkaloids, glycosides, steroids, and terpenoids	Antibacterial activities	El-Amier et al. (2021)
35	*Myrsine africana*	Leaves, Fruits	Terpenoids, steroids, flavonoids, carbohydrates, tannins, and saponins	Antitumour, antimicrobial and antioxidant activities	Laraib et al. (2021)
36	*Hibiscus rosa sinensis*	Leaves	Phenols, alkaloids, tannins, flavonoids, carbohydrates/ reducing sugars, terpenoids, phlobatanin, cardiac glycoside, and saponins	Antimicrobial activity	Priya and Sharma (2021)
37	*Chassalia kolly*	Leaves	Glycosides, alkaloids, flavonoids, anthocynes, reducing compound, mucilages, and saponosids	Anti-inflammatory and antioxidant activities	Alain et al. (2021)
38	*Pogostemon bengalensis*	Leaves	Flavonoids, phenols, alkaloids, saponins, tannins, and glycosides	Red blood cells haemolysis assay, antibacterial and antioxidant activities	Pimpliskar et al. (2021)
39	*Eryngium pyramidale Boiss.* and *Hausskn*	Aerial parts	Flavonoids, steroids, terpenoids, glycosides, and phenols	Antibacterial activity	Nejati et al. (2021)
40	*Gentiana cruciata* L.	Flowers	Polyphenols and tannins	Antibacterial and antifungal activity	Budniak et al. (2021)

and Rahaman 2020). The hydroxide (–OH), carbonyl (–C=O), aldehyde (–CHO), ethoxy (–C–O–C), and carboxylic acid (–COOH) functional groups in the skeleton of flavonoid compounds are present in the reduction and stabilization of flavonoid-based nanoparticles (Khatamiet al. 2015; Hussain et al. 2019). The existence of the aforementioned and other functional groups in flavonoids enables the active chelation of the precursor ions and allows the desired nanoparticle formation (El-Seedi et al. 2019).

1.3.2 Roles of Terpenoids in Nanoparticles Synthesis

Terpenoids are a large class of naturally occurring compounds derived from the five-carbon compound isoprene, and isoprene polymers called terpenes (Perveen 2021). They are important phytochemicals for the biosynthesis of various nanoparticles. Plant terpenoids including monoterpenes, sesquiterpenes, diterpenes, sesquiterpenes, and triterpenes have found beneficial uses in medicine and their existence in plant biomass form a good platform for green synthesis of nanomaterials (Bergman et al. 2019). Medicinal plant terpenoids attached to the surface of nanoparticles during controlled synthesis are important molecules in reducing the precursor ions to required zerovalent forms and act as capping and stabilizing agents. According to Khan et al. (2016), terpenoids as major chemical constituents of various plant essential oils involved in the biosynthesis of AgNPs and their derivatives may play a crucial role as the surface-active molecule that may participate in reducing Ag^+ to Ag^0 and stabilizing the nanoparticles. Parlinska-Wojtan et al. (2016), on the other hand, justified that terpenoids present in the leaf extracts of chamomile are major reductants in the green synthesis of aqueous colloidal solutions of AgNPs, and also act as a capping and stabilizing agent to prevent undesired agglomerations and aggregations of the particles. In the other work, Kavitha et al. (2017) isolated terpenoid compounds from medicinal plant leaf extracts of *Andrographis paniculate* and participated in reducing the divalent (Zn^{+2}) to zerovalent (Zn^0) form by anchoring to the surface of zinc oxide nanoparticles (ZnONPs) via –C=O functional groups

1.3.3 Roles of Proteins in Nanoparticles Synthesis

Proteins, the primary plant metabolites, are macromolecules consisting of one or more long-chain amino acid units linked together by peptide bondage (Godeto et al. 2021). Recently, the reports revealed that biosynthesized protein nanoparticles are widely active in nanomedicines, mainly in drug delivery. This is mainly because the protein molecules have added advantages over other biomolecules, such as biocompatibility, biodegradability, ease of availability and preparation, high drug loading efficiencies, they are non-immunogenic, increase cellular uptake, and a high number of functional groups can be modified for targeting (Jain et al. 2018; Kianfar 2021; Reddy and Rapisarda 2021). In their study, Ahmad et al. (2013) depicted the presence of different –C=O groups attached to the surface of biosynthesized nanoparticles by Fourier-transform infrared spectroscopy (FT-IR) spectrophotometer, and the –C=O groups of amino acid residues tend to act as capping ligands for nanoparticles, thereby stabilizing the particles in aqueous solution by preventing unexpected agglomerations. It was sufficiently documented that the different functional groups such as thiols (SH), carbonyls (–C=O), hydroxyls (–OH), amines (–NH2), and amides (–C=ONH2) in amino acid residues of protein molecules are extensively useful components in the biosynthesis of various nanoparticles.

Protein molecules containing a disulphide bridge (S–S) as of in cysteine–cysteine linkage and thiol surface (S–H) in cysteine and methionine amino acids forms a catalytic site for the reduction of nanoparticles and act as reducing and capping agents (Durán et al. 2015). When they come into contact with the nanoparticles' surfaces, protein molecules can bind immediately to their surfaces (Elechiguerra et al. 2005). They can also facilitate the transportation of drug or generic materials into human cells (Hu et al. 2011). Moreover, the uptake and retention of nanoparticles inside human cells increase when proteins are used as natural capping agents (Rodriguez et al. 2013). Therefore there is no need for a separate capping step when peptides are present in nanoparticles as both reducing and capping agents, which is very significant for most therapeutic applications of nanoparticles (Chowdhury et al. 2014).

1.3.4 Roles of Phenolic Acids in Nanoparticles Synthesis

Phenolic acids are a class of polyphenolic compounds abundant in various medicinal plant sources. They refer to prominent bioactive secondary metabolites with –COOH functional groups with two subcategories, hydroxybenzoic acids and hydroxycinnamic acids, based on the attachment of –COOH to the benzene ring (Kumar and Goel 2019; Al Mamari 2021). The –OH and –COOH functional groups in phenolic acids, especially those of caffeic acid, gallic acid, ellagic acid, and protocatechuic acid are bioreductants of metal nanoparticles (Dauthal and Mukhopadhyay 2016; Amini and Akbari 2019). Natural phenolic acids present their reducing potentials by chelating to metal surfaces via –OH and –COOH groups. They facilitate the nucleation of nanoparticles by providing an electron (e^-) to the apparent metal ions and mediate the growth through oxidized phenolic acid anchored to the nanoparticles' surface (Amini and Akbari 2019). According to Kim and Han (2016), the catechol functional group in caffeic acid (3,4-dihydroxycinnamic acid) allows its strong chelation to the metal or metal oxides, thereby reducing the metal/metal oxide ions to corresponding nanoparticles.

1.4 PLANT METABOLITES-BASED SYNTHESIS PATHWAYS

As discussed in the sections above, metabolites sourced from different parts a of plant actively participate in the biosynthesis of various nanoparticles. In addition to reducing the precursor compound to corresponding nanoparticles and stabilizing them, biomolecules from plants also functionalize the surface of the nanoparticles (Küünal et al. 2018), which allows further modification to nanoparticle surfaces and induces synergetic effect for multiple applications.

1.4.1 Medicinal Plant Metabolites in Metal and Metal Oxide Nanoparticles Synthesis

Green synthesis of nanoparticles using metallic precursors was extensively studied through intracellular and extracellular reduction with the help of biomolecules from different microorganisms (bacteria, fungi, algae, yeast, etc.) and plant sources. Metallic nanoparticles are submicron scale entities with a metal/metal oxide core usually covered with a shell made up of organic or inorganic material or metal oxide (Khan 2020). Their extensive engagement in synthesizing various nanomaterials is due to their conspicuous properties such as good electrical and optical properties, antibacterial activity, chemical properties, and biocompatibility (Jamkhande et al. 2019; Wijesinghe et al. 2020). In addition, the flexibility to prepare particles with varied shapes, sizes, and morphology and the possibility for further modification of the surface broaden their application in different fields.

Primary and secondary metabolites of several medicinal plants have been widely consumed for metal and metal oxide nanoparticles synthesis through green chemistry approaches. Metallic nanoparticles synthesis involves the addition of prepared plant extracts to metal salt solutions at optimized reaction conditions (such as temperature, pH, reaction time, precursor, and extract dose, etc.) and observing a colour change that represents the first signal for metal nanoparticle formation (Dikshit et al. 2021). The general route for the formation of metal and metal oxide nanoparticles with the help of medicinal plant metabolites is presented in Figure 1.2. The phytochemicals in the extracts of medicinal plants are responsible for reducing metal ions to the corresponding nanoparticles. The progress of formation can be monitored with the help of UV-visible spectroscopy (Anandalakshmi et al. 2016). Heavy metals such as silver (Ag), gold (Au), zinc (Zn), palladium (Pd), lead (Pb), copper (Cu), iron (Fe), nickel (Ni), aluminium (Al), cobalt (Co), platinum (Pt) and metallic oxides like iron oxide (Fe_3O_4), zinc oxide (ZnO), titanium dioxide (TiO_2), copper oxide (CuO),

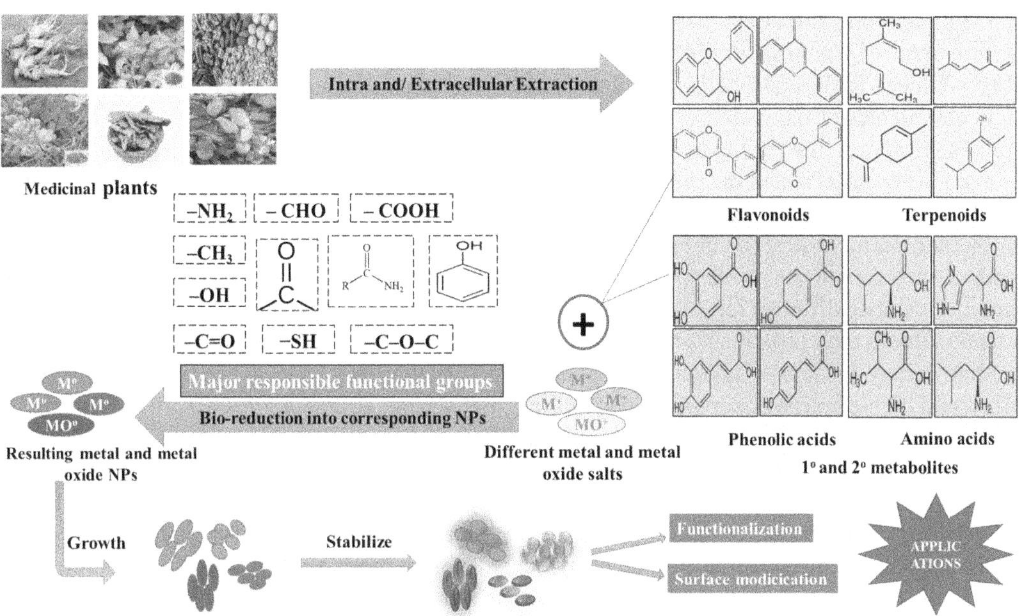

FIGURE 1.2 Major route for synthesis of metal and metal oxide nanoparticles from medicinal plant metabolites.

magnesium oxide (MgO), and aluminium oxide (Al_2O_3) are biologically reduced into corresponding nanoparticles with the help of plant biomolecules (Jeevanandam et al. 2016). Currently, Ag and Au have received more attention among widely reported metal nanoparticles due to their outstanding properties, viz, their inert nature, stability, high disparity, non-cytotoxicity, and biocompatibility (Alaqad and Saleh. 2016). They have been biologically synthesized with different shapes, sizes, and structures for different intended applications.

Recently, Garibo et al. (2020) synthesized a spherical and quasi-spherical shaped AgNPs with an average diameter size of 5 nm using aqueous stem and root extract of *Lysiloma acapulcensis*, a Mexican medicinal plant. Accordingly, secondary metabolites of the plant-like alkyl halides act as reducing agents and protein compounds and ethylene groups detected by FT-IR could act as stabilizing/capping agents. Moreover, AgNPs with the average size of 7 nm were synthesized biologically using chamomile extract for antibacterial applications (Parlinska-Wojtan et al. 2016). The other work by Khatami et al. (2015) reported spherically shaped antifungal AgNPs with an average size of 14 nm synthesized using the seed exudate of *Sinapis arvensis*, an Iranian medicinal crop. In this case, the –C=O, –OH, andamine (N–H) groups in seed exudates confirmed by FT-IR spectra mainly participate in the reduction of Ag^+ ions to Ag^0 nanoparticles. Furthermore, a phenolic compound-rich hot water extract of *calendula officinalis* with strong antioxidant properties was used by Baghizadeh, et al. (2015) to reduce $AgNO_3$ precursor to AgNPs.

The other extensively studied metal nanoparticles after Ag are Au nanoparticles (AuNPs). They have been synthesized extracellularly and intracellularly using medicinal plants' living and dead biomass. For instance, AuNPs for antimicrobial application have been synthesized extracellularly using leaf extract of *Simarouba glauca* in an aqueous medium (Thangamani and Bhuvaneshwari 2019). In this report, the plant extracts possess different biomolecules such as aldehyde, carboxylic acids, alcohol, alkaloids, flavonoids, and other phenolic groups as confirmed by FT-IR and gas chromatography-mass spectrometry (GC-MS) spectrophotometer. Among these, the oxidation of aldehydes facilitated the reduction of metal ions to AuNPs to carboxylic acid within the solutions.

In the other work, Liu et al. (2019) reported a green route for the synthesis of AuNPs using ethanolic extract of *Euphrasia officinalis* as a major anti-inflammatory agent in lipopolysaccharide (LPS) induced RAW 264.7 cells. The phenolic acid, especially caffeic acid, luteolin-glucoside, and rutin compounds of the plant extract present their reduction and stabilization potential in forming AuNPs. Moreover, the fruit extract of *Dillenia indica* rich in phenolic groups is reported for reduction of Au^+ to AuNPs, and the resulting nanoparticle was tested for its in-vitro cytotoxicity (Sett et al. 2016).

In addition to various metal nanomaterials, metal oxide nanoparticles are also promising contenders, especially those synthesized through green technology, which is of interest to research communities. Among different reported metal oxide nanoparticles, green synthesized ZnONPs received attention due to their prominent properties such as a wide bandgap, better electrochemical activities, binding energy, low-cost synthesis, non-toxicity, biocompatibility, chemical and photochemical stability, and high-electron communication features (Mehta et al. 2012; George et al. 2018; Selim et al. 2020). Currently, several medicinal plant metabolites are being engaged in synthesizing ZnONPs for industrial, environmental, and even domestic applications. Ogunyemi et al. (2019) reported the green synthesis of ZnONPs employing an extract of medicinal plants such as chamomile flower (*Matricaria chamomilla*), olive leaf (*Olea europaea*), and red tomato fruit (*Lycopersicon esculentum*). The study reveals that different phytochemicals in these plant extracts such as terpenes, flavonoids, tannins, and glycosides are facilitators in reducing the ZnO precursor to its nanoparticle involved in capping and stabilizing the nanoparticles. Furthermore, medicinal plant metabolites of the leaf extracts of *Abutilon indicum*, *Melia azedarach*, and *Indigofera tinctoria* have been used to reduce $Zn(NO_3)_2.6H_2O$ precursor to ZnONPs to be used in anticancer activity against DU-145 and Calu-6 cancer cells (Prashanth et al. 2018). Their work reveals that various metabolites such as phenolic compounds, glycosides, proteins, terpenoids, etc. present in the leaf extracts of the plants are major biofuels responsible for reducing, stabilizing, and capping the nanoparticles.

The other most important metal oxide nanoparticles that have been receiving current research attention owing to their optical, electrical, and magnetic properties are Fe_3O_4 nanoparticles. Due to these and other promising properties, Fe_3O_4 nanoparticles, especially those of magnetite (Fe_3O_4) are widely engaged in various application fields such as magnetic storage media, ferrofluids, biosensors, catalysts, substrates in cancer treatment, targeted drug delivery in clinical trials, separation processes, and environmental remediation (Balamurugan 2014; Niraimathee et al. 2016). Today, various medicinal plant metabolites have been employed for the synthesis of Fe_3O_4 nanoparticles as an efficient and eco-friendly substitute for chemical or physical synthesis methods. For instance, the leaf extract of *Phyllanthus niruri*, an important Indian Ayurveda, was employed as a green reducing agent for synthesis of Fe_3O_4 for antibacterial activities (Kumar et al. 2018). The plant active phytoconstituents such as flavonoids, tannins, terpenoids, alkaloids, lignans, polyphenols, coumarins, and saponins found in *Phyllanthus niruri* have a therapeutic effect and are good candidates for capping and reducing Fe_3O_4 nanoparticles (Kumar et al. 2018). In addition, spherically shaped maghemite (γ -Fe_2O_3) was successfully synthesized by Demirezen et al. (2019) using *Ficus carica* (common fig) dried fruit extract as capping and reducing agents. Their FT-IR report reveals that the phytochemicals in the fruit extract of the plant, mainly those of –OH and –C=O-based compounds, are important reductant and capping components.

1.4.2 Medicinal Plant Metabolites in Carbon-Based Nanoparticles Synthesis

Carbon-based nanomaterials are among the current research focus and considered a promising material for various applications owing to their interesting properties such as having the highest surface area, better mechanical and thermal strength, and electrochemical activities (Patel et al. 2020). Different architectures of carbon-based nanomaterials with varied structures, shapes, and physicochemical properties have been generated since carbon atoms are able to bond with each other in various ways to form different allotropes of carbon (Rauti et al. 2019). Accordingly, 0D fullerene, nano-diamonds (NDs), carbon dots (CDs), graphene quantum dots (GQDs), 1D carbon nanotubes (CNTs), 2D graphene and its derivatives, and a nitrogen-rich graphene-like nanostructure – graphitic carbon nitride (g-CN) are the most studied categories of carbon-based nanomaterials (Maiti et al. 2019; Xin et al. 2019). Carbon-based nanomaterials have been extensively exploited for various environmental, agricultural, biological, and medical applications. Such materials have been synthesized via non-green pathways (i.e., physically or chemically), which are reported as hazardous and not environmentally benign.

Hence, biological pathways comprising plants, microorganisms, and agricultural and industrial wastes have been implemented as green alternatives to synthesize carbon-based nanomaterial in recent years. Just like they do in the synthesis of metal and metal oxide nanoparticles, medicinal plant phytochemicals participate in the formation of carbon-based nanomaterials, especially in those of fullerene and graphene-based architectures as capping and reducing agents. Some medicinal plant phytochemicals that have been involved in the reducing and capping of carbon-based nanomaterials are given in Table 1.2. For instance, more recently, reduced graphene oxide (rGO) nanosheet was synthesized biologically using leaf extracts of *Tridax procumbens*, an Indian medicinal plant for antioxidant and antibacterial applications (Thiyagarajulu et al. 2020). In addition, an antimicrobial and anticancer rGO nanoparticle was successfully synthesized employing vegetable extract, *Chenopodium album*, as a reducing and stabilizing agent (Umar et al. 2020).

According to Hamedani et al. (2019), the leaf extract of *Pistacia atlantica* is used to modify the surface of multi-walled carbon nanotubes (MWCNTs) for in-situ reduction and immobilization of AgNPs. In this case, the phytochemicals in the *Pistacia atlantica* extracts are first allowed to adsorb on the surface of MWCNTs and coat it completely, and then the adsorbed compounds reduce Ag ions to AgNPs and stabilize them. The other recent work by Vinay (2021) reveals that fullerene [C_{60}]-linked AGNPs were successfully synthesized employing neem gum originated from the neem tree (*Melia azedarach*), which is a complex polysaccharide acid salt that has been medicinally used in India as a fuel for many centuries. The synthesized nanocomposites have effective cytotoxic activity against the lung cancer cell line (A549) with IC_{50} value of 87.85 μg/mL. Therefore, metabolites of medicinal plants not only engage in the synthesis of metal and metal oxide nanoparticles but also participate in reducing, capping, and stabilizing carbon-based nanomaterials and their hybrid forms.

1.5 APPLICATIONS OF MEDICINAL PLANT METABOLITE-MEDIATED NANOPARTICLES

1.5.1 Antimicrobial Applications

Nowadays, microbes are showing high resistance to antibiotics. This issue needs to be resolved through the production of novel, more advanced platforms for the study and development of more potent antimicrobial agents against multidrug-resistant strains (Chung et al. 2016). Currently, research communities are investigating various novel works, especially green synthesized nanomaterials, and advanced products to be used as antimicrobial alternatives for antibiotic resistant microbes. In this regard, medicinal plant metabolites have been employed for the synthesis of several antimicrobial nanoparticles. Phytoconstituents in *Rumex nepalensis* leaves, *Phytolacca dodecandra* fruits, *Grewia ferruginea* bark and leaves are reported to have prominent antibacterial activities against different organisms, probably due to abundant availability of bioactive compounds (Tura et al. 2017).

It has been reported that plants and their extracts rich in tannins, flavonoids, and saponins are effective against resistant bacterial species (Tura et al. 2017). Recently, a wide variety of nanoparticles synthesized employing medicinal plant metabolites has been engaged in antimicrobial activities against several pathogenic microbes. In particular, pharmacologically important phytochemicals in *Viola betonicifolia* leaf extract used for AuNPs synthesis (VB-AuNPs) has the capability of killing

TABLE 1.2
Medicinal Plant Metabolites Involved in Synthesis of Different Nanoparticles

Medicinal plants	Parts used	Major metabolites participated in reducing, capping and/ stabilizing the nanoparticles	Synthesized nanoparticles	Average Size and shapes	Characterization techniques	Medicinal and other applications of synthesized nanoparticles	References
Lysiloma acapulcensis	Roots, Stems	Alkyl halides, proteins, phenolic, and other aromatic compounds	Ag	5 nm, spherical and quasi-spherical	FT-IR, XRD, TEM, LC–MS, XPS, and EDS	Antimicrobial activities	Garibo et al. (2020)
Sinapis arvensis	Seeds	Alcohols, phenols, carbohydrates, and amino acids	Ag	14 nm, spherical	UV-vis, XRD, TEM, FT-IR, and ICP	Antifungal activities	Khatami et al. (2015)
Helicteres isora	Roots	Carbohydrates, proteins, fibres, calcium, phosphorus, and iron	Ag	30–40 nm, crystalline and spherical	UV-vis, SEM, TEM, FT-IR, and XRD	Antibacterial and antioxidant activities	Bhakya et al. (2016)
Rheum palmatum	Roots	Phenolic compounds, esters, anthraquinone, carbohydrates, aldehydes, and ketones	Ag	121 nm, cubic, spherical, and hexagonal	UV-vis, XRD, GC-MS, DLS, SEM, TEM, and FT-IR	Antibacterial activities	Arokiyaraj et al. (2017)
Excoecaria agallocha	Leaves	Phenolic classes, amides, flavonoids, methylene groups, and carboxylate groups	Ag	20 nm, crystalline, hexagonal and spherical	UV-vis, FT-IR, XRD, FE-SEM, and EDX	Antibacterial, antioxidant, and anticancer ctivities	Bhuvaneswari et al. (2017)
Terminalia arjuna	Bark	Polyphenols, phenolic acids, and proteins	Ag	64.6 ± 1.8 nm, spherical	FT-IR, XRD, FE-SEM, UV-vis, and DLS	Antibacterial ctivities	Ahmed et al. (2017)
Croton caudatus Geisel	Leaf	Polyphenol-based secondary metabolites	Au	20–50 nm, spherical	UV-vis, FT-IR,XRD, SEM-EDAX, and TEM	Antimicrobial, antioxidant activities	Kumar et al. (2019)

(Continued)

TABLE 1.2 (CONTINUED)
Medicinal Plant Metabolites Involved in Synthesis of Different Nanoparticles

Medicinal plants	Parts used	Major metabolites participated in reducing, capping and/ stabilizing the nanoparticles	Synthesized nanoparticles	Average Size and shapes	Characterization techniques	Medicinal and other applications of synthesized nanoparticles	References
Simarouba glauca	Leaves	Alkaloids, flavonoids, and phenolic compounds	Au	<10 nm, prism and spherical	UV-vis, XRD, HR-TEM, FT-IR, and GC-MS	Antimicrobial activities	Thangamania and Bhuvaneshwari (2019)
Euphrasia officinalis	Leaves	Phenolic compounds (phenolic acids and flavonoids)	Au	5–30 nm, spherical or hexagonal and a few triangular-shaped	UV-vis, FT-IR, XRD, TEM, EDX, and SAED	Anti-inflammatory activities	Liu et al. (2019)
Dillenia indica	Fruits	Alcohol, carboxylic acid, esters, and ethers	Au	5–50 nm, spherical, triangular, tetragonal, and pentagonal	FT-IR, XRD, TEM, EDX, HR-TEM, SAED, and TGA/ DSC		Sett et al. (2016)
Corchorus olitorius	Leaves	Phenolic compounds (such as chlorogenic acid, α-tocopherol, and quercetin 3-galactoside derivatives), ascorbic acid, and proteins	Au	37–50 nm, quasi-spherical	UV-vis, FT-IR, XRD, TEM, and TGA	Anticancer activities	Ismail et al. (2018)
Sumac	Fruits	Flavonoid, tannins and other phenolic compounds	Au	20.83 nm, spherical	UV-vis, FT-IR, XRD, TEM, and zeta potential	Antioxidant activities	Shabestarian et al. (2016)
Dendropanax morbifera	Leaves	Polysaccharides and terpenoids	Ag and Au	100–150 nm, polygonal for Ag and 10–20 nm, hexagonal for Au	UV-vis, XRD, EDX, FE-TEM, and DLS	Anticancer activities	Wang et al. (2016)
Acer pentapomicum	Leaves	Phenolic and alcoholic compounds	Au	19–24 nm, spherical	SEM, XRD, EDX, UV-vis, and FT-IR	Antioxidant, antibacterial and antifungal activities	Khan et al. (2018)

(Continued)

TABLE 1.2 (CONTINUED)
Medicinal Plant Metabolites Involved in Synthesis of Different Nanoparticles

Medicinal plants	Parts used	Major metabolites participated in reducing, capping and/ stabilizing the nanoparticles	Synthesized nanoparticles	Average Size and shapes	Characterization techniques	Medicinal and other applications of synthesized nanoparticles	References
Xanthium strumarium	Leaves	Monoterpene such as limonene and borneol	Pt	22 nm, cubic to rectangular shape	UV-vis, FT-IR, PXRD, SEM-EDAX, and TEM	Antimicrobial and anticancer activities	Kumar et al. (2019)
Taraxacum laevigatum	Aerial parts	Proteins, flavonoids, and saponins	Pt	2–7 nm, spherical	UV-vis, XRD, TEM, SEM, EDX, DLS, and FT-IR	Antibacterial activities	Tahir et al. (2017)
Punica granatum	Fruits	Alkaloids, glycosides, flavonoids, phenolic compounds, reducing sugars, resins, and tannins	Pt	20.12 nm, spherical	UV-vis, XRD, TEM, FE-SEM, and FT-IR	Antitumour activities	Şahin et al. (2018)
Filicium decipiens	Leaves	Saponins, tannins, flavonoids, steroids, alkaloids, and other phenolics	Pd	6.36 nm, spherical	UV-vis, FT-IR, XRD, and TEM	Antibacterial activities	Sharmila et al. (2017)
Withania coagulans	Leaves	Flavonoid and other phenolics	Pd	<15 nm, spherical	XRD, FE-SEM, FT-IR, EDS, TEM, and VSM	Catalytic activities	Atarod et al. (2016)
Andrographis paniculata	Leaves	Terpenoids	ZnO	20.23 nm, hexagonal nanorod	UV-vis, FT-IR, XRD, SEM, ^{1}H-NMR, and DLS	Drug delivery	Kavitha et al. (2017)

(*Continued*)

TABLE 1.2 (CONTINUED)
Medicinal Plant Metabolites Involved in Synthesis of Different Nanoparticles

Medicinal plants	Parts used	Major metabolites participated in reducing, capping and/stabilizing the nanoparticles	Synthesized nanoparticles	Average Size and shapes	Characterization techniques	Medicinal and other applications of synthesized nanoparticles	References
Matricaria chamomilla L.	Flowers	Terpenes, flavonoids, alkaloids, glycosides, and tannins	ZnO	51.2 ± 3.2 nm, cubic	SEM, TEM, UV-vis, FT-IR, EDS, and XRD	Antimicrobial activities	Ogunyemi et al. (2019)
Olea europaea	Leaves	Terpenes, flavonoids, glycosides, saponins, and tannins		41.0 ± 2.0 nm, cubic			
Lycopersicon esculentum M.	Fruits	Terpenes, flavonoids, glycosides saponins, tannins, and alkaloids		51.6 ± 3.6 nm, cubic			
Deverra tortuosa	Aerial parts	Flavonoids, terpenoids, glycosides, alkaloids, coumarin, steroids, tannin, and other polyphenolic compounds	ZnO	15.22 nm, hexagonal	UV-vis, FT-IR, XRD, HR-TEM, and HPLC	Anticancer activities	Selim et al. (2020)
Abutilon indicum	Leaves	Glycoside, tannin, and phenolic compounds	ZnO	15 nm, spherical	XRD, SEM, TEM, XPS, GC-MS, and TGA	Anticancer activities	Prashanth et al. (2018)
Melia azedarach		Proteins, terpenoids, steroids, glycosides, tannins, and phenolic compounds		12 nm, spherical			
Indigofera tinctoria		Saponins		21 nm, spherical			

(Continued)

TABLE 1.2 (CONTINUED)
Medicinal Plant Metabolites Involved in Synthesis of Different Nanoparticles

Medicinal plants	Parts used	Major metabolites participated in reducing, capping and/ stabilizing the nanoparticles	Synthesized nanoparticles	Average Size and shapes	Characterization techniques	Medicinal and other applications of synthesized nanoparticles	References
Lippia adoensis (koseret)	Leaves	Alcohols, ketones, aldehydes, and phenols	ZnO	19.78 nm, spherical, and hexagonal wurtzite	XRD, SEM-EDX, FT-IR, TEM, UV-vis, and TGA	Antibacterial activities	Demissie et al. (2020)
Beta vulgaris, *Cinnamomum tamala*	–	Alcohols, ketones, aldehydes,alkanes, alkenes, and proteins.	ZnO	20 ± 2 nm, spherical 30 ± 3 nm, rod-shaped	XRD, SEM, FT-IR, TEM, and UV-vis	Antibacterial and antifungal activities	Pillai et al. (2020)
Cinnamomum verum				46 ± 2 nm, spherical			
Brassica oleracea var. Italica				47 ± 2 nm, spherical			
Catharanthus roseus	Leaves	Aliphatic amines, alcohols, carbohydrates, and proteins	TiO_2	65 nm, clustered and irregular shapes	XRD, FT-IR, SEM, and AFM	Antiparasitic activities	Velayutham et al. (2012)
Hibiscus rosasenansis	Flowers	Phenolic groups, proteins, and amines	TiO_2	7 nm, spherical and monodispersed	FT-IR, SEM, and XRD	Antibacterial activities	Kumar et al. (2014)
Psidium guajava	Leaves	Alcohols, carboxylic acids, and alkenes	TiO_2	32.58 nm, clusters and spherical	XRD, FT-IR, FE-SEM, and EDX	Antibacterial and antioxidant activities	Santhoshkumar et al. (2014)
Solanum trilobatum	Leaves	Vinyl ethers, alkanes, aldehydes, alkynes, beta lactones, and aliphatic amines	TiO_2	70 nm, oval and uneven spherical	FT-IR, SEM, XRD, EDX, and AFM	Larvicidal and pediculicidal activities	Rajakumar et al. (2014)

(Continued)

TABLE 1.2 (CONTINUED)
Medicinal Plant Metabolites Involved in Synthesis of Different Nanoparticles

Medicinal plants	Parts used	Major metabolites participated in reducing, capping and/ stabilizing the nanoparticles	Synthesized nanoparticles	Average Size and shapes	Characterization techniques	Medicinal and other applications of synthesized nanoparticles	References
Curcuma longa		Terpenoids, flavonoids, and proteins	TiO_2	50–110 nm, spherical	UV-vis, XRD, AFM, and SEM	Antifungal activities	Jalill et al. (2016)
Kalopanax pictus	Leaves	–	MnO_2	19.2 nm, spherical	UV-vis, FT-IR, XPS, TEM, and EDX	Dye removal activities	Moon et al. (2015)
Camellia japonica	Leaves	Phenolic acid, flavonoids, terpenoids and protein molecules	CuO ZnO	15 nm, spherical 20 nm, spherical	UV-vis, FT-IR, EDS, SEM, XRD, and TEM	Optical sensing of metals ions like Ag^+ and Li^+	Maruthupandy et al. (2017)
Phyllanthus niruri	Leaves	Terpenoids, flavonoids, polyphenols, alkaloids, coumarins and saponins	Fe_3O_4	10 nm, square	XRD, FT-IR, TEM, SEM, and UV-vis	Antimicrobial activities	Kumar et al. (2018)
Eucalyptus globulus	Leaves	Limonene, 1,8-cineole, α-pinene, p-cymene, γ-terpinene, and α-terpineol	β-Fe_2O_3	100 nm, agglomerated cluster	UV-vis, SEM-EDX, TEM, XRD, and FT-IR	–	Balamurugan et al. (2014)
Mimosa pudica	Roots	Tannin, calcium oxalate crystals, and mimosine	Fe_3O_4	67 nm, spherical	SEM, XRD, FT-IR, UV-vis, and VSM	–	Niraimathee et al. (2016)
Ficus carica	Fruits	Vitamins, minerals, sugars, carbohydrates, organic acids, and phenolic compounds	γ-Fe_2O_3	9 ± 4 nm, spherical	EDX, TEM, XRD, FT-IR, DLS, and UV-vis	–	Demirezen et al. (2019)
Euphorbia herita	Leaves	Polyphenols, flavonoids, and alcoholic compounds	Fe_3O_4	25–80 nm, irregular	SEM, XRD, FT-IR, and UV-vis	Antimicrobial activities	Ahmad et al. (2021)

(Continued)

TABLE 1.2 (CONTINUED)
Medicinal Plant Metabolites Involved in Synthesis of Different Nanoparticles

Medicinal plants	Parts used	Major metabolites participated in reducing, capping and/ stabilizing the nanoparticles	Synthesized nanoparticles	Average Size and shapes	Characterization techniques	Medicinal and other applications of synthesized nanoparticles	References
Laurus nobilis L.	Leaves	Alcohols, flavonoids, and polyphenols	α-Fe_2O_3	8.03 ± 8.99 nm, spherical like and partly as a hexagonal	FE-SEM, TEM, FT-IR, XRD, EDS, and UV-vis	Antimicrobial activities	Jamzad and Bidkorpeh (2020)
Terminalia bellirica	Leaves	Polyphenols, alcohols, and carboxylic acid	α-Fe_2O_3	21.32 nm, spherical	XRD, FT-IR, TEM, SEM, and UV-vis	Antioxidant, antibacterial and thermoacoustic activities	Jegadeesan et al. (2019)
Moringa oleifera	Fruits, Leaves			45 nm, irregular shaped			
Platanus orientalis	Leaves	Quercetins, flavonoids, and glycosides	α-Fe_2O_3 and γ-Fe_2O_3	38 nm, spherical	XRD, FT-IR, SEM, TEM, EDX, DLS, and UV-vis	Antifungal activities	Devi et al. (2019)
Psoralea corylifolia	Seeds	Carboxylic acids, aliphatic amines, alcohols, and phenols	α-Fe2O3	39 nm, spherical, rod-like and uneven shapes	FT-IR, UV-vis, XRD, Raman spectrophotometer, SEM, and HR-TEM	Anticancer and catalytic activities	Nagajyothi et al. (2017)
Tridax procumbens	Leaves	Alkaloids, flavonoids, phenols, saponins, steroids, and tannins	Reduced graphene oxide (rGO) nanosheet	Non-uniform sheet-like structure	Raman spectrophotometer, UV-vis, XRD, FT-IR, FE-SEM, TEM, and EDS	Antioxidant and antibacterial activities	Thiyagarajulu et al. (2020)
Pistacia atlantica	Leaves	Phenolic compounds and flavonoids	Ag-NP/CNT	Quasi-spherical	XRD, FT-IR, SEM, TEM, and EDAX	Heterogeneous nano-catalyst for degradation of organic dyes	Hamedani et al. (2019)
Gum of *Melia azadirachta*	-	-	Ag@C_{60} NPs	30 nm, bore quasi-spherical	XRD, EDAX, SEM, and TEM	Anticancer activities	Vinay (2021)

(Continued)

TABLE 1.2 (CONTINUED)
Medicinal Plant Metabolites Involved in Synthesis of Different Nanoparticles

Medicinal plants	Parts used	Major metabolites participated in reducing, capping and/ stabilizing the nanoparticles	Synthesized nanoparticles	Average Size and shapes	Characterization techniques	Medicinal and other applications of synthesized nanoparticles	References
Chenopodium album	Leaves	Vitamin C, casein, caffeic acid, and polyphenols	rGO	Scrolled shape and lateral corrugations	FT-IR, UV-vis, SEM, and TEM	Anticancer, antibacterial, antifungal and antibiofilm activities	Umar et al. (2020)
Citrus limon L. (Lemon juice)	Fruits	Ascorbic acid, citric acid, sugars, and polyphenols	rGO	Exfoliated nanosheets	Raman spectrophotometer, FT-IR, XRD, UV-vis, SEM, and TEM	Dye removal activities	Mahiuddin and Ochiai (2021)
Lotus garcinii	Leaves	Carbohydrates, triterpenoids saponins, steroids, coumarins, tannins, flavonoids, glycosides, proteins, nucleic acid, and carotenoids	Ag/rGO/Fe_3O_4	Spherical, 7–20 nm	FT-IR, FE-SEM, EDX, XRD, TEM, and UV-vis	Heterogeneous catalyst for reduction of organic pollutants	Maham et al. (2017)
Eucalyptus twigs	Stems	Flavonoids, eucalyptone, alkanoids, tannins, triterpenes, and several terpenoids.	Full-colour fluorescent carbon nanoparticles (CNPs)	Spherical or oval, 15–50 nm	UV-vis, XPS, FT-IR, FE-SEM, and TEM	Synthetic food colorant sensing and bioimaging	Damera et al. (2020)

Abbreviations: AFM: atomic force microscopy; DLS: dynamic light scattering; DSC: differential scanning calorimetry; EDAX: energy dispersive X-ray; FT-IR: Fourier-transform infrared spectroscopy; FE-SEM: field emission scanning electron microscopy; HE-TEM: high-resolution transmission electron microscopy; HPLC: high performance liquid chromatography; ICP: inductively coupled plasma; LCMS: liquid chromatography mass spectrometry; SAED: selected area electron diffraction; SEM: scanning electron microscope; TEM: transmission electron microscopes; TGA: thermogravimetric analyzer; UV-vis: ultraviolet–visible spectroscopy; XPS: X-ray photoelectron spectroscopy; XRD: X-ray crystallography

S. aureus, *Bacillus subtilis*, *Escherichia coli*, and *Pseudomonas aeruginosa* bacterial strains effectively (Wang et al. 2021). Flower-based nanoparticles from *Rosa floribunda charisma* possessed promising antibacterial activity against three skin pathogens that pose a significant threat to public health, as *Staphylococcus epidermidis, Streptococcus pyogenes, and Pseudomonas aeruginosa* (Younis et al. 2021).

1.5.2 Applications as Antioxidants

Natural antioxidants derived from various medicinal plants have great potential to reduce different diseases induced by reactive oxygen species (ROS) in the body. Most compounds with antioxidant activities in the human diet typically originate from plant sources (Rauf et al. 2021). Both in vitro and in vivo studies have shown that nanoparticles, especially those of metals and metal oxides exhibit antioxidant activity by scavenging ROS and reactive nitrogen species (RNS), causing unwanted side reactions, promoting degenerative and age-related diseases (Kumar et al. 2021). Plant secondary metabolites, especially those of flavonoid and phenolic compounds in cells and plant tissue have major antioxidant activities (Hussain et al. 2019). Such metabolites have been repeatedly employed for synthesis of various nanomaterials as a green alternative to be used as an antioxidant agent against numerous ailments. In particular, flower-based AuNPs from *Rosa floribunda charisma* showed good radical scavenging activity, demonstrated by inhibition of ROS and RNS (such as superoxide, nitric oxide, hydroxyl radical, and xanthine oxidase) (Younis et al. 2021). Seed-based *Mangifera indica* aqueous extract of AgNPs showed 2,2-diphenyl-1-picrylhydrazyl (DPPH) activity in the concentration range of 160–960 μg/m leading to inhibition of 16–90% with IC_{50} value of 544 μg/m (Donga and Chanda 2021). Thus, it has been widely reported that various nanoparticles synthesized biologically using medicinal plant phytochemicals as reducing and stabilizing agents have been tested for antioxidant activities in vivo and in vitro showed promising effectiveness.

1.5.3 Hepatoprotective Applications

Liver damage induced by hepatotoxicity is common in living organisms, which is mainly caused by a medicine, chemical, or herbal or dietary supplement, and it can be the side effect of medications such as HIV drugs. Nanoparticles carefully synthesized employing medicinal plant metabolites have been employed in hepatoprotective applications as a fascinating alternative to chemical-based drugs. In addition to being an alternative to toxic chemical treatments, the plant phytochemicals involved during nanoparticle synthesis can lead to synergetic effects, including antioxidant activities (Zhang et al. 2019). Plant-mediated nanoparticles of selenium, Ag, and Au have been widely used to prevent hepatotoxicity, a common form of liver damage. According to Kumar et al. (2021) report, rats intoxicated by carbon tetrachloride (CCl_4) were treated with aqueous leaf extract of *Punica granatum* and generated AgNPs. The results obtained obviously suggested that the aqueous extract of *Punica granatum* had a hepatoprotective effect, as the liver profile is affected by CCl_4 toxicity. On the other hand, AGNPs synthesized from *Rhizophora apiculata* have been shown to protect against hepatotoxin-induced liver damage caused by carbon tetrachloride (Zhang et al. 2019).

1.5.4 Anticancer Therapeutic Potential

Numerous therapeutic plant extracts and active components have been reported to have anticancer properties since ancient times. Medicinal herbs have been shown to have anticancer and cytotoxic properties in numerous studies. Plant-mediated nanoparticles have been reported to have improved anticancer activities against different cancer cell lines. In particular, those synthesized using secondary metabolites and other non-metallic components of various medicinal plant extracts have shown great effect against various cancer cell lines such as human breast cancer (KB) cell line, Hep

2, HCT 116, and Hela cell lines, and able to control tumour cell growth (Kuppusamy et al. 2016). For instance, medicinal plants containing various metabolites such as terpenoids, alkaloids, glycosides, flavonoids, and other phenolic compounds have been involved in the synthesis of nanoparticles, especially those of Ag and Au for the treatment of cancer (Kathiravanet al. 2014; Wang et al. 2016). The properties and structure of nanoparticles, mainly size, shape, and surface composition determine the effectivities of synthesized nanoparticles against cancer cells.

1.5.5 Applications in Drug Delivery

Nanoparticles have been widely involved in drug delivery, such as in anticancer drugs and therapeutic proteins in the field of nanomedicine as drug nanocarriers. Scientific reports reveal that they have the ability to improve the stability and solubility of encapsulated cargos, facilitate transport across membranes, and extend circulation times for better safety and efficacy (Mitchell et al. 2021). Among widely reported nanoparticles, magnetite nanoparticles (Fe_3O_4-NPs) biosynthesized using various plant extracts are gaining popularity in drug delivery systems due to their excellent magnetic properties. In addition to superparamagnetic property, the shape, size, surface modification, and stability of the nanoparticles determine the ability of the particles in delivering drugs to specific sites safely and effectively (Yew et al. 2020). Besides, AuNPs are considered one of the most suitable carrier systems in drug delivery carriers and macromolecular carriers, due to their improved biocompatibility, stability, and oxidation resistance (Sengani et al. 2017).

1.5.6 Catalytic Applications

Nanoparticles synthesized using medicinal plant phytochemicals have been extensively studied as promising alternatives in the reduction of various organic pollutants such as organic dyes and facilitate catalyst-induced reactions. Currently, nanoparticles are synthesized biologically using metals, and their oxides showed promising catalytic activities. For instance, AuNPS synthesized biologically using a traditional medicinal plant, *Piper longum* fruit, showed potent degradation capacity against the four organic dyes – methylene blue (MB), methyl red, crystal violet, and acridine orange (Nakkala et al. 2016). The other work by Ahmad et al. (2015) reveals that AuNPs synthesized using *Fagonia indica* as a reducing agent presented strong catalytic activity for the photocatalytic reduction of methylene blue (about 80% of MB was photodegraded under visible-light irradiation after 80 min) and chemical reduction of 4-nitrophenol. In another work, biosynthesized ZnONPs showed about 96% photocatalytic degradation of MB dye (5×10^{-5} M MB at pH 12) (Kumar et al. 2018).

In their work, Rostami-Vartooni et al. (2019) biosynthesized Ag and Pd nanoparticles supported on Fe_3O_4/bentonite using an aqueous *Salix aegyptiaca* leaf extract. Their experiment reveals that the magnetically recoverable Ag or Pd nanoparticles/Fe_3O_4/bentonite nanocomposites exhibited remarkable catalytic activity for the reduction reaction of rhodamine B (RhB), MB, and methyl orange (MO) azo dyes in the presence of aqueous $NaBH_4$ with good reusability. Moreover, AgNPs synthesized employing *Dalbergia spinosa* leaf extract as a biological reducing agent exhibited prominent catalytic activities in reduction of 4-nitrophenol (4-NP) into 4-aminophenol (4-AP) (Muniyappan, and Nagarajan 2014) (Figure 1.3).

1.6 CONCLUSION

Green synthesis of nanomaterials is a promising area for various industrial and environmental applications owing to its energy efficiency, low production cost, ease of synthesis, use of few or no toxic chemicals, and reduced environmental challenges. Since our world is blessed with the grace of nature, several natural entities have been utilized to synthesize and modify nanomaterials. Medicinal plants have been employed for modern and folk medicines among various natural resources directly

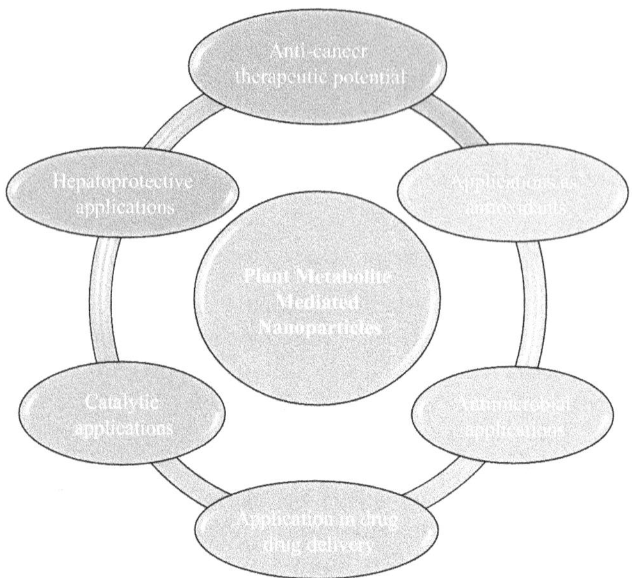

FIGURE 1.3 Applications of plant metabolite-mediated nanoparticles.

or indirectly throughout human history. Currently, more focus has been given to medicinal plants and their biomass to synthesize nanoparticles, mainly due to ease of availability, the flexibility to choose any parts, and because they encompass several natural compounds that are directly involved in reducing, capping, and stabilizing the resulting nanoparticles. Different metabolites of plants, such as phenolics, flavonoids, terpenes, tannins, glycosides, steroids, proteins, etc., possess several functional groups including hydroxides, carbonyls, thiols, amines, amides, etc. and play a vital role in reducing precursor compounds to intended nanoparticles. The resulting nanoparticles with controlled size, shape, and surface properties can participate in industrial, agricultural, environmental, and health application areas. Even though the properties, characteristics, and applicability of nanomaterials have been thoroughly studied and proved by various scientific communities, their environmental challenge and toxicity and health impact are not well documented. Hence, in addition to synthesis and characterization studies, the attention of future research should focus on the environmental challenges posed by the usage of nanomaterials for various applications. Considering this, nanoparticles critically synthesized from medicinal plant metabolites with extreme caution can be a modern solution to the existing problems encountered by our societies.

REFERENCES

Abate, Limenew, A. Bachheti, R.K. Bachheti, and A. Husen. 2021. Antibacterial Properties of Medicinal Plants: Recent Trends, Progress, and Challenges. In *Traditional Herbal Therapy for the Human Immune System*. Boca Raton, FL: CRC Press, pp. 13–54.

Abate, Limenew, M.G. Tadesse, A. Bachheti, and Rakesh Kumar Bachheti. 2022. Traditional and Phytochemical Bases of Herbs, Shrubs, Climbers, and Trees from Ethiopia for Their Anticancer Response. *BioMed Research International*, 2022. https://doi.org/10.1155/2022/1589877.

Ahmad, A., Y. Wei, F. Syed, M. Imran, Z.U.H. Khan, K. Tahir, A.U. Khan, M. Raza, Q. Khan, and Q. Yuan. 2015. Size Dependent Catalytic Activities of Green Synthesized Gold Nanoparticles and Electro-catalytic Oxidation of Catechol on Gold Nanoparticles Modified Electrode. *RSC Advances*, 5(120), 99364–99377. https://doi.org/10.1039/C5RA20096B.

Ahmad, T., I.A. Wani, N. Manzoor, J. Ahmed, and A.M. Asiri. 2013. Biosynthesis, Structural Characterization and Antimicrobial Activity of Gold and Silver Nanoparticles. *Colloids and Surfaces B: Biointerfaces*, 107, 227–234. http://doi.org/10.1016/j.colsurfb.2013.02.004.

Ahmad, W., J.K. Kumar, and M. Amjad. 2021. Euphorbia Herita Leaf Extract as a Reducing Agent in a Facile Green Synthesis of Iron Oxide Nanoparticles and Antimicrobial Activity Evaluation. *Inorganic and Nano-Metal Chemistry*, 51(9), 1147–1154. https://doi.org/10.1080/24701556.2020.1815062.

Ahmed, Q., N. Gupta, A. Kumar, and S. Nimesh. 2017. Antibacterial Efficacy of Silver Nanoparticles Synthesized Employing Terminalia Arjuna Bark Extract. *Artificial Cells, Nanomedicine, and Biotechnology*, 45(6), 1192–1200. https://doi.org/10.1080/21691401.2016.1215328.

Alain, K.Y., A.J. Morand, B.D. Andreea, O. Théophile, A.D.C. Pascal, A.G. Alain, A. Félicien, D.R. Mihaela, and S.C.K. Dominique. 2021. Phytochemical Analysis, Antioxidant and Anti-Inflammatory Activities of Chassalia Kolly Leaves Extract, a Plant Used in Benin to Treat Skin Illness. *GSC Biological and Pharmaceutical Sciences*, 15(3), 63–72. https://doi.org/10.30574/gscbps.2021.15.3.0148.

Alamgir, A.N.M. 2017. Pharmacognostical Botany: Classification of Medicinal and Aromatic Plants (Maps), Botanical Taxonomy, Morphology, and Anatomy of Drug Plants. In *Therapeutic Use of Medicinal Plants and Their Extracts*. Cham: Springer, 1, pp. 177–293. https://doi.org/10.1007/978-3-319-63862-1_6.

Alanazi, A.D., and R. Hesham. 2016. *Calendula officinalis*-Mediated Biosynthesis of Silver Nanoparticles and Their Electrochemical and Optical Characterization. *International Journal of Electrochemical Science*, 11, 10795–10805. https://doi.org/10.20964/2016.12.88.

Alaqad, K., and T.A. Saleh. 2016. Gold and Silver Nanoparticles: Synthesis Methods, Characterization Routes and Applications Towards Drugs. *Journal of Environmental and Analytical Toxicology*, 6(4), 525-2161. https://doi.org/10.4172/2161-0525.1000384.

Al Mamari, H.H. 2021. Phenolic Compounds: Classification, Chemistry, and Updated Techniques of Analysis and Synthesis. In *Phenolic Compounds*. IntechOpen. https://doi.org/10.5772/intechopen.98958.

Alzandi, A.A., E.A. Taher, N.A. Al-Sagheer, A.W. Al-Khulaidi, M. Azizi, and D.M. Naguib. 2021. Phytochemical Components, Antioxidant and Anticancer Activity of 18 Major Medicinal Plants in Albaha Region, Saudi Arabia. *Biocatalysis and Agricultural Biotechnology*, 34, 102020. https://doi.org/10.1016/j.bcab.2021.102020.

Amini, S.M.A., and A. Akbari. 2019. Metal Nanoparticles Synthesis Through Natural Phenolic Acids. *IET Nanobiotechnology*, 13(8), 771–777. https://doi.org/10.1049/iet-nbt.2018.5386.

Anand, U., N. Jacobo-Herrera, A. Altemimi, and N. Lakhssassi. 2019. A Comprehensive Review on Medicinal Plants as Antimicrobial Therapeutics: Potential Avenues of Biocompatible Drug Discovery. *Metabolites*, 9(11), 258. https://doi.org/10.3390/metabo9110258.

Anandalakshmi, K., J. Venugobal, and V. Ramasamy. 2016. Characterization of Silver Nanoparticles by Green Synthesis Method Using *Pedalium Murex* Leaf Extract and their Antibacterial Activity. *Applied Nanoscience*, 6(3), 399–408. https://doi.org/10.1007/s13204-015-0449-z.

Arib, C., J. Spadavecchia, and M.L. Chapelle. 2021. Enzyme Mediated Synthesis of Hybrid Polyedric Gold Nanoparticles. *Scientific Reports*, 11(1), 1–8. https://doi.org/10.1038/s41598-021-81751-1.

Aritonang, H.F., H. Koleangan, and A.D. Wuntu. 2019. Synthesis of Silver Nanoparticles Using Aqueous Extract of Medicinal Plants' (*Impatiens balsamina* and *Lantana camara*) fresh leaves and analysis of antimicrobial activity. *International Journal of Microbiology*, 2019. https://doi.org/10.1155/2019/8642303.

Arokiyaraj, S., S. Vincent, M. Saravanan, Y. Lee, Y.K. Oh, and K.H. Kim. 2017. Green Synthesis of Silver Nanoparticles Using *Rheum Palmatum* Root Extract and their Antibacterial Activity Against *Staphylococcus Aureus* and *Pseudomonas Aeruginosa*. *Artificial Cells, Nanomedicine, and Biotechnology*, 45(2), 372–379. https://doi.org/10.3109/21691401.2016.1160403.

Asaduzzaman, M., and T. Asao. 2018. Introductory Chapter: Phytochemicals and Disease Prevention. In *Phytochemicals-Source of Antioxidants and Role in Disease Prevention*. IntechOpen. https://doi.org/10.5772/intechopen.81877.

Asfaw, T.B., T.B. Esho, R.K. Bachheti, D.P. Pandey, and A. Husen. 2022. Exploring Important Herbs, Shrubs, and Trees for Their Traditional Knowledge, Chemical Derivatives, and Potential Benefits. In *Herbs, Shrubs, and Trees of Potential Medicinal Benefits*. Boca Raton, FL: CRC Press, pp. 1–26.

Atarod, M., M. Nasrollahzadeh, and S.M. Sajadi. 2016. Green Synthesis of Pd/RGO/Fe_3O_4 Nanocomposite Using *Withania coagulans* Leaf Extract and Its Application as Magnetically Separable and Reusable Catalyst for the Reduction of 4-Nitrophenol. *Journal of Colloid and Interface Science*, 465, 249–258. https://doi.org/10.1016/j.jcis.2015.11.060.

Bachheti, A., R.K. Bachheti, L. Abate, and A. Husen. 2021a. Current Status of Aloe-Based Nanoparticle Fabrication, Characterization and Their Application in Some Cutting-Edge Areas. *South African Journal of Botany*. https://doi.org/10.1016/j.sajb.2021.08.021.

Bachheti, R.K., A. Fikadu, A. Bachheti, and A. Husen. 2020. Biogenic Fabrication of Nanomaterials from Flower-Based Chemical Compounds, Characterization and Their Various Applications: A Review. *Saudi Journal of Biological Sciences*, 27(10), 2551–2562.

Bachheti, R.K., A. Sharma, A. Bachheti, A.Husen, G.M. Shanka, and D.P. Pandey. 2020b. Nanomaterials from Various Forest Tree Species and Their Biomedical Applications. In Husen A., and Jawaid M. (eds) *Nanomaterials for Agriculture and Forestry Applications*. Elsevier, pp. 81–106. https://doi.org/10.1016/B978-0-12-817852-2.00004-4.

Bachheti, A., A. Sharma, R.K. Bachheti, A. Husen, and V.K. Mishra. 2019a. Plant-Mediated Synthesis of Copper Oxide Nanoparticles and Their Biological Applications. In *Nanomaterials and Plant Potential*. Cham: Springer, pp. 221–237. https://doi.org/10.1007/978-3-030-05569-1_8.

Bachheti, R.K., L. Abate, A. Bachheti, A. Madhusudhan, and A. Husen. 2021. Algae-, Fungi-, and Yeast-Mediated Biological Synthesis of Nanoparticles and Their Various Biomedical Applications. In *Handbook of Greener Synthesis of Nanomaterials and Compounds*. Elsevier, pp. 701–734. https://doi.org/10.1016/B978-0-12-821938-6.00022-0.

Bachheti, R.K., R. Konwarh, V. Gupta, A. Husen, and Archana Joshi. 2019. Green Synthesis of Iron Oxide Nanoparticles: Cutting Edge Technology and Multifaceted Applications. In*Nanomaterials and Plant Potential*. Cham: Springer, pp. 239–259. https://doi.org/10.1007/978-3-030-05569-1_9.

Bachheti, R.K., Y. Godebo, A. Bachheti, M.O. Yassin, and Azamal Husen. 2020a. Root-Based Fabrication of Metal/Metal-Oxide Nanomaterials and Their Various Applications. In Husen A., and Jawaid M. (eds) *Nanomaterials for Agriculture and Forestry Applications*. Elsevier, pp. 135–166. https://doi.org/10.1016/B978-0-12-817852-2.00006-8.

Baghizadeh, A., S. Ranjbar, V.K. Gupta, M. Asif, S. Pourseyedi, M.J. Karimi, and R. Mohammadinejad. 2015. Green Synthesis of Silver Nanoparticles Using Seed Extract of Calendula officinalis in Liquid Phase. *Journal of Molecular Liquids*, 207, 159–163. https://doi.org/10.1016/j.molliq.2015.03.029.

Balamurugan, M., S. Saravanan, and T. Soga. 2014. Synthesis of Iron Oxide Nanoparticles by Using *Eucalyptus Globulus* Plant Extract. *e-Journal of Surface Science and Nanotechnology*, 12, 363–367. https://doi.org/10.1380/ejssnt.2014.363.

Barata, A.M., F. Rocha, V. Lopes, and A.M. Carvalho. 2016. Conservation and Sustainable Uses of Medicinal and Aromatic Plants Genetic Resources on the Worldwide for Human Welfare. *Industrial Crops and Products*, 88, 8–11. https://doi.org/10.1016/j.indcrop.2016.02.035.

Begum, H.A., M. Hamayun, T. Yaseen, S. Akhter, and M. Shakeel. 2021. Phytochemical Analysis, Antifungal Bioassay and Folklore Uses of Selected Medicinal Plants of Family Rosaceae. *Pure and Applied Biology (PAB)*, 5(2), 183–192. http://doi.org/10.19045/bspab.2016.50024.

Bergman, M.E., B. Davis, and M.A. Phillips. 2019. Medically Useful Plant Terpenoids: Biosynthesis, Occurrence, and Mechanism of Action. *Molecules*, 24(21), 3961. https://doi.org/10.3390/molecules24213961.

Beshah, F., Y. Hunde, M. Getachew, R.K. Bachheti, A. Husen, and A. Bachheti. 2020. Ethnopharmacological, Phytochemistry and Other Potential Applications of *Dodonaea* Genus: A Comprehensive Review. *Current Research in Biotechnology*, 2, 103–119.

Bhakya, S., S. Muthukrishnan, M. Sukumaran, and M. Muthukumar. 2016. Biogenic Synthesis of Silver Nanoparticles and Their Antioxidant and Antibacterial Activity. *Applied Nanoscience*, 6(5), 755–766. https://doi.org/10.1007/s13204-015-0473-z.

Bhuvaneswari, R., R.J. Xavier, and M. Arumugam. 2017. Facile Synthesis of Multifunctional Silver Nanoparticles Using Mangrove Plant *Excoecaria Agallocha* L. for its Antibacterial, Antioxidant and Cytotoxic Effects. *Journal of Parasitic Diseases*, 41(1), 180–187. https://doi.org/10.1007/s12639-016-0773-6.

Budniak, L., L. Slobodianiuk, S. Marchyshyn, R. Basaraba, and A. Banadyga. 2021. The Antibacterial and Antifungal Activities of the Extract of *Gentiana cruciata* L. Herb. *Pharmacology Online*, 2, 188–197.

Bundschuh, M., J. Filser, S. Lüderwald, M.S. McKee, G. Metreveli, G.E. Schaumann, R. Schulz, and S. Wagner. 2018. Nanoparticles in the Environment: Where Do We Come from, Where Do We Go To? *Environmental Sciences Europe*, 30(1), 1–17. https://doi.org/10.1186/s12302-018-0132-6.

Cele, T. 2020. Preparation of Nanoparticles. In *Engineered Nanomaterials-Health and Safety*. London, UK: IntechOpen. https://doi.org/10.5772/intechopen.90771.

Chandra, H., P. Kumari, E. Bontempi, and S. Yadav. 2020. Medicinal Plants: Treasure Trove for Green Synthesis of Metallic Nanoparticles and Their Biomedical Applications. *Biocatalysis and Agricultural Biotechnology*, 24, 101518. https://doi.org/10.1016/j.bcab.2020.101518.

Chowdhury, S., A. Basu, and S. Kundu. 2014. Green Synthesis of Protein Capped Silver Nanoparticles from Phytopathogenic Fungus *Macrophomina Phaseolina* (Tassi) Goid with Antimicrobial Properties Against Multidrug-Resistant Bacteria. *Nanoscale Research Letters*, 9(1), 1–11. https://doi.org/10.1186/1556-276X-9-365.

Chung, I.M., I. Park, K. Seung-Hyun, M. Thiruvengadam, and G. Rajakumar. 2016. Plant-Mediated Synthesis of Silver Nanoparticles: Their Characteristic Properties and Therapeutic Applications. *Nanoscale Research Letters*, 11(1), 1–14. https://doi.org/10.1186/s11671-016-1257-4.

Cox-Georgian, D., N. Ramadoss, C. Dona, and C. Basu. 2019. Therapeutic and Medicinal Uses of Terpenes. In *Medicinal Plants*. Cham: Springer, pp. 333–359.

Dai, J., and R.J. Mumper. 2010. Plant Phenolics: Extraction, Analysis and Their Antioxidant and Anticancer Properties. *Molecules*, 15(10), 7313–7352. https://doi.org/10.3390/molecules15107313.

Damera, D.P., R. Manimaran, V.V. Krishna, and A. Nag. 2020. Green Synthesis of Full-Color Fluorescent Carbon Nanoparticles from Eucalyptus Twigs for Sensing the Synthetic Food Colorant and Bioimaging. *ACS Omega*, 5(31), 19905–19918. https://doi.org/10.1021/acsomega.0c03148.

Dauthal, P., and M. Mukhopadhyay. 2016. Noble Metal Nanoparticles: Plant-Mediated Synthesis, Mechanistic Aspects of Synthesis, and Applications. *Industrial and Engineering Chemistry Research*, 55(36), 9557–9577. https://doi.org/10.1021/acs.iecr.6b00861.

Demirezen, D.A., Y.Ş. Yıldız, Ş. Yılmaz, and D.D. Yılmaz. 2019. Green Synthesis and Characterization of Iron Oxide Nanoparticles Using *Ficus Carica* (Common Fig) Dried Fruit Extract. *Journal of Bioscience and Bioengineering*, 127(2), 241–245. https://doi.org/10.1016/j.jbiosc.2018.07.024.

Demissie, M.G., F.K. Sabir, G.D. Edossa, and B.A. Gonfa. 2020. Synthesis of Zinc Oxide Nanoparticles Using Leaf Extract of *Lippia Adoensis* (Koseret) and Evaluation of Its Antibacterial Activity. *Journal of Chemistry*, 2020. https://doi.org/10.1155/2020/7459042.

Devi, H.S., M.A. Boda, M.A. Shah, S. Parveen, and A.H. Wani. 2019. Green Synthesis of Iron Oxide Nanoparticles Using *Platanus Orientalis* Leaf Extract for Antifungal Activity. *Green Processing and Synthesis*, 8(1), 38–45. https://doi.org/10.1515/gps-2017-0145.

Dikshit, P.K., J. Kumar, A.K. Das, S. Sadhu, S. Sharma, S. Singh, P.K. Gupta, and B.S. Kim. 2021. Green Synthesis of Metallic Nanoparticles: Applications and Limitations. *Catalysts*, 11(8), 902. https://doi.org/10.3390/catal11080902.

Donga, S., and S. Chanda. 2021. Facile Green Synthesis of Silver Nanoparticles Using *Mangifera Indica* Seed Aqueous Extract and its Antimicrobial, Antioxidant and Cytotoxic Potential (3-In-1 System). *Artificial Cells, Nanomedicine, and Biotechnology*, 49(1), 292–302. https://doi.org/10.1080/21691401.2021.1899193.

Durán, M., C.P. Silveira, and N. Durán. 2015. Catalytic Role of Traditional Enzymes for Biosynthesis of Biogenic Metallic Nanoparticles: A Mini-Review. *IET Nanobiotechnology*, 9(5), 314–323. http://doi.org/10.1049/iet-nbt.2014.0054.

Ebenezer, T.E., M. Zoltner, A. Burrell, A. Nenarokova, A.M. Novák, B. Prasad, P. Soukal, C. Santana-Molina, E. O'Neill, N.N. Nankissoor, and N. Vadakedath. 2019. Transcriptome, Proteome and Draft Genome of *Euglena Gracilis*. *BMC Biology*, 17(1), 1–23. https://doi.org/10.1186/s12915-019-0626-8.

El-Amier, Y.A., O.N. Al-hadithy, A.A. Fahmy, and M.M. El-Zayat. 2021. Phytochemical Analysis and Biological Activities of Three Wild Mesembryanthemum Species Growing in Heterogeneous Habitats. *Journal of Phytology*, 13, 1–8. https://doi.org/10.25081/jp.2021.v13.6403.

Elechiguerra, J.L., J.L. Burt, J.R. Morones, A. Camacho-Bragado, X. Gao, H.H. Lara, and M.J. Yacaman. 2005. Interaction of Silver Nanoparticles With HIV-1. *Journal of Nanobiotechnology*, 3(1), 1–10. https://doi.org/10.1186/1477-3155-3-6.

El-Seedi, H.R., R.M. El-Shabasy, S.A. Khalifa, A. Saeed, A. Shah, R. Shah, F.J. Iftikhar, M.M. Abdel-Daim, A. Omri, N.H. Hajrahand, and J.S. Sabir. 2019. Metal Nanoparticles Fabricated by Green Chemistry Using Natural Extracts: Biosynthesis, Mechanisms, and Applications. *RSC Advances*, 9(42), 24539–24559. https://doi.org/10.1039/C9RA02225B.

El-Zayat, M.M., M.M. Eraqi, F.A. Alfaiz, and M.M. Elshaer. 2021. Antibacterial and Antioxidant Potential of Some Egyptian Medicinal Plants Used in Traditional Medicine. *Journal of King Saud University – Science*, 33(5), 101466. https://doi.org/10.1016/j.jksus.2021.101466.

Fernandes, F., M. Dias-Teixeira, C. Delerue-Matos, and C. Grosso. 2021. Critical Review of Lipid-Based Nanoparticles as Carriers of Neuroprotective Drugs and Extracts. *Nanomaterials*, 11(3), 563. https://doi.org/10.3390/nano11030563.

Gagare, S.B., S.L. Chavan, and A.B. Sagade. 2021. Antibacterial Potential and Phytochemical Screening of *Physalis Angulata* and *Solanum Virgianum*. *International Journal of Researches in Biosciences, Agriculture and Technology*, 1(9), 36–40.

Garibo, D., H.A. Borbón-Nuñez, J.N.D. León, M.E. García, I. Estrada, Y. Toledano-Magaña, H. Tiznado, M. Ovalle-Marroquin, A.G. Soto-Ramos, A. Blanco, and J.A. Rodríguez. 2020. Green Synthesis of Silver Nanoparticles Using *Lysilomaa capulcensis* Exhibit High-Antimicrobial Activity. *Scientific Reports*, 10(1), 1–11. https://doi.org/10.1038/s41598-020-69606-7.

George, J.M., A. Antony, and B. Mathew. 2018. Metal Oxide Nanoparticles in Electrochemical Sensing and Biosensing: A Review. *Microchimica Acta*, 185(7), 1–26. https://doi.org/10.1007/s00604-018-2894-3.

Godeto, Y.G., A. Bachheti, A. Husen, D.P. Pandey, and R.K. Bachheti. 2021. Forest-Based Edible Seeds and Nuts for Health Care and Disease Control. In *Non-Timber Forest Products*. Cham: Springer, pp. 145–174. https://doi.org/10.1007/978-3-030-73077-2_7.

Golbarg, H., and M.J. Moghaddam. 2021. Antibacterial Potency of Medicinal Plants Including *Artemisia Annua* and *Oxalis Corniculata* Against Multi-Drug Resistance *Escherichia Coli*. *BioMed Research International*, 2021. https://doi.org/10.1155/2021/9981915.

Hamedani, Y.P., and M. Hekmati. 2019. Green Biosynthesis of Silver Nanoparticles Decorated on Multi-walled Carbon Nanotubes Using the Extract of *Pistacia atlantica* Leaves as a Recyclable Heterogeneous Nano-Catalyst for Degradation of Organic Dyes in Water. *Polyhedron*, 164, 1–6. https://doi.org/10.1016/j.poly.2019.02.010.

Hilou, A., O.G. Nacoulma, and T.R. Guiguemde. 2006. In Vivo Antimalarial Activities of Extracts from *Amaranthus Spinosus* L. and *Boerhaaviaerecta* L. in Mice. *Journal of Ethnopharmacology*, 103(2), 236–240. https://doi.org/10.1016/j.jep.2005.08.006.

Hong, S., D.W. Choi, H.N. Kim, C.G. Park, W. Lee, and H.H. Park. 2020. Protein-Based Nanoparticles as Drug Delivery Systems. *Pharmaceutics*, 12(7), 604. https://doi.org/10.3390/pharmaceutics12070604.

Hu, C.M.J., L. Zhang, S. Aryal, C. Cheung, R.H. Fang, and L. Zhang. 2011. Erythrocyte Membrane-Camouflaged Polymeric Nanoparticles as a Biomimetic Delivery Platform. *Proceedings of the National Academy of Sciences*, 108(27), 10980–10985. https://doi.org/10.1073/pnas.1106634108.

Husen, A., Q.I. Rahman, M. Iqbal, M.O. Yassin, and R.K. Bachheti. 2019. Plant-Mediated Fabrication of Gold Nanoparticles and Their Applications. In *Nanomaterials and Plant Potential*. Cham: Springer, pp. 71–110. https://doi.org/10.1007/978-3-030-05569-1_3.

Hussain, M., N.I. Raja, M. Iqbal, and S. Aslam. 2019. Applications of Plant Flavonoids in the Green Synthesis of Colloidal Silver Nanoparticles and Impacts on Human Health. *Iranian Journal of Science and Technology, Transactions A: Science*, 43(3), 1381–1392. https://doi.org/10.1007/s40995-017-0431-6.

Hussein, R.A., and A.A. El-Anssary. 2019. Plants Secondary Metabolites: The Key Drivers of the Pharmacological Actions of Medicinal Plants. *Herbal Medicine*, 1, 13. https://doi.org/10.5772/intechopen.76139.

Ijaz, I., E. Gilani, A. Nazir, and A. Bukhari. 2020. Detail Review on Chemical, Physical and Green Synthesis, Classification, Characterizations and Applications of Nanoparticles. *Green Chemistry Letters and Reviews*, 13(3), 223–245. https://doi.org/10.1080/17518253.2020.1802517.

Ismail, E.H., A. Saqer, E. Assirey, A. Naqvi, and R.M. Okasha. 2018. Successful Green Synthesis of Gold Nanoparticles Using a *Corchorus Olitorius* Extract and Their Antiproliferative Effect In Cancer Cells. *International Journal of Molecular Sciences*, 19(9), 2612. https://doi.org/10.3390/ijms19092612.

Jacob, J.M., R. Ravindran, M. Narayanan, S.M. Samuel, A. Pugazhendhi, and G. Kumar. 2021. Microalgae: A Prospective Low-Cost Green Alternative for Nanoparticle Synthesis. *Current Opinion in Environmental Science and Health*, 20, 00163. https://doi.org/10.1016/j.coesh.2019.12.005.

Jadoun, S., R. Arif, N.K. Jangid, and R.K. Meena. 2021. Green Synthesis of Nanoparticles Using Plant Extracts: A Review. *Environmental Chemistry Letters*, 19(1), 355–374. https://doi.org/10.1007/s10311-020-01074-x.

Jain, A., S.K. Singh, S.K. Arya, S.C. Kundu, and S. Kapoor. 2018. Protein Nanoparticles: Promising Platforms for Drug Delivery Applications. *ACS Biomaterials Science and Engineering*, 4(12), 3939–3961. https://doi.org/10.1021/acsbiomaterials.8b01098.

Jalill, A., D.H. Raghad, R.S. Nuaman, and A.N. Abd. 2016. Biological Synthesis of Titanium Dioxide Nanoparticles by *Curcuma longa* Plant Extract and Study Its Biological Properties. *World Scientific News*, 49(2), 204–222. bwmeta1.element.psjd-b2915893-0c75-4c7a-a51f-b29b6f76db03.

Jamkhande, P.G., N.W. Ghule, A.H. Bamer, and M.G. Kalaskar. 2019. Metal Nanoparticles Synthesis: An Overview on Methods of Preparation, Advantages and Disadvantages, and Applications. *Journal of Drug Delivery Science and Technology*, 53, 101174. https://doi.org/10.1016/j.jddst.2019.101174.

Jamzad, M., and B.M. Kamari. 2020. Green Synthesis of Iron Oxide Nanoparticles by the Aqueous Extract of *Laurus nobilis* L. Leaves and Evaluation of the Antimicrobial Activity. *Journal of Nanostructure in Chemistry*, 10(3), 193–201. https://doi.org/10.1007/s40097-020-00341-1.

Jeevanandam, J., Y.S. Chan, and M.K. Danquah. 2016. Biosynthesis of Metal and Metal Oxide Nanoparticles. *ChemBioEng Reviews*, 3(2), 55–67. https://doi.org/10.1002/cben.201500018.

Jegadeesan, G.B., K. Srimathi, N.S. Srinivas, S. Manishkanna, and D. Vignesh. 2019. Green Synthesis of Iron Oxide Nanoparticles Using *Terminalia Bellirica* and *Moringa Oleifera* Fruit and Leaf Extracts: Antioxidant, Antibacterial and Thermoacoustic Properties. *Biocatalysis and Agricultural Biotechnology*, 21, 101354. https://doi.org/10.1016/j.bcab.2019.101354.

Jyothiprabha, V., and P. Venkatachalam. 2016. Preliminary Phytochemical Screening of Different Solvent Extracts of Selected Indian Spices. *International Journal of Current Microbiology and Applied Sciences*, 5(2), 116–122. http://doi.org/10.20546/ijcmas.2016.502.013.

Karak, P. 2019. Biological Activities of Flavonoids: An Overview. *International Journal of Pharmaceutical Sciences and Research*, 10(4), 1567–1574. http://doi.org/10.13040/IJPSR.0975-8232.10(4).1567-74.

Kathiravan, V., S. Ravi, and S. Ashokkumar. 2014. Synthesis of Silver Nanoparticles from *Melia Dubia* Leaf Extract and Their In-Vitro Anticancer Activity. *Spectrochimica Acta Part A: Molecular and Biomolecular Spectroscopy*, 130, 116–121. https://doi.org/10.1016/j.saa.2014.03.107.

Kavitha, S., M. Dhamodaran, R. Prasad, and M. Ganesan. 2017. Synthesis and Characterization of Zinc Oxide Nanoparticles Using Terpenoid Fractions of *Andrographis paniculata* Leaves. *International Nano Letters*, 7(2), 141–147. https://doi.org/10.1007/s40089-017-0207-1.

Khan, I., K. Saeed, and I. Khan. 2019. Nanoparticles: Properties, Applications and Toxicities. *Arabian Journal of Chemistry*, 12(7), 908–931. http://doi.org/10.1016/j.arabjc.2017.05.011.

Khan, M.A., T. Khan, and A. Nadhman. 2016. Applications of Plant Terpenoids in the Synthesis of Colloidal Silver Nanoparticles. *Advances in Colloid and Interface Science*, 234, 132–141. https://doi.org/10.1016/j.cis.2016.04.008.

Khan, S.A. 2020. Metal Nanoparticles Toxicity: Role of Physicochemical Aspects. In *Metal Nanoparticles for Drug Delivery and Diagnostic Applications*. Elsevier, pp. 1–11. https://doi.org/10.1016/B978-0-12-816960-5.00001-X.

Khan, S., J. Bakht, and F. Syed. 2018. Green Synthesis of Gold Nanoparticles Using *Acer Pentapomicum* Leaves Extract its Characterization, Antibacterial, Antifungal and Antioxidant Bioassay. *Digest Journal of Nanomaterials and Biostructures*, 13, 579–589. Https://Www.Chalcogen.Ro/579_Khans.Pdf.

Khatami, M., S. Pourseyedi, M. Khatami, H. Hamidi, M. Zaeifi, and L. Soltani. 2015. Synthesis of Silver Nanoparticles Using Seed Exudates of *Sinapis Arvensis* as a Novel Bioresource, and Evaluation of Their Antifungal Activity. *Bioresources and Bioprocessing*, 2(1), 1–7. https://doi.org/10.1186/s40643-015-0043-y.

Kianfar, E. 2021. Protein Nanoparticles in Drug Delivery: Animal Protein, Plant Proteins and Protein Cages, Albumin Nanoparticles. *Journal of Nanobiotechnology*, 19(1), 1–32. https://doi.org/10.1186/s12951-021-00896-3.

Kim, K., and J.W. Han. 2016. Effect of Caffeic Acid Adsorption in Controlling the Morphology of Gold Nanoparticles: Role of Surface Coverage and Functional Groups. *Physical Chemistry Chemical Physics*, 18(40), 27775–27783. https://doi.org/10.1039/C6CP04122A.

Koul, A., A. Kumar, V.K. Singh, D.K. Tripathi, and S. Mallubhotla. 2018. Exploring Plant-Mediated Copper, Iron, Titanium, and Cerium Oxide Nanoparticles and Their Impacts. In *Nanomaterials in Plants, Algae*, and *Microorganisms*. San Diego: Academic Press, pp. 175–194. http://doi.org/10.1016/B978-0-12-811487-2.00008-6.

Kumar, K.S., N. Dhananjaya, and L.R. Yadav. 2018. *Euphorbia Tirucalli* Plant Latex Mediated Green Combustion Synthesis of ZnO Nanoparticles: Structure, Photoluminescence and Photo-catalytic Activities. *Journal of Science: Advanced Materials and Devices*, 3(3), 303–309. https://doi.org/10.1016/j.jsamd.2018.07.005.

Kumar, M., R. Ranjan, A. Kumar, M.P. Sinha, R. Srivastava, S. Subarna, and M.S. Kumar. 2021. Hepatoprotective Activity of Silver Nanoparticles Synthesized Using Aqueous Leaf Extract of *Punica Granatum* Against Induced Hepatotoxicity in Rats. *Nova BiologicaReperta*, 7(4), 381–389.

Kumar, M., V. Pratap, A.K. Nigam, B.K. Sinha, M. Kumar, and J.K.G. Singh. 2021. Plants as a Source of Potential Antioxidants and Their Effective Nanoformulations. *Journal of Scientific Research*, 65(3), 57–72.

Kumar, N., and N. Goel. 2019. Phenolic Acids: Natural Versatile Molecules with Promising Therapeutic Applications. *Biotechnology Reports*, 24, e00370. https://doi.org/10.1016/j.btre.2019.e00370.

Kumar, P.S.M., A.P. Francis, and T. Devasena. 2014. Biosynthesized and Chemically Synthesized Titania Nanoparticles: Comparative Analysis of Antibacterial Activity. *Journal of Environmental Nanotechnology*, 3(3), 73–81. https://doi.org/10.13074/jent.2014.09.143098.

Kumar, P.V., S.M.J. Kala, and K.S. Prakash. 2019a. Green Synthesis of Gold Nanoparticles Using *Croton Caudatus Geisel* Leaf Extract and their Biological Studies. *Materials Letters*, 236, 19–22. https://doi.org/10.1016/j.matlet.2018.10.025.

Kumar, P.V., S.M.J. Kala, and K.S. Prakash. 2019b. Green Synthesis Derived Pt-Nanoparticles Using *Xanthium Strumarium* Leaf Extract and their Biological Studies. *Journal of Environmental Chemical Engineering*, 7(3), 103146. https://doi.org/10.1016/j.jece.2019.103146.

Kuppusamy, P., M.M. Yusoff, G.P. Maniam, and N. Govindan. 2016. Biosynthesis of Metallic Nanoparticles Using Plant Derivatives and Their New Avenues in Pharmacological Applications–an Updated Report. *Saudi Pharmaceutical Journal*, 24(4), 473–484. http://doi.org/10.1016/j.jsps.2014.11.013.

Küünal, S., P. Rauwel, and E. Rauwel. 2018. Plant Extract Mediated Synthesis of Nanoparticles. In *Emerging Applications of Nanoparticles and Architecture Nanostructures*. Elsevier, pp. 411–446. https://doi.org/10.1016/B978-0-323-51254-1.00014-2.

Laraib, S., S. Sharif, Y. Bibi, S. Nisa, R. Aziz, and A. Qayyum. 2021. Phytochemical Analysis and Some Bioactivities of Leaves and Fruits of *Myrsine Africanalinn. Arabian Journal for Science and Engineering*, 46(1), 53–63. https://doi.org/10.1007/s13369-020-04710-4.

Liu, Y., S. Kim, Y.J. Kim, H. Perumalsamy, S. Lee, E. Hwang, and T.H. Yi. 2019. Green Synthesis of Gold Nanoparticles Using *Euphrasia Officinalis* Leaf Extract to Inhibit Lipopolysaccharide-Induced Inflammation Through NF-κB and JAK/STAT pathways in RAW 264.7 macrophages. *International Journal of Nanomedicine*, 14, 2945. https://doi.org/10.2147/IJN.S199781.

Maham, M., M. Nasrollahzadeh, S.M. Sajadi, and M. Nekoei. 2017. Biosynthesis of Ag/Reduced Graphene Oxide/Fe3O4 Using *Lotus Garcinii* Leaf Extract and its Application as a Recyclable Nanocatalyst for the Reduction of 4-Nitrophenol and Organic Dyes. *Journal of Colloid and Interface Science*, 497, 33–42. http://doi.org/10.1016/j.jcis.2017.02.064.

Mahiuddin, M.B., and B. Ochiai. 2021. Lemon Juice Assisted Green Synthesis of Reduced Graphene Oxide and Its Application for Adsorption of Methylene Blue. *Technologies*, 9(4), 96. https://doi.org/10.3390/technologies9040096.

Maiti, D., X. Tong, X. Mou, and K. Yang. 2019. Carbon-Based Nanomaterials for Biomedical Applications: A Recent Study. *Frontiers in Pharmacology*, 1401. https://doi.org/10.3389/fphar.2018.01401.

Makkar, H.P.S., T. Norvsambuu, S. Lkhagvatseren, and K. Becker. 2009. Plant Secondary Metabolites in Some Medicinal Plants of Mongolia Used for Enhancing Animal Health and Production. *Tropicultura*, 27(3), 159–167.

Mardani, S., H. Nasri, S. Hajian, A. Ahmadi, R. Kazemi, and M. Rafieian-Kopaei. 2014. Impact of *Momordica Charantia* Extract on Kidney Function and Structure in Mice. *Journal of Nephropathology*, 3(1), 35. https://doi.org/10.12860%2Fjnp.2014.08.

Marrelli, M., F. Conforti, F. Araniti, and G.A. Statti. 2016. Effects of Saponins on Lipid Metabolism: A Review of Potential Health Benefits in the Treatment of Obesity. *Molecules*, 21(10), 1404. https://doi.org/10.3390/molecules21101404.

Marslin, G., K. Siram, Q. Maqbool, R.K. Selvakesavan, D. Kruszka, P. Kachlicki, and G. Franklin. 2018. Secondary Metabolites in the Green Synthesis of Metallic Nanoparticles. *Materials*, 11(6), 940. http://doi.org/10.3390/ma11060940.

Maruthupandy, M., Y. Zuo, J.S. Chen, J.M. Song, H.L. Niu, C.J. Mao, S.Y. Zhang, and Y.H. Shen. 2017. Synthesis of Metal Oxide Nanoparticles (CuO and ZnO NPs) via Biological Template and Their Optical Sensor Applications. *Applied Surface Science*, 397, 167–174. https://doi.org/10.1016/j.apsusc.2016.11.118.

Mbuni, Y.M., S. Wang, B.N. Mwangi, N.J. Mbari, P.M. Musili, N.O. Walter, G. Hu, Y. Zhou, and Q. Wang. 2020. Medicinal Plants and Their Traditional Uses in Local Communities Around Cherangani Hills, Western Kenya. *Plants*, 9(3), 331. https://doi.org/10.3390/plants9030331

Mehta, S.K., K. Singh, A. Umar, G.R. Chaudhary, and S. Singh. 2012. Ultra-high Sensitive Hydrazine Chemical Sensor Based on Low-Temperature Grown ZnO Nanoparticles. *Electrochimica Acta*, 69, 128–133. https://doi.org/10.1016/j.electacta.2012.02.091.

Mitchell, M.J., M.M. Billingsley, R.M. Haley, M.E. Wechsler, N.A. Peppas, and R. Langer. 2021. Engineering Precision Nanoparticles for Drug Delivery. *Nature Reviews Drug Discovery*, 20(2), 101–124. https://doi.org/10.1038/s41573-020-0090-8.

Mondal, S., and S.T. Rahaman. 2020. Flavonoids: A Vital Resource in Healthcare and Medicine. *Pharmacy & Pharmacology International Journal*, 8(2), 91–104. https://doi.org/10.15406/ppij.2020.08.00285.

Moon, S.A., B.K. Salunke, B. Alkotaini, E. Sathiyamoorthi, and B.S. Kim. 2015. Biological Synthesis of Manganese Dioxide Nanoparticles by *Kalopanax Pictus* Plant Extract. *IET Nanobiotechnology*, 9(4), 220–225. https://doi.org/10.1049/iet-nbt.2014.0051.

Mukherjee, S., and S.K. Nethi. 2019. Biological Synthesis of Nanoparticles Using Bacteria. In *Nanotechnology for Agriculture*. Singapore: Springer, pp. 37–51. https://doi.org/10.1007/978-981-32-9370-0_3.

Muniyappan, N., and N.S. Nagarajan. 2014. Green Synthesis of Silver Nanoparticles with *Dalbergia spinosa* Leaves and Their Applications in Biological and Catalytic Activities. *Process Biochemistry*, 49(6), 1054–1061. https://doi.org/10.1016/j.procbio.2014.03.015.

Mustafa, G., R. Arif, A. Atta, S. Sharif, and A. Jamil. 2017. Bioactive Compounds from Medicinal Plants and Their Importance in Drug Discovery in Pakistan. *Matrix Science Pharma (MSP)*, 1(1), 17–26.

Nagajyothi, P.C., M. Pandurangan, D.H. Kim, T.V.M. Sreekanth, and J. Shim. 2017. Green Synthesis of Iron Oxide Nanoparticles and Their Catalytic and In Vitro Anticancer Activities. *Journal of Cluster Science*, 28(1), 245–257. https://doi.org/10.1007/s10876-016-1082-z.

Nakkala, J.R., R. Mata, and S.R. Sadras. 2016. The Antioxidant and Catalytic Activities of Green Synthesized Gold Nanoparticles from *Piper Longum* Fruit Extract. *Process Safety and Environmental Protection*, 100, 288–294. https://doi.org/10.1016/j.psep.2016.02.007.

Neethu, S., E.K. Radhakrishnan, and M. Jyothis. 2019. Biofabrication of Nanoparticles Using Fungi. In *Nanotechnology for Agriculture*. Singapore: Springer, pp. 53–73. https://doi.org/10.1007/978-981-32-9370-0_4.

Nejati, M., S. Masoudi, D. Dastan, and N. Masnabadi. 2021. Phytochemical Analysis and Antibacterial Activity of *Eryngium Pyramidaleboiss* and *Hausskn. Journal of the Chilean Chemical Society*, 66(2), 5230–5236. http://doi.org/10.4067/S0717-97072021000205230.

Niraimathee, V.A., V. Subha, R.E. Ravindran, and S. Renganathan. 2016. Green Synthesis of Iron Oxide Nanoparticles from *Mimosa Pudica* Root Extract. *International Journal of Environment and Sustainable Development*, 15(3), 227–240. https://www.inderscienceonline.com/doi/abs/10.1504/IJESD.2016.077370.

Nurudeen, Q.O., and M.B. Falana. 2021. Identification and Quantification of Secondary Metabolites and the Antimicrobial Efficacy of Leaves Extracts of Some Medicinal Plants. *Zanco Journal of Pure and Applied Sciences*, 33(1), 91–106. https://doi.org/10.21271/ZJPAS.33.1.10.

Ogunyemi, S.O., Y. Abdallah, M. Zhang, H. Fouad, X. Hong, E. Ibrahim, M.M.I. Masum, A. Hossain, J. Mo, and B. Li. 2019. Green Synthesis of Zinc Oxide Nanoparticles Using Different Plant Extracts and Their Antibacterial Activity Against *Xanthomonas* Oryzaepv. Oryzae. *Artificial Cells, Nanomedicine, and Biotechnology*, 47(1), 341–352. https://doi.org/10.1080/21691401.2018.1557671.

Okigbo, R.N., C.L. Anuagasi, and J.E. Amadi. 2009. Advances in Selected Medicinal and Aromatic Plants Indigenous to Africa. *Journal of Medicinal Plants Research*, 3(2), 86–95.

Ovais, M., A.T. Khalil, N.U. Islam, I. Ahmad, M. Ayaz, M. Saravanan, Z.K. Shinwari, and S. Mukherjee. 2018. Role of Plant Phytochemicals and Microbial Enzymes in Biosynthesis of Metallic Nanoparticles. *Applied Microbiology and Biotechnology*, 102(16), 6799–6814. https://doi.org/10.1007/s00253-018-9146-7.

Painuli, S., P. Semwal, A. Bacheti, R.K. Bachheti, and A. Husen. 2020. Nanomaterials from Nonwood Forest Products and Their Applications. In Husen A., Jawaid M. (eds) *Nanomaterials for Agriculture and Forestry Applications*. Elsevier, pp. 15–40. https://doi.org/10.1016/B978-0-12-817852-2.00002-0.

Pan, S.Y., G. Litscher, S.H. Gao, S.F. Zhou, Z.L. Yu, H.Q. Chen, S.F. Zhang, M.K. Tang, J.N. Sun, and K.M. Ko. 2014. Historical Perspective of Traditional Indigenous Medical Practices: The Current Renaissance and Conservation of Herbal Resources. *Evidence-Based Complementary and Alternative Medicine*, 2014. https://doi.org/10.1155/2014/525340.

Pang, Z., J. Chen, T. Wang, C. Gao, Z. Li, L. Guo, Z. Li, L. Guo, and J.X.Y. Cheng. 2021. Linking Plant Secondary Metabolites and Plant Microbiomes: A Review. *Frontiers in Plant Science*, 12, 300. https://doi.org/10.3389/fpls.2021.621276.

Parekh, J., N. Karathia, and S. Chanda. 2006. Evaluation of Antibacterial Activity and Phytochemical Analysis of *Bauhinia Variegata* L. Bark. *African Journal of Biomedical Research*, 9(1), 53–56.

Parlinska-Wojtan, M., J. Depciuch, B. Fryc, and M. Kus-Liskiewicz. 2018. Green Synthesis and Antibacterial Effects of Aqueous Colloidal Solutions of Silver Nanoparticles Using Clove Eugenol. *Applied Organometallic Chemistry*, 32(4), e4276. https://doi.org/10.1002/aoc.4276.

Parlinska-Wojtan, M., M. Kus-Liskiewicz, J. Depciuch, and O. Sadik. 2016. Green Synthesis and Antibacterial Effects of Aqueous Colloidal Solutions of Silver Nanoparticles Using Chamomile Terpenoids as a Combined Reducing and Capping Agent. *Bioprocess and Biosystems Engineering*, 39(8), 1213–1223. https://doi.org/10.1007/s00449-016-1599-4.

Patel, D.K., H.B. Kim, S.D. Dutta, K. Ganguly, and K.T. Lim. 2020. Carbon Nanotubes-Based Nanomaterials and Their Agricultural and Biotechnological Applications. *Materials*, 13(7), 1679. https://doi.org/10.3390/ma13071679.

Pathmanathan, M.K., K. Uthayarasa, J.P. Jeyadevan, and E.C. Jeyaseelan. 2010. In Vitro Antibacterial Activity and Phytochemical Analysis of Some Selected Medicinal Plants. *International Journal of Pharmaceutical and Biological Archives*, 1(3), 291–299.

Perveen, S. 2021. Introductory Chapter: Terpenes and Terpenoids. In *Terpenes and Terpenoids-Recent Advances*. IntechOpen. https://doi.org/10.5772/intechopen.98261.

Pillai, A.M., V.S. Sivasankarapillai, A. Rahdar, J. Joseph, F. Sadeghfar, K. Rajesh, and G.Z. Kyzas. 2020. Green Synthesis and Characterization of Zinc Oxide Nanoparticles with Antibacterial and Antifungal Activity. *Journal of Molecular Structure*, 1211, 128107. https://doi.org/10.1016/j.molstruc.2020.128107.

Pimpliskar, M.R., R. Jadhav, Y. Ughade, and R.N. Jadhav. 2021. Preliminary Phytochemical and Pharmacological Screening of *Pogostemon benghalensis* for Antioxidant and Antibacterial Activity. *Asian Journal of Pharmacy and Pharmacology*, 7(1), 28–32. https://doi.org/10.31024/ajpp.2021.7.1.7.

Pott, D.M., S. Osorio, and J.G. Vallarino. 2019. From Central to Specialized Metabolism: An Overview of Some Secondary Compounds Derived from the Primary Metabolism for Their Role in Conferring Nutritional and Organoleptic Characteristics to Fruit. *Frontiers in Plant Science*, 835. https://doi.org/10.3389/fpls.2019.00835.

Prashanth, G.K., P.A. Prashanth, B.M. Nagabhushana, S. Ananda, G.M. Krishnaiah, H.G. Nagendra, H.M. Sathyananda, S.C. Rajendra, S. Yogisha, S. Anand, and Y. Tejabhiram. 2018. Comparison of Anticancer Activity of Biocompatible ZnO Nanoparticles Prepared by Solution Combustion Synthesis Using Aqueous Leaf Extracts of *Abutilon indicum*, *Melia azedarach* and *Indigofera tinctoria* as Biofuels. *Artificial Cells, Nanomedicine, and Biotechnology*, 46(5), 968–979. https://doi.org/10.1080/21691401.2017.1351982.

Priya, K., and H.P. Sharma. 2021. Phytochemical Analysis and Antimicrobial Activity of *Hibiscus rosa sinensis*. *European Journal of Biotechnology and Bioscience*, 9(1), 21–26.

Rafique, M., I. Sadaf, M.S. Rafique, and M.B. Tahir. 2017. A Review on Green Synthesis of Silver Nanoparticles and Their Applications. *Artificial Cells, Nanomedicine, and Biotechnology*, 45(7), 1272–1291. https://doi.org/10.1080/21691401.2016.1241792.

Rajakumar, G., A.A. Rahuman, C. Jayaseelan, K.T. Santhosh, S. Marimuthu, C. Kamaraj, A. Bagavan, A.A. Zahir, A.V. Kirthi, G. Elango, and P. Arora. 2014. *Solanum trilobatum* Extract-Mediated Synthesis of Titanium Dioxide Nanoparticles to Control *Pediculus Humanus Capitis*, *Hyalomma Anatolicum* and *Anopheles Subpictus*. *Parasitology Research*, 113(2), 469–479. https://doi.org/10.1007/s00436-013-3676-9.

Rauf, A., T. Ahmad, A. Khan, G. Uddin, B. Ahmad, Y.N. Mabkhot, S. Bawazeer, N. Riaz, B.K. Malikovna, and Z.M. Almarhoon. 2021. Green Synthesis and Biomedicinal Applications of Silver and Gold Nanoparticles Functionalized with Methanolic Extract of *Mentha longifolia*. *Artificial Cells, Nanomedicine, and Biotechnology*, 49(1), 194–203. https://doi.org/10.1080/21691401.2021.1890099.

Rauti, R., M. Musto, S. Bosi, M. Prato, and L. Ballerin. 2019. Properties and Behavior of Carbon Nanomaterials When Interfacing Neuronal Cells: How Far Have We Come? *Carbon*, 143, 430–446. https://doi.org/10.1016/j.carbon.2018.11.026.

Reddy, N., and M. Rapisarda. 2021. Properties and Applications of Nanoparticles from Plant Proteins. *Materials*, 14(13), 3607. https://doi.org/10.3390/ma14133607.

Rehman, J.U., A. Iqbal, A. Mahmood, H.M. Asif, E. Mohiuddin, and M. Akram. 2021. Phytochemical Analysis, Antioxidant and Antibacterial Potential of Some Selected Medicinal Plants Traditionally Utilized for the Management of Urinary Tract Infection. *Pakistan Journal of Pharmaceutical Sciences*, 34(3). https://doi.org/1056-1062. https://doi.org/10.36721/PJPS.2021.34.3.SUP.1057-1062.1.

Roby, M.H.H., M.A. Sarhan, K.A.H. Selim, and K.I. Khalel. 2013. Antioxidant and Antimicrobial Activities of Essential Oil and Extracts of Fennel (*Foeniculum vulgare* L.) and Chamomile (*Matricaria Chamomilla* L.). *Industrial Crops and Products*, 44, 437–445. https://doi.org/10.1016/j.indcrop.2012.10.012.

Rodriguez, P.L., T. Harada, D.A. Christian, D.A. Pantano, R.K. Tsai, and D.E. Discher. 2013. Minimal "Self" Peptides That Inhibit Phagocytic Clearance and Enhance Delivery of Nanoparticles. *Science*, 339(6122), 971–975. https://doi.org/10.1126/science.1229568.

Rostami-Vartooni, A., L. Rostami, and M. Bagherzadeh. 2019. Green Synthesis of Fe_3O_4/Bentonite-Supported Ag and Pd Nanoparticles and Investigation of Their Catalytic Activities for the Reduction of Azo Dyes. *Journal of Materials Science: Materials in Electronics*, 30(24), 21377–21387. https://doi.org/10.1007/s10854-019-02514-3.

Roy, B., S. Mukherjee, N. Mukherjee, P. Chowdhury, and S.P.S. Babu. 2014. Design and Green Synthesis of Polymer Inspired Nanoparticles for the Evaluation of Their Antimicrobial and Anti-filarial Efficiency. *RSC Advances*, 4(65), 34487–34499. https://doi.org/10.1039/c4ra03732d.

Şahin, B., A. Aygün, H. Gündüz, K. Şahin, E. Demir, S. Akocak, and F. Şen. 2018. Cytotoxic Effects of Platinum Nanoparticles Obtained from Pomegranate Extract by the Green Synthesis Method on the MCF-7 Cell Line. *Colloids and Surfaces B: Biointerfaces*, 163, 119–124. https://doi.org/10.1016/j.colsurfb.2017.12.042.

Santhoshkumar, T., A.A. Rahuman, C. Jayaseelan, G. Rajakumar, S. Marimuthu, A.V. Kirthi, K. Velayutham, J. Thomas, J. Venkatesan, and S.K. Kim. 2014. Green Synthesis of Titanium Dioxide Nanoparticles Using *Psidium Guajava* Extract and its Antibacterial and Antioxidant Properties. *Asian Pacific Journal of Tropical Medicine*, 7(12), 968–976. https://doi.org/10.1016/S1995-7645(14)60171-1.

Šantić, Ž., N. Pravdić, M. Bevanda, and K. Galić. 2017. The Historical Use of Medicinal Plants in Traditional and Scientific Medicine. *Psychiatria Danubina*, 29(4), 69–74.

Saxena, M., J. Saxena, R. Nema, D. Singh, and A. Gupta. 2013. Phytochemistry of Medicinal Plants. *Journal of Pharmacognosy and Phytochemistry*, 1(6), 168–182.

Selim, Y.A., M.A. Azb, I. Ragab, and M.H. Abd El-Azim. 2020. Green Synthesis of Zinc Oxide Nanoparticles Using Aqueous Extract of *Deverra tortuosa* and Their Cytotoxic Activities. *Scientific Reports*, 10(1), 1–9. https://doi.org/10.1038/s41598-020-60541-1.

Sengani, M., A.M. Grumezescu, and V.D. Rajeswari. 2017. Recent Trends and Methodologies in Gold Nanoparticle Synthesis–A Prospective Review on Drug Delivery Aspect. *OpenNano*, 2, 37–46. https://doi.org/10.1016/j.onano.2017.07.001.

Sett, A., M. Gadewar, P. Sharma, M. Deka, and U. Bora. 2016. Green Synthesis of Gold Nanoparticles Using Aqueous Extract of *Dillenia indica*. *Advances in Natural Sciences: Nanoscience and Nanotechnology*, 7(2), 025005. https://doi.org/10.1088/2043-6262/7/2/025005.

Shabestarian, H., M. Homayouni-Tabrizi, M. Soltani, F. Namvar, S. Azizi, R. Mohamad, and H. Shabestarian. 2016. Green Synthesis of Gold Nanoparticles Using Sumac Aqueous Extract and Their Antioxidant Activity. *Materials Research*, 20(1), 264–270. http://doi.org/10.1590/1980-5373-MR-2015-0694.

Sharara, D.T., A.H. Al-Marzoqi, and H.J. Hussein. 2021. In Vitro Antibacterial Efficacy of the Secondary Metabolites Extracted from *Myrtus communis* L. Against Some Pathogenic Bacteria Isolated from Hemodialysis Fluid. *Annals of the Romanian Society for Cell Biology*, 25(6), 9267–9274. https://annalsofrscb.ro/index.php/journal/article/view/7197.

Sharma, D., S. Kanchi, and K. Bisetty. 2019. Biogenic Synthesis of Nanoparticles: A Review. *Arabian Journal of Chemistry*, 12(8), 3576–3600. http://doi.org/10.1016/j.arabjc.2015.11.002.

Sharmila, G., M.F. Fathima, S. Haries, S. Geetha, N.M. Kumar, and C. Muthukumaran. 2017. Green Synthesis, Characterization and Antibacterial Efficacy of Palladium Nanoparticles Synthesized Using *Filicium Decipiens* Leaf Extract. *Journal of Molecular Structure*, 1138, 35–40. https://doi.org/10.1016/j.molstruc.2017.02.097.

Singh, O., Z. Khanam, N. Misra, and M.K. Srivastava. 2011. Chamomile (*Matricaria Chamomilla* L.): An Overview. *Pharmacognosy Reviews*, 5(9), 82. https://doi.org/10.4103/0973-7847.79103.

Skalickova, S., M. Baron, and J. Sochor. 2017. Nanoparticles Biosynthesized by Yeast: A Review of Their Application. *Kvasny Prumysl*, 63(6), 290–292. https://doi.org/10.18832/kp201727.

Tahir, K., S. Nazir, A. Ahmad, B. Li, A.U. Khan, Z.U.H. Khan, F.U. Khan, Q.U. Khan, A. Khan, and A.U. Rahman. 2017. Facile and Green Synthesis of Phytochemicals Capped Platinum Nanoparticles and In Vitro Their Superior Antibacterial Activity. *Journal of Photochemistry and Photobiology B: Biology*, 166, 246–251. https://doi.org/10.1016/j.jphotobiol.2016.12.016.

Thangamani, N., and N. Bhuvaneshwari. 2019. Green Synthesis of Gold Nanoparticles Using *Simarouba Glauca* Leaf Extract and their Biological Activity of Micro-Organism. *Chemical Physics Letters*, 732, 136587. https://doi.org/10.1016/j.cplett.2019.07.015.

Thiyagarajulu, N., S. Arumugam, A.L. Narayanan, T. Mathivanan, and R.R. Renuka. 2020. Green Synthesis of Reduced Graphene Nanosheets Using Leaf Extract of *Tridax procumbens* and Its Potential In Vitro Biological Activities. *Biointerface Research in Applied Chemistry*, 11(3), 9975–9984. https://doi.org/10.33263/BRIAC113.99759984.

Tiwari, R., and C.S. Rana. 2015. Plant Secondary Metabolites: A Review. *International Journal of Engineering Research and General Science*, 3(5), 661–670.

Tiwari, R., R. Baral, N. Parajuli, R. Shrestha, S. Pun, A. Pahari, and S. Gurung. 2021. Phytochemical Screening, Free Radical Scavenging and In-Vitro Anti-bacterial Activity of Ethanolic Extracts of Selected Medicinal Plants of Nepal and Effort Towards Formulation of Antibacterial Cream from the Extracts. *International Journal of Herbal Medicine*, 9(3), 39–47. http://www.florajournal.com/.

Tripathi, S.K., A.K. Mahakud, and B.K. Biswal. 2019. A Green Approach Towards Formulation, Characterization, and Antimicrobial Activity of Poly (Lactic-Co-glycolic) Acid-*Alstonia Scholaris* Based Nanoparticle. *Materials Research Express*, 6(9), 095325. https://doi.org/10.1088/2053-1591/ab30d1.

Tumbde, S., F. Rehman, A. Ghosh, and A. Dedhe. 2021. Current Approaches in In-Vitro Production of Secondary Metabolites from Medicinal Plants. *EPRA International Journal of Research and Development*, 6(3), 67–71.

Tura, G.T., W.B. Eshete, and G.T. Tucho. 2017. Antibacterial Efficacy of Local Plants and Their Contribution to Public Health in Rural Ethiopia. *Antimicrobial Resistance and Infection Control*, 6(1), 1–7. https://doi.org/10.1186/s13756-017-0236-6.

Umar, M.F., F. Ahmad, H. Saeed, S.A. Usmani, M. Owais, and M. Rafatullah. 2020. Bio-mediated Synthesis of Reduced Graphene Oxide Nanoparticles from *Chenopodium album*: Their Antimicrobial and Anticancer Activities. *Nanomaterials*, 10(6), 1096. https://doi.org/10.3390/nano10061096.

Umashankar, D.D. 2020. Plant Secondary Metabolites as Potential Usage in Regenerative Medicine. *The Journal of Phytopharmacology*, 9(4), 270–273. https://doi.org/10.31254/phyto.2020.9410.

Velayutham, K., A.A. Rahuman, G. Rajakumar, T. Santhoshkumar, S. Marimuthu, C. Jayaseelan, A. Bagavan, A.V. Kirthi, C. Kamaraj, A.A. Zahir, and G. Elango. 2012. Evaluation of Catharanthus roseus Leaf Extract-Mediated Biosynthesis of Titanium Dioxide Nanoparticles Against *Hippobosca Maculata* and *Bovicolaovis*. *Parasitology Research*, 111(6), 2329–2337. https://doi.org/10.1007/s00436-011-2676-x.

Vinay, S.P. 2021. Synthesis of Fullerene (C60)-Silver Nanoparticles Using Neem Gum Extract Under Microwave Irradiation. *BioNanoScience*, 11(1), 1–7. https://doi.org/10.1007/s12668-020-00799-x.

Wang, C., R. Mathiyalagan, Y.J. Kim, V. Castro-Aceituno, P. Singh, S. Ahn, D. Wang, and D.C. Yang. 2016. Rapid Green Synthesis of Silver and Gold Nanoparticles Using *Dendropanax Morbiferus* Leaf Extract and their Anticancer Activities. *International Journal of Nanomedicine*, 11, 3691. http://doi.org/10.2147/IJN.S97181.

Wang, M., Y. Meng, H. Zhu, Y. Hu, C.P. Xu, X. Chao, W. Li, C. Li, and C. Pan. 2021. Green Synthesized Gold Nanoparticles Using *Viola Betonicifolia* Leaves Extract: Characterization, Antimicrobial, Antioxidant, and Cytobiocompatible Activities. *International Journal of Nanomedicine*, 16, 7319. https://doi.org/10.2147%2FIJN.S323524.

Wanjohi, B.K., V. Sudoi, E.W. Njenga, and W.K. Kipkore. 2020. An Ethnobotanical Study of Traditional Knowledge and Uses of Medicinal Wild Plants Among the Marakwet Community in Kenya. *Evidence-Based Complementary and Alternative Medicine*, 2020. https://doi.org/10.1155/2020/3208634.

Wijesinghe, W.P.S.L., M.M.M.G.P.G. Mantilaka, K.A.A. Ruparathna, R.B.S.D. Rajapakshe, S.A.L. Sameera, and M.G.G.S.N. Thilakarathna. 2020. Filler Matrix Interfaces of Inorganic/Biopolymer Composites and Their Applications. In *Interfaces in Particle and Fibre Reinforced Composites*, 95–112. https://doi.org/10.1016/B978-0-08-102665-6.00004-2.

Xin, Q., H. Shah, A. Nawaz, W. Xie, M.Z. Akram, A. Batool, L. Tian, S.U. Jan, R. Boddula, B. Guo, and Q. Liu. 2019. Antibacterial Carbon-Based Nanomaterials. *Advanced Materials*, 31(45), 1804838. https://doi.org/10.1002/adma.201804838.

Yadav, R.N.S., and M. Agarwala. 2011. Phytochemical Analysis of Some Medicinal Plants. *Journal of Phytology*, 3(12), 10–14. http://journal-phytology.com/.

Yadi, M., E. Mostafavi, B. Saleh, S. Davaran, I. Aliyeva, R. Khalilov, M. Nikzamir, N. Nikzamir, A. Akbarzadeh, Y. Panahi, and M. Milani. 2018. Current Developments in Green Synthesis of Metallic Nanoparticles Using Plant Extracts: A Review. *Artificial Cells, Nanomedicine, and Biotechnology*, 46(3), S336–S343. https://doi.org/10.1080/21691401.2018.1492931.

Yew, Y.P., K. Shameli, M. Miyake, N.B.B.A. Khairudin, S.E.B. Mohamad, T. Naiki, and K.X. Lee. 2020. Green Biosynthesis of Superparamagnetic Magnetite Fe_3O_4 Nanoparticles and Biomedical Applications in Targeted Anticancer Drug Delivery System: A Review. *Arabian Journal of Chemistry*, 13(1), 2287–2308. https://doi.org/10.1016/j.arabjc.2018.04.013.

Yonekura-Sakakibara, K., Y. Higashi, and R. Nakabayashi. 2019. The Origin and Evolution of Plant Flavonoid Metabolism. *Frontiers in Plant Science*, 10, 943. https://doi.org/10.3389/fpls.2019.00943.

Younis, I.Y., S.S. El-Hawary, O.A. Eldahshan, M.M. Abdel-Aziz, and Z.Y. Ali. 2021. Green Synthesis of Magnesium Nanoparticles Mediated from *Rosa floribunda Charisma* Extract and Its Antioxidant, Antiaging and Antibiofilm Activities. *Scientific Reports*, 11(1), 1–15. https://doi.org/10.1038/s41598-021-96377-6.

Zare, E.N., V.V. Padil, B. Mokhtari, A. Venkateshaiah, S. Wacławek, M. Černík, F.R. Tay, R.S. Varma, and P. Makvandi. 2020. Advances in Biogenically Synthesized Shaped Metal-And Carbon-Based Nanoarchitectures and Their Medicinal Applications. *Advances in Colloid and Interface Science*, 283, 102236. https://doi.org/10.1016/j.cis.2020.102236.

Zaynab, M., M. Fatima, Y. Sharif, M.H. Zafar, H. Ali, and K.A. Khan. 2019. Role of Primary Metabolites in Plant Defense Against Pathogens. *Microbial Pathogenesis*, 137, 103728. https://doi.org/10.1016/j.micpath.2019.103728.

Zhang, H., J.A. Jacob, Z. Jiang, S. Xu, K. Sun, Z. Zhong, N. Varadharaju, and A. Shanmugam. 2019. Hepatoprotective Effect of Silver Nanoparticles Synthesized Using Aqueous Leaf Extract of *Rhizophora apiculata*. *International Journal of Nanomedicine*, 14, 3517. https://doi.org/10.2147/IJN.S198895.

2 Medicinal Plant-Based Flavonoid-Mediated Nanoparticles Synthesis, Characterization, and Applications

Abrha Mengstu, Siraye Esubalew, Limenew Abate, Rakesh Kumar Bachheti, Azamal Husen, and Archana Bachheti

CONTENTS

2.1 INTRODUCTION

Nanotechnology is a large multidisciplinary subject of research, development, and economic activity that has gained prominence around the world in recent years. It is a term that refers to the design, synthesis, development, characterization, exploration, and application of nanomaterials for the advancement of science. Nanometres are used to determine at least one typical measurement of nanostructured materials (Bayda et al. 2020). The green chemistry approach is critical for the longer-term prospect of nanoparticles (NPs) synthesis and supplies safe, eco-friendly, and environmentally benign NPs (Varma 2012). NPs are mainly classified as inorganic and organic NPs. Organic NPs comprise carbon NPs such as carbon nanotubes, quantum dots, and fullerenes. In contrast, inorganic NPs comprise metallic NPs such as silver, gold, copper, and aluminium, semiconductor NPs like zinc oxide (ZnO), zinc sulphide (ZnS), cadmium sulphide (CdS), and magnetic NPs of nickel, iron, cobalt, and silver. Gold and silver NPs are attracting attention these days because they have superior properties and useful stability (Bachheti et al. 2020). They have electrical, magnetic,

DOI: 10.1201/9781003213727-2

optical, physical, and chemical properties. They have a wide range of uses in the medical, electronics, biosensor, cosmetics, and textiles industries due to their higher surface area (Ahmed et al. 2010).

The 'top-down' or breakdown and 'bottom-up' or build-up approaches are the techniques that are used to synthesize NPs. The top-down approach is that in which bulk materials are broken into fine particles using various methods such as pulse wire discharge, ball milling, evaporation–condensation, and pulsed laser ablation. On the other hand, the bottom-up strategy is the self-assembly of atoms into new nuclei, which grow into nanoscale particles. These two approaches comprise three methods of NPs synthesis: biological, chemical, and physical methods. The physical method of NPs synthesis mainly follows a top-down approach. The chemical and biological methods of NPs synthesis are classified as the bottom-up approach.

NPs have been made using a variety of secondary metabolites, which are abundant in medicinal essential plant species and could be used as capping and reducing agents, implying that they could be effective agents for NPs biosynthesis (Mashwani et al. 2015). To date, there has not been a comprehensive review of the role of flavonoids in the synthesis of metal oxide and metal NPs in a green manner. This chapter emphasizes and discusses the molecular role of flavonoids as capping and reducing agents in producing metal oxide and metal NPs.

2.2 MAJOR PLANT-BASED FLAVONOIDS AND THEIR STRUCTURE

Flavonoids are a natural product with a complex phenolic structure that may exist in vegetables and fruits, cereals, flowers, woods, roots, teas, and wines. These natural compounds' health advantages are well recognized, and attempts are being undertaken to extract flavonoids from the components. Flavonoids are now widely identified as a vital constituent in many medical and cosmetic products. They're becoming more popular due to their antimutagenic, anti-inflammatory, antioxidant, and anticarcinogenic properties, as well as their ability to change the activity of key cellular enzymes. These compounds are involved in a wide range of biological activities in animals, plants, and microbes, and they have long been known to be produced in certain plants. Flavonoids are essential to the production of the colour and scent of flowers and fruit dispersion to aid in the germination of seeds and spores and the development and production of young plants (Griesbach 2005). Flavonoids are used as antimicrobial compounds, detoxifying agents, phytoalexins, allopathic compounds, signal molecules, and UV filters to protect plants from abiotic and biotic stresses (Takahashi and Ohnishi 2004).

Flavonoids have a structure made from two aromatic rings (A and B) connected by an oxygenated heterocyclic ring (C) that is hydroxylated in many positions (Figure 2.1). Flavonoids with the B ring connected in position 3 of the C ring are known as isoflavones. In neoflavonoids, the B ring is attached to the C ring at position 4, however, those with the B ring attached at position 2 can be divided into subgroups based on the C ring's structural features. Based on the carbon of the C ring to which the B ring is attached and the degree of unsaturation and oxidation of the C ring, flavonoids are classified as isoflavones, neoflavonoids, flavones, chalcones, flavanonols, flavanols, and flavones. Flavonoids tend to lose electrons or hydrogen atoms to absorb activated oxygen and become the intermediate products of metal ions. So, flavonoids are responsible for reducing metal ions.

2.3 FLAVONOID-BASED NANOPARTICLES SYNTHESIS AND CHARACTERIZATION

2.3.1 Metal Nanoparticles

Metal NPs are used in a variety of industries due to their unique physicochemical properties, including a higher surface-to-volume ratio and small size compared to bulk materials (Kumar et al. 2013). Metal NPs can be made using standard methods, but they are costly and need strong chemical reagents that are highly hazardous to human health (Cushing et al. 2004). As a result, researchers

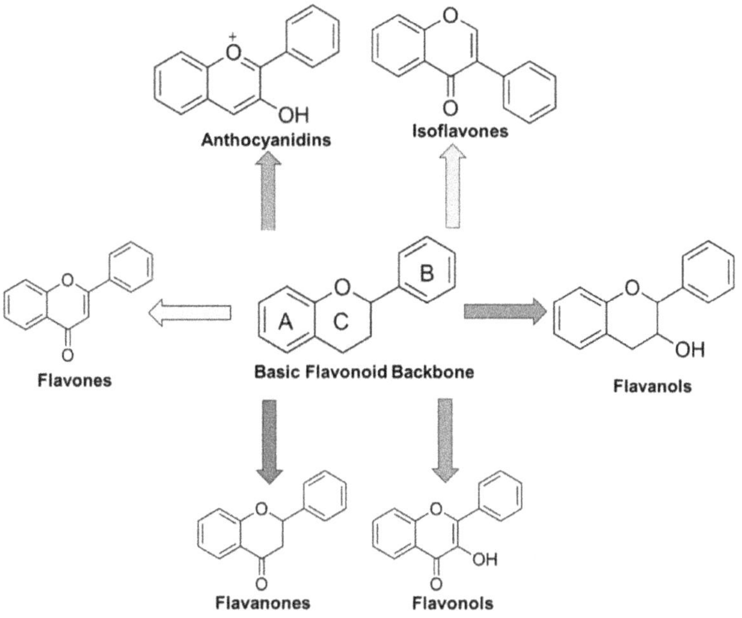

FIGURE 2.1 Basic skeleton of flavonoids and their subclasses (Adapted from Li et al. 2018).

use green or biological methods to synthesize metal NPs. Secondary metabolites like flavonoids are currently significant in synthesizing metal NPs via capping and reducing agents. Flavonoids such as tricetin, kaempferol, quercetin, hesperidin, naringin and diosmin, dihydromyricetin (DMY), apiin, baicalein, etc., have been shown in recent studies to have anti-inflammatory properties. They play a major role in the bioreduction of metal ions and capping of the synthesized NPs (Table 2.1). Zeta potential, ATR-FTIR, TGA, AFM, DLS, EDS, SEM, TEM, XRD, and UV-Vis spectroscopy are the most commonly used techniques for the characterization of NPs.

Alsamhary et al. (2020) stated that gold nanoparticles (AuNPs) could be produced at optimal values of pH 8, temperature 30°C, 250 μM chloroauric acid, and 125 μM tricetinutilizing flavonoid tricetin as a capping and reducing agent. ATR-FTIR, TEM, XRD, and UV-Vis spectroscopy analyzed the AuNPs. The carbonyl (C=O) and hydroxyl (–OH) groups of tricetin were predominantly involved in creating AuNPs, and the NPs were observed to be spherical with a 12 nm average size. To synthesize AuNPs, Hou et al. (2020) employed procyanidins (PCs) as a reducing and stabilizing agent. According to a TEM analysis, the NP is spherical. Oueslati et al. (2020) used kaempferol 3-O-b-D-apiofuranosyl-7-O-a-L-rhamnopyranoside (KG) from the *Lotus leguminosae* plant to make AuNPs. The synthesis parameters (mass ratio, pH, incubation time) were varied to produce stable colloids of AuNPs with tuned size and morphology. By using a fixed pH of 10 and three different values of 1/4, 1/8, and 1/16, the influence of the mass ratio (HAuCl4/KG) was investigated. The pH and HAuCl4/KG were optimized to produce exceptionally stable colloids of spherical-shaped AuNPs with a diameter in the range of around 37 nm. The synthesized NPs were characterized by TGA, TEM, FTIR, and UV–Vis spectrophotometry.

Jain and Mehata (2017) employed FTIR, XRD, and TEM to evaluate silver nanoparticles (AgNPs) made from *Ocimum sanctum* leaf extract and its derivative quercetin (flavonoid) as precursors under various conditions concentration of reactant, time of reaction, pH, and temperature. These analysis techniques have shown that AgNPs produced from both extracts and quercetin have the same morphological, optical, and antimicrobial properties, suggesting that quercetin is primarily responsible for converting metal NPs. Ameen et al. (2018) utilized DMY to make AgNPs, and the they were characterized by XRD, FTIR, TEM, FE-SEM, and UV-Vis. They had a spherical shape,

TABLE 2.1
Flavonoids-Based Metal Nanoparticles Synthesis, Particle Sizes, Shapes, and Their Applications

Plants/flavonoid	NP	Size (nm)	Shape	Characterization	Application	Reference
Tricetin	Au	12	Spherical	TEM, FTIR, UV-Vis, and ATR-XRD	Antibacterial activity	Alsamhary et al. (2020)
Procyanidins	Au	19	Spherical	UV—Vis, FTIR, TEM, and XRD,	Light-controlled drug release	Hou et al. (2020)
Ocimum sanctum (quercetin)	Ag	10–20	–	TEM, XRD, FTIR, and photoluminescence	Antibacterial activity	Jain and Mehata (2017)
Dihydromyricetin	Ag	34	Spherical	UV-Vis, FESEM-EDX, TEM, FTIR, and XRD	Antifungal activity	Ameen et al. (2018)
Quercetin and rutin	Ag	3–22 and 4–17	Spherical	SEM, UV-Vis, TEM, XRD, and CV	Antimicrobial activity	Zhou and Tang (2018)
Pongamia pinnata L. (Karanjin)	Ag	20	Spherical	UV-Vis, TEM, FTIR, XRD, EDS, and AFM	Antifungal activity	Naveena et al. (2018)
Lilium casa blanca (flavonoids)	Ag	12.7	Spherical	UV-Vis, TEM, FTIR, DLS, and zeta potential	Antibacterial and catalytic activity	Luo et al. (2018)
Citrus flavonoids (hesperetin)	Au	22	Spherical	HR-TEM, DLS, XRP, andUV-Vis	–	Sierra et al. (2016)
Quercetin	Ag, Au	53, 27	–	UV-Vis, SEM, DLS, EDX, and HR-SEM	Anti-inflammatory activity	Ozdal et al. (2019)
Pinocembrin andgalangin	Au, Ag	50, 50	Spherical	UV-Vis, TEM, FTIR, and CV	–	Roy et al. (2010)
Quercetin and gallic acid	Ag/Se	30–35	Spherical	Zeta-potential, EDX, XRD, HR-TEM, and UV-Vis	Antitumour, antimicrobial, and antioxidant activities	Mittal et al. (2014)
Lotus leguminosae (kaempferol)	Au	37	Spherical	UV-Vis, NMR, TEM, TGA, and FTIR	Antioxidant, catalytic and anticancer activities	Oueslati et al. (2020)
Quercetin phosphate	Ag	11.89–30.6	Spherical	TEM and UV-Vis	Antifungal and antibacterial activities	Osonga et al. (2018)
Elaeisguineensis (flavonoid)	Au	13.3–15.22	Spherical	XPS, TEM, FTIR and XRD, EDAX, UV-Vis, and DLS	–	Irfan et al. (2017)
Psidium guajava (flavonoids)	Ag	15–20	Spherical	XRD, FTIR, TEM, UV-Vis, and SEM-EDX	Photo-catalytic and antimicrobial activities	Wang et al. (2018)
Quercetin	Au	20–45	Spherical	TEM, UV-Vis, AFM, and DLS	–	Das et al. (2013)

(Continued)

TABLE 2.1 (CONTINUED)
Flavonoids-Based Metal Nanoparticles Synthesis, Particle Sizes, Shapes, and Their Applications

Plants/flavonoid	NP	Size (nm)	Shape	Characterization	Application	Reference
Radix Hedysari (Formononetin)	Au		Spherical	DLS, FTIR, TEM, and UV-Vis	Real sample detection	Chen et al. (2020)
Hesperidin, naringin, and diosmin	Ag	10–80	Oval and hexagonal	TEM, FTIR, XRD, and UV-Vis	Cytotoxic and antibacterial activities	Sahu et al. (2016)
Dihydromyricetin	Au	25–43	Spherical	UV-Vis, TEM, FTIR, and particle size analyzer	–	Guo et al. (2014)
Kaempferol	Au	18.24	Spherical	UV-Vis, zeta potential, HR-TEM, EDAX, and FTIR	Antileishmanial activity	Halder et al. (2017)
Quercetin (capped)	Au	<100	Spherical	TEM, FTIR, and UV-Vis	Antioxidant, antimicrobial, and cytotoxic activities	Milanezi et al. (2019)
Baicalein	Au	26.5	Spherical	HR-TEM, SEM, DLS, EDAX, FTIR, and UV-Vis,	Anti-biofilm activity	Rajkumari et al. (2017)
Apiin	Au, Ag	21 and 39	Spherical, quasi-spherical	XRD, TEM, FTIR, and UV-Vis	IR-absorbing optical coatings	Kasthuri et al. (2009)
Kaempferol	Au	16.5 $\pm$ 2.5	Spherical	DLS, ICP-OES, XRD, FTIR, TEM, TGA, and zeta-potential	Anticancer, antioxidant, and antiangiogenic activities	Raghavan et al. (2015)
Dihydromyricetin	Ag	114	Spherical	XRD, XPs, FITR, DLS, TEM, SEM, and UV-Vis	Anticancer, antibacterial, and antioxidant activities	Li et al. (2021)
Quercetin	Ag, Au	32.4 and 20.2	Spherical	FE-TEM, HR-XRD, and UV-Vis	Anticancer and catalytic activities	Lee and Park (2020)
Myricetin (MY)	Ag	20–50	Spherical	XRD, TEM, SEM, FTIR and UV-Vis	Antioxidant and antibacterial activities	Li et al. (2020)
Elaeisguineensis (flavonoid)	Au	26.41 $\pm$ 7.95	Spherical	UV-Vis, FTIR, GCMS, TEM, and FE-SEM	–	Ahmad et al. (2019)
Myricetin (MY)	Au	6	Spherical	TEM, FTIR and XRD	–	Sathishkumar et al. (2019)
Genistein	Au	64.64	Spherical	UV-Vis, NMR, TEM, DLS, and TGA	Cytotoxic activity	Stolarczyk et al. (2017)

with a mean particle size of 34 nm. According to FTIR studies, the C=O and –OH groups were responsible for NP production. Latif et al. (2018) investigated the influence of temperature on the preparation of AuNPs using crude flavonoids extract (CACrF) from *Centella asiatica* as a reducing agent. AuNP synthesis is directly related to temperature; according to the findings, there is a blue shift in the UV-Vis spectra as temperature rises. The size of the AuNPs decreases as the temperature of the reaction media rises. The broadening of the peak demonstrates the polydispersity of NPs. As evidenced by their sharp and narrow maximal absorption, the polydispersity of AuNPs synthesized at high temperatures is higher than that of AuNPs synthesized at low temperatures. AuNP concentration rises as the temperature of the reaction medium rises since their maximal absorption value is directly proportional to their concentration.

AgNPs were produced in an environmentally friendly and non-toxic manner using a natural flavonoid called karanjin, obtained from *Pongamia pinnata* L. seed and assisted by a microwave technique (Naveena et al. 2018). AFM, XRD, FTIR, UV-Vis, TEM, EDS and have characterized the synthesized NPs. According to UV-visible spectroscopy analysis, the peak absorbance of plasmon resonance was found at 424 nm. The NP has a spherical shape with an approximately 20 nm average size. Yusuf et al. (2020) synthesized AgNPs from *Clinacanthus nutans* leaf and stem extracts, suggesting that flavonoids and phenolics are excellent silver nitrate ($AgNO_3$) reducers. The method is low-cost, non-toxic, and non-polluting, and the produced AgNP-S and AgNP-L were found to be spherical, with 129.9 and 114.7 nm average diameters, respectively. The following are the best conditions for making AgNPs with aqueous extract from *Clinacanthus nutans* leaf and stem with the following parameters: 15% (v/v) extract, 15 mM $AgNO_3$, 60°C, and a 24-hour incubation period. The polydispersity index, the shape of the NP, and the particle size distribution are all influenced by these parameters. The existence of functional groups like –COOH, –C=O and –OH in plant extracts was identified using FTIR, showing that flavonoids and phenolics were conjugated with silver ions. Luo et al. (2018) synthesized AgNPs using flavonoid extracted from *Lilium casablanca* leaves. The reaction parameters, such as reaction time, reaction temperature, the concentration of $AgNO_3$, type of flavonoids, and pH control the synthesis of AgNPs. FTIR, DLS, TEM, UV-Vis, and zeta potential were used to characterize the AgNPs formed. According to the data, the best conditions were 10 mL of 20 mM $AgNO_3$, 1 mL of flavonoid extracts, pH 10, and a temperature of 70°C. According to the UV-Vis investigation, the NP has a maximum absorption peak of 404 nm, and the NPs formed were spherical and 12.7 nm in size.

Bimetallic (Ag/Se) NPs with high yield, that are mono-dispersed and stable, were produced using flavonoids such as gallic acid and quercetin. The method is green, easy, eco-friendly, bioreactive, cost effective, and occurs at ambient conditions (Mittal et al. 2014). Various reaction parameters such as reaction time, temperature, pH, concentration of Ag/Se salt, gallic acid, quercetin, gallic acid, and Ag/Se salt concentration were tuned. The characterization of NPs was carried out by zeta potential, EDX, XRD, HR-TEM, and UV-Vis. They were determined to be 30–35 nm in size. Flavonoids and phenolics were shown to be responsible for both the reduction and stabilization of NPs. Ahmad et al. (2019) revealed that polyphenolic and flavonoid components are significant in the production and stability of AuNPs using aqueous *Elaeis guineensis* leaf extract. The findings revealed that flavonoids and the –OH groups of phenolic compounds were oxidized to C=O groups during the synthesis step, which were then linked to the surface of AuNPs to prevent long periods of time aggregation.

Chen et al. (2020) used a non-toxic, environmentally safe, effective, and renewable technique to make AuNPs from an ethyl acetate *Radix hedysari* extract. Using HPLC, the extract's components were identified, and formononetin accounted for more than 90% of the total material. According to the study, formononetin is used for the green synthesis of AuNPs. Sahu et al. (2016) utilized three distinct flavonoids such as diosmin, naringin, and hesperidin for AgNP synthesis. The NPs produced were characterized by TEM, FTIR, XRD, and UV-Vis. According to TEM analysis, the size of the NPs was found to be in the range of 20–80, 5–40, and 5–50 nm, respectively, and naringin- and hesperidin-derived AuNPs were polydispersed and oval shaped and whereas diosmin-derived AgNPs were hexagonal shaped (Figure 2.2).

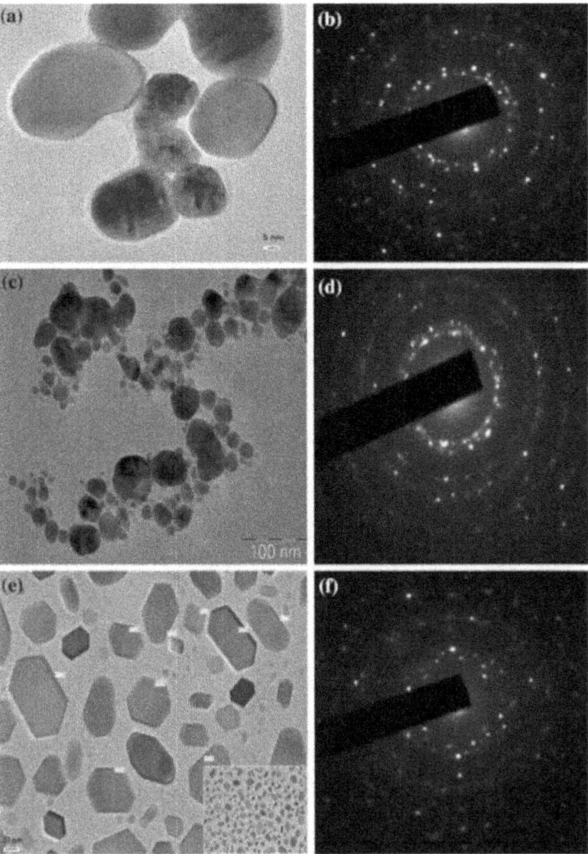

FIGURE 2.2 Transmission electron micrographs of silver nanoparticles synthesized by hesperidin (a), naringin (c), and diosmin (e), and their respective electron diffraction pattern (b, d, f) (Sahu et al. 2016).

The XRD pattern of the AgNPs produced revealed the existence of many peaks of varying intensities over the whole XRD spectrum (Figure 2.3), confirming their polydispersed character. Hesperidin > naringin > diosmin was the order of polydispersity of AgNPs. For diosmin, naringin, and hesperid in the synthesized NPs had the maximum intensity peak at two values of 29.52, 29.260, and 35.280 respectively. XRD results confirmed the crystalline character of the AgNPs produced.

Milanezi et al. (2019) synthesized quercetin-capped AuNPs (Qct-AuNPs), which were verified by TEM, FTIR, and UV-vis. The integrity of the quercetin molecules on the NP surface was shown by FTIR analysis. A slight spherical structure with a diameter of less than 100 nm was discovered using TEM. Rajkumari et al. (2017) utilized flavonoids called baicalein as capping and reducing agents to make AuNPs in a green way. HR-TEM, FTIR, SEM, UV-Vis, DLS, and EDAX were used to examine the baicalein-based AuNPs (BCL-AuNPs). The biosynthesized SCL-AuNPs were discovered to have a morphology of spherical shape and 26.5 nm average size. In a simple and uncomplicated technique to generate Q-AuNPs and Q-AgNPs, quercetin was used as a stabilizing and reducing agent (Lee and Park 2020). For spherical-shaped Q-AuNPs and Q-AgNPs, strong SPR was found at 527 and 401 nm, respectively. AuNPs had a mean size of 20.24 nm. However, AgNPs had a mean size of 32.414 nm, according to FE-TEM pictures. According to XRD patterns, the NPs exhibit a face-centred cubic structure.

AgNPs were effectively synthesized from dietary myricetin (MY) using a cost-effective, non-toxic, simple, and ecologically acceptable green method (Li et al. 2020). XRD, FTIR, TEM, SEM, and UV-Vis were used to characterize MY-AgNPs. According to the findings, the maximum UV-Vis

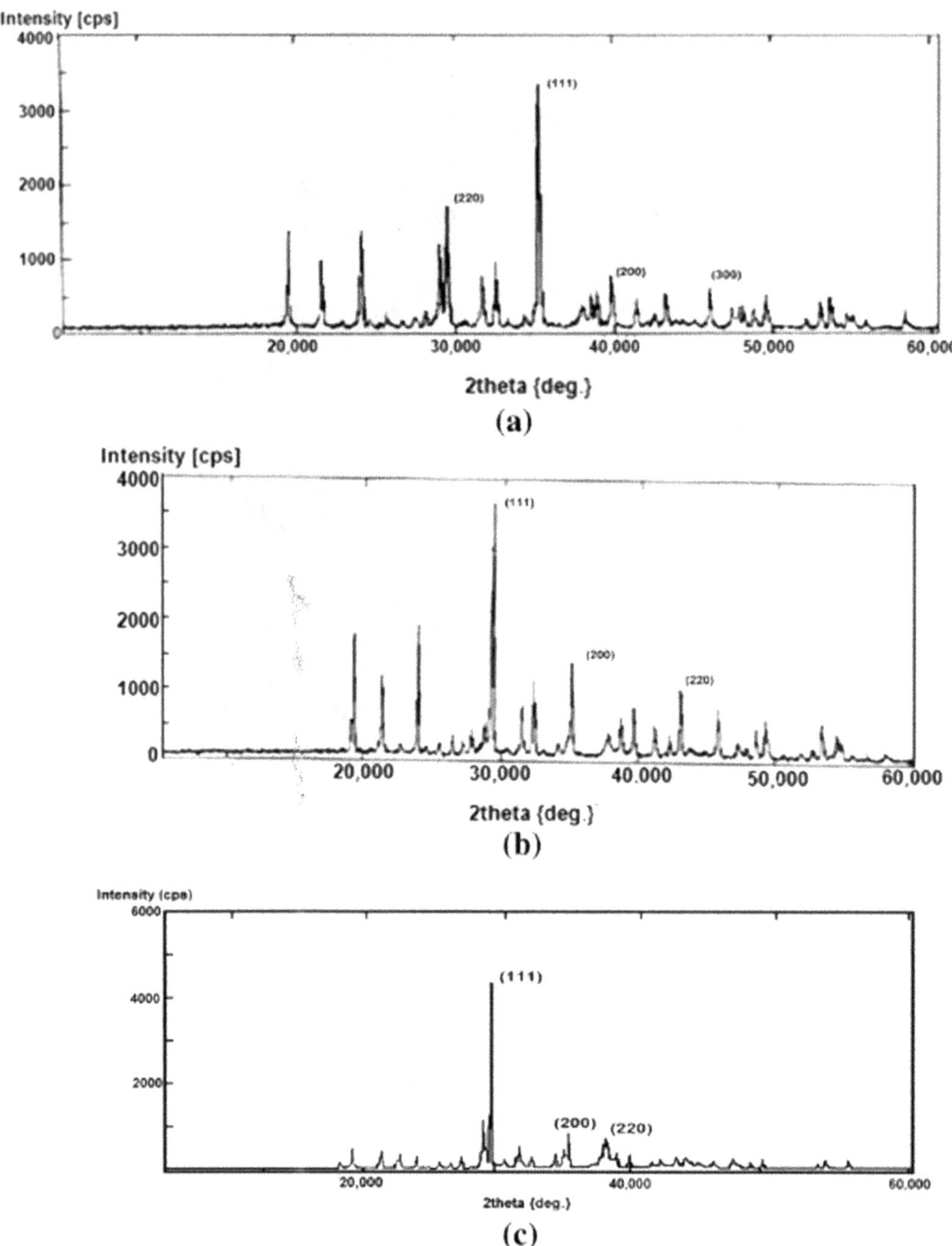

FIGURE 2.3 X-ray diffraction pattern of silver nanoparticles synthesized by a hesperidin, (b) naringin, and (c) diosmin (Sahu et al. 2016).

absorption of AgNPs dispersion was at 410 nm, almost spherical. Sathishkumar et al. (2019) utilized MY flavonoid as a reducing agent to produce AuNPs with a particle size of 6 nm and an approximately spherical shape. The physicochemical characteristics of the produced MY-AuNPs were investigated using XRD, FTIR TEM, FTIR, and XRD analyses. According to FTIR measurements, the –OH groups of MY interact with Au^{3+} ions during nanoparticle formation. The pH, temperature, and storage stability of the MY-AuNPs generated were also investigated. In addition, MY-AuNPs were proven to be safe for therapeutic application in a biocompatibility investigation.

Li et al. (2021) synthesized AgNPs using the natural flavonoid DMY as a reducing and stabilizing agent. To characterize DMY-mediated AgNPs, SEM, TEM, DLS, FTIR, XPS, UV-Vis, and XRD were utilized. According to the study, AgNPs exhibited the highest absorption of UV-visible at around 410 nm. The NPs that resulted were virtually spherical. FTIR was used to identify functional groups that could aid in the transition of metal ions into NPs. Riaz et al. (2020) created AuNPs using flavonoids isolated from the *Berberis lyceum* Royle leaves. XRD, FTIR, UV-Vis, STEM, and EDS were used to describe them, revealing spherical AuNPs with a size of 23 nm. The optimum conditions for the synthesis of AuNPs were found to be a 1:3 ratio of Au to flavonoid, temperature of 70°C and a pH of 4.

Furthermore, details of flavonoid-based metal NP fabrication and their characterization with their shape, size, and application are presented in Table 2.1. In general, flavonoids are used to fabricate Au, Ag, and Ag/Se NPs and the synthesized NPs are mainly characterized by SEM, UV-Vis, FTIR, TEM, EDX, DLS, and zeta potential. The NPs that were synthesized are virtually spherical with size <100 nm and have various properties including antibacterial, antioxidant, catalytic, antifungal, anticancer, antitumour, light-controlled drug release, photo-degradation and so forth.

2.3.2 Metal Oxide Nanoparticles

Metal oxide NPs have unique characteristics that allow them to be employed in a variety of industrial applications, including solar energy conversion, pharmaceuticals, magnetic storage media, sensors, electronics, catalytic processes, and wastewater treatment (Khan et al. 2019; Singh et al. 2021). Flavonoids from diverse plants have an important role in converting metal precursors to useful NPs and stabilizing the synthesized NPs (Suresh et al. 2015; Nibras et al. 2020). Jeyaleela et al. (2020) used a simple homogeneous precipitation process to make zinc oxide nanoparticles (ZnONPs) from the isolated flavonoid quercetin obtained from the medicinal plant *Combretum ovalifolium*. Zinc acetate dehydrate was used as a precursor, and quercetin was used as a reducing agent to create the NP. The quercetin-mediated NPs were investigated using SEM, EDX, XRD, UV-Vis, FTIR, and SEM. According to SEM and XRD investigations, the Q-ZnONPs are fibre-shaped and single crystalline, with a diameter of 31.24 nm. Before and after adding the reducing agent (quercetin), colour changes were used to determine the formation of ZnONPs. The elemental composition of biosynthesized quercetin-mediated ZnONPs was determined using EDX.

SEM imaging was used to examine synthesized quercetin-mediated ZnONPs morphology. Figure 2.4(a) indicates that oxygen and zinc elements mainly exist in the quercetin-mediated ZnONPs, with some small impurities. Figure 2.4(b) demonstrates the morphology of ZnONPs mediated by quercetin.

Generally, flavonoids are used as reducing, capping, and stabilizing agents in the preparation of ZnO, TiO_2, and Fe_3O_4 NPs. They have several applications such as an antioxidant, catalytic, antimicrobial, and for removal of heavy metals. Furthermore, details of flavonoid-based metal oxide NPs syntheses, their characterization with their size, shape, and application are shown in Table 2.2. The synthesized metal oxide NPs were characterized by TEM, SEM, EDX, XRD, FTIR, UV-Vis, and AFM and generally had spherical shapes.

2.4 MECHANISM OF NANOPARTICLES SYNTHESIS FROM FLAVONOIDS

Chemical reduction of organic and inorganic reducing agents is the most prevalent method for producing NPs. Various reducing agents, like poly (ethylene glycol)-block copolymers, N, N-dimethylformamide, Tollens' reagent, polyol process, elemental hydrogen, sodium borohydride, ascorbate, and sodium citrate are important to reduce metal ions in aqueous or non-aqueous liquids.

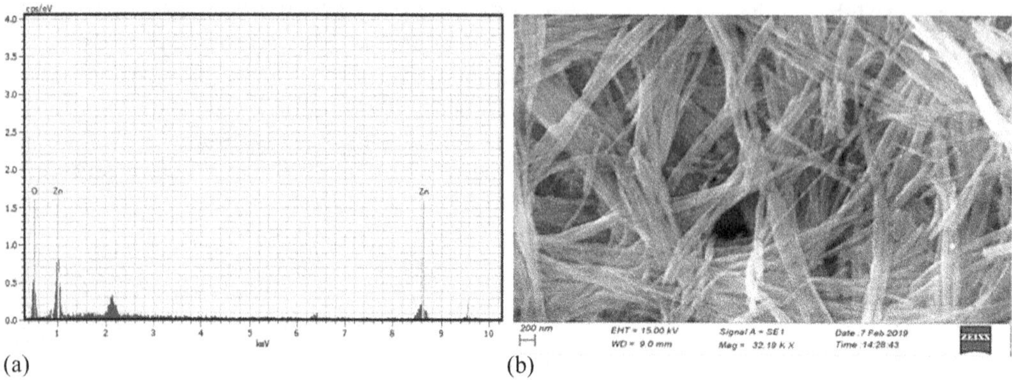

FIGURE 2.4 (a) EDX of biosynthesized Q-ZnONPs and (b) SEM magnification image of biosynthesized Q-ZnONPs (Jeyaleela et al. 2020.)

These reducing agents cause metal ions to be reduced, resulting in the formation of metallic atoms, which then agglomerate to form oligomeric clusters. Metallic colloidal particles are produced as a result of these clusters. Protective agents should be used to prepare metal NPs to stabilize dispersive NPs and safeguard NPs that may absorb on or cling to NP surfaces, preventing agglomeration (Oliveira et al. 2005). Surfactants with functionalities for interactions with particle surfaces such as alcohol, acid, amines, and thiol can stabilize particle growth and protect particles from agglomeration, sedimentation, or losing their surface properties. Such stabilizing and reducing agents are harmful to the environment, toxic, and costly. It is preferable to use green (biological) methods of NP synthesis to overcome these issues.

Synthesis of NPs from metal and metal oxide is a simple and facile process of combining $AgNO_3$, chloroauric acid ($HAuCl_4$), and zinc acetate dehydrate with biological substances commonly known as microorganisms (fungi, bacterial, and algae), plant extract, or isolated phytochemicals (flavonoids) used as stabilizing and capping agents (Vadlapudi and Kaladhar 2014). The stabilization and reduction from Ag^+/Au^{3+} to Ag^0/Au^0 have been attributed to the existence of some functional groups such as –OH, C=O, –COOH flavonoids such as tricetin, baicalein, quercetin, DMY, myricetin, apiin, kaempferol, hesperidin, naringin, diosmin, pinocembrin, galangin, etc. (Alsamhary et al. 2020; Li et al. 2020; Ozdal et al. 2019; Li et al. 2021; Halder et al. 2017). Flavonoid reduction and the stability of AgNPs are not well understood mechanisms. NP formation has three stages: ion reduction, clustering, and subsequent NP growth. The nature of the reducing agent and its concentration, pH, and $AgNO_3$ concentration determine the characteristics of each step (Borodina and Mirgorod 2014). A flavonoid is a potent antioxidant biomolecule that comprises phenolic groups and exhibits keto-enol tautomerism. Few studies have postulated a mechanism for the interaction of flavonoids with metal ions that leads to the creation of nanomaterials (Guo et al. 2014; Raghavan et al. 2015; Zhou and Tang 2018; Jain and Mehata 2017; Naveena et al. 2018; Roy et al. 2010; Oueslati et al. 2020; Ahmad et al. 2019).

2.5 APPLICATION OF FLAVONOID-BASED METAL AND METAL OXIDE NANOPARTICLES

The applications of metal and metal oxide-derived NPs derived from different phytochemical sources, such as flavonoids, have been noted in various studies. These NPs have gained much attention because they have unique properties, including electrical, magnetic, optical, and catalytic properties. Thus, they are used in medicinal (antimicrobial, anticancer, antioxidant, antifungal, antioxidant, drug delivery), environmental remediation (catalytic degradation, heavy metal trapping),

TABLE 2.2
Flavonoids-Based Metal Oxide Nanoparticles Synthesis, Particle Sizes, Shapes, and Their Applications

Plants	NP	Size (nm)	Shape	Characterization	Application	Reference
Artemisia herba alba (flavonoid)	ZnO	18	Spherical	AFM and FTIR	Antioxidant activities	Nibras et al. (2020)
Combretum ovalifolium (Quercetin)	ZnO	31.24	Fibre shaped	UV-Vis, FTIR, XRD, EDX, and SEM	Antioxidant and catalytic activities	Jeyaleela et al. (2020)
Cassia fistula (flavonoid and polyphenol)	ZnO	5–15	Hexagonal	UV-Vis, PXRD, and TEM	Antioxidant and antimicrobial activities	Suresh et al. (2015)
Proanthocyanidins	Fe_3O_4	47 ± 7.3	Sheets and rods	FTIR, SEM, TEM, XRD, EDS and AAS	Removal of heavy metal ions	Shi et al. (2020)
Azadirachta indica (flavonoid, terpenoid)	TiO_2	25–87	Spherical	TEM, SEM, XRD, and FTIR	Antibacterial activities	Thakur et al. (2019)
Sesbania grandiflora (flavonoids)	TiO_2	43–56	Spherical and triangular	XRD, TEM, SEM-EDX, FTIR, and UV-Vis	Acute toxicity to zebrafish	Srinivasan et al. (2019)
Zingiber officinale (flavonoids)	ZnO	30–50	Spherical	FTIR, SEM, and EDX	Antibiofilm activity	Raj and Jayalakshmy (2015)
Glycosmis cochinchinensis (flavonoid)	TiO_2	40 ± 5	Spherical	TEM, SEM-EDX, UV-Vis, FTIR, and XRD	Photocatalytic and antimicrobial activities	Rosi et al. (2018)

industrial (catalysis), and agricultural applications (Suresh et al. 2015; Shi et al. 2020; Srinivasan et al. 2019; Jeyaleela et al. 2020; Lee and Park 2020; Riaz et al. 2020; Li et al. 2021) as presented in Figure 2.5 and Tables 2.1 and 2.2.

2.5.1 Anticancer Activities

NPs have anticancer properties and can disrupt the mitochondrial respiratory chain, causing reactive oxygen species (ROS) to be produced, ATP synthesis to be disrupted, and severe DNA damage (Asharani et al. 2009). Stolarczyk et al. (2017) used genistein to make AuNPs and tested their cytotoxicity. Compared to free genistein, the MTT results show that genistein conjugated with AuNPs has the most significant degree of cytotoxicity. The findings indicate that AuNPs-GE may boost genistein's anticancer properties. The cytotoxicity of quercetin-derived AgNPs and AuNPs against cancer cells was studied by Lee and Park (2020). The cytotoxicity of both types of NPs on cancer cells showed that they could be good delivery routes for biologically active molecules like anticancer drugs. Li et al. (2021) utilized DMY to biogenically generate AgNPs and assessed their cytotoxicity against the cancer cell lines MDA-MB-231, HepG2, and HeLa. Oueslati et al. (2020) examined AuNPs' cytotoxicity effect made from the plant *Lotus leguminosae*'s kaempferol KG towards breast cancer (MCF-7). The results showed that the produced AuNPs had moderate to low cytotoxicity in MCF-7 cells at increasing doses. The mechanism through which NPs cause cytotoxicity is depicted in Figure 2.6.

2.5.2 Antimicrobial Activities

The antimicrobial properties of metal/metal oxide NPs fabricated from various flavonoids have been reported in several studies (Tables 2.1 and 2.2). The antimicrobial efficacy of NPs is influenced by their size and shape and the substance used to make them (Sorbiun et al. 2018). NPs have a large surface area, which helps them to make close interaction with a microorganism's cell membrane.

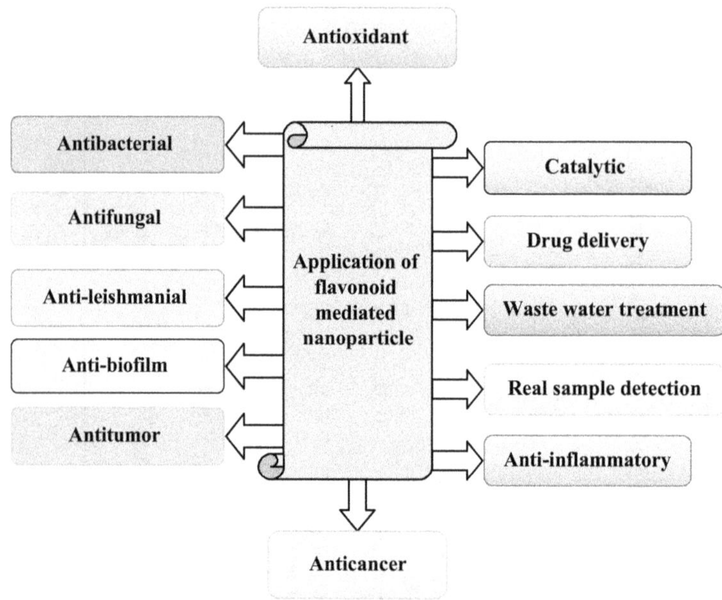

FIGURE 2.5 Applications of flavonoid-mediated nanoparticles.

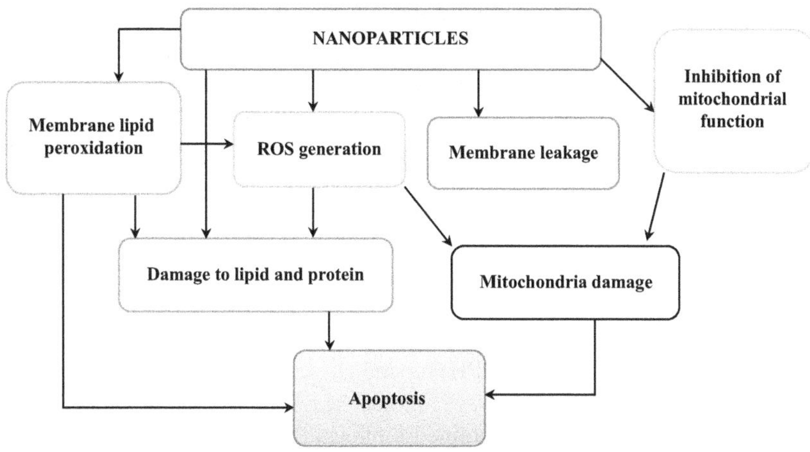

FIGURE 2.6 Possible mechanism of nanoparticles to induce cytotoxicity.

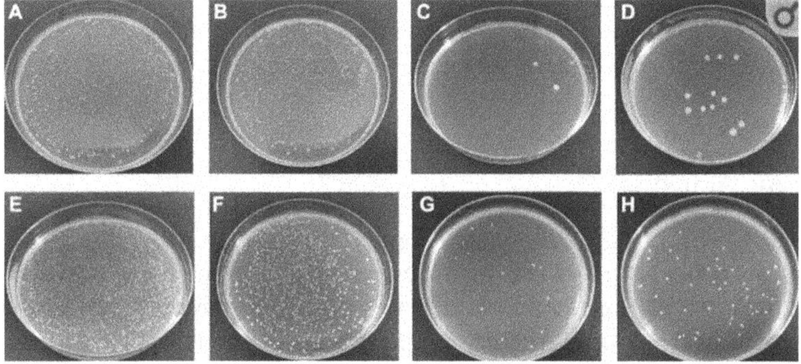

FIGURE 2.7 (A) *Escherichia coli*, (B) *Escherichia coli* treated with 0.1 mg/mL dihydromyricetin, (C) *Escherichia coli* treated with 0.1 mg/mL dihydromyricetin-silver nanoparticles, (D) *Escherichia coli* treated with 0.1 mg/mL tetracycline, (E) *Salmonella*, (F) *Salmonella* treated with 0.1 mg/mL dihydromyricetin, (G) *Salmonella* treated with 0.1 mg/mL dihydromyricetin-silver nanoparticles, (H) *Salmonella* treated with 0.1 mg/mL tetracycline (Adapted from Li et al. 2021).

Metal oxide NPs and metal NPs stop bacteria, fungi, and viruses from growing by cell division, cell membrane growth, generating metal ions that cause DNA replication to be disrupted that interfere with DNA replication, and other processes (Sorbiun et al. 2018; Siddiqi et al. 2018). Because of their high effectiveness against different microorganisms, metal NPs such as Au and Ag cover the most significant portion of the most widely recognized metal NPs. For example, Li et al. (2021) recently biosynthesized AgNPs using DMY stabilizing and reducing agents. They assessed their potential against food-borne bacteria and human pathogenic such as *Salmonella* and *Escherichia coli*. The biosynthesized AgNPs significantly inhibited bacterial growth. Figure 2.7 illustrates the number of live bacteria in the culture medium at the same concentration in different samples, showing that DMY-AgNPs have a superior and relevant antibacterial activity compared to DMY and tetracycline.

2.5.3 Antioxidant Activities

Several investigations have revealed the antioxidant properties of metal/metal oxide NPs generated from different flavonoids (Tables 2.1 and 2.2). ROS are involved in cardiovascular disease, the development of cancer, asthma, and age-related diseases. Antioxidant-active materials decrease the formation of ROS (Obrenovich et al. 2011). The capacity of phenolics and flavonoids to transfer a hydrogen atom to a free radical has been thoroughly investigated using the free radical scavenging test (Mohan et al. 2012). The ABTS, DPPH (2,2-diphenyl-1-picryl-hydrazyl), DPPH (1,1-diphenyl-2-picrylhydrazile), and MTT assays are used to determine the radical reduction. The redox potential of flavonoid and phenolic bioactive compounds is linked to NP antioxidant activity, allowing them to serve as powerful singlet oxygen scavengers, hydrogen donors, and reducing agents. These molecules have a strong antioxidant capacity, and their consumption reduces the risk of illness. The antioxidant efficiency of AuNPs made from kaempferol KG from the *Lotus leguminosae* plant was investigated by Oueslati et al. (2020), who discovered that AuNPs had a much higher DPPH radical scavenging activity than flavonoid extract, with an IC_{50} of 30.56 μg/mL. Biofabricated AuNPs transmit an electron or H to the DPPH free radical, transforming it to the light yellow stable DPPH-H, according to the findings. The DPPH radical scavenging assay method was important to assess the AgNPs' antioxidant activity obtained biologically using MY, a dietary flavonoid, and the results demonstrated that the ability of MY-AgNPs to scavenge free radicals grew continuously as the concentration was raised (Li et al. 2020). Li et al. (2021) recently synthesized AgNPs from DMY and used the DPPH scavenging method to evaluate the antioxidant potential of DMY-AgNPs. Furthermore, the antioxidant effectivity of AuNPs was investigated using nitric oxide free radical scavenging methods, ABTS and DPPH, and using the nitric oxide free radical scavenging method; it was discovered that Qct-AuNPs have stronger antioxidant activity than free radicals (Milanezi et al. 2019).

2.5.4 Catalytic Activities

Metal oxide and/or metal NPs are commonly employed for catalysis because of their high stability, more significant activity, and surface area. Several types of research on the catalytic activity of metal oxide and metal NPs generated from different flavonoids have been undertaken (Tables 2.1 and 2.2). Luo et al. (2018) produced AgNPs from flavonoids isolated from *Lilium casablanca* petals and demonstrated that the biosynthesized AgNPs displayed significant catalytic activity via hydroboration of p-nitrophenol. Lee and Park (2020) evaluated the catalytic activity of quercetin-based AuNPs and AgNPs for 4-nitrophenol reduction and methyl orange degradation. The researchers showed that raising the number of AuNPs or AgNPs employed as a catalyst enhanced the rate constant of the catalytic reaction. Additionally, Oueslati et al. (2020) synthesized AuNPs from Lotus *leguminosae kaempferol* glucoside. They explored the catalytic effectiveness of the AuNPs for the catalytic reduction of p-NP to p-AP by $NaBH_4$.

2.5.5 Other Applications

Metal/metal oxide NPs mediated by flavonoids have a variety of other uses. AuNPs were synthesized using the polyphenolic compound procyanidins in an environmentally safe manner, and the synthesized NPs had a spherical structure, good stability, and excellent light-controlled drug release (Hou et al. 2020). Sharma et al. (2019) used flavonoids isolated from the seed plant of *Madhuca longifolia* (Ml) as a reducing and stabilizing agent to synthesize Ag, Au, and bimetallic Ag/Au NPs, which were then tested for their bioefficacy in wound healing. As compared to native seed extract, Mlf@AgNps outperformed all other biofabricated NPs in terms of wound healing bioefficacy (30.40%). The inherent antimicrobial ability of Ag, the zeta potential variation, the large surface area of NPs, and the coating of medicinally significant flavonoid content on the NPs were all attributed to Mlf@AgNps' enhanced wound-healing capacity. The electron deficiency of oxidized flavonoids (quinone moiety) confers further antioxidant activities. Flavonoid-loaded AgNPs have a bright future ahead of them, and they open up a new avenue for the production of useful complementary herbal nanomedicine. In another study, Chen et al. (2020) synthesized AuNPs using an ethyl acetate extract of

Radix Hedysari in a non-toxic and efficient method. The authors predicted that formononetin was responsible for manufacturing NPs and were used to identify real samples.

2.6 CONCLUSION

The NPs synthesized from flavonoids have shown potential antimicrobial, anti-inflammatory, anti-tumour, antioxidant, antibiofilm, catalytic, and cytotoxic activity. Quercetin, kaempferol, apigenin, genistein, naringenin, hesperetin, and other flavonoids served as stabilizing and reducing agents in the flavonoid-mediated green synthesis NPs. Tricetin, kaempferol, formononetin, kaempferol, and genistein also used for the preparation of spherical AuNPs for antibacterial, anticancer, anti-angiogenic activities. The synthesized NPs are characterized using a variety of characterization techniques such as TEM, SEM, XRD, FTIR, and UV-Vis, to examine crystalline, size, and shape of nanocellulose. It has been determined that flavonoids are essential as medication globally; thus, using them to synthesize NPs paves the way for cheap/simple and most effective green technology, which can be used in various fields. Flavonoids have a potential for NP synthesis in medicine and consumer products in the future. It is crucial to consider how bioactive groups bond to the NP surface and which chemical groups are responsible for making NPs more efficient. Simultaneously, comprehensive research into the bioavailability and biocompatibility of NPs is needed.

REFERENCES

Ahmad, T., M. A. Bustam, M. Irfan, M. Moniruzzaman, H. M. Asghar and S. Bhattacharjee. 2019. Mechanistic investigation of phytochemicals involved in green synthesis of gold nanoparticles using aqueous *Elaeis guineensis* leaves extract: Role of phenolic compounds and flavonoids. *Biotechnology and Applied Biochemistry* 66(4): 698–708.

Ahmed, M., M. S. AlSalhi and M. K. J. Siddiqui. 2010. Silver nanoparticle applications and human health. *Clinica Chimica Acta* 411: 1841–1848.

Alsamhary, K., N. Al-Enazi, W. A. Alshehri and F. Ameen. 2020. Gold nanoparticles synthesized by flavonoid tricetin as a potential antibacterial nanomedicine to treat respiratory infections causing opportunistic bacterial pathogens. *Microbial Pathogenesis* 139: 103928.

Ameen, F., S. A. AlYahya, M. A. Bakhrebah, M. S. Nassar and A. Aljuraifani. 2018. Flavonoid dihydromyricetin-mediated silver nanoparticles as potential nanomedicine for biomedical treatment of infections caused by opportunistic fungal pathogens. *Research on Chemical Intermediates* 44(9): 5063–5073.

AshaRani, P. V., K. M. G. Low, M. P. Hande and S. Valiyaveettil. 2009. Cytotoxicity and genotoxicity of silver nanoparticles in human cells. *ACS Nano* 3(2): 279–290.

Bachheti, R. K., A. Fikadu, A. Bachheti and A. Husen. 2020. Biogenic fabrication of nanomaterials from flower-based chemical compounds, characterization and their various applications: A review. *Saudi Journal of Biological Sciences* 27(10): 2551–2562.

Bayda, S., M. Adeel, T. Tuccinardi, M. Cordani and F. Rizzolio. 2020. The history of nanoscience and nanotechnology: From chemical-physical applications to nanomedicine. *Molecules* 25(1): 112.

Borodina, V. G. and Y.A. Mirgorod. 2014. Kinetics and mechanism of the interaction between $HAuCl_4$ and rutin. *Kinetics and Catalysis* 55(6): 683–687.

Chen, X., J. Ji, G. Shi, Z. Xue, X. Zhou, L. Zhao and S. Feng. 2020. Formononetin in Radix Hedysari extract-mediated green synthesis of gold nanoparticles for colorimetric detection of ferrous ions in tap water. *RSC Advances* 10(54): 32897–32905.

Cushing, B. L., V. L. Kolesnichenko and C. J. O'connor. 2004. Recent advances in the liquid-phase syntheses of inorganic nanoparticles. *Chemical Reviews* 104(9): 3893–3946.

Das, D. K., A. Chakraborty, S. Bhattacharjee and S. Dey. 2013. Biosynthesis of stabilised gold nanoparticle using an aglycone flavonoid, quercetin. *Journal of Experimental Nanoscience* 8(4): 649–655.

Griesbach, R. J. 2005. Biochemistry and genetics of flower color. *Plant Breeding Reviews* 25: 89–114.

Guo, Q., Q. Guo, J. Yuan and J. Zeng. 2014. Biosynthesis of gold nanoparticles using a kind of flavonol: Dihydromyricetin. *Colloids and Surfaces A: Physicochemical and Engineering Aspects* 441: 127–132.

Halder, A., S. Das, T. Bera and A. Mukherjee. 2017. Rapid synthesis for monodispersed gold nanoparticles in kaempferol and anti-leishmanial efficacy against wild and drug-resistant strains. *RSC Advances* 7(23): 14159–14167.

Hou, K., M. Bao, C. Xin, L. Wang, H. Zhang, H. Zhao and Z. Wang. 2020. Green synthesis of gold nanoparticles coated doxorubicin liposomes using procyanidins for light-controlled drug release. *Advanced Powder Technology* 31(8): 3640–3649.

Irfan, M., M. Moniruzzaman, T. Ahmad, P. C. Mandal, S. Bhattacharjee and B. Abdullah. 2017. Ionic liquid-based extraction of flavonoids from *Elaeisguineensis* leaves and their applications for gold nanoparticles synthesis. *Journal of Molecular Liquids* 241: 270–278.

Jain, S. and M. S. Mehata. 2017. Medicinal plant leaf extract and pure flavonoid mediated green synthesis of silver nanoparticles and their enhanced antibacterial property. *Scientific Reports* 7(1): 1–3.

Jeyaleela, G. D., J. R. Vimala, S. M. Sheela, A. Agila, M. S. Bharathy and M. Divya. 2020. Biofabrication of zinc oxide Nanoparticles using the Isolated Flavonoid from *Combretum ovalifolium* and its anti-oxidative Ability and Catalytic degradation of methylene blue Dye. *Oriental Journal of Chemistry* 36(4): 655–664.

Kasthuri, J., S. Veerapandia and N. Rajendiran. 2009. Biological synthesis of silver and gold nanoparticles using apiin as reducing agent. *Colloids and Surfaces B: Biointerfaces* 68(1): 55–60.

Khan, I., K. Saeed and I. Khan. 2019. Nanoparticles: Properties, applications and toxicities. *Arabian Journal of Chemistry* 12(7): 908–931.

Kumar, P., M. Govindaraju, S. Senthamilselvi and K. Premkumar. 2013. Photocatalytic degradation of methyl orange dye using silver (Ag) nanoparticles synthesized from *Ulva lactuca. Colloids and Surfaces B: Biointerfaces* 103: 658–661.

Latif, M. S., F. Kormin, M. K. Mustafa, I. I. Mohamad, M. Khan, S. Abbas, M. I. Ghazali, N. S. Shafie, M. F. Bakar, S. F. Sabran and S. F. Fuzi. 2018. Effect of temperature on the synthesis of *Centella asiatica* flavonoids extract-mediated gold nanoparticles: UV-visible spectra analyses. *AIP Conference Proceedings* 1: 020071.

Lee, Y. J. and Y. Park. 2020. Green synthetic Nanoarchitectonics of gold and silver nanoparticles prepared using quercetin and their cytotoxicity and catalytic applications. *Journal of Nanoscience and Nanotechnology* 20(5): 2781–2790.

Li, Z., I. Ali, J. Qiu, H. Zhao, W. Ma, A. Bai, D. Wang and J. Li. 2021. Eco-friendly and facile synthesis of antioxidant, antibacterial and anticancer dihydromyricetin-mediated silver nanoparticles. *International Journal of Nanomedicine* 16: 481.

Li, Z., W. Ma, I. Ali, H. Zhao, D. Wang and J. Qiu. 2020. Green and facile synthesis and antioxidant and antibacterial evaluation of dietary myricetin-mediated silver nanoparticles. *ACS Omega* 50: 32632–32640.

Li, Y., T. Zhang and G.Y. Chen. 2018. Flavonoids and colorectal cancer prevention. *Antioxidants* 7(12): p. 187.

Luo, Q., W. Su, H. Li, J. Xiong, W. Wang, W. Yang and J. Du. 2018. Antibacterial activity and catalytic activity of biosynthesized silver nanoparticles by flavonoids from petals of *Lilium casa blanca. Micro and Nano Letters* 13(6): 824–828.

Mashwani, Z. U., T. Khan, M. A. Khan and A. Nadhman. 2015. Synthesis in plants and plant extracts of silver nanoparticles with potent antimicrobial properties: Current status and future prospects. *Applied Microbiology and Biotechnology* 99: 9923–9934.

Mat Yusuf, S. N. A., C. N. A. Che Mood, N. H. Ahmad, D. Sandai, C. K. Lee and V. Lim. 2020. Optimization of biogenic synthesis of silver nanoparticles from flavonoid-rich Clinacanthus nutans leaf and stem aqueous extracts. *Royal Society Open Science* 7(7): 20006.

Milanezi, F. G., L. M. Meireles, M. M. de Christo Scherer, J. P. de Oliveira, A. R. da Silva, M. L. de Araujo, D. C. Endringer, M. Fronza, M. C. Guimarães and R. Scherer. 2019. Antioxidant, antimicrobial and cytotoxic activities of gold nanoparticles capped with quercetin. *Saudi Pharmaceutical Journal* 27(7): 968–974.

Mittal, A. K., S. Kumar and U. C. Banerjee. 2014. Quercetin and gallic acid mediated synthesis of bimetallic (silver and selenium) nanoparticles and their antitumor and antimicrobial potential. *Journal of Colloid and Interface Science* 431: 194–199.

Mohan, R., R. Birari, A. Karmase, S. Jagtap and K. K. Bhutani. 2012. Antioxidant activity of a new phenolic glycoside from *Lagenaria siceraria* stand fruits. *Food Chemistry* 132(1): 244–251.

Naveena, N. L., R. Naik, R. Pratap and S. A. Shivashankar. 2018. Microwave-assisted greener synthesis of silver nanoparticles using Karanjin and their antifungal activity. *Journal of Materials Nanoscience* 5(1): 23–28.

Nibras, A. L., R. M. Hasan and K. Alslman. 2020. Effect of zinc oxide nanoparticles on the oxidative stress (malonaldehyde MDA, lipid peroxidation level LPO) and antioxidants (GSH glutation). *Medico Legal Update* 20(1): 882–888.

Obrenovich, E. M., Y. Li, K. Parvathaneni, B. Yendluri, H. Palacios, J. Leszek and G. Aliev. 2011. Antioxidants in health, disease and aging. *CNS and Neurological Disorders – Drug Targets* 10(2): 192–207.

Oliveira, M. M., D. Ugarte, D. Zanchet and A. J. Zarbin. 2005. Influence of synthetic parameters on the size, structure, and stability of dodecanethiol-stabilized silver nanoparticles. *Journal of Colloid and Interface Science* 292(2): 429–435.

Osonga, F. J., A. Akgul, I. Yazgan, A. Akgul, R. Ontman, V. M. Kariuki, G. B. Eshun and O. A. Sadik. 2018. Flavonoid-derived anisotropic silver nanoparticles inhibit growth and change the expression of virulence genes in *Escherichia coli* SM10. *RSC Advances* 8(9): 4649–4661.

Oueslati, M. H., L. B. Tahar and A. H. Harrath. 2020. Catalytic, antioxidant and anticancer activities of gold nanoparticles synthesized by kaempferol glucoside from *Lotus leguminosae. Arabian Journal of Chemistry* 13(1): 3112–3122.

Ozdal, Z. D., E. Sahmetlioglu, I. Narin and A. Cumaoglu. 2019. Synthesis of gold and silver nanoparticles using flavonoid quercetin and their effects on lipopolysaccharide-induced inflammatory response in microglial cells. *3 Biotech* 9(6): 1–8.

Raghavan, B. S., S. Kondath, R. Anantanarayanan and R. Rajaram. 2015 k.Kaempferol mediated synthesis of gold nanoparticles and their cytotoxic effects on MCF-7 cancer cell line. *Process Biochemistry* 50(11): 1966–1976.

Raj, L. F. and E. Jayalakshmy. 2015. Biosynthesis and characterization of zinc oxide nanoparticles using root extract of *Zingiber officinale. Oriental Journal of Chemistry* 31(1): 51–56.

Rajkumari, J., S. Busi, A. C. Vasu and P. Reddy. 2017. Facile green synthesis of baicalein fabricated gold nanoparticles and their antibiofilm activity against *Pseudomonas aeruginosa* PAO1. *Microbial Pathogenesis* 107: 261–269.

Riaz, S., R. N. Fatima, I. Hussain, T. Tanweer, A. Nawaz, F. Menaa, H. A. Janjua, T. Alam, A. Batool, A. Naeem, M. Hameed and S. M. Ali. 2020. Effect of flavonoid-coated gold nanoparticles on bacterial colonization in mice organs. *Nanomaterials* 10(9): 1769.

Rosi, H. and S. Kalyanasundaram. 2018. Synthesis, characterization, structural and optical properties of titanium dioxide nanoparticles using *Glycosmis cochinchinensis* Leaf extract and its photocatalytic evaluation and antimicrobial properties. *World News of Natural Sciences International Science Journal* 17: 1–15.

Roy, N., S. Mondal, R. A. Laskar, S. Basu, D. Mandal and N. A. Begum. 2010. Biogenic synthesis of Au and Ag nanoparticles by *Indian propolis* and its constituents. *Colloids and Surfaces B: Biointerfaces* 76(1): 317–325.

Sahu, N., D. Soni, B. Chandrashekhar, D. B. Satpute, S. Saravanadevi, B. K. Sarangi and R. A. Pandey. 2016. Synthesis of silver nanoparticles using flavonoids: Hesperidin, naringin and diosmin, and their antibacterial effects and cytotoxicity. *International Nano Letters* 6(3): 173–181.

Sathishkuma, P., Z. Li, B. Huang, X. Guo, Q. Zhan, C. Wang and F. L. Gu. 2019. Understanding the surface functionalization of myricetin-mediated gold nanoparticles: Experimental and theoretical approaches. *Applied Surface Science* 493: 634–644.

Sharma, M., S. Yadav, N. Ganesh, M. M. Srivastava and S. Srivastava. 2019. Biofabrication and characterization of flavonoid-loaded Ag, Au, Au–Ag bimetallic nanoparticles using seed extract of the plant Madhuca longifolia for the enhancement in wound healing bio-efficacy. *Progress in Biomaterials* 8: 51–63.

Shi, Y., Y. Xing, S. Deng, B. Zhao, Y. Fu and Z. Liu. 2020. Synthesis of proanthocyanidins-functionalized Fe_3O_4 magnetic nanoparticles with high solubility for removal of heavy metal ions. *Chemical Physics Letters* 753: 137600.

Siddiqi, K. S., A. Husen and R. A. Rao. 2018. A review on biosynthesis of silver nanoparticles and their biocidal properties. *Journal of Nanobiotechnology* 16(1): 1–28.

Sierra, J. A., C. R. Vanoni, M. A. Tumelero, C. C. Cid, R. Faccio, D. F. Franceschini, T. B. Creczynski-Pasa and A. A. Pasa. 2016. Biogenic approaches using citrus extracts for the synthesis of metal nanoparticles: The role of flavonoids in gold reduction and stabilization. *New Journal of Chemistry* 40(2): 1420–1429.

Singh, N., J. Bhagat, E. Tiwari, N. Khandelwal, G. K. Darbha and S. K. Shyama. 2021. Metal oxide nanoparticles and polycyclic aromatic hydrocarbons alter nanoplastic's stability and toxicity to zebrafish. *Journal of Hazardous Materials* 407: 124382.

Sorbiun, M., E. Shayegan-Mehr, A. Ramazani and M. A. Mashhadi. 2018. Biosynthesis of metallic nanoparticles using plant extracts and evaluation of their antibacterial properties. *Nanochemistry Research* 3(1): 1–6.

Srinivasan, M., M. Venkatesan, V. Arumugam, G. Natesan, N. Saravanan, S. Murugesan, S. Ramachandran, R. Ayyasamy and A. Pugazhendhi. 2019. Green synthesis and characterization of titanium dioxide nanoparticles (TiO_2 NPs) using *Sesbania grandiflora* and evaluation of toxicity in zebrafish embryos. *Process Biochemistry* 80: 197–202.

Stolarczyk, E. U., K. Stolarczyk, M. Łaszcz, M. Kubiszewski, W. Maruszak, W. Olejarz and D. Bryk. 2017. Synthesis and characterization of genistein conjugated with gold nanoparticles and the study of their cytotoxic properties. *European Journal of Pharmaceutical Sciences* 96: 176–185.

Suresh, D., P. C. Nethravathi, H. Rajanaika, H. Nagabhushana and S. C. Sharma. 2015. Green synthesis of multifunctional zinc oxide (ZnO) nanoparticles using *Cassia fistula* plant extract and their photodegradative, antioxidant and antibacterial activities. *Materials Science in Semiconductor Processing* 31: 446–454.

Takahashi, A. and T. Ohnishi. 2004. The significance of the study about the biological effects of solar ultraviolet radiation using the exposed facility on the international space station. *Biological Sciences Space* 18(4): 255–260.

Thakur, B. K., A. Kumar and D. Kumar. 2019. Green synthesis of titanium dioxide nanoparticles using *Azadirachta indica* leaf extract and evaluation of their antibacterial activity. *South African Journal of Botany* 124: 223–227.

Vadlapudi, V. and D. S. Kaladhar. 2014. Review: Green synthesis of silver and gold nanoparticles. *Middle East Journal of Scientific Research* 19(6): 834–842.

Varma, R. 2012. Greener approach to nanomaterials and their sustainable applications. *Current Opinion in Chemical Engineering* 1(2): 123–128.

Wang, L., F. Lu, Y. Liu, Y. Wu and Z. Wu. 2018. Photocatalytic degradation of organic dyes and antimicrobial activity of silver nanoparticles fast synthesized by flavonoids fraction of *Psidium guajava* L. leaves. *Journal of Molecular Liquids* 263: 187–192.

Zhou, Y. and R. C. Tang. 2018. Facile and eco-friendly fabrication of colored and bioactive silk materials using silver nanoparticles synthesized by two flavonoids. *Polymers* 10(4): 404.

3 Medicinal Plant-Based Terpenoids in Nanoparticles Synthesis, Characterization, and Their Applications

Sepideh Khoee and Mozhdeh Madadi

CONTENTS

DOI: 10.1201/9781003213727-3

LIST OF ABBREVIATIONS

(LA) (CA-PLGA-PEG-LA)	Cholic acid poly (lactic-co-glycolic acid) (PLGA)-polyethylene glycol (PEG)-lacto-bionic acid (LA)
2-MeTHF	2-Methyltetrahydrofuran
5F	Ent-11a-hydroxy-15-oxo-kaur-16-en-19-oic-acid
Ach	Acetylcholine
AChE	Acetylcholinesterase
AD	Alzheimer's disease
Ag	Silver
AMR	Antimicrobial resistance
ART	Artemisinin
Au	Gold
Aβ	Amyloid-β
BuChE	Butyrylcholinesterase
CA	Cholic acid
car	Carveol
CDs	Carbon dots
Ce	Cerium
CNS	Central nerve systems
COVID-19	Coronavirus infectious disease 2019
CQDs	Carbon quantum dots
CS	Chitosan
CTAB	Cetyl trimethyl ammonium bromide
CTCs	Circulating tumour cells
Cu	Copper
DBU	1,8-diazabicyclo [5.4.0] undec-7-ene
DMA	Dynamic mechanical analysis
DMAPP	Dimethylallyl pyrophosphate
DSC	Differential scanning calorimetry
EDC	1-ethyl-3-(3-dimethylaminopropyl)-carbodiimide hydrochloride
EDX	Energy dispersive X-ray analysis
EM	Exosome membrane
EMPCs	Exosome-like sequential-bioactivating paclitaxel prodrug nanoplatform
EVs	Exosome vesicles
F	Fresh
FD	Freeze-dried
Fe	Iron
FTIR	Fourier-transform infrared spectroscopy
GANPs	Glycyrrhizic acid nanoparticles
GC-CS	Galactosylated-chitosan
ger	Geraniol
GRAS	Generally recognized as safe
HCC	Hepatocellular carcinoma
H-NMR	Proton nuclear magnetic resonance
HRTEM	High-resolution TEM
IC_{50}	Half maximal inhibitory concentration
IPP	Isopentenyl pyrophosphate
JOLE	*Jasminum officinale* L. leaf extract
JO	Jasmine oil

LA	Lactide
Lim-OH	Limonene-OH
MEP	2-C-methyl-D-erythriol 4-phosphate
Mg	Magnesium
MIC	Minimum inhibitory concentration
Mn	Manganese
MNPs	Metallic nanoparticles
MRI	Magnetic resonance imaging
MVA	Mevalonate
NHS	N-hydroxysuccinimide
NSCLC	Non-small cell lung cancer
NPs	Nanoparticles
PCL–PEG–PCL	poly (ε-caprolactone)–poly (ethylene glycol)–poly (ε-caprolactone
PEG–PCL	polyethylene glycol–poly E-caprolactone
PDI	Polydispersity index
PDLLA	Poly (D,L -lactide)
PdNCs	Palladium nanocubes
PLA	Poly (lactic acid)
PTX	Paclitaxel
PVA	Polyvinyl alcohol
PWD	Pulsed wire discharge
QDs	Quantum dots
RME	Receptor-mediated endocytosis
ROCOP	Ring-opening copolymerization
ROP	Ring-opening polymerization
ROS	Reactive oxygen species
SEM	Scanning electron microscope
SPR	Surface plasmon resonance
STM	Scanning tunneling microscopy
TAP	Terpenoid fraction
Ti	Titanium
TP	Triptolide
TSE	*Tamarindus indica* shell-husk extract
t-sob	Trans-sobrerol
UA	Ursolic acid
UV-Vis	Ultraviolet-visible spectroscopy
WHO	World Health Organization
XRD	X-ray powder diffraction
Zn	Zinc

3.1 INTRODUCTION

Nanotechnology is a ground-breaking, interdisciplinary area with a flourishing role in a variety of fields such as material science, engineering, environment, medicine, pharmaceutical, and so on (Rabiee et al. 2020). Without being specifically labelled, the general concept of nanotechnology was established by Professor Richard Feynman in 1959 in his lecture at the American Physical Society meeting. Later on, in 1974, Professor Norio Taniguchi of Tokyo University introduced the term 'nanotechnology' to elucidate an extremely thin film of materials (Taniguchi 1974). The discovery of scanning tunnelling microscopy (STM) in 1980 initiated great breakthroughs in the field of nanotechnology (*Nature Nanotechnology* 2010). Given the phenomenal growth in this field over the past few decades, nanotechnology is regarded as frontier technology and a 21st-century revolution in all aspects of human lives. The word 'nano' is derived from the ancient Greek for 'dwarf' and denotes a factor of 10^{-9} m. Nanotechnology is the science of manipulation and handling materials with at least one dimension of a size less than 100 nm (Thakkar, Mhatre, and Parikh 2010). Nanomaterials are of significant importance due to their remarkable physical and chemical (physicochemical) properties over their bulk version. Some distinguishing properties of materials, such as emission, thermal and electrical conductivity, and optical behaviour, alter as their size approach the nanoscales (Patra, Kwon, and Baek 2016). Another distinctive feature of nanoparticles (NPs) is their relatively higher surface area to volume ratio. The large surface area makes nanomaterials ideal candidates for applications ranging from drug delivery to electronics (Lin et al. 2021; Zahin et al. 2020).

Generally, NPs are synthesized through physical and chemical methods. Laser ablation (Menazea 2020), vacuum vapour deposition (Yu et al. 2018), lithography (Gupta et al. 2019), and pulsed wire discharge (PWD) (Chu et al. 2019) are some instances of physical processes. Furthermore, chemical methods such as microemulsion (Rane et al. 2018), sol-gel (Thiagarajan, Sanmugam, and Vikraman 2017), photochemical reduction (Meader et al. 2017) are utilized for NP fabrications. However, all the methods mentioned require high energy and are costly. Moreover, besides aprotic solvents, hazardous chemicals are environmentally damaging and toxic, which restrict their biomedical and clinical applications. To resolve these problems, the biological synthesis method burgeoned as an alternative route for NP production (Ovais et al. 2018). Biosynthesis brings about a paradigm shift in nanomaterials considering their cost-effective, convenient, eco-friendly nature (Rauwel 2017; Iravani 2011; Kumar and Yadav 2009). Green synthesis of NPs is carried out with various natural sources such as yeast (Faramarzi, Anzabi, and Jafarizadeh-Malmiri 2020), fungi (Molnár et al. 2018), algae (Senthilkumar et al. 2019), bacteria (Tsekhmistrenko et al. 2020), viruses (Salem and Fouda 2021), and plants (Jadoun et al. 2021). Among the bio-based suggested precursors, plant extract and essence appear to be superior candidates over other organisms owing to their numerous advantages, including:

(1) higher synthetic rate, (2) easy process, (3) elimination of cell culturing and special storage conditions, (4) more stable NPs, (5) extra cost-effective, (6) exclusion infection possibilities, (7) readily available, and therefore (8) suitable for larger scale (Yadi et al. 2018). In addition, these NPs are more varied in size and shape (Chung et al. 2016).

The broad applications of plants in different areas of science are inherently based on their phytochemical constituents. According to the role that takes part in the metabolic process, these compounds are characterized in two main groups: primary metabolites and secondary metabolites. Primary metabolites are approximately the same in all animals and plants and are responsible for vital metabolic activities like growth, reproduction, nutrition. Sugars, amino acids, nucleic acids, lipids, etc., are primary metabolites (Chen and Wang 2017). On the other hand, secondary metabolites are not directly essential for the life of plants but associated with their survival (Barrios-González 2018). They often involve plant protection against pathogens and predators (Boncan et al. 2020; Mumm, Posthumus, and Dicke 2008), and a few of them can produce UV-absorbing compounds for preserving leaves (Nunn et al. 2020). It is also known that these secondary metabolites attract insects for pollination (Abbas et al. 2017). Overall secondary metabolites adjust plants for coevolution. The major classes of plants' secondary metabolites are phenolic, alkaloids, and terpenoids.

3.2 TERPENOIDS

Terpenoids, oxygenated derivatives of terpenes, among all the plants' secondary metabolites, are the largest and have the most structural diversity. Approximately, one-third of all natural compounds belong to the terpenomes family including steroids, carotenoids, terpenoids, and their derivatives (Christianson 2017). Terpenoids are usually odorous, which clarifies their wide applications in the fragrance, flavour, perfume, and cosmetics industries. They also display great potential for medical and pharmaceutical purposes (Jahangeer et al. 2021). All terpenoids are derived structurally from the isoprene five carbons unit (Markus Lange and Ahkami 2013; S and C 2008; Cheng et al. 2007).

Based on the number of isoprene units in the terpenoids structures, they are categorized as follows: hemiterpenoids, monoterpenoids, sesquiterpenoids, diterpenoids, sesterterpenoids, triterpenoids, triterpenoids, and polyterpenoids.

Isopentenyl pyrophosphate (IPP) and its allylic isomer dimethylallyl pyrophosphate (DMAPP) serve as the isoprene precursor in the biosynthesis steps of terpenoids through two different pathways: (i) mevalonate (MVA) and (ii) 2-C-methyl-D-erythriol 4-phosphate (MEP). The MVA pathway that occurs in the cytoplasm is mainly responsible for the synthesis of sesquiterpenoids and triterpenoids. While the MEP pathway takes place in the plastids and produces hemi-, mono-, di-, and triterpenoids (Dubey, Bhalla, and Luthra 2003) (Lichtenthaler 1999; Lichtenthaler, Rohmer, and Schwender 1997). The details of these two pathways are precisely described in several reviews (Eisenreich et al. 1998; Kuzuyama 2002; Dubey, Bhalla, and Luthra 2003; Vranová, Coman, and Gruissem 2013).

3.3 TYPES OF TERPENOIDS

3.3.1 Hemiterpenoids

Hemiterpenoids are considered the simplest form of terpenoid. They consist of one isoprene unit. Isoprene, as the only hemiterpene, with an important role in atmospheric chemistry, is emitted from hard-stem herbs like oaks, poplars, eucalyptus. Other well-known examples of hemiterpenoids are tiglic acid, isovaleric acid, and angelic acid (Figure 3.1).

3.3.2 Monoterpenoids

Monoterpenoids are composed of two isoprene units, with the formula $C_{10}H_{16}$. They have an intense aroma (and are a major component of a plant's essential oils), making them a prime constituent of the perfume industry. Based on the number of rings in their structure, monoterpenoids can be classified into three generic subgroups: acyclic, monocyclic, and bicyclic. Myrcene, geraniol, lavandulol, linalool, and citral are some prominent acyclic monoterpenoids/terpenes. Monocyclic monoterpenoids/terpenes comprise limonone, menthol, thymol, and so on. The most representative of bicyclic monoterpenoids/terpenes are α- and β-pinene, iridoids, camphor, and thujone as shown in Figure 3.2.

3.3.3 Sesquiterpenoids

Sesquiterpenoids are composed of three isoprene units with the formula $C_{15}H_{24}$. They are the most varied and the most extensive group of terpenoids. Although they are less volatile than monoterpenoids, they are a valuable part of essential oils. Moreover, sesquiterpenoids are widely applied in biological activities including anti-inflammatory, antimicrobial, and a few more. Sesquiterpenoids will also be sorted into different groups of acyclic, monocyclic, bicyclic, and tricyclic. Farnesol is a typical case of acyclic sesquiterpene. α-zingiberene, β-bisabolene, and α-humulene are good illustrations of monocyclic. Bicyclic sesquiterpenes include β-santalol, δ-cadinene, chamazulene, etc. Thujopsene and β-cedrol are exemplars of tricyclics (Figure 3.3).

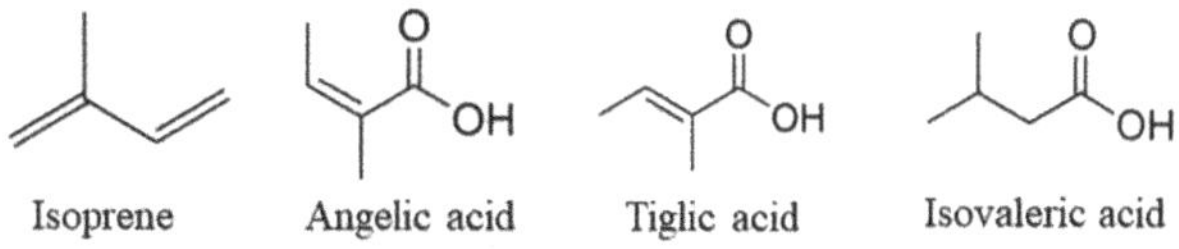

FIGURE 3.1 Chemical structures of hemiterpenoids.

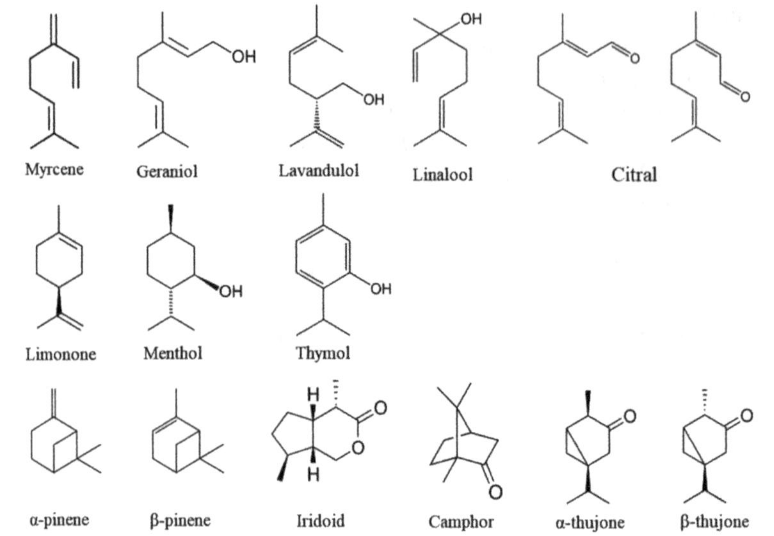

OH
OH
OH
OH
O
O
Myrcene
Geraniol
Lavandulol
Linalool
Citral
OH
OH
Limonone
Menthol
Thymol
H
O
O
H
O
O
O
α-pinene
β-pinene
Iridoid
Camphor
α-thujone
β-thujone

FIGURE 3.2 Chemical structures of monoterpenoids.

CH2OH
Farnesol
α-Zingiberene
β-Bisabolene
α-Humulene
OH
H
HO
H
β-Santalol
δ-Cadinene
Chamazulene
Thujopsene
β-Cedrol

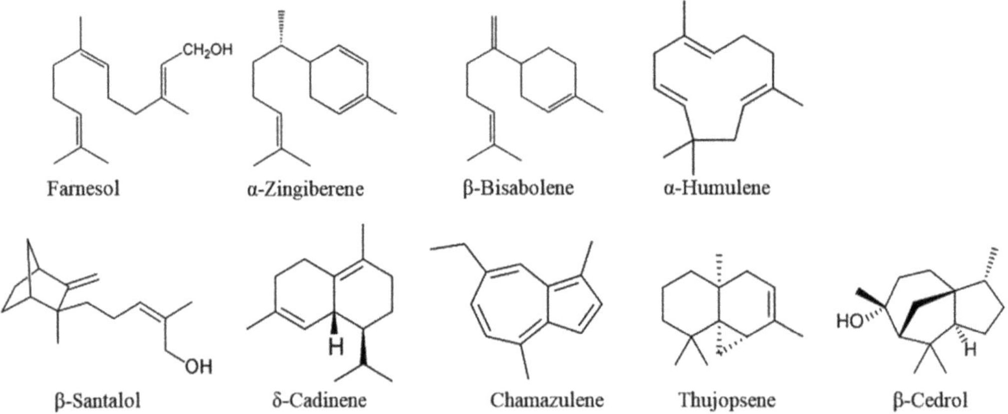

FIGURE 3.3 Chemical structures of sesquiterpenoids.

3.3.4 Diterpenoids

Diterpenoids consist of four isoprene units with the formula $C_{20}H_{32}$. They are less volatile than mono- and sesquiterpenoids; therefore, their role in essential oils is very small. Diterpenoids and terpenes are categorized as linear (e.g., phytol), monocyclic (e.g., retinol), bicyclic (e.g., sclareol), tricyclic (e.g., carnosic acid), tetracyclic (e.g., steviol) (Figure 3.4). Some diterpenes are the precursor of fundamental biological compounds, namely vitamins E and K1 (phytol) and vitamin A (retinol) compounds. They also demonstrate therapeutic properties, viz. antibacterial, antifungal, and anti-inflammatory activities.

3.3.5 Sesterterpenoids

They are built from five isoprene units with the formula $C_{25}H_{40}$ (with 25 carbons in their structure). They are considered an atypical class of terpenoid. The chief source of sesterterpenoids is marine

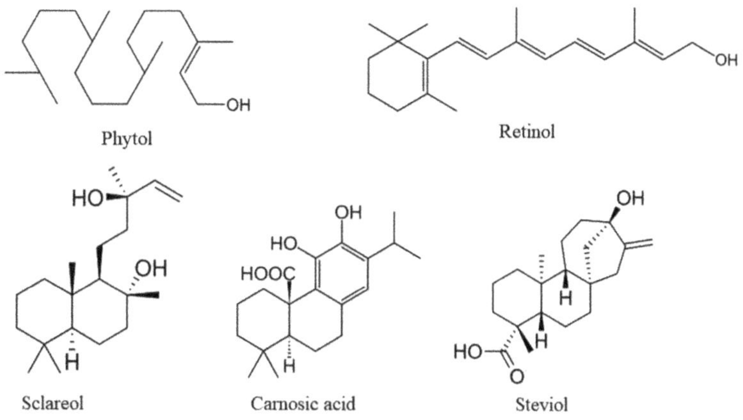

Phytol Retinol

Sclareol Carnosic acid Steviol

FIGURE 3.4 Chemical structures of diterpenoids.

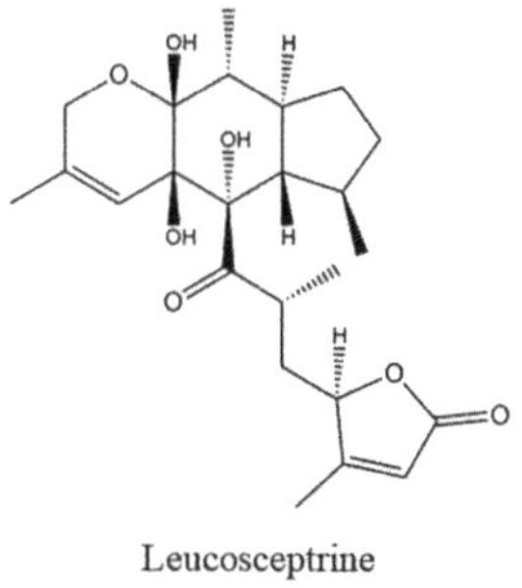

Leucosceptrine

FIGURE 3.5 Chemical structure of plant-sourced sesterterpenoids.

organisms, in particular sponges, and an approximate amount of 15% is extracted from plants such as Lamiaceae (Guo, Liu, and Li 2021). Sesterterpenoids exist in a variety of forms: acyclic, mono-, bi-, tri-, tetra-, and macrocyclic structures. Leucosceptrine is part of the sestertepenoids family (Figure 3.5).

3.3.6 Triterpenoids

Triterpenoids are based on six isoprene units with 30 carbons and complex cyclic structure (mostly tetra- and pentacyclic) in their backbone. The linear triterpene, squalene, is the precursor to all steroids in plants and animals. Sterols are tetracyclic triterpenoid derivatives with cyclopentane perhydrophenanthrene ring. Sitosterol, stigmasterol, and α-spinasterol are examples of phytosterol in vascular plants. Ursolic acid, oleanolic acid, and betulinic acid are among the well-known triterpenoids in plant extracts (Figure 3.6). Saponins are a remarkable subgroup of triterpenes that accounts for a large group in plants.

3.3.7 Tetraterpenoids

Tetraterpenoids are composed of eight isoprene units with the molecular formula $C_{40}H_{64}$. Carotenoids are the most prominent tetraterpenoids in charge of the pigment of compounds (due to their conjugated systems) ranging from pale yellow to deep red. Lycopene, the simplest tetraterpene, has a bright red colour and is found in red fruits and vegetables like tomatoes, watermelons, berries, etc. Tetraterpenoids are either basic unsaturated hydrocarbon like α-carotene, β-carotene known as

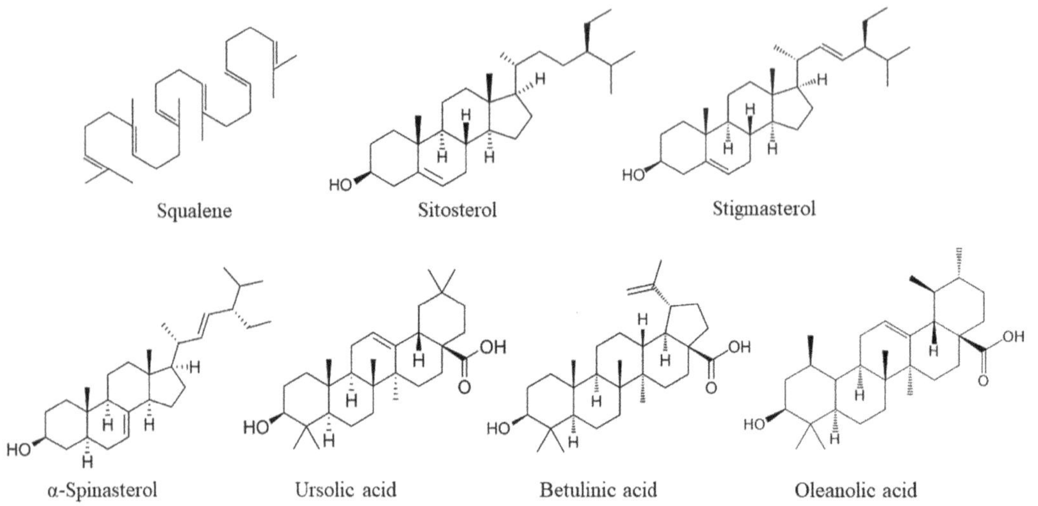

Squalene Sitosterol Stigmasterol

α-Spinasterol Ursolic acid Betulinic acid Oleanolic acid

FIGURE 3.6 Chemical structures of triterpenoids.

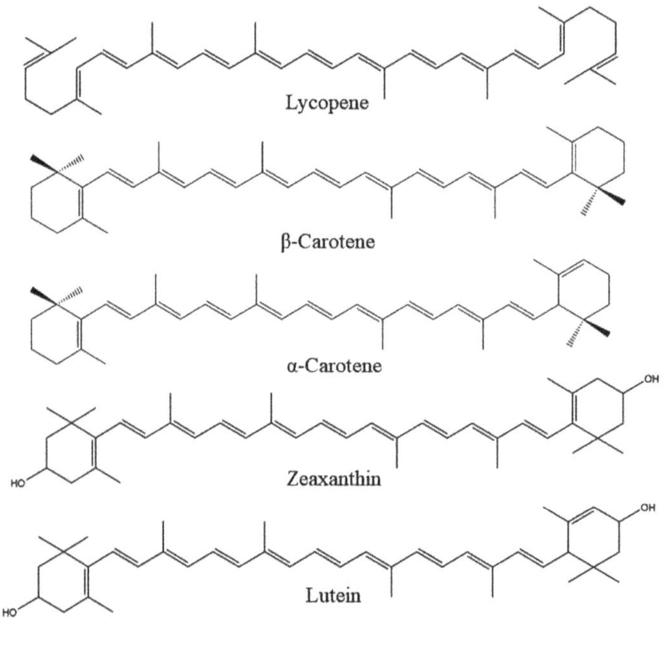

Lycopene

β-Carotene

α-Carotene

Zeaxanthin

Lutein

FIGURE 3.7 Chemical structures of tetraterpenoids.

carotenes, or their oxygenated analogous as lutein, zeaxanthin known as xanthophylls (Figure 3.7). Carotene subdivision is an orange pigment. Sweet potato, orange cantaloupe, and melon contain carotenes. Xanthophylls are yellow, and fruits with a yellow and orange colour and dark green leafy vegetables like spinach, broccoli, chlorella, and parsley are the best source of xanthophylls.

3.3.8 Polyterpenoids

Polyterpenoids consist of more than eight isoprene units and are polymeric isoprenoids. Natural rubber is categorized in this class. They are high molecular weight polymers of isoprene units with

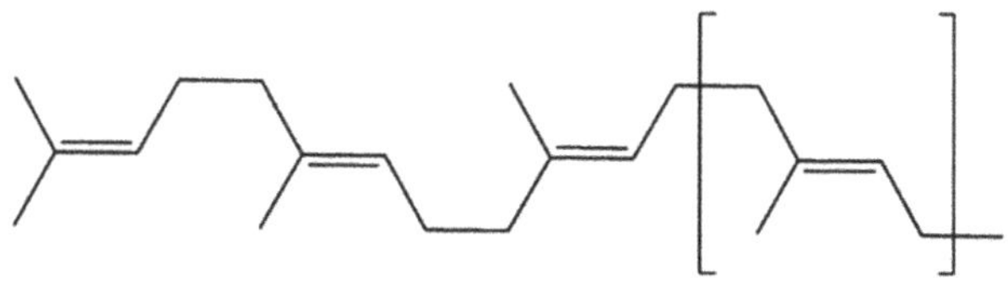

Gutta-percha

FIGURE 3.8 Chemical structure of polyterpenoids.

cis double bonds. Common latex elastomers are examples. Some plants produce polyisoprenes with trans configuration known as gutta percha (Figure 3.8) from *Palaquium gutta* (Sapotaceae).

This chapter focuses on three aspects of terpenoid-NPs in (1) loading terpenoids in NPs and their applications in drug delivery, (2) using terpenoid for synthesis of metallic and metal oxide NPs and highlighting their biomedical applications, and (3) utilizing terpenoids as precursors for polymer synthesis.

3.4 TERPENOIDS AS LOADED MEDICINE IN NANOCARRIERS

Since the dawn of history, humankind has tried to find a method in nature for a cure for diseases. Medicinal plants have played a cardinal role in traditional medicine all around the world for centuries and carry on providing new remedies. Furthermore, numerous phytochemicals with potential biomedical activities such as flavonoids (Khan et al. 2021; Nouri and Khoee 2020), alkaloids (Zheng et al. 2018), and terpenoids have been identified from them. Among the active compounds in medicinal plants, terpenoids have great potential in treating many chronic diseases such as neurodegenerative diseases (González-Cofrade et al. 2019), bacterial and viral infections (Nurdin et al. 2020), and various types of cancers. Nowadays, the demand for natural medicine has increased significantly because of its unique features, such as fewer side effects and toxicity, low cost, and easy availability. An ideal medicine must enable delivery of therapeutics to the required sites and offer physical and chemical stability with good bioavailability. Despite the proven potential of terpenoids in pharmaceutics, targeted delivery of hydrophobic terpenoids hampers limitations due to their low aqueous solubility, low bioavailability, and low permeability. To overcome these obstacles, novel delivery systems, including NPs and nanocapsules, have developed. Encapsulation of terpenoids improves solubility and obtains controlled delivery; moreover, it solves the problem of stability, protects the drug from degradation, and enhances bioavailability (Lasoń 2020; Serrano-Marín et al. 2020; Hafez et al. 2020). Currently, it is worth mentioning that over half of the approved anticancer agents are derived from plant sources. Ent-11a-hydroxy-15-oxo-kaur-16-en-19-oic-acid (5F) is a diterpenoid purified from *Pteris semipinnata* – a Chinese herbal medicine – with a potent anticancer effect. Due to its low bioavailability, 5F encounters application limitations. Aiming to enhance the bioavailability of 5F, Gong et al. (2019) suggested 5F-loaded cholic acid (CA)-functionalized star-shaped poly (lactic-co-glycolic acid) (PLGA)-polyethylene glycol (PEG)-lacto-bionic acid (LA) (CA-PLGA-PEG-LA) NPs synthesized via nanoprecipitation method as a novel targeting drug delivery system for liver cancer therapy. The general process is displayed as Figure 3.9(a). The encapsulated 5F was produced using the nanoprecipitation method by dissolving a mixture of 5F and copolymer in an organic solvent such as dichloromethane, then adding the organic phase to the aqueous phase. CA-based copolymers displayed higher drug-loading content and encapsulation efficiency. PEGylation of the surface of NPs improved their hydrophilicity and stabilized the micelle. The NP size is a critical property for drug delivery systems. The average hydrodynamic size of the NPs was reported to be 120 nm by dynamic light scattering (DLS) which is shown in Figure 3.9(b). The proper size and narrow size distribution <0.2 made it a good candidate for drug delivery systems. As exhibited in transmission electron microscope (TEM) images (Figure 3.9(c))

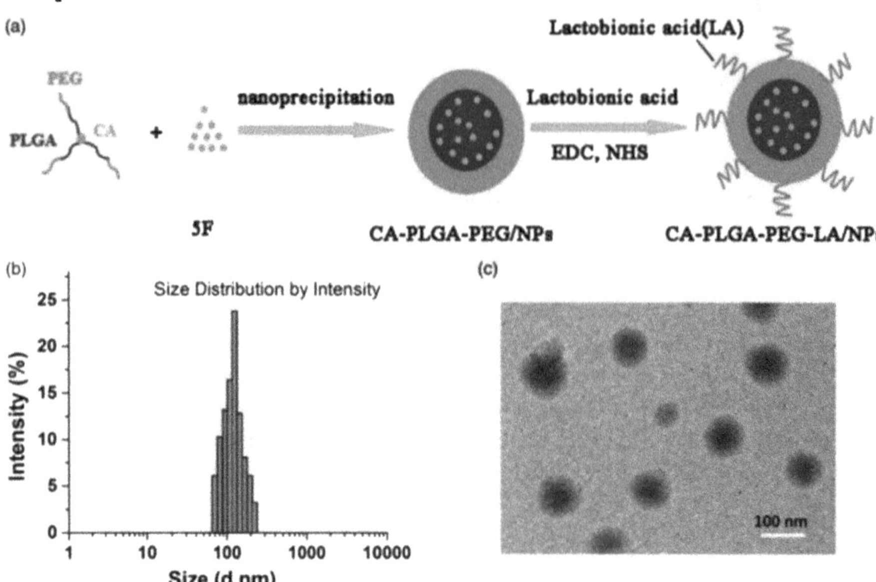

FIGURE 3.9 Preparation and characterization of nanoparticles. (a) Schematic representation of nanoparticle synthesis. (b) Dynamic light scattering size distribution. (c) Transmission electron microscope image of nanoparticles (Gong et al. 2019).

the NPs are well-dispersed and of spherical shape. The in vivo and in vitro results showed great anticancer activity and interestingly demonstrated that 5F-loaded NPs show higher cytotoxicity than 5F against HepG2 (liver cancer) cell lines.

Paclitaxel (PTX), a diterpenoid extracted from the bark of the Pacific yew tree, is one of the oldest chemotherapeutic agents in use. Despite being Food and Drug Administration (FDA) approved, a notable disadvantage of PTX is its poor solubility. Targeted chemotherapeutic delivery, including PTX, is preferable to overcome low water solubility and other drawbacks. Accordingly, Engelberg et al.'s (2021) study proposed a co-encapsulation of PTX and a terpenoid jasmine oil (JO) within biocompatible block copolymer of polyethylene glycol (PEG)–poly E-caprolactone (PCL) (PEG-PCL) NPs, then decorated with S15-APT, an non-small cell lung cancer (NSCLC)-specific aptamer, and were labelled with a diagnostic fluorescent tracer, Cy5 (Figure 3.10). Due to the structural similarity of JO and PTX, the JO core stabilized PTX and greatly improved encapsulation efficiency and reduced premature drug release. Eventually, the adhesive core JO enhances cytotoxic activity on NSCLC cells. The DLS results demonstrate the NPs average size <50 nm, which revealed a desirable prospect for cell entry via receptor-mediated endocytosis (RME).

In another study, for a cancer antimetastasis treatment, a co-encapsulation of a pro-drug reactive oxygen species (ROS)-responsive thioether-linked paclitaxel-linoleic acid conjugates (PTX-S-LA) and cucurbitacin B (CuB), a tetracyclic triterpenoid from the plant of the Cucurbitaceae family, into polyethylene glycol (PEG)–poly ε-caprolactone (PCL) (PEG-PCL) micelles (PCNP) through the emulsion evaporation technique was investigated. The PCNPs were later decorated with an exosome membrane (EM) in order to enhance capture circulating tumour cells (CTCs). CTCs, which are carried around the body with blood circulation, are the seeds of metastasis – additional tumour growth in distant organs. Metastasis is responsible for 90% of tumour death related to cancer; therefore, the early detection of CTCs is of great importance. Exosomes – the 40–180 nm sized extracellular vesicles (EVs) – can escape phagocytosis and achieve long-term circulation by the wide range of cellular adhesion molecules on their membranes, like CD44 and CD47. These surface adhesion molecules can also facilitate homotypic targeting. Herein, decorating the nanocarrier with exosome-membrane (EM) induced immune evasion and achieved CTC recognition. Furthermore,

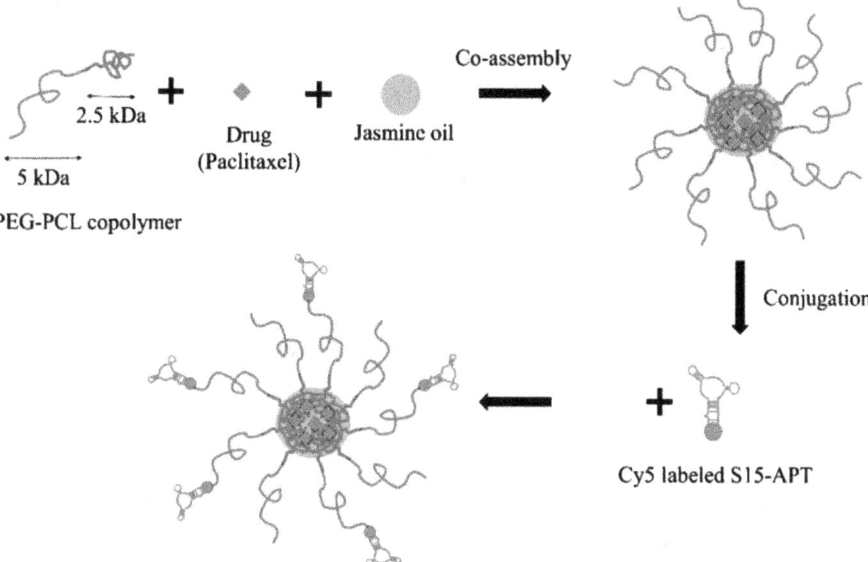

FIGURE 3.10 Schematic diagram of co-encapsulation of jasmine oil and paclitaxel in nanoparticles (Engelberg et al. 2021).

the presence of the CD44 on EM facilitates their penetration into the same source cancer cells. After the cellular uptake, at the initial stage CuB would release and inhibit tumour metastasis and increase intracellular ROS levels, which follows in the bioactivation of ROS-responsiveness of PTX-S-LA and contributes to the effective PTX release from PTX-S-LA. The release of PTX has been stated to be ~2.9 times greater, verifying CuB-mediated prodrug activation. It has also been manifested that the EM-coating significantly decreased the PTX leakage (Wang et al. 2020).

Triptolide (TP) is another diterpenoid isolated from *Tripterygium wilfordii* with anticancer effects, including hepatocellular carcinoma (HCC). However, the low soluble and highly toxic nature of triptolide restrains its application. Therefore, a galactosylated-chitosan (GC-CS) nanocarrier was examined for TP delivery as a hepatocyte-targeted therapeutic (GC-TP-NP). Chitosan (CS) is a polycationic polymer with great biocompatibility, biodegradability, and, more importantly, low toxicity. Due to its pKa (6.2–7.0), CS can only be protonated and soluble in acidic environments. So, in order to use CS as a nanocarrier in hepatocyte-targeted delivery, modification was done on this multi-purpose natural polymer. GC-CS has been carried out by the tripolyphosphate cross-linking method to incorporate galactose groups with lactobionic acid onto CS. Lactobionic acid was anchored to CS via the reaction of amine groups of CS and carboxylic acid, activated by 1-ethyl-3-(3-dimethylaminopropyl)-carbodiimide hydrochloride (EDC) and N-hydroxysuccinimide (NHS) of lactobiontic acid. Galactosylation of CS was confirmed by the number of hydrogen atoms in the proton nuclear magnetic resonance (H-NMR) spectra. The TEM results displayed a near-spherical shape for GC-TP-NP. The results indicated that GC-TP-NP minimized the toxic effect of TP because of the encapsulation of TP. Moreover, it revealed that GC-TP-NP induced apoptosis (Zhang et al. 2019).

Ursolic acid (UA) is a natural pentacyclic triterpenoid derived from different medicinal plants such as ginseng (*Panax ginseng*) and rosemary (*Rosmarinus officinalis*) with a strong anticancer effect. Nevertheless, its poor water solubility hinders its oral bioavailability. Thus, Mainardes suggested CS-modified poly (lactic acid) (PLA) NPs loaded with UA to overcome the limitations. PLA is a biodegradable synthetic polymer with a negative zeta potential. CS has a positively charged amino group which brings about its ability to interact with the human membrane. CS is electrostatically coated on the surface of the negatively charged PLA NPs to enhance its oral delivery. Preparation of CS/PLA NPs containing UA was via the oil-in-water emulsion approach in which the organic phase consist of PLA and UA in dichloromethane is added to the aqueous phase of CS and polyvinyl alcohol (PVA). In the Fourier-transform infrared spectroscopy (FTIR) spectra of the UA-loaded CS/PLA NPs, no peak elimination or specific shift of the UA's functional groups was detected, which elucidates that the drug (UA) hasn't been degraded and didn't have any chemical (covalent) interaction with polymer. CS-coating on UA/PLA NPs was indicated by the positive zeta potential of the NPs and triggers its mucoadhesive properties that is, due to the electrostatic interaction of CS and the negatively charged sialic acid and sugar molecules in mucin. Moreover, the higher the zeta potential, the more it prevents agglomeration, which means more stable NPs and prevents premature UA release. UA-loaded CS/PLA NPs showed concentration-dependent cytotoxicity, but it was lower than the free UA, possibly because of the slow drug release (Antonio et al. 2021).

Artemisinin (ART) is a sesquiterpenoid lactone with various biological applications. Despite the effectiveness of ART, it is limited by its poor bioavailability and solubility, and toxicity. For this purpose, Manjili et al. (2018) incorporated ART into poly (ε-caprolactone)–poly (ethylene glycol)–poly (ε-caprolactone) (PCL–PEG–PCL) tri-block copolymers by a nanoprecipitation method. The tri-block copolymers (PCL–PEG–PCL) were synthesized by ring-opening polymerization of ε-caprolactone in the presence of PEG as molecule macroinitiator and $Sn(Oct)_2$ as the catalyst. DLS reported the average size of the ART/PCL–PEG–PCL NPs to be around 70 nm. The zeta potential of the surface was slightly negative, which contributes to drug circulation time increase. ART slowly release from ART/PCL–PEG–PCL NPs and inhibit breast cancer cell growth.

Glycyrrhizic acid, a triterpenoid isolated from the roots of liquorice, exhibits great potential against inflammatory and viral diseases for a long time. COVID-19 has been one of the grievous world threats which cause excessive inflammation, and until 2021, there hadn't been any specific drugs for it. Keeping in view the antiviral background of glycyrrhizic acid, Zhao et al. (2021) proposed glycyrrhizic acid nanoparticles (GANPs) to combat SARS-CoV-2. GANPs had been prepared via the hydrothermal procedure in which the solution of glycyrrhizic acid was incubated at 185°C for 6 h. Later on, for further investigation, the GANP was modified by PEG-Cy5 coupled with amide bonds through activation by EDS and NHS. The results revealed an excellent inhibitory effect on the proliferation of coronavirus and decreased proinflammation. Moreover, it has been shown that GANP, by targeting severe inflammation areas via EPR effect, enhanced GANPs' accumulation in the lungs and liver, leading to more efficient treatment. The remarkable in vivo and in vitro antiviral and anti-inflammatory results of GANPs brought hope to a therapeutic solution for the pandemic and other hyperinflammation.

3.5 SYNTHESIS OF NANOPARTICLES BY USING TERPENOID-BASED MATERIALS

Due to the mentioned advantages of plant extract-based synthesis of metallic NPs (MNPs), in the recent era phyto-mediated approach of MNP fabrication has attracted tremendous attention among researchers. Various parts of plants from roots (Rashid et al. 2019), stem (Balachandar et al. 2019), bark (Burlacu et al. 2019), leaves (Naseer et al. 2020), and flowers (Kumar et al. 2020) can be envisaged as MNP synthesis sources. The process of NP synthesis using plant extracts is simple. It starts with getting the extract of plants using standard methods such as maceration, infusion, decoction, digestion, percolation, and Sohxlet extraction (the advanced form of digestion and decoction methods).

Maceration: in this process, the powdered crude is mixed with water or organic solvents and stand for at least three days.

Infusion: in this procedure, the powdered crude is infused in cold or boiling water.

Decoction: in this method, the powdered crude is boiled in water.
Digestion: this method is a form of maceration with gentle heat.
Percolation: the extraction of the macerated mixture with a percolator.
Sohxlet extraction: a repeated extraction cycle by refluxing the mixture through the thimble using a condenser and a siphon sidearm.

The different extraction methods of medicinal plants have been widely studied (Belwal et al. 2018; Zhang, Lin, and Ye 2018). As well as, applying these extracts in the production of NPs known as 'green nanotechnology' is also interesting for many researchers. For this purpose, the plant extract and the metallic salt solution are mixed in particular ratios. The reaction usually takes hours or minutes to complete, and the formation of NPs occurs as the biomolecule metabolites in extracts reduce metallic ions precursor. The colour change of the mixture denotes the MNP formation followed by collecting the fabricated MNPs after centrifuging, washing, and drying them consecutively. In some cases, calcination will be needed at the final stage too. The process of reduction of metal ions mediated by plants is called bioreduction. The diverse metabolites in plant extracts act not only as reducing agents but also as stabilizing agents. The detailed mechanism for bioreduction of MNPs hasn't been fully understood because of the diverse nature of the engaged phytochemicals. An instance of a biosynthesized (ZnO) NP from the separated terpenoid fraction is discussed later in this chapter. However, a general route has been assumed; the biomolecules in plants such as flavonoids, alkaloids, and terpenoids first take part in the activation phase, reducing the metallic ion precursor to metals following the nucleation phase which the MNPs are formed. Next is the growth phase; during this stage, NPs agglomerate to form oligomeric clusters with different morphologies, the final size and shape of the MNPs determined in this step. As the growth phase proceeds, to limit the NPs' agglomeration and, therefore, size increase, the biometabolites cover the NPs and restrict further agglomeration (Peng et al. 2018). Accordingly, biomolecules stabilize NPs through capping.

Plants and plant extracts display an essential role in the synthesis of NPs and here we classify them based on the type of NPs that they could produce, i.e., noble metal, metal oxide, quantum dot, graphene, and carbon quantum dots NPs.

3.5.1 Noble Metal Nanoparticle Synthesis

On account of the unique characteristics of noble metals, such as their chemical inertness, surface oxidation resistance, relative stability, and superior shape control contrary to other metals, they have attracted tremendous attention; therefore, green synthesis of these metals has been explored extensively. Among them, silver (Ag) and gold (Au) have been used from ancient times due to their extraordinary biomedical properties to the point that more than half of the publications address silver nanoparticle (AgNP) synthesis and their application (Khan et al. 2019; Santhosh et al. 2019).

Eugenol, a phenolic monoterpenoid, is the prime element of *Syzygium aromaticum* L. (clove) essential oil. Besides that, triterpenoids and several other sesquiterpenoids are also contained in clove essential oil (Mittal et al. 2014; Batiha et al. 2020). *Syzygium aromaticum* L. essential oils were used as a reducing and capping agent to rapidly synthesize mostly spherical AgNPs. The green synthesis was completed within 30 min, which was confirmed by changing the colourless silver nitrate ($AgNO_3$) solution to brown-yellow. Colour change and the excitation of the surface plasmon resonance (SPR) bands of the AgNPs were the rationale for the reduction of Ag^+ to Ag^0. It has been noticed that eugenol plays a critical role in the procedure by deprotonating eugenol OH groups and forming an ionic structure. The ionic structure then shifts into the stable resonance, making it capable of two-electron release; eventually Ag^+ can take these two electrons and produce Ag^0 (Singh et al. 2010). Besides, in this study, the effect of different pH (7, 8, 9, 10) on the AgNPs' formation was also examined. Ultraviolet-visible spectroscopy (UV-Vis) monitored, and the results are displayed in Figure 3.11. The band of SPR with a higher adsorption peak was determined at pH

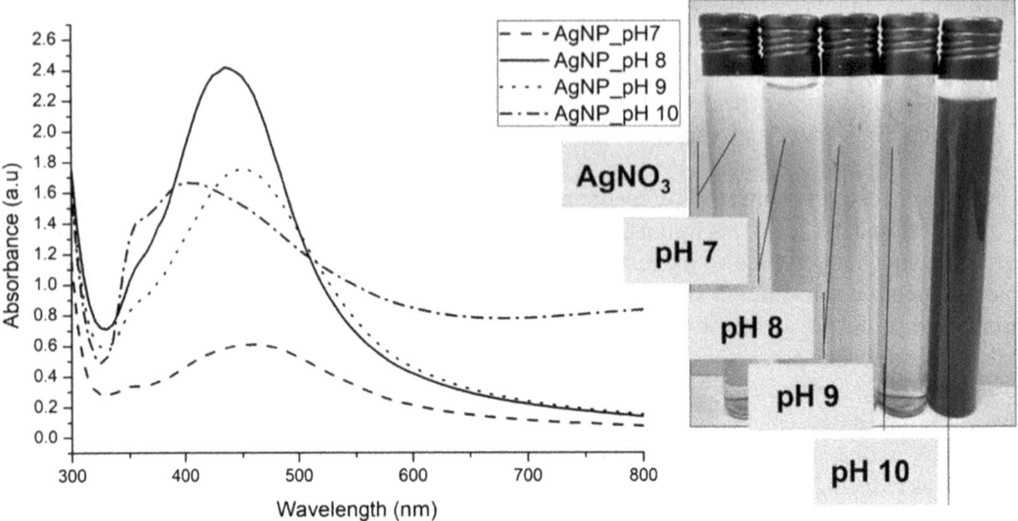

FIGURE 3.11 The surface plasmon resonance band for silver nanoparticles by different pH condition and the respective colour (Maciel et al. 2019).

8, which indicated the maximum formation of AgNPs, whereas wider bands were recognized for pH 7 and 10, representing NPs with a higher polydispersity index.

DLS and TEM analysis also approved the acquired results from UV-Vis. TEM results displayed irregular-shaped AgNPs, including triangles, squares, hexagons, and spheres. The spherical shape was predominant in all pHs, and needle and pentagon shapes were also observed in pH 10 and 7, respectively. The zeta potential values for all AgNPs indicate stability against agglomeration (Maciel et al. 2019).

Monoterepenoids such as linalool and geraniol are among the terpenoids existing in *Jasminum officinale* L. leaf extract (JOLE). Leaf extracts of *Jasminum officinale* have been used to reduce Ag ions into AgNPs with ~9.22 nm diameter and mostly spherical shaped. The JOLE-synthesized AgNPs were obtained after 24 h of incubation at room temperature in a dark place. The HPLC-PDA-MS/MS analysis was carried out in order to evaluate the phenolic phytochemical of JOLE; the results demonstrated that from the 51 identified compounds, the majority of them belongs to secoiridoid glycoside (a structural derivation of monoterpenoids). After incubation, the appearance of a yellowish-brown colour was the sign of AgNP formation. This was also endorsed by the characteristic peak that appeared at 363 nm in UV-Vis spectra. The JOLE FTIR spectra exhibited featured OH peaks, and C=O functional groups stand for secoiridoid glycoside (along with the aromatic ring C-O-C and secondary OH groups' peaks). The deviation of these peaks in the JOLE-AgNPs' IR spectra proposed their role in the biosynthesis of AgNPs. The results for the hydrodynamic size of the AgNPs from DLS showed monodispersed NPs with a size of ~87.6 nm that differs from the TEM, in which only the metallic core has been measured. Furthermore, the negative charge on the AgNPs' surface created good stability. JOLE-AgNPs exhibited a crystalline nature with an fcc structure by comparing the obtained peaks of X-ray powder diffraction (XRD) with the JCPDS data (Elhawary et al. 2020).

Ziziphus jujube fruit is plentiful in triterpenoid acids such as betulinic acid, alphitolic acid, oleanolic acid, and more (Gao, Wu, and Wang 2013; Bai et al. 2016). The facile green synthesis of AuNPs via *Ziziphus jujube* fruit aqueous extract was examined. The composition of the *Ziziphus jujube* fruit was analyzed by GC-MS and manifested a significant number of triterpenoids. The GC results also assume the participation of OH groups in AuNP formation as reducing agents. The appearance of the purple colour after an hour was the formation evidence. DLS reported particle size to be 25.5 nm with a narrow distribution (with a 0.7 polydispersity index (PDI) value), and the zeta potential (in PBS and pH 7) revealed high stability. Similar outcomes were reinforced by UV-Vis in which the adsorption peak of NPs didn't change from 550 nm in 12 days, or with the temperature change (Jafarizad, Safaee, and Ekinci 2017).

While the extracts of various parts of plants, from flowers, leaves, bark, and roots, have been a matter of interest for NPs synthesis. The idea of employing plant waste such as peel, seed, and husk without harming the plant for NP fabrication has been an enticing topic among researchers.

Tamarindus indica shell-husk extract (TSE) has been utilized with the purpose of reducing agents for a short genuinely green synthesis of AgNPs. The synthesis was carried out at 85–90°C for 30 min and was completed with the sign of a dark-brown mixture (Figure 3.12).

The Salkowski test – a qualitative terpenoid-specific test – was performed to identify the TSE's terpenoids by adding concentrated H_2SO_4 to the plant extract's chloroform solution; a reddish-brown colour was formed. The reddish colour is an indication of the presence of terpenoids. This study concluded that the OH groups of the compounds oxidized to C=O groups; the released electrons are gained by Ag^+ and form AgNPs. Moreover, it has been observed that the variation of pH dramatically influences the size of the AgNPs. DLS has attained the average NP size to be 98.1 nm at pH 8.5 (slightly basic condition), and by increasing the pH to 9.5, a two-fold increase in size has been demonstrated (from 98.1 to 186.2 nm without a significant PDI change) (Tade, Nangare, and Patil 2020).

As it is clear from the name of the well-known component of monoterpenoids, α-pinene is the principal constituent of pine needle leaves. Other than α-pinene, β-pinene, eugenol, p-cymene, and many other terpenoids have also been detected (Berta, Supuka, and Chladná 1997). The pine needle extract (as a reducing agent) has been used to synthesize one of the other important noble metals palladium; palladium nanocubes (PdNCs) with an average size of 11.18 nm. In this synthesis, cetyl trimethyl ammonium bromide (CTAB) has been added to the mixture of Na_2PdCl_4 and extract as the capping agent. UV-Vis, XRD, TEM characterized the formed palladium nanoparticles (PdNPs). FTIR spectra illustrated that the reductive groups of the phytochemicals act as reducing agents by a significant decrease in the intensity of C=C and OH functional

FIGURE 3.12 Schematic representation of silver nanoparticle synthesis from *Tamarindus indica* shell-husk extract (Tade, Nangare, and Patil 2020).

groups peaks. Based on the peaks' decrease, the reductive activity of the phytochemicals has been suggested to be like terpenoids and polyhydric alcohols > esters and flavonoids. During the bioreduction process, CTAB forms $[PdCl_nBr_{4-n}]^{2-}$, a coordination compound of Na_2PdCl_4 and CTAB, because of the higher stability of $[PdBr_4]^{2-}$ than $[PdCl_4]^{2-}$. Further, the separated Br^- acts as a stabilizer and adsorb selectively on the Pd (100) plane and decrease the growth rate on them, resulting in the eventual fabrication of NPs with surrounding (100) faces. This phenomenon was also verified by UV-Vis analysis; the sharp absorption peak at 283 nm during the first 30 min displayed the rapid reduction step of Pd^{2+} to PdNPs, and the broad peak after t = 30 was due to the decrease in the reduction rate Pd^{2+} and slower growth pf PdNCs brought by the selective adsorbed Br^-. It has been stated that during different stages of bioreduction, different morphologies have appeared. This was validated via TEM analysis. During the nucleation phase, the Pd^{2+} was reduced to Pd^0 at t = 10 min; many of these seeds had been produced, and at t = 30 min, the seeds continued to grow from 3.93 to 8.96 nm. The TEM showed PdNPs with spherical-like morphology. As Br^- adsorbed, the PdNCs started growing. XRD, selected area electron diffraction (SAED) announced the crystallinity of the PdNCs and high-resolution TEM (HRTEM) as well in which the fringes on the surface were consistent with (100) of the fcc structure of Pd. Also, in TEM images among the majority of PdNCs, some nanorod and nanotriangle NPs were also observed (Peng et al. 2018).

3.5.2 Metal Oxide Nanoparticle Synthesis

Metal oxide NPs like metallic NPs are among the most used nanomaterials due to their distinct properties. Metal oxide NPs in particular are gaining lots of attention in the fields of semiconductors and have promising potential in biological activities as well. On the basis of the growing demand for green synthesis of NPs, metal oxide NPs from zinc (Zn), titanium (Ti), cerium (Ce), copper (Cu), iron (Fe), manganese (Mn), and magnesium (Mg) have also been produced from plant extracts.

Manganese oxide (Mn_3O_4) NPs are critical NPs due to their wide application range such as energy, electrochemical, catalysis, sensor, and therapeutic uses (Sobańska et al. 2021). Moreover, Mn itself plays a crucial role in cellular redox status, neurotransmitter synthesis, metabolism, etc. Further, the free radicals from Mn have a great link with diseases like Alzheimer's, diabetes, cancer, and more (Shaik et al. 2021). Furthermore, within biomedical applications, Mn_3O_4 and iron oxide (Fe_2O_3) have the potential to enhance the contrast of the images taken by magnetic resonance imaging (MRI). The magnetic properties of iron oxides (superparamagnetic) make them an excellent candidate for biomedical applications such as cell separation and targeted drug delivery, as well as MRI. Other than biomedical applications, iron oxides are used for catalysis, sensors, water treatment, etc. (Das and Patra 2021).

Triterpenoids are the main compound of the *Chaenomeles* species (Yao et al. 2020). Meanwhile, a novel approach of Mn_3O_4 and Fe_2O_3 NP synthesis with the flower extracts of *Chaenomeles* has been published. The NPs synthesized under a moderate condition without using elevated temperature with the ratio of 1:1 the precursor $Fe(NO_3)_3.9H_2O$ and $Mn(NO_3)_2.6H_2O$ solution and the extract. After 24 h, the brown solution appeared as the indicator of the NPs' formation. XRD results of the Mn_3O_4 and Fe_2O_3 NPs have been notified tetragonal and rhombohedral crystalline structures, respectively. TEM and scanning electron microscope (SEM) images unveiled irregular cubic and spherical shapes with an average size of 110 nm for Mn_3O_4 and spherical and rod shapes with an average size of 39 nm for Fe_2O_3 individually (Karunakaran et al. 2017).

The wide bandgap and considerable binding energy of zinc oxide nanoparticles (ZnONPs) make them widely applicable in photocatalysis, coating, sunscreen, and antimicrobial applications (Selim et al. 2020). Additionally, having FDA approval enhances their antimicrobial and biological applications in surgical tapes, antiseptic creams, pharmaceutics, cancer therapy, drug delivery, and many more (Akintelu and Folorunso 2020). The high content of sesquiterpenoids like β-copaene and monoterpenoids such as terpinene-4-ol, γ-terpinene in *Deverra tortuosa* leaf extract has made it a prominent candidate for ZnONP synthesis (Biol 2020). Selim et al.'s (2020) study will be the first report examining *Deverra tortuosa* ZnONPs for anticancer properties. The presence of triterpenoids and other terpenoids in the *Deverra tortuosa* extract was tested by Liebermann–Burchard reaction which the chloroform solution of the extract is treated with acetic anhydride and a few drops of concentrated H_2SO_4. The formation of the green-blue colour inhibits the presence of terpenoids. The synthesis of ZnONPs was determined with the formation of a creamy (yellowish)-white paste after 24 h of incubation at 60°C. The ZnONPs exhibited a prominent peak at 374 nm in UV-Vis spectra, ascribing the bandgap of ZnO, which is due to the oscillation of the electrons to the conductive band from the valence band. The bandgap of the ZnONPs plays a significant part in its biological application, like cytotoxic effect. Indication of the C=C functional groups in the IR spectra of the ZnONP accounts for the stability of the NPs. TEM and XRD techniques identified the average size of 15.22 nm with a hexagonal structure.

As previously mentioned, due to the broad spectrum of biomolecules from proteins to secondary metabolites as terpenoids, flavonoids, and alkaloids implicated in the bioreduction mechanism, there is no precise understanding of each compound's role. Nevertheless, as a case in point, Kavitha et al. (2017) stated that the bitter taste of the leaves of *Andrographis paniculata* points to its high terpenoid content. With this in mind, they proposed synthesizing ZnONPs from the terpenoid fraction of *Andrographis paniculata* leaves. The terpenoid fraction (TAP) was separated from *Andrographis paniculata* leaf extract by column chromatography and affirmed by Salkowski test followed by 3 h mixing of TAP, $Zn(NO_3)_2$, and 0.1 N NaOH (in a time gap manner). FTIR characterized the resultant cloudy-white precipitate (ZnONPs) to elucidate the biocompounds' role. The IR spectra of TAP exhibited peaks corresponded to C–H, C–O, –C=C–, and a strong one for –OH, while the ZnONP spectra displayed a peak for –CH=O group in addition to the observed –OH, C–H, C–O. The remaining –C=C– was a sign for the surrounding aromatic rings in ZnONPs. Additionally, XRD and SEM findings supported each other on hexagonal-shaped ZnONPs with an average size of 20.23 nm. It has been suggested that the +17.6 mV zeta potential value of ZnONPs made them an excellent applicant for absorption to negatively charged cells and further drug delivery applications.

Titanium dioxides (TiO_2s) like ZnO are considered inorganic substances with excellent oxidation strength, photostable, and nontoxic properties. Due to this, they have been utilized greatly in material sciences and medicine, such as drug delivery, medical implants, etc. (Jafari et al. 2020). A single-step green synthesis using *Azadirachta indica* leaf extract to fabricate TiO_2NPs was investigated in another experiment. The white-u TiO_2NPs were found to have crystalline anatase and rutile structures examined by XRD and SAED. Further, the SEM and TEM depicted that the TiO_2NPs are spheres with a 15–50 nm size range (Thakur, Kumar, and Kumar 2019). Finally, it is worth mentioning that *Azadirachta indica* (Indian neem tree) compose a separate subclass of terpenoids called tetranotriterpenoids with a chemical core of furanolactones. Limonoid and azadirachtin are famous examples of this class (Githua et al. 2010).

Mg is a vital mineral for the body, and among the other inorganic metal oxides, is greatly considered nontoxic. On the other hand, they have been classified by the FDA as generally recognized as safe (GRAS) that open new windows for biomedical applications, including scaffold repairing, as a hyperthermia agent in cancer therapy, etc. (Cai et al. 2018). A rapid, facile green synthesis of magnesium oxide nanoparticles (MgONPs) using both solid and liquid extracts of *Aloe barbadensis* has been reported. Phytochemical analysis of the *Aloe barbadensis* obtained a great number of triterpenoids such as lupeol (Gangadharan et al. 2019), linoleic acid, b-sitosterol, and campesterol (Hamman 2008). In the extract preparation step after filtration with a vacuum filter, the residue (solid state) and the filtrate (aqueous state) have been separately purified via rotary vacuum evaporator, and the result was used for further MgONPs synthesis. Reaction parameters such as reaction time, extract ratio, temperature, and precursor concentration have been optimally evaluated to obtain smaller MgONPs. The result of experiments demonstrated that the optimum reaction time

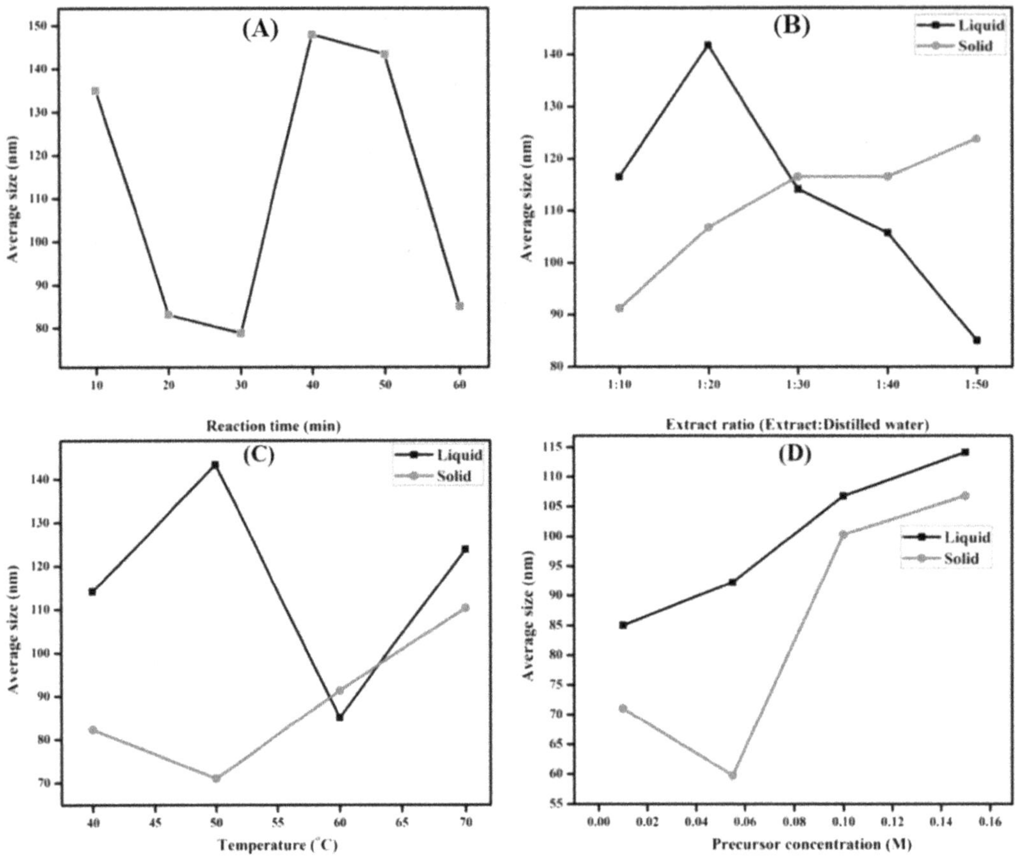

FIGURE 3.13 Optimization of magnesium oxide nanoparticle synthesis parameters, (A) reaction time, (B) extract ratio, (C) temperature, and (D) precursor concentration (Jeevanandam et al. 2020).

for both solid and liquid extracts was 30 min, because the smallest NPs with the size of 78.82 nm produced (Figure 3.13(A)). With changing the plant extract concentration, liquid and solid phases exhibit different mechanisms in the formation of NPs. As it is known, with the decrease of concentration, the reduction rate increases, which leads to smaller sized particles, so as it is observed in Figure 3.13(B) that the extract ratio of 1:50 (liquid extract: aqueous extract) yields smaller MgONPs (85.5 nm), and on the contrary, the smallest MgONP (91.28 nm) obtained from the solid extract was with the 1:10 ratio.

These findings indicate that the liquid extract and solid extract are different in the quantity of the phytochemicals. UV-Vis illustrated that $Mg(OH)_2$ instead of MgO has been obtained from the solid extract thus, further calcination needs to be done – which the optimum temperature for it was based on the obtained data from thermogravimetric analysis (TGA). Generally, as the reaction temperature increased, the size of the NPs tends to increase due to the reduction of supersaturation which, in this situation, the growth rate increases and is expected to have larger-sized NPs. It has been reported that temperatures below 60°C don't have enough energy to stabilize the NPs. Despite obtaining smaller NPs at 50°C, the 60°C temperature was chosen as the optimum temperature for both solid and liquid extracts; due to the lowest PDI value (broad distribution range) that was observed at this temperature (Figure 3.13(C)). It is also a matter of fact that the increase in precursor concentration could induce agglomeration and enhance the NPs size as demonstrated in Figure 3.13(D). Similarly, the absence of characteristic peaks of the extract represents the formation of MgONPs in the FTIR spectrum. TEM results manifest spherical MgONPs in the range of 25–35 nm and 39–60 nm for liquid and solid extracts, respectively (Jeevanandam et al. 2020).

The other interesting metal oxide NP is cerium oxide (CeO). CeOs are semiconductor oxides with broad application in solar cells, surface polishing, automotive exhaust treatment, etc. Besides this they have the unique ability to mimic the activities of some enzymes as oxidase, catalase, phosphatase, and many more, which because of this capacity they have potential in drug delivery, neurodegenerative diseases, diabetes, and wound healing (Dhall and Self 2018; Abuid et al. 2019).

CuONPs are nowadays mainly used as antimicrobial agents. For non-medicinal applications, they can be referred to as catalysts, superconductors, and sensing materials (Rani et al. 2020; Prajapati, Yadav, and Gauri 2020). *Holoptelea integrifolia* possessing a variety of triterpenoids such as Friedlin, β-sitosterol has been utilized to synthesize mostly spherical CuNPs. The shape and the involvement of the phytochemicals of the extract were analyzed by SEM and FTIR separately (Khalid et al. 2013; Anju et al. 2020).

3.5.3 Quantum Dots

Quantum dots (QDs) are the newest class of NPs. The reason behind their name goes back to the optical properties of nanocrystals which quantum mechanics dedicates. Due to their superior electronic and optical characteristics such as high photostability, broad absorption spectrum, high quantum yields (up to 85%), etc., they are assessed in various biological applications such as bioimaging, biosensing, drug delivery, tissue engineering, and many others as well as material sciences (Rani et al. 2020). QDs are generally comprised of metal and transition metals with a core-shell structure in which the capping ligand shell on the semiconductor core enhances their water solubility (Frasco and Chaniotakis 2010). Acknowledging the developing application of QDs, emphasizing the green synthesis of QDs from different biological sources, including bacteria, fungi, algae, plant, and plant extract, has earned significant consideration (Omran, Whitehead, and Baek 2021).

Herein, Begum and Ahmaruzzaman (2018) had applied *Parkia speciosa* Hassk. pods extract for green synthesis of SnO_2QDs. *Parkia speciosa* Hassk. contains a great portion of sesquiterpenoids and triterpenoids such as β-sitosterol, campesterol, stigmasterol, and lupeol (Izzah Ahmad et al. 2019). The synthesis of rounded SnO_2QDs from *Parkia speciosa* Hassk. extract was adopted by the addition of the leaf extract to $SnCl_4$ aqueous solution under ultrasonic; after drying at 60°C in a hot oven, the yellowish-brown SnO_2QDs were produced. The EDX analysis confirms the formation of SnO_2 with the noticeable peaks of Sn and O (3.4 and 0.5 keV). The appearance of the terminal Sn–O and the binding O–Sn–O bonds in FTIR supported the production of QDs. Moreover, the indication of C–C, CH_3, and OH peaks are attributed to the *Parkia speciosa* Hassk. extract. XRD, SAED, and HRTEM shared the results and revealed a tetragonal rutile crystalline structure. The size and shape of the SnO_2QDs were examined via TEM to be 1.9 nm-sized spherical NPs.

3.5.4 Carbon and Graphene Quantum Dots

Despite the wide application of QDs, due to the toxic nature and high cost of the heavy metals in their composition, they should be tailored to suit more benign choices. Carbon dots (CDs) or carbon quantum dots (CQDs) are an excellent alternative for QDs. CDs represent a new class of nanocarbon materials with a size less than 10 nm, which were discovered in 2004 by Xu and his co-workers (Xu et al. 2004). They have lower toxicity, more biocompatibility, an abundant carbon precursor source; besides, the unique optical properties of QDs, had made them the centre of attention (Shahshahanipour et al. 2019). Using the phytochemicals in plant extract as the carbon source is an innovative approach to synthesize CDs ecologically (Nasseri et al. 2020). Ramanarayanan and Swaminathan (2019) reported utilizing guava leaf extract for synthesizing CDs via a hydrothermal method. In this procedure, the guava leaf extract was heated at 160°C in a Teflon autoclave. After 1 h the dark brown CDs were obtained. The characterization peak affirmed the synthesis of CDs at 360 nm in UV-Vis. TEM revealed nearly spherical CDs with a narrow distribution from 4 to 7 nm. The peaks of the functional groups of OH, CH, C=C, C-O-C, and C-O of the extract were identified by FTIR. Notably, guava leaf extract used in this study as a carbon source consisted mainly of caryopbyllene (95%), a bicyclic sesquiterpenoid, followed by percentages of pinene, limonene, humulene, and many other terpenoids (Wilson and Shaw 1978).

Graphene quantum dots (GQDs) are categorized as one type of CD with better crystallinity (Li et al. 2015). In this regard, Anooj and Praseetha (2020) proposed synthesizing GQDs from *Rosa gallica. Rosa gallica* is a well-known element of the perfume industry. Besides that, it has excellent biological activities such as anti-inflammatory, antioxidant, antidiabetic, and cytotoxic effects. The major component of *Rosa gallica* is the well-known monoterpenoid, geraniol. Limonene, citronellol, eugenol, and nerol are its other monoterpenoidal constituents (Mileva et al. 2021). The presence of terpenoids in this work was also confirmed by the Salkowski test. They synthesized GQDs via two different approaches, pyrolysis and the hydrothermal method. In the pyrolysis procedure, the solution of the *Rosa gallica* extract in water was heated to 100°C for 90 min. At this point, the change of the citric acid in the extract results in an orange solution. However, in the latter, first, the mixture of extract, hydrazine hydrate, and water was boiled in an ultrasonic bath, and then the solution was heated at 200°C for 6–10 h in a Teflon reactor. The sharp peak at the 2θ of 38.63 in XRD denotes the graphitic of GQDs. The IR peaks of the –O-C=O, aromatic C=C, $-CH_3$, and C–O bonds confirm the strong interaction of the extract of *Rosa gallica* and the component.

3.6 BIOMEDICAL APPLICATIONS OF TERPENOID-SYNTHESIZED NANOPARTICLES

3.6.1 Anticancer Application

Despite the remarkable progress in the field of cancer treatment, there is still a need for more effective therapies with fewer side effects. Thus recently, plants and phytochemicals from plant extracts have been considered as a great potential against a variety of cancers (Abdulridha et al. 2020).

Sesquiterpenoids and monoterpenoids such as caryophyllene, β-elemene, β–pinene, and linalool are just some of the major components of *Piper longum* L. leaf extract (Anwar 2008). The spherical AgNPs from *Piper longum* L. aqueous extract have been tested for the first time against cervical cancer cells and showed remarkable cytotoxicity against HeLa cell lines using MTT assay. MTT assay is a colorimetric assay to determine cell viability and cytotoxicity. It is based on the reduction of the yellow tetrazolium salt solution to purple formazan crystals in metabolically active cells by a mitochondrial dehydrogenase enzyme. The darker the solution colour, the greater the number of viable cells. The AgNPs displayed a dose and time-dependent inhibition of HeLa cells. The maximum IC_{50} value of the AgNPs on HeLa cells was 5.27 µg/mL which means that 5.27 of the as-synthesized AgNPs are enough for 50% inhibition of the viability of the human cervical cell lines. In order to compare the cytotoxic effect of the AgNPs with the leaf extract, the *P. longum* extract was separately dissolved in different solvents of hexane, chloroform, ethyl acetate, methanol, aqueous, and the observed IC_{50} were 221.4, 35.6, 17.8, 12.9, and 8.8 µg/mL, respectively. Furthermore, it has been observed that the IC_{50} value of the piperlongumine, a critical phytocompound of *P. longum* with therapeutic effects, for HeLa cells was 174.6 µg/m. Comparing the IC_{50} values of piperlongumine and AgNPs demonstrates that the greater cytotoxic effect of AgNPs is due to the synergistic effect of the biomolecules of the plants, moreover; they display minimum toxic effects on normal cells (HEK 293). The AgNPs, by producing ROS and causing DNA degradation, eventually induce apoptosis in HeLa cells at a minimum concentration (Yadav et al. 2019).

Generally, jasmine species comprises a broad range of monoterpenoids, iridoids, and secoiridoid. Among them, *Jasminum auriculatum* extensively possesses a great source of the triterpenoid jasmonil and pentacyclic triterpenoid lupeol (Scientific and Sciences 2018). Considering the great source of the phytochemicals of *Jasminum auriculatum* leaves, it has been used as a reducing and stabilizing agent to prepare AuNPs. The Au^{3+} was reduced to Au^0 within 1 h. The *Jasminum auriculatum*-based AuNPs have been manifested to strongly inhibit the growth of HeLa cells with a dose and time-dependent manner; up to 69.28% suppression of the viability of HeLa cells was observed with the 200 μg/mL concentrated AuNPs, and the half maximal inhibitory concentration (IC_{50}) has been described to be 104 μg/mL. The anticancer effect of the prepared AuNPs was tested by MTT assay. The amphiphilic feature of AuNPs could be the basis of the cytotoxic effect, which causes the AuNPs to penetrate into cancerous cells and, by inducing ROS, results in cell death (Balasubramanian, Kala, and Pushparaj 2020).

Although the major constituent of *Coleus aromaticus* varies in different studies between carcavol and thymol, it is shown that in both cases, monoterpenoids are the main components of *C. aromaticus* (Govindaraju and Arulselvi 2018). TiO_2NPs by leaf extract of *C. aromaticus* have been successfully tested for their apoptosis effect on HeLa cell lines. It has been suggested that the cytotoxicity was due to the provided excess electrons to TiO_2NPs by plant extract, which resulted in the increase of ROS on the HeLa cells' surface. It has been exhibited that the TiO_2NPs decreased the viability of HeLa cells via an increase in their concentration (about 92.37% and 81.24% inhibitory effect at 200 and 100 μg/mL, respectively) (Narayanan et al. 2021).

3.6.2 Antimicrobial Application

Antimicrobial resistance (AMR) is a global health crisis; moreover, according to the World Health Organization (WHO), it is considered one of the top ten global threats (World Health Organization 2020). Therefore, with the increase in drug resistance and viral infections, there is a call for a need for alternative therapeutics. Metal and metal oxide NPs have grown to be a perfect aid against microbial resistance.

3.6.3 Antibacterial Application

Antibiotic resistance to life-threatening infections caused by *Escherichia coli*, *Staphylococcus aureus*, and *Klebsiella pneumonia* have been rapidly increasing. The Coronavirus infectious disease (COVID-19) outbreaks caused by SARS-CoV-2 arose from Wuhan, China in 2019. It was one of the fatal global pandemics after the 1918 Spanish flu. Secondary bacterial infections in COVID-patients can significantly escalate morbidity and mortality. About 62% of the fatalities among COVID-19 patients was diagnosed with bacterial infections (Shafran et al. 2021). Some of the common pathogens in hospitalized COVID-patients are *E. coli*, *S. aureus*, and *Klebsiella* sp. (Chen et al. 2020; Hoque et al. 2021; Zhou et al. 2020). Terpenoid-based NPs have been reported to have a good antibacterial effect.

Terpenoids such as lutein, luteoxanthin, β-carotene are presented in *Moringa oleifera* extracts (Zahirah et al. 2018). The green synthesis of spherical-shaped AgNPs was observed from freeze-dried (FD) and fresh (F) *Moringa oleifera* leaf extract. The primary energy source for the NPs' formation was direct sunlight irradiation. This is the first reporting of sunlight mediated AgNP synthesis using *M. oleifera* for their antibacterial activity. The AgNPs declared a broad antibacterial spectrum range against *Staphylococcus aureus*, *Enterococcus faecalis*, *Escherichia coli*, *Pseudomonas aeruginosa*, and *Klebsiella pneumonia*, which are famous among well-known pathogens with antimicrobial resistance. The minimum inhibitory concentration (MIC) of AgNP from both FD and F leaf extract was reported to be 25 μg/mL for *K. pneumoniae*, *P. aeruginosa*, and *S. aureus* strains, while as with a decrease in half, this number reduced to 12.5 μg/mL for *E. coli* and *E. faecalis*. Comparing these data with the MIC of neomycin, a commercially available antibiotic, in the range of 6.25–25 μg/mL confirmed that besides being more effective at the minimum concentration, they have a wider susceptible bacterial range. It has been stated that *E. faecalis*, *P. aeruginosa*, and *S. aureus* strains showed no inhibition for neomycin, which means they were insensitive to it; however, they were susceptible toward both of the synthesized AgNPs. It is noteworthy that the AgNPs derived from *Moringa oleifera* showed antibacterial against *Klebsiella pneumoniae* due to its size being six times smaller than other reports (Moodley et al. 2018).

Boswellia carterii extract is a rich source of various types of terpenoids, from monoterpenoids to sesquiterpenoids. Aside from that, Boswellic acid, a pentacyclic triterpenoid, is a distinguishing compound in it (Mannino et al. 2016). In this specific work, the *Boswellia carterii*-derived AgNPs have been found to have a prominent inhibitory effect against *Streptococcus mutans*, *Pseudomonas aeruginosa*, and *Enterococcus faecalis*. At the same concentration of *Boswellia carterii*-derived AgNPs (200 μg/mL), *S. mutans* was found to have the largest inhibitory zone rather than *Pseudomonas aeruginosa*, and *Enterococcus faecalis* with the reported diameter of 22 ± 0.6 mm, ±0.9 mm, and 12.0 ± 1.4 mm respectively. Moreover, different MIC values have been illustrated for the mentioned bacteria at agar dilution and broth dilution. In agar dilution, *Streptococcus mutans, Pseudomonas aeruginosa*, and *Enterococcus faecalis* at a concentration of 25.0, 34.0, and 40 μg/mL respectively, were inhibited while these concentration changes to 10.4, 15.2, and 18.2 μg/mL respectively in broth dilution (Al-Dahmash et al. 2021).

Peel extract of *Actinidia deliciosa* (kiwi fruit) is the reservoir of terpenoids like carotenoids. Using *Actinidia deliciosa* peel extract as a reducing and capping agent, 8.2 nm truly green ZnONPs with a crystalline structure were synthesized. SEM and TEM analysis indicated that the ZnONPs produced were spherical and hexagonal shapes with just a little aggregation. Furthermore, the zeta potential results showed a strong negative surface charge, which eased the chemical interaction of the NPs with the bacterial membrane. For the antibactericidal activity of the ZnONPs they were tested against *Staphylococcus aureus* (*S. aureus*), and SEM images scrutinized the mechanism of cell death. The images of the bacteria treated with ZnONPs indicated cell membrane damage. In contrast, there was no damage in uninfected bacteria cells, indicating two probable routes: the NPs' interaction and penetration ability with the bacteria, the other; increase of the reactive oxygen species like •OH, H_2O_2, and O_2• that could interact with the cell which in both cases leads to the bacterial death (Li et al. 2021).

3.6.4 Antifungal Application

Antifungal resistance is another important class of antimicrobial resistance that is becoming a concern, particularly for patients at high risk. Recent studies have demonstrated that considerable numbers of co-fungal infections with COVID-19 include *Aspergillus* species (especially *Aspergillus flavus*), *Candida albicans*, and *Candida glabrata* (van de Veerdonk et al. 2021; Hoque et al. 2021).

Hence, the antifungal potential of AgNPs synthesis from *Kigelia pinnata* aqueous bark extract has been examined. *Kigelia pinnata* is a source of terpenoids including limonene, α-pinene, α-terpineol, monoterpenoids, phytol diterpenoid, squalene triterpene, and caryophyllene sesquiterpenoid (Bello et al. 2016). The *Kigelia pinnata*-mediated AgNPs have been experimented on against three fungal pathogens, *Aspergillus niger, Aspergillus flavus*, and *Candida albicans*, and demonstrated great MIC values of 62.4, 31.2, and 15.6 μg/mL, respectively with the highest antifungal activity against *C. Albicans.* As exhibited, the synthesized AgNPs could be a great antifungal agent (Ravi and Kannabiran 2021).

3.6.5 Anti-Inflammatory Application

Neuroinflammation is a serious condition in central nervous system (CNS) disorders like neurodegenerative diseases. Alzheimer's disease (AD) is the most common age-related neurodegenerative disease with immense worldwide progression, which is caused by multiple factors such as oxidative stress and depletion of acetylcholine (ACh), and aggregation of amyloid-β (Aβ) in the brain. acetylcholinesterase (AChE) and butyrylcholinesterase (BuChE) enzyme hydrolysis acetylcholine in the brain. Inhibition of these enzymes, thus increasing the acetylcholine level, is one of the most effective treatment approaches to AD. Moreover, it has been reported that BuChE is closely associated with Aβ-plaque forming. Accumulation of Aβ, which is toxic to neurons and induces inflammation, is the main contributor to the pathogenesis of AD; thus, suppressing the Aβ fibrils aggregation is the principal strategy. Common drugs for AD treatment have severe side effects; therefore, the implication of NPs in these diseases has attracted great interest. Herein, the neuroprotective and anti-Alzheimer effects of the green synthesized AuNP from *Terminalia arjuna* bark extract have been studied. As published before, the oleanane-type triterpenoids such as arjunic acid, arjunolic acid, and arjungenin are the main biocompound of *Terminalia arjuna* bark extract. The arjunolic acid itself has an AChE inhibitory effect. The as-synthesized AuNPs manifested great concentration-dependent inhibitory effect of AChE (98.23 ± 0.015% at 50 μg/mL), and BuChE (99.4 ± 0.13% at 50 μg/mL) with the IC_{50} of 4.25 ± 0.02 and 5.05 ± 0.02 μg/mL, respectively. The mechanism of inhibitory activity of AuNPs against AChE was recommended to be due to interaction and adsorption of the lipophilic AuNPs with the hydrophobic environment of the enzymes (ChEs) and by changing the conformation results in the inhibition of enzymes. In supplement to this, the synthesized AuNPs have been suggested to prevent Aβ oligomers aggregation and destroy the formed fibrils. The Aβ fibrils formation was probed by a Thioflavin (Th-T) assay. Th-T is a fluorescent dye that gives strong fluorescence upon binding to amyloid fibrils. The decrease in fluorescence with AuNPs (50 μg/mL) indicates β-sheet growth is decreased, which leads to the inhibition of fibril formation. The IR results are in accordance with the Th-T and describe that the sharp peak at 1630 cm^{-1}, which accounts for β-sheets, decreases after AuNP treatment (Suganthy et al. 2018).

3.6.6 Antiviral Application

Linalool and nerolidol are among the terpenoids in strawberry and ginger and are well-known rich sources of terpenoids – from monoterpenoids to sesquiterpenoids. Very recently, in an exciting study, green synthesized AgNPs from *Fragaria ananassa* Duch. (strawberry) and *Zingiber officinale* (ginger) have been successfully explored against SARS-CoV-2. The *Fragaria ananassa* Duch. and *Zingiber officinale* derived AgNPs have been synthesized extremely rapidly from $AgNO_3$ precursor and the extracts –separately – with the addition of NaOH in the reaction medium at 60°C for 10 min. To investigate the anti-SARS-CoV-2 activities of the AgNPs, MTT assay was used on vero cells and exhibited great antiviral activity. The results demonstrated great anti-SARS potential with an IC_{50} value of 0.0989 and 0.034 μg/mL for strawberry and ginger, respectively. Moreover, the interaction of AgNPs with viruses can be in two possible processes: AgNPs could either bind to the virus outer layer and inhibit the further attachment of the virus or bind to the DNA or RNA of the virus and prohibit the virus's reproduction. The use of AgNPs as an anti-COVID agent can be due to the adjunct of the AgNPs to the glycoprotein spike of the SARS-CoV-2 and prohibit their ability to bind to living cells. Further, by releasing Ag ions, the pH of the respiratory epithelium in which coronavirus usually lives tends to decrease. The acidic condition is a hostile environment for coronavirus. Therefore, by this mechanism, AgNPs contribute to control SARS-CoV-2. MD was carried out for a deeper insight; however, the MD studies suggest the anti-SARS mechanism through the inhibitory effect of the neohesperidin on human AAK1 protein and SARS-CoV-2 NSP16 protein. These findings indicate it is a promising candidate for future therapeutic studies against the SARS-CoV-2 pandemic (Al-Sanea et al. 2021).

On the same path, Al-Dahmash et al. (2021) stated that the negative zeta value of the synthesized AgNPs not only stabilizes them but also influences the NPs' attachment to viruses such as SARS-CoV-2. It has been proclaimed that NPs with the size of <30 nm can act as an anti-COVID agent and attach to positive charge glycoprotein spike receptors of coronavirus and target it. In this case, the

TABLE 3.1
Non-Green Synthesized Nanoparticles and Their Biological Activities

Application	Nanoparticle	Analyte	Toxicity threshold	Reference
Anticancer	AgNP	HeLa cells	2.0 μg/mL	Kaba and Egorova (2015)
	AuNP	HeLa cells	>5 μg/mL	Shanei and Akbari-Zadeh (2019)
	TiO_2NP	HeLa cells	≥100 μg/mL	Geng et al. (2020)
Antibacterial	AgNP	*Escherichia coli*	IC_{50} = 11.89 mg/L	Kaweeteerawat et al. (2017)
	AgNP	*Staphylococcus aureus*	IC_{50} = 6.98 mg/L	Kaweeteerawat et al. (2017)
	AgNP	*Klebsiella pneumonia*	MIC_{90} ≥ 512 μg/mL	Khatoon et al. (2019)
	AgNP	*Enterococcus faecalis*	19.2 μg/mL	Schwass et al. (2018)
	AgNP	*Pseudomonas aeruginosa*	MIC_{90} ≥ 640 μg/mL	Khatoon et al. (2019)
	ZnONP	*Staphylococcus aureus*	200 μg/mL	Siddiqi et al. (2018)
Antifungal	AgNP	*Aspergillus flavus*	80 μg/mL	Villamizar-Gallardo, Cruz, and Ortíz (2016)
	AgNP	*Aspergillus niger*	2 mmol/L	Shuaib et al. (2020)
	AgNP	*Candida albicans*	60 μg/mL	Vazquez-Muñoz, Avalos-Borja, and Castro-Longoria (2014)
Anti-inflammatory	AuNP	Aβ42 monomer	110 nmol/L	Hou et al. (2020)
Antiviral (SARS-CoV-2)	AgNP	Vero cells	IC_{50} = 0.002190%	Almanza-Reyes et al. (2021)

average size of the AgNPs has been reported to be 14 nm. However, this suggestion and the efficacy of SARS-CoV-2 inhibitory activity still needs to be clinically tested.

In the following, the NPs mentioned here are compared with some of their corresponding non-green synthesized NPs with the identical biological application in Table 3.1.

3.7 TERPENOIDS AS RENEWABLE PRECURSORS FOR POLYMER SYNTHESIS

Since the beginning of the 20th century, on account of the ubiquitous applications of polymers in various fields, there has been an increasing demand for such versatile material. Nevertheless, polymer production is currently mainly dependent on fossil-derived resources, resulting in severe environmental concerns regarding global warming, depletion of oil resources, and economic issues. Therefore, reducing the dependence on fossil fuels and seeking out renewable alternatives is of great importance. There are two different approaches to obtaining renewable polymers: 1) the typical process is utilizing natural polymers such as hemicellulose, lignin, and chitin for polymer production. The limitation of these biopolymers is their molecular weight and industrial applications. 2) Using the renewable monomers like plant oil, rosin, terpene, terpenoids for polymerization (O'Brien et al. 2020). The advantage of this approach over the former is the adjustment of molecular weight distribution. Among the renewable feedstocks, the family of terpenes and terpenoids has attracted great attention. Monoterpenes and monoterpenoids are the most studied terpenes. Limonene is a well-known monoterpenoid with significance in terpenoids' polymerization. Limonene-derived monomers can polymerize to polyesters by ring-opening polymerization (ROP); likewise, limonene oxide can react with CO_2 in a ring-opening copolymerization (ROCOP) and form polycarbonate (Stößer et al. 2017). Detailed examples have been much reviewed (Della Monica and Kleij 2020). Other types of terpenes such as β-myrcene, β-ocimene, and β-farnesene, which β-myrcene and β-ocimene regard to monoterpenes and β-farnesene to sesquiterpenes, can undergo a radical polymerization and produce isoprene like their polymeric homologs (Sahu and Bhowmick 2019). Similarly, there have been articles about the modification of monoterpenoids such as tetrahydrogeraniol, citronellol, menthol, and isoborneol into (meth)acrylates (Droesbeke et al. 2020). Citronellol, geraniol, and farnesol have also been reported to prepare sulphur-terpenoid composites by inverse vulcanization (Maladeniya et al. 2020). However, the polyterpenes explored until now are mainly elastomers with industrial applications, and deeper insight into this subject exceeds our chapter's scope. Unfortunately, to date, only one report has been found reporting biocompatible terpenoid-based polymer NPs with potential biological application. In this approach, O'Brien et al. (2020) proposed a totally green, metal-free ROP of lactide with geraniol (ger), carveol (car), limonene-OH (Lim-OH), and trans-sobrerol (t-sob) terpenoids as initiator, 1,8-diazabicyclo [5.4.0]undec-7-ene (DBU) as a naturally occurring catalyst, and 2-methyltetrahydrofuran (2-MeTHF) solvent as a greener replacement for dichloromethane. The precursors used are all biomass-derived; additionally, 2-MeTHF is not only sustainable but also enables the dissolve of t-sob, which was insoluble in dichloromethane. By dissolving the lactide (LA) and initiator with the ratio of 1:10 in 2-MeTHF at 60°C and adding DBU, the ROP started, and the reaction carried on for 15 min to produce poly (D,L-lactide) (PDLLA) oligomers with 10 LA units. After that the nanoprecipitation method was carried out to obtain polymeric NPs. The presence of terpenoid-head was confirmed with the distinguished double-bond of the terpenoid by 1H-NMR and IR spectroscopies. The terpenoid-head directly affects glass transition temperature (Tg) values. Differential scanning calorimetry (DSC) and dynamic mechanical analysis (DMA) techniques demonstrated that Lim-LA10 had the highest Tg, while the lowest was for ger-LA10. Water contact angle values of terpenoid-LA10 evaluated the hydrophilic/hydrophobic properties, in which, $\theta > 90°$ is considered hydrophobic, while $\theta < 90°$ is hydrophilic. The reported θ was in the range of 55° to 72°, which confirmed the amphiphilic structure of the oligomers. The $\theta = 55°$ was attributed to t-sob-LA10, declaring it the most hydrophilic terpenoid due to its hydroxyl group. The highest amphiphilic balance of the LA with the t-sob also causes the t-sob-LA10 to have the smallest NP size among other terpenoid initiators, which was affirmed by DLS results demonstrating a range of 175–230 nm size for the self-assembled NPs. Moreover, it is reported that the negative zeta potential value was due to the NPs' surface hydroxyl groups. Furthermore, t-sob-LA20 (with the 20:1 ratio of [M]:[I]) has been synthesized and demonstrated more well-defined 103 nm NPs (70 nm smaller than t-sob-LA10 NPs), which is due to the control of the hydrophilic/hydrophobic balance of the polymer. Moreover, the t-sob-LA20 was more stable. The terpenoid-initiated PDLLA oligomers' cytotoxicity on Caco-2 intestinal, A549 lung, and A431 skin cells were tested by PrestoBlue metabolic activity and LDH release to indicate membrane damage. The results demonstrate no cytotoxicity with NPs, showing great potential for further biomedical applications. The authors suggest that the negative zeta potential of NPs results in their electrostatic repulsion with cells and higher blood circulation, so the uptake of NPs by the liver reduces; therefore, they can be exploited as nanocarriers for drug delivery applications.

The area of terpene/terpenoid-based polymers can have a broader range of potentialities within both synthesis approach and application, mainly when applied in nanosized dimensions.

3.8 CONCLUSION AND FUTURE PERSPECTIVES

Reflecting the immense importance of green chemistry in science, much attention has been drawn to the green synthesis of NPs. Using plant extracts as precursors for NP formation has been studied extensively, and their utilization in various applications has been examined. This process can offer an environmentally friendly, safe, and cost-effective alternative for NP synthesis. Terpenoids constitute the largest class of secondary metabolites in plants and have been successfully proven to be valuable medicinal chemistry candidates. Terpenoids themselves are a rich reservoir for therapeutic properties; therefore, their essential aspects such as pharmacokinetics should be considered for terpenoid-loaded NPs as the drug delivery systems. Apart from that, respecting the popularity of terpenoids in biomedical sciences, recent researchers have been delved into the development of terpenoid-base NPs. Despite the increasing interests in this field, there are still some issues about the exact procedure of these phytochemicals' reduction and capping process. Thus, the future of terpenoid-based NPs remains open for more investigation with the specific need to focus on their mechanism. For this purpose, instead of the whole extract, more examples of pure terpenoids can be used to eliminate the interaction of phytochemicals with each other and give us more precise insight into the mechanism. In addition to this, new and advanced characterization techniques can help our understanding of the mechanism of action. NPs are featured by their size, shape, and dispersity. Having control over these properties is also critical for their potential biomedical application. To obtain this, changing experiment conditions such as temperature, incubation time, pH, mixing speed, etc., can be applied to find the optimized conditions. Taken together, although there are current deficiencies in this field, more advances with deeper research into terpenoid-mediated NPs can discover new avenues and opportunities.

REFERENCES

Abbas, Farhat, Yanguo Ke, Rangcai Yu, Yuechong Yue, Sikandar Amanullah, Muhammad Muzammil Jahangir, and Yanping Fan. 2017. "Volatile Terpenoids: Multiple Functions, Biosynthesis, Modulation and Manipulation by Genetic Engineering." *Planta* 246 (5). Springer Berlin Heidelberg: 803–816. doi:10.1007/s00425-017-2749-x.

Abdulridha, Manal Khalid, Ali H. Al-Marzoqi, Ghaidaa Raheem Lateef Al-awsi, Shaden M.H. Mubarak, Maryam Heidarifard, and Abdolmajid Ghasemian. 2020. "Anticancer Effects of Herbal Medicine Compounds and Novel Formulations: A Literature Review." *Journal of Gastrointestinal Cancer* 51 (3): 765–773. doi:10.1007/s12029-020-00385-0.

Abuid, Nicholas J., Kerim M. Gattás-Asfura, Daniel J. LaShoto, Alexia M. Poulos, and Cherie L. Stabler. 2019. "Biomedical Applications of Cerium Oxide Nanoparticles: A Potent Redox Modulator and Drug Delivery Agent." *Nanoparticles for Biomedical Applications: Fundamental Concepts, Biological Interactions and Clinical Applications*, 283–301. doi:10.1016/B978-0-12-816662-8.00017-5.

Akintelu, Sunday Adewale, and Aderonke Similoluwa Folorunso. 2020. "A Review on Green Synthesis of Zinc Oxide Nanoparticles Using Plant Extracts and Its Biomedical Applications." *BioNanoScience* 10 (4): 848–863. doi:10.1007/s12668-020-00774-6.

Al-Dahmash, Nora D., Mysoon M. Al-Ansari, Fatimah O. Al-Otibi, and A.J.A. Ranjith Singh. 2021. "Frankincense, an Aromatic Medicinal Exudate of Boswellia Carterii Used to Mediate Silver Nanoparticle Synthesis: Evaluation of Bacterial Molecular Inhibition and Its Pathway." *Journal of Drug Delivery Science and Technology* 61 (November 2020). Elsevier B.V.: 102337. doi:10.1016/j.jddst.2021.102337.

Almanza-Reyes, Horacio, Sandra Moreno, Ismael Plascencia-López, Martha Alvarado-Vera, Leslie Patrón-Romero, Belén Borrego, Alberto Reyes-Escamilla, et al. 2021. "Evaluation of Silver Nanoparticles for the Prevention of SARS-CoV-2 Infection in Health Workers: In Vitro and in Vivo." *PLoS ONE* 16 (8 August): 1–14. doi:10.1371/journal.pone.0256401.

Al-Sanea, Mohammad M., Narek Abelyan, Mohamed A. Abdelgawad, Arafa Musa, Mohammed M. Ghoneim, Tarfah Al-Warhi, Nada Aljaeed, et al. 2021. "Strawberry and Ginger Silver Nanoparticles as Potential Inhibitors for Sars-Cov-2 Assisted by in Silico Modeling and Metabolic Profiling." *Antibiotics* 10 (7). doi:10.3390/antibiotics10070824.

Anju, Shruti Sharma, Hari Ram Dhanetia, and Alka Sharma. 2020. "Green Synthesis of Copper Nanoparticles using *Holoptelea integrifolia* Fruit Extract." *Rasayan Journal of Chemistry* 13 (4): 2664–2671. doi:10.31788/RJC.2020.1346306.

Anooj, E S, and P K Praseetha. 2020. "Green Synthesis and Characterization of Graphene Quantum Dots from *Rosa gallica* Petal Extract." *Plant Archives* 20 (2): 6151–6155.

Antonio, Emilli, Osmar dos Reis Antunes Junior, Rossana Gabriela Del Jesús Vásquez Marcano, Camila Diedrich, Juliane da Silva Santos, Christiane Schineider Machado, Najeh Maissar Khalil, and Rubiana Mara Mainardes. 2021. "Chitosan Modified Poly (Lactic Acid) Nanoparticles Increased the Ursolic Acid Oral Bioavailability." *International Journal of Biological Macromolecules* 172. Elsevier B.V.: 133–142. doi:10.1016/j.ijbiomac.2021.01.041.

Anwar, M N. 2008. "Volatile Constituents of Essential Oils Isolated from Leaf and Inflorescences of *Piper longum* Linn." *Chittagong University Journal of Biological Sciences* 3: 77–85.

Bai, Lu, Hai Zhang, Qingchao Liu, Yong Zhao, Xueqin Cui, Sen Guo, Li Zhang, Chi Tang Ho, and Naisheng Bai. 2016. "Chemical Characterization of the Main Bioactive Constituents from Fruits of *Ziziphus jujuba*." *Food and Function* 7 (6): 2870–2877. doi:10.1039/c6fo00613b.

Balachandar, Ramalingam, Paramasivam Gurumoorthy, Natchimuthu Karmegam, Hamed Barabadi, Ramasamy Subbaiya, Krishnan Anand, Pandi Boomi, and Muthupandian Saravanan. 2019. "Plant-mediated Synthesis, Characterization and Bactericidal Potential of Emerging Silver Nanoparticles using Stem Extract of *Phyllanthus pinnatus*: A Recent Advance in Phytonanotechnology." *Journal of Cluster Science* 30 (6). Springer US: 1481–1488. doi:10.1007/s10876-019-01591-y.

Balasubramanian, S., S. Mary Jelastin Kala, and T. Lurthu Pushparaj. 2020. "Biogenic Synthesis of Gold Nanoparticles Using Jasminum Auriculatum Leaf Extract and Their Catalytic, Antimicrobial and Anticancer Activities." *Journal of Drug Delivery Science and Technology* 57 (May 2019). Elsevier: 101620. doi:10.1016/j.jddst.2020.101620.

Barrios-González, Javier. 2018. "Secondary Metabolites Production." *Current Developments in Biotechnology and Bioengineering*: 257–283. doi:10.1016/b978-0-444-63990-5.00013-x.

Batiha, Gaber El Saber, Luay M. Alkazmi, Lamiaa G. Wasef, Amany Magdy Beshbishy, Eman H. Nadwa, and Eman K. Rashwan. 2020. "Syzygium Aromaticum l. (Myrtaceae): Traditional Uses, Bioactive Chemical Constituents, Pharmacological and Toxicological Activities." *Biomolecules* 10 (2): 1–17. doi:10.3390/biom10020202.

Begum, Shamima, and Md Ahmaruzzaman. 2018. "Green Synthesis of SnO2 Quantum Dots Using Parkia Speciosa Hassk Pods Extract for the Evaluation of Anti-Oxidant and Photocatalytic Properties." *Journal of Photochemistry and Photobiology B: Biology* 184. Elsevier B.V: 44–53. doi:10.1016/j.jphotobiol.2018.04.041.

Bello, Idris, Mustapha W. Shehu, Mustapha Musa, Mohd Zaini Asmawi, and Roziahanim Mahmud. 2016. "Kigelia Africana (Lam.) Benth. (Sausage Tree): Phytochemistry and Pharmacological Review of a Quintessential African Traditional Medicinal Plant." *Journal of Ethnopharmacology* 189. Elsevier: 253–276. doi:10.1016/j.jep.2016.05.049.

Belwal, Tarun, Shahira M. Ezzat, Luca Rastrelli, Indra D. Bhatt, Maria Daglia, Alessandra Baldi, Hari Prasad Devkota, et al. 2018. "A Critical Analysis of Extraction Techniques Used for Botanicals: Trends, Priorities, Industrial Uses and Optimization Strategies." *TrAC – Trends in Analytical Chemistry* 100 (2018). Elsevier B.V.: 82–102. doi:10.1016/j.trac.2017.12.018.

Berta, Frantisek, Ján Supuka, and Anna Chladná. 1997. "The Composition of Terpenes in Needles of Pinus Sylvestris in a Relatively Clear and in a City Environment." *Biologia* 52 (1): 71–78.

Boncan, Delbert Almerick T., Stacey S.K. Tsang, Chade Li, Ivy H.T. Lee, Hon Ming Lam, Ting Fung Chan, and Jerome H.L. Hui. 2020. "Terpenes and Terpenoids in Plants: Interactions with Environment and Insects." *International Journal of Molecular Sciences* 21 (19): 1–19. doi:10.3390/ijms21197382.

Burlacu, Ema, Corneliu Tanase, Nastaca Alina Coman, and Lavinia Berta. 2019. "A Review of Bark-Extract-Mediated Green Synthesis of Metallic Nanoparticles and Their Applications." *Molecules* 24 (23): 1–18. doi:10.3390/molecules24234354.

Cai, Lin, Juanni Chen, Zhongwei Liu, Hancheng Wang, Huikuan Yang, and Wei Ding. 2018. "Magnesium Oxide Nanoparticles: Effective Agricultural Antibacterial Agent against Ralstonia Solanacearum." *Frontiers in Microbiology* 9 (APR): 1–19. doi:10.3389/fmicb.2018.00790.

Chen, Chiy Rong, Li Hui Chao, Min Hsiung Pan, Yun Wen Liao, and Chi I. Chang. 2007. "Tocopherols and Triterpenoids from Sida Acuta." *Journal of the Chinese Chemical Society* 54 (1): 41–45. doi:10.1002/jccs.200700008.

Chen, Hongzhang, and Lan Wang. 2017. *Sugar Strategies for Biomass Biochemical Conversion. Technologies for Biochemical Conversion of Biomass*. Cambridge, MA: Metallurgical Industry Press. doi:10.1016/b978-0-12-802417-1.00006-5.

Chen, Nanshan, Min Zhou, Xuan Dong, Jieming Qu, Fengyun Gong, Yang Han, Yang Qiu, et al. 2020. "Epidemiological and Clinical Characteristics of 99 Cases of 2019 Novel Coronavirus Pneumonia in Wuhan , China : A Descriptive Study." *The Lancet* 395 (10223). Elsevier Ltd: 507–513. doi:10.1016/S0140-6736(20)30211-7.

Cheng, Ai Xia, Yong Gen Lou, Ying Bo Mao, Shan Lu, Ling Jian Wang, and Xiao Ya Chen. 2007. "Plant Terpenoids: Biosynthesis and Ecological Functions." *Journal of Integrative Plant Biology* 49 (2): 179–186. doi:10.1111/j.1744-7909.2007.00395.x.

Christianson, David W. 2017. "Structural and Chemical Biology of Terpenoid Cyclases." *Chemical Reviews* 117 (17): 11570–11648. doi:10.1021/acs.chemrev.7b00287.

Chu, Ngo Minh, Nguyen Duy Hieu, Dung Thi Mai Do, Ramanujam Sarathi, Tadachika Nakayama, and Hisayuki Suematsu. 2019. "Synthesis of Molybdenum Carbide Nanoparticles Using Pulsed Wire Discharge in Mixed Atmosphere of Kerosene and Argon." *Journal of the American Ceramic Society* 102 (12): 7108–7115. doi:10.1111/jace.16621.

Chung, Ill Min, Inmyoung Park, Kim Seung-Hyun, Muthu Thiruvengadam, and Govindasamy Rajakumar. 2016. "Plant-Mediated Synthesis of Silver Nanoparticles: Their Characteristic Properties and Therapeutic Applications." *Nanoscale Research Letters* 11 (1): 1–14. doi:10.1186/s11671-016-1257-4.

Das, Sourav, and Chitta Ranjan Patra. 2021. "Green Synthesis of Iron Oxide Nanoparticles Using Plant Extracts and Its Biological Application." In Boris Kharisov and Oxana Kharissova (eds), *Handbook of Greener Synthesis of Nanomaterials and Compounds*. Amsterdam, Netherlands: Elsevier Inc. doi:10.1016/C2019-0-04948-5.

Della Monica, Francesco, and Arjan W. Kleij. 2020. "From Terpenes to Sustainable and Functional Polymers." *Polymer Chemistry* 11 (32): 5109–5127. doi:10.1039/d0py00817f.

Dhall, Atul, and William Self. 2018. "Cerium Oxide Nanoparticles: A Brief Review of Their Synthesis Methods and Biomedical Applications." *Antioxidants* 7 (8): 1–13. doi:10.3390/antiox7080097.

Droesbeke, Martijn A., Alexandre Simula, José M. Asua, and Filip E. Du Prez. 2020. "Biosourced Terpenoids for the Development of Sustainable Acrylic Pressure-Sensitive Adhesives: Via Emulsion Polymerisation." *Green Chemistry* 22 (14): 4561–4569. doi:10.1039/d0gc01350a.

Eisenreich, Wolfgang, Matthias Schwarz, Alain Cartayrade, Duilio Arigoni, Meinhart H. Zenk, and Adelbert Bacher. 1998. "The Deoxyxylulose Phosphate Pathway of Terpenoid Biosynthesis in Plants and Microorganisms." *Chemistry and Biology* 5 (9). doi:10.1016/S1074-5521(98)90002-3.

Elhawary, Seham, Hala El-Hefnawy, Fatma Alzahraa Mokhtar, Mansour Sobeh, Eman Mostafa, Samir Osman, and Mohamed El-Raey. 2020. "Green Synthesis of Silver Nanoparticles Using Extract of Jasminum Officinal l. Leaves and Evaluation of Cytotoxic Activity towards Bladder (5637) and Breast Cancer (Mcf-7) Cell Lines." *International Journal of Nanomedicine* 15: 9771–9781. doi:10.2147/IJN.S269880.

Engelberg, Shira, Yuexi Lin, Yehuda G. Assaraf, and Yoav D. Livney. 2021. "Targeted Nanoparticles Harboring Jasmine-Oil-Entrapped Paclitaxel for Elimination of Lung Cancer Cells." *Int J Mol Sci* 22 (3): 1019.

Faramarzi, Sara, Younes Anzabi, and Hoda Jafarizadeh-Malmiri. 2020. "Nanobiotechnology Approach in Intracellular Selenium Nanoparticle Synthesis Using Saccharomyces Cerevisiae—Fabrication and Characterization." *Archives of Microbiology* 202 (5). Springer Berlin Heidelberg: 1203–1209. doi:10.1007/s00203-020-01831-0.

Frasco, Manuela F., and Nikos Chaniotakis. 2010. "Bioconjugated Quantum Dots as Fluorescent Probes for Bioanalytical Applications." *Analytical and Bioanalytical Chemistry* 396 (1): 229–240. doi:10.1007/s00216-009-3033-0.

Gangadharan, C., M. Arthanareeswari, R. Pandiyan, K. Ilango, and R. MohanKumar. 2019. "Enhancing the Bioactivity of Lupeol, Isolated from Aloe Vera Leaf via Targeted Semi - Synthetic Modifications of the Olefinic Bond." *Materials Today: Proceedings* 14. Elsevier Ltd: 296–301. doi:10.1016/j.matpr.2019.04.150.

Gao, Qing Han, Chun Sen Wu, and Min Wang. 2013. "The Jujube (Ziziphus Jujuba Mill.) Fruit: A Review of Current Knowledge of Fruit Composition and Health Benefits." *Journal of Agricultural and Food Chemistry* 61 (14): 3351–3363. doi:10.1021/jf4007032.

Geng, Runqing, Yuanyuan Ren, Rong Rao, Xi Tan, Hong Zhou, Xiangliang Yang, Wei Liu, and Qunwei Lu. 2020. "Titanium Dioxide Nanoparticles Induced Hela Cell Necrosis under Uva Radiation through the Ros-Mptp Pathway." *Nanomaterials* 10 (10): 1–15. doi:10.3390/nano10102029.

Githua, Mercy, Ahmed Hassanali, Joseph Keriko, Grace Murilla, Mary Ndungu, and Gathu Nyagah. 2010. "Newantitrypanosomal Tetranotriterpenoids from Azadirachta Indica." *African Journal of Traditional, Complementary and Alternative Medicines* 7 (3): 207–213. doi:10.4314/ajtcam.v7i3.54776.

Gong, Xianling, Yi Zheng, Guangzhi He, Kebing Chen, and Xiaowei Zeng. 2019. "Multifunctional Nanoplatform Based on Star- Shaped Copolymer for Liver Cancer Targeting Therapy." *Drug Delivery* 26 (1): 595–603. doi:10.1080/10717544.2019.1625467.

González-Cofrade, Laura, Beatriz De Las Heras, Luis Apaza Ticona, and Olga M. Palomino. 2019. "Molecular Targets Involved in the Neuroprotection Mediated by Terpenoids." *Planta Medica* 85 (17): 1304–1315. doi:10.1055/a-0953-6738.

Govindaraju, S., and P. Indra Arulselvi. 2018. "Characterization of Coleus Aromaticus Essential Oil and Its Major Constituent Carvacrol for in Vitro Antidiabetic and Antiproliferative Activities." *Journal of Herbs, Spices & Medicinal Plants* 24 (1). Taylor & Francis: 37–51. doi:10.1080/10496475.2017.1369483.

Guo, Kai, Yan Liu, and Sheng-Hong Li. 2021. "The Untapped Potential of Plant Sesterterpenoids: Chemistry, Biological Activities and Biosynthesis." *Natural Product Reports*. Royal Society of Chemistry. doi:10.1039/d1np00021g.

Gupta, Vaibhav, Patrick T. Probst, Fabian R. Goßler, Anja Maria Steiner, Jonas Schubert, Yannic Brasse, Tobias A.F. König, and Andreas Fery. 2019. "Mechanotunable Surface Lattice Resonances in the Visible

Optical Range by Soft Lithography Templates and Directed Self-Assembly." *ACS Applied Materials and Interfaces* 11 (31): 28189–28196. doi:10.1021/acsami.9b08871.

Hafez, Dina A., Kadria A. Elkhodairy, Mohamed Teleb, and Ahmed O. Elzoghby. 2020. "Nanomedicine-Based Approaches for Improved Delivery of Phyto-Therapeutics for Cancer Therapy." *Expert Opinion on Drug Delivery* 17 (3). Taylor & Francis: 279–285. doi:10.1080/17425247.2020.1723542.

Hamman, Josias H. 2008. "Composition and Applications of Aloe Vera Leaf Gel." *Molecules* 13 (8): 1599–1616. doi:10.3390/molecules13081599.

Hoque, M. Nazmul, Salma Akter, Israt Dilruba, M. Rafiul Islam, M. Shaminur Rahman, Masuda Akhter, Israt Islam, et al. 2021. "Microbial Pathogenesis Microbial Co-Infections in COVID-19 : Associated Microbiota and Underlying Mechanisms of Pathogenesis." *Microbial Pathogenesis* 156 (April). Elsevier Ltd: 104941. doi:10.1016/j.micpath.2021.104941.

Hou, Ke, Jing Zhao, Hui Wang, Bin Li, Kexin Li, Xinghua Shi, Kaiwei Wan, et al. 2020. "Chiral Gold Nanoparticles Enantioselectively Rescue Memory Deficits in a Mouse Model of Alzheimer's Disease." *Nature Communications* 11 (1). Springer US: 1–11. doi:10.1038/s41467-020-18525-2.

Iravani, Siavash. 2011. "Green Synthesis of Metal Nanoparticles Using Plants." *Green Chemistry* 13 (10): 2638–2650. doi:10.1039/c1gc15386b.

Izzah Ahmad, Nurul, Salina Abdul Rahman, Yin-Hui Leong, and Nur Hayati Azizul. 2019. "A Review on the Phytochemicals of Parkia Speciosa, Stinky Beans as Potential Phytomedicine." *Journal of Food Science and Nutrition Research* 2 (3): 151–173. doi:10.26502/jfsnr.2642-11000017.

Jadoun, Sapana, Rizwan Arif, Nirmala Kumari Jangid, and Rajesh Kumar Meena. 2021. "Green Synthesis of Nanoparticles Using Plant Extracts: A Review." *Environmental Chemistry Letters* 19 (1). Springer International Publishing: 355–374. doi:10.1007/s10311-020-01074-x.

Jafari, S., B. Mahyad, H. Hashemzadeh, S. Janfaza, T. Gholikhani, and L. Tayebi. 2020. "Biomedical Applications of TiO2 Nanostructures: Recent Advances." *International Journal of Nanomedicine* 15: 3447–3470.

Jafarizad, Abbas, Khadijeh Safaee, and Duygu Ekinci. 2017. "Green Synthesis of Gold Nanoparticles Using Aqueous Extracts of Ziziphus Jujuba and Gum Arabic." *Journal of Cluster Science* 28 (5). Springer US: 2765–2777. doi:10.1007/s10876-017-1258-1.

Jahangeer, Muhammad, Rameen Fatima, Mehvish Ashiq, Aneela Basharat, Sarmad Ahmad Qamar, Muhammad Bilal, and Hafiz M.N. Iqbal. 2021. "Therapeutic and Biomedical Potentialities of Terpenoids-A Review." *Journal of Pure and Applied Microbiology* 15 (2): 471–483. doi:10.22207/JPAM.15.2.04.

Jeevanandam, Jaison, Yen San Chan, Yee Jing Wong, and Yiik Siang Hii. 2020. "Biogenic Synthesis of Magnesium Oxide Nanoparticles Using Aloe Barbadensis Leaf Latex Extract." *IOP Conference Series: Materials Science and Engineering* 943 (1): 1–14. doi:10.1088/1757-899X/943/1/012030.

Kaba, Said I., and Elena M. Egorova. 2015. "In Vitro Studies of the Toxic Effects of Silver Nanoparticles on HeLa and U937 Cells." *Nanotechnology, Science and Applications* 8: 19–29. doi:10.2147/NSA.S78134.

Karunakaran, Gopalu, Matheswaran Jagathambal, Evgeny Kolesnikov, Arkhipov Dmitry, Artur Ishteev, Alexander Gusev, and Denis Kuznetsov. 2017. "Floral Biosynthesis of Mn3O4 and Fe2O3 Nanoparticles Using Chaenomeles Sp. Flower Extracts for Efficient Medicinal Applications." *Jom* 69 (8): 1325–1333. doi:10.1007/s11837-017-2349-z.

Kavitha, S., M. Dhamodaran, Rajendra Prasad, and M. Ganesan. 2017. "Synthesis and Characterisation of Zinc Oxide Nanoparticles Using Terpenoid Fractions of Andrographis Paniculata Leaves." *International Nano Letters* 7 (2). Springer Berlin Heidelberg: 141–147. doi:10.1007/s40089-017-0207-1.

Kaweeteerawat, Chitrada, Preeyawis Na Ubol, Sanirat Sangmuang, Sasitorn Aueviriyavit, and Rawiwan Maniratanachote. 2017. "Mechanisms of Antibiotic Resistance in Bacteria Mediated by Silver Nanoparticles." *Journal of Toxicology and Environmental Health – Part A: Current Issues* 80 (23–24). Taylor & Francis: 1276–1289. doi:10.1080/15287394.2017.1376727.

Khalid, Shaukat, Ghazala H. Rizwan, Hina Yasin, Rehana Perveen, Hina Abrar, Huma Shareef, Kaneez Fatima, and Maryam Ahmed. 2013. "Medicinal Importance of Holoptelea Integrifolia (Roxb). Planch – Its Biological and Pharmacological Activities." *Natural Products Chemistry & Research* 2 (1): 2–5.

Khan, Haroon, Hammad Ullah, Miquel Martorell, Susana Esteban Valdes, Tarun Belwal, Silvia Tejada, Antoni Sureda, and Mohammad Amjad Kamal. 2021. "Flavonoids Nanoparticles in Cancer: Treatment, Prevention and Clinical Prospects." *Seminars in Cancer Biology* 69 (July). Elsevier: 200–211. doi:10.1016/j.semcancer.2019.07.023.

Khan, Tariq, Nazif Ullah, Mubarak Ali Khan, Zia ur Rehman Mashwani, and Akhtar Nadhman. 2019. "Plant-Based Gold Nanoparticles; a Comprehensive Review of the Decade-Long Research on Synthesis, Mechanistic Aspects and Diverse Applications." *Advances in Colloid and Interface Science* 272. Elsevier B.V.: 102017. doi:10.1016/j.cis.2019.102017.

Khatoon, Nafeesa, Hammad Alam, Afreen Khan, Khalid Raza, and Meryam Sardar. 2019. "Ampicillin Silver Nanoformulations against Multidrug Resistant Bacteria." *Scientific Reports* 9 (1). Springer US: 1–10. doi:10.1038/s41598-019-43309-0.

Kumar, Harsh, Kanchan Bhardwaj, Kamil Kuča, Anu Kalia, Eugenie Nepovimova, Rachna Verma, and Dinesh Kumar. 2020. "Flower-Based Green Synthesis of Metallic Nanoparticles: Applications beyond Fragrance." *Nanomaterials* 10 (4). doi:10.3390/nano10040766.

Kumar, Vineet, and Sudesh Kumar Yadav. 2009. "Plant-Mediated Synthesis of Silver and Gold Nanoparticles and Their Applications." *Journal of Chemical Technology and Biotechnology* 84 (2): 151–157. doi:10.1002/jctb.2023.

Kuzuyama, Tomohisa. 2002. "Mevalonate and Nonmevalonate Pathways for the Biosynthesis of Isoprene Units." *Bioscience, Biotechnology and Biochemistry* 66 (8): 1619–1627. doi:10.1271/bbb.66.1619.

Lasoń, Elwira. 2020. "Topical Administration of Terpenes Encapsulated in Nanostructured Lipid-Based Systems." *Molecules* 25 (23). doi:10.3390/molecules25235758.

Li, Jie, Yi Li, Haisuo Wu, Saraschandra Naraginti, and Yunbo Wu. 2021. "Facile Synthesis of ZnO Nanoparticles by Actinidia Deliciosa Fruit Peel Extract: Bactericidal, Anticancer and Detoxification Properties." *Environmental Research* 200 (May). doi:10.1016/j.envres.2021.111433.

Li, Xiaoming, Muchen Rui, Jizhong Song, Zihan Shen, and Haibo Zeng. 2015. "Carbon and Graphene Quantum Dots for Optoelectronic and Energy Devices: A Review." *Advanced Functional Materials* 25 (31): 4929–4947. doi:10.1002/adfm.201501250.

Lichtenthaler, Hartmut K. 1999. "The 1-Deoxy-D-Xylulose-5-Phosphate Pathway of Isoprenoid Biosynthesis in Plants." *Annual Review of Plant Biology* 50: 47–65. doi:10.1146/annurev.arplant.50.1.47.

Lichtenthaler, Hartmut K., Michel Rohmer, and Jörg Schwender. 1997. "Two Independent Biochemical Pathways for Isopentenyl Diphosphate and Isoprenoid Biosynthesis in Higher Plants." *Physiologia Plantarum* 101 (3): 643–652. doi:10.1034/j.1399-3054.1997.1010327.x.

Lin, Neil, Daksh Verma, Nikhil Saini, Ramis Arbi, Muhammad Munir, Marko Jovic, and Ayse Turak. 2021. "Antiviral Nanoparticles for Sanitizing Surfaces: A Roadmap to Self-Sterilizing against COVID-19." *Nano Today* 40. Elsevier: 101267. doi:10.1016/j.nantod.2021.101267.

Maciel, Matheus Vinicius de Oliveira Brisola, Aline da Rosa Almeida, Michelle Heck Machado, Ana Paula Zapelini de Melo, Cleonice Gonçalves da Rosa, Daniele Ziglia de Freitas, Carolina Montanheiro Noronha, Gerson Lopes Teixeira, Rafael Dutra de Armas, and Pedro Luiz Manique Barreto. 2019. "*Syzygium Aromaticum* L. (Clove) Essential Oil as a Reducing Agent for the Green Synthesis of Silver Nanoparticles." *Open Journal of Applied Sciences* 9 (2): 45–54. doi:10.4236/ojapps.2019.92005.

Maladeniya, Charini P., Menisha S. Karunarathna, Moira K. Lauer, Claudia V. Lopez, Timmy Thiounn, and Rhett C. Smith. 2020. "A Role for Terpenoid Cyclization in the Atom Economical Polymerization of Terpenoids with Sulfur to Yield Durable Composites." *Materials Advances* 1 (6). Royal Society of Chemistry: 1665–1674. doi:10.1039/d0ma00474j.

Manjili, Hamidreza Kheiri, Hojjat Malvandi, Mir Sajjad Mousavi, Elahe Attari, and Hossein Danafar. 2018. "In Vitro and in Vivo Delivery of Artemisinin Loaded PCL–PEG–PCL Micelles and Its Pharmacokinetic Study." *Artificial Cells, Nanomedicine and Biotechnology* 46 (5): 926–936. doi:10.1080/21691401.2017.1347880.

Mannino, G., A. Occhipinti, and M.E. Maffei. 2016. "Quantitative Determination of 3-O-acetyl-11-keto-β-boswellic Acid (AKBA) and Other Boswellic Acids in Boswellia sacra Flueck (syn. B. carteri Birdw) and Boswellia serrata Roxb." *Molecules* 21 (10): 1329.

Markus Lange, B., and Amirhossein Ahkami. 2013. "Metabolic Engineering of Plant Monoterpenes, Sesquiterpenes and Diterpenes-Current Status and Future Opportunities." *Plant Biotechnology Journal* 11 (2): 169–196. doi:10.1111/pbi.12022.

Meader, Victoria Kathryn, Mallory G. John, Collin J. Rodrigues, and Katharine Moore Tibbetts. 2017. "Roles of Free Electrons and H2O2 in the Optical Breakdown-Induced Photochemical Reduction of Aqueous [AuCl4]-." *Journal of Physical Chemistry A* 121 (36): 6742–6754. doi:10.1021/acs.jpca.7b05370.

Menazea, A. A. 2020. "Femtosecond Laser Ablation-Assisted Synthesis of Silver Nanoparticles in Organic and Inorganic Liquids Medium and Their Antibacterial Efficiency." *Radiation Physics and Chemistry* 168. Elsevier Ltd: 108616. doi:10.1016/j.radphyschem.2019.108616.

Mileva, Milka, Yana Ilieva, Gabriele Jovtchev, Svetla Gateva, Maya Margaritova Zaharieva, Almira Georgieva, Lyudmila Dimitrova, et al. 2021. "Rose Flowers—A Delicate Perfume or a Natural Healer?" *Biomolecules* 11 (1): 1–32. doi:10.3390/biom11010127.

Mittal, Monika, Nomita Gupta, Palak Parashar, Varsha Mehra, and Manisha Khatri. 2014. "Phytochemical Evaluation and Pharmacological Activity of Syzygium Aromaticum: A Comprehensive Review." *International Journal of Pharmacy and Pharmaceutical Sciences* 6 (8): 67–72.

Molnár, Zsófia, Viktória Bódai, George Szakacs, Balázs Erdélyi, Zsolt Fogarassy, György Sáfrán, Tamás Varga, et al. 2018. "Green Synthesis of Gold Nanoparticles by Thermophilic Filamentous Fungi." *Scientific Reports* 8 (1): 1–12. doi:10.1038/s41598-018-22112-3.

Moodley, Jerushka S., Suresh Babu Naidu Krishna, Karen Pillay, Sershen, and Patrick Govender. 2018. "Green Synthesis of Silver Nanoparticles from Moringa Oleifera Leaf Extracts and Its Antimicrobial Potential." *Advances in Natural Sciences: Nanoscience and Nanotechnology* 9 (1). IOP Publishing. doi:10.1088/2043-6254/aaabb2.

Mumm, Roland, Maarten A. Posthumus, and Marcel Dicke. 2008. "Significance of Terpenoids in Induced Indirect Plant Defence against Herbivorous Arthropods." *Plant, Cell and Environment* 31 (4): 575–585. doi:10.1111/j.1365-3040.2008.01783.x.

Narayanan, Mathiyazhagan, Paramasivam Vigneshwari, Devarajan Natarajan, Sabariswaran Kandasamy, Mishal Alsehli, Ashraf Elfasakhany, and Arivalagan Pugazhendhi. 2021. "Synthesis and Characterization of TiO2 NPs by Aqueous Leaf Extract of Coleus Aromaticus and Assess Their Antibacterial, Larvicidal, and Anticancer Potential." *Environmental Research* 200 (May). Elsevier Inc.: 111335. doi:10.1016/j.envres.2021.111335.

Naseer, Minha, Usman Aslam, Bushra Khalid, and Bin Chen. 2020. "Green Route to Synthesize Zinc Oxide Nanoparticles Using Leaf Extracts of Cassia Fistula and Melia Azadarach and Their Antibacterial Potential." *Scientific Reports*. Springer US: 1–10. doi:10.1038/s41598-020-65949-3.

Nasseri, Mohammad Ali, Hamideh Keshtkar, Milad Kazemnejadi, and Ali Allahresani. 2020. "Phytochemical Properties and Antioxidant Activity of Echinops Persicus Plant Extract: Green Synthesis of Carbon Quantum Dots from the Plant Extract." *SN Applied Sciences* 2 (4). Springer International Publishing: 1–12. doi:10.1007/s42452-020-2466-0.

Nature Nanotechnology. 2010. "Editorial: A Brief History of Some Landmark Papers." 5 (4): 237. doi:10.1038/nnano.2010.80.

Nouri, Akram, and Sepideh Khoee. 2020. "Preparation of Amylose-Poly(Methyl Methacrylate) Inclusion Complex as a Smart Nanocarrier with Switchable Surface Hydrophilicity." *Carbohydrate Polymers* 246 (May). Elsevier: 116662. doi:10.1016/j.carbpol.2020.116662.

Nunn, Alistair V.W., Geoffrey W. Guy, Stanley W. Botchway, and Jimmy D. Bell. 2020. "From Sunscreens to Medicines: Can a Dissipation Hypothesis Explain the Beneficial Aspects of Many Plant Compounds?" *Phytotherapy Research* 34 (8): 1868–1888. doi:10.1002/ptr.6654.

Nurdin, Denny, Andri Hardiansyah, Elsy Rahimi Chaldun, Anti Khoerul Fikkriyah, H. D. Adhita Dharsono, Dikdik Kurnia, and Mieke Hemiawati Satari. 2020. "Preparation and Characterization of Terpenoid-Encapsulated PLGA Microparticles and Its Antibacterial Activity Against Enterococcus Faecalis." *Key Engineering Materials* 829: 263–269. doi:10.4028/www.scientific.net/KEM.829.263.

O'Brien, Dara M., Rachel L. Atkinson, Robert Cavanagh, Ana A.C. Pacheco, Ryan Larder, Kristoffer Kortsen, Eduards Krumins, et al. 2020. "A "Greener" One-Pot Synthesis of Monoterpene-Functionalised Lactide Oligomers." *European Polymer Journal* 125 (January). doi:10.1016/j.eurpolymj.2020.109516.

Omran, Basma A., Kathryn A. Whitehead, and Kwang Hyun Baek. 2021. "One-Pot Bioinspired Synthesis of Fluorescent Metal Chalcogenide and Carbon Quantum Dots: Applications and Potential Biotoxicity." *Colloids and Surfaces B: Biointerfaces* 200 (January). Elsevier B.V.: 111578. doi:10.1016/j.colsurfb.2021.111578.

Ovais, Muhammad, Ali Talha Khalil, Nazar Ul Islam, Irshad Ahmad, Muhamamd Ayaz, Muthupandian Saravanan, Zabta Khan Shinwari, and Sudip Mukherjee. 2018. "Role of Plant Phytochemicals and Microbial Enzymes in Biosynthesis of Metallic Nanoparticles." *Applied Microbiology and Biotechnology* 102 (16): 6799–6814. doi:10.1007/s00253-018-9146-7.

Patra, Jayanta Kumar, Yongseok Kwon, and Kwang Hyun Baek. 2016. "Green Biosynthesis of Gold Nanoparticles by Onion Peel Extract: Synthesis, Characterization and Biological Activities." *Advanced Powder Technology* 27 (5). The Society of Powder Technology Japan: 2204–2213. doi:10.1016/j.apt.2016.08.005.

Peng, Xuwen, Zelin Cui, Xuefeng Bai, and Hongfei Lv. 2018. "Bio-Synthesis of Palladium Nanocubes and Their Electrocatalytic Properties." *IET Nanobiotechnology* 12 (8): 1031–1036. doi:10.1049/iet-nbt.2018.5159.

Prajapati, Pradeep K., Pramod Yadav, and Aleena Gauri. 2020. "Possible Potential of Tamra Bhasma (Calcined Copper) in COVID-19 Management." *Journal of Research in Ayurvedic Sciences* 4 (3): 113–120. doi:10.5005/jras-10064-0111.

Rabiee, Navid, Mojtaba Bagherzadeh, Mahsa Kiani, and Amir Mohammad Ghadiri. 2020. "Rosmarinus Officinalis Directed Palladium Nanoparticle Synthesis: Investigation of Potential Anti-Bacterial, Anti-Fungal and Mizoroki-Heck Catalytic Activities." *Advanced Powder Technology* 31 (4). The Society of Powder Technology Japan: 1402–1411. doi:10.1016/j.apt.2020.01.024.

Ramanarayanan, Rajita, and Sindhu Swaminathan. 2019. "Synthesis and Characterisation of Green Luminescent Carbon Dots from Guava Leaf Extract." *Materials Today: Proceedings* 33 (xxxx). Elsevier Ltd: 2223–2227. doi:10.1016/j.matpr.2020.03.805.

Rane, Ajay Vasudeo, Krishnan Kanny, V.K. Abitha, and Sabu Thomas. 2018. "Methods for Synthesis of Nanoparticles and Fabrication of Nanocomposites." In *Synthesis of Inorganic Nanomaterials*. Elsevier Ltd. doi:10.1016/b978-0-08-101975-7.00005-1.

Rani, Humaira, Satarudra Prakash Singh, Thakur Prasad Yadav, Mohd Sajid Khan, Mohammad Israil Ansari, and Akhilesh Kumar Singh. 2020. "In-Vitro Catalytic, Antimicrobial and Antioxidant Activities of Bioengineered Copper Quantum Dots Using Mangifera Indica (L.) Leaf Extract." *Materials Chemistry and Physics* 239 (August 2019). Elsevier B.V.: 122052. doi:10.1016/j.matchemphys.2019.122052.

Rashid, Saddaf, Muhammad Azeem, Sabaz Ali, and Mohammad Maroof. 2019. "Characterization and Synergistic Antibacterial Potential of Green Synthesized Silver Nanoparticles Using Aqueous Root Extracts of Important Medicinal Plants of Pakistan." *Colloids and Surfaces B : Biointerfaces* 179 (December 2018): 317–325. doi:10.1016/j.colsurfb.2019.04.016.

Rauwel, Protima. 2017. "Emerging Trends in Nanoparticle Synthesis Using Plant Extracts for Biomedical Applications." *Global Journal of Nanomedicine* 1 (3): 55–57. doi:10.19080/gjn.2017.01.555562.

Ravi, Lokesh, and Krishnan Kannabiran. 2021. "Antifungal Potential of Green Synthesized Silver Nanoparticles (Agnps) from the Stem Bark Extract of Kigelia Pinnata." *Research Journal of Pharmacy and Technology* 14 (4): 1842–1846. doi:10.52711/0974-360X.2021.00326.

Sahu, Pranabesh, and Anil K. Bhowmick. 2019. "Redox Emulsion Polymerization of Terpenes: Mapping the Effect of the System, Structure, and Reactivity." *Industrial and Engineering Chemistry Research* 58 (46): 20946–20960. doi:10.1021/acs.iecr.9b02001.

Salem, Salem S., and Amr Fouda. 2021. "Green Synthesis of Metallic Nanoparticles and Their Prospective Biotechnological Applications: An Overview." *Biological Trace Element Research* 199 (1): 344–370. doi:10.1007/s12011-020-02138-3.

Santhosh, Abhirami, V. Theertha, Priyanka Prakash, and S. Smitha Chandran. 2019. "From Waste to a Value Added Product: Green Synthesis of Silver Nanoparticles from Onion Peels Together with Its Diverse Applications." *Materials Today: Proceedings* 46. Elsevier Ltd.: 4460–4463. doi:10.1016/j.matpr.2020.09.680.

Schwass, D. R., K. M. Lyons, R. Love, G. R. Tompkins, and C. J. Meledandri. 2018. "Antimicrobial Activity of a Colloidal AgNP Suspension Demonstrated In Vitro against Monoculture Biofilms: Toward a Novel Tooth Disinfectant for Treating Dental Caries." *Advances in Dental Research* 29 (1): 117–123. doi:10.1177/0022034517736495.

Selim, Yasser A., Maha A. Azb, Islam Ragab, and Mohamed H. M. Abd El-Azim. 2020. "Green Synthesis of Zinc Oxide Nanoparticles Using Aqueous Extract of Deverra Tortuosa and Their Cytotoxic Activities." *Scientific Reports* 10 (1). Springer US: 1–9. doi:10.1038/s41598-020-60541-1.

Senthilkumar, P., L. Surendran, B. Sudhagar, and D. S. Ranjith Santhosh Kumar. 2019. "Facile Green Synthesis of Gold Nanoparticles from Marine Algae Gelidiella Acerosa and Evaluation of Its Biological Potential." *SN Applied Sciences* 1 (4). Springer International Publishing. doi:10.1007/s42452-019-0284-z.

Serrano-Marín, Joan, Irene Reyes-Resina, Eva Martínez-Pinilla, Gemma Navarro, and Rafael Franco. 2020. "Natural Compounds as Guides for the Discovery of Drugs Targeting G-Protein-Coupled Receptors." *Molecules (Basel, Switzerland)* 25 (21): 1–14. doi:10.3390/molecules25215060.

Shafran, Noa, Inbal Shafran, Haim Ben Zvi, Summer Sofer, Liron Sheena, and Ilan Krause. 2021. "Secondary Bacterial Infection in COVID 19 Patients Is a Stronger Predictor for Death Compared to Influenza Patients." *Scientific Reports*. Nature Publishing Group UK: 1–8. doi:10.1038/s41598-021-92220-0.

Shahshahanipour, M., B. Rezaei, Ali A. Ensafi, and Zahra Etemadifar. 2019. "An Ancient Plant for the Synthesis of a Novel Carbon Dot and Its Applications as an Antibacterial Agent and Probe for Sensing of an Anti-Cancer Drug." *Materials Science and Engineering C* 98 (January): 826–833. doi:10.1016/j.msec.2019.01.041.

Shaik, Mohammed Rafi, Rabbani Syed, Syed Farooq Adil, Mufsir Kuniyil, Mujeeb Khan, Mohammed S. Alqahtani, Jilani P. Shaik, et al. 2021. "Mn3O4 Nanoparticles: Synthesis, Characterization and Their Antimicrobial and Anticancer Activity against A549 and MCF-7 Cell Lines." *Saudi Journal of Biological Sciences* 28 (2). The Author(s): 1196–1202. doi:10.1016/j.sjbs.2020.11.087.

Shanei, Ahmad, and Hadi Akbari-Zadeh. 2019. "Investigating the Sonodynamic-Radiosensitivity Effect of Gold Nanoparticles on HeLa Cervical Cancer Cells." *Journal of Korean Medical Science* 34 (37): 1–15. doi:10.3346/jkms.2019.34.e243.

Shanker Dubey, Vinod, Ritu Bhalla, and Rajesh Luthra. 2003. "テルペン生合成(非メバロン酸経路) An Overview of the Non-Mevalonate Pathway for Terpenoid Biosynthesis in Plants." *Journal of Biosciences* 28 (5): 637–646. https://www.ias.ac.in/article/fulltext/jbsc/028/05/0637-0646.

Shuaib, Urooj, Tousif Hussain, Riaz Ahmad, Muhammad Zakaullah, Farrukh Ehtesham Mubarik, Sidra Tul Muntaha, and Sana Ashraf. 2020. "Plasma-Liquid Synthesis of Silver Nanoparticles and Their Antibacterial and Antifungal Applications." *Materials Research Express* 7 (3). IOP Publishing. doi:10.1088/2053-1591/ab7cb6.

Siddiqi, Khwaja Salahuddin, Aziz ur Rahman, Tajuddin, and Azamal Husen. 2018. "Properties of Zinc Oxide Nanoparticles and Their Activity Against Microbes." *Nanoscale Research Letters* 13. doi:10.1186/s11671-018-2532-3.

Singh, Ashwani Kumar, Mahe Talat, D. P. Singh, and O. N. Srivastava. 2010. "Biosynthesis of Gold and Silver Nanoparticles by Natural Precursor Clove and Their Functionalization with Amine Group." *Journal of Nanoparticle Research* 12 (5): 1667–1675. doi:10.1007/s11051-009-9835-3.

Sobańska, Zuzanna, Joanna Roszak, Kornelia Kowalczyk, and Maciej Stępnik. 2021. "Applications and Biological Activity of Nanoparticles of Manganese and Manganese Oxides in in Vitro and in Vivo Models." *Nanomaterials* 11 (5). doi:10.3390/nano11051084.

Stößer, Tim, Chunliang Li, Junjuda Unruangsri, Prabhjot K. Saini, Rafaël J. Sablong, Michael A.R. Meier, Charlotte K. Williams, and Cor Koning. 2017. "Bio-Derived Polymers for Coating Applications: Comparing Poly(Limonene Carbonate) and Poly(Cyclohexadiene Carbonate)." *Polymer Chemistry* 8 (39). Royal Society of Chemistry: 6099–6105. doi:10.1039/c7py01223c.

Suganthy, Natarajan, Vijayan Sri Ramkumar, Arivalagan Pugazhendhi, Giovanni Benelli, and Govindaraju Archunan. 2018. "Biogenic Synthesis of Gold Nanoparticles from Terminalia Arjuna Bark Extract: Assessment of Safety Aspects and Neuroprotective Potential via Antioxidant, Anticholinesterase, and Antiamyloidogenic Effects." *Environmental Science and Pollution Research* 25 (11): 10418–10433. doi:10.1007/s11356-017-9789-4.

Sultana, Shahnaz, Mohammed Ali, and Showkat Rassol Mir. 2018. "Chemical Constituents from the Aerial Parts of Jasminum Auriculatum Vahl and Seeds of Holarrhena Pubescens Wall. ex G. Don." *Acta Scientific Pharmaceutical Sciences* 2 (3): 16–21.

Tade, Rahul Shankar, Sopan Namdev Nangare, and Pravin Onkar Patil. 2020. "Agro-Industrial Waste-Mediated Green Synthesis of Silver Nanoparticles and Evaluation of Its Antibacterial Activity." *Nano Biomedicine and Engineering* 12 (1): 57–66. doi:10.5101/nbe.v12i1.p57-66.

Taniguchi, N. 1974. "On the Basic Concept of Nanotechnology." *Proceeding of the ICPE*. https://ci.nii.ac.jp/naid/10008480916/.

Thakkar, Kaushik N., Snehit S. Mhatre, and Rasesh Y. Parikh. 2010. "Biological Synthesis of Metallic Nanoparticles." *Nanomedicine: Nanotechnology, Biology, and Medicine* 6 (2). Elsevier Inc.: 257–262. doi:10.1016/j.nano.2009.07.002.

Thakur, B. K., A. Kumar, and D. Kumar. 2019. "Green Synthesis of Titanium Dioxide Nanoparticles Using Azadirachta Indica Leaf Extract and Evaluation of Their Antibacterial Activity." *South African Journal of Botany* 124. South African Association of Botanists: 223–227. doi:10.1016/j.sajb.2019.05.024.

Thiagarajan, Shrividhya, Anandhavelu Sanmugam, and Dhanasekaran Vikraman. 2017. "Facile Methodology of Sol-Gel Synthesis for Metal Oxide Nanostructures." *Recent Applications in Sol-Gel Synthesis*, 1–16. doi:10.5772/intechopen.68708.

Tsekhmistrenko, S. I., V. S. Bityutskyy, O. S. Tsekhmistrenko, L. P. Horalskyi, N. O. Tymoshok, and M. Y. Spivak. 2020. "Bacterial Synthesis of Nanoparticles: A Green Approach." *Biosystems Diversity* 28 (1): 9–17. doi:10.15421/012002.

van de Veerdonk, Frank L., Roger J.M. Brüggemann, Shoko Vos, Gert De Hertogh, Joost Wauters, Monique H.E. Reijers, Mihai G. Netea, Jeroen A. Schouten, and Paul E. Verweij. 2021. "COVID-19-Associated Aspergillus Tracheobronchitis: The Interplay between Viral Tropism, Host Defence, and Fungal Invasion." *The Lancet Respiratory Medicine* 9 (7): 795–802. doi:10.1016/S2213-2600(21)00138-7.

Vazquez-Muñoz, Roberto, Miguel Avalos-Borja, and Ernestina Castro-Longoria. 2014. "Ultrastructural Analysis of Candida Albicans When Exposed to Silver Nanoparticles." *PLoS ONE* 9 (10): 1–10. doi:10.1371/journal.pone.0108876.

Villamizar-Gallardo, Raquel, Johann Faccelo Osma Cruz, and Oscar O. Ortíz. 2016. "Fungicidal Effect of Silver Nanoparticles on Toxigenic Fungi in Cocoa." *Pesquisa Agropecuaria Brasileira* 51 (12): 1929–1936. doi:10.1590/S0100-204X2016001200003.

Vranová, Eva, Diana Coman, and Wilhelm Gruissem. 2013. "Network Analysis of the MVA and MEP Pathways for Isoprenoid Synthesis." *Annual Review of Plant Biology* 64 (February): 665–700. doi:10.1146/annurev-arplant-050312-120116.

Wang, Kaiyuan, Hao Ye, Xuanbo Zhang, Xia Wang, Bin Yang, Cong Luo, Zhiqiang Zhao, et al. 2020. "Biomaterials An Exosome-like Programmable-Bioactivating Paclitaxel Prodrug Nanoplatform for Enhanced Breast Cancer Metastasis Inhibition." 257 (July). doi:10.1016/j.biomaterials.2020.120224.

Wilson III, C.W., and Shaw, P.E. 1978. "Terpene Hydrocarbons from Psidium guajava." *Phytochemistry* 17 (8): 1435–1436.

World Health Organization. 2020, October. "Antimicrobial Resistance." https://www.who.int/news-room/fact-sheets/detail/antimicrobial-resistance.

Xu, Xiaoyou, Robert Ray, Yunlong Gu, Harry J. Ploehn, Latha Gearheart, Kyle Raker, and Walter A. Scrivens. 2004. "Electrophoretic Analysis and Purification of Fluorescent Single-Walled Carbon Nanotube Fragments." *Journal of the American Chemical Society* 126 (40): 12736–12737. doi:10.1021/ja040082h.

Yadav, Renuka, Himanshu Saini, Dinesh Kumar, Shweta Pasi, and Veena Agrawal. 2019. "Bioengineering of Piper Longum L. Extract Mediated Silver Nanoparticles and Their Potential Biomedical Applications." *Materials Science and Engineering C* 104 (June). Elsevier. doi:10.1016/j.msec.2019.109984.

Yadi, Morteza, Ebrahim Mostafavi, Bahram Saleh, Soodabeh Davaran, Immi Aliyeva, Rovshan Khalilov, Mohammad Nikzamir, et al. 2018. "Current Developments in Green Synthesis of Metallic Nanoparticles Using Plant Extracts: A Review." *Artificial Cells, Nanomedicine and Biotechnology* 46 (sup3). Taylor & Francis: S336–S343. doi:10.1080/21691401.2018.1492931.

Yao, Liangliang, Suyou Zhu, Ziyi Hu, Lin Chen, Muhammad Farrukh Nisar, and Chunpeng Wan. 2020. "Anti-Inflammatory Constituents From Chaenomeles Speciosa." *Natural Product Communications* 15 (3). doi:10.1177/1934578X20913691.

Yu, Fan, Jiale Xu, Huiqin Li, Zizhao Wang, Limin Sun, Tao Deng, Peng Tao, and Qi Liang. 2018. "Ga-In Liquid Metal Nanoparticles Prepared by Physical Vapor Deposition." *Progress in Natural Science: Materials International* 28 (1). Elsevier B.V.: 28–33. doi:10.1016/j.pnsc.2017.12.004.

Zahin, Nuzhat, Raihanatul Anwar, Devesh Tewari, Md Tanvir Kabir, Amin Sajid, Bijo Mathew, Md Sahab Uddin, Lotfi Aleya, and Mohamed M. Abdel-Daim. 2020. "Nanoparticles and Its Biomedical Applications in Health and Diseases: Special Focus on Drug Delivery." *Environmental Science and Pollution Research* 27 (16): 19151–19168. doi:10.1007/s11356-019-05211-0.

Zahirah, Nur, Abd Rani, Khairana Husain, and Endang Kumolosasi. 2018. "Moringa Genus: A Review of Phytochemistry and Pharmacology." *Frontiers in Pharmacology* 9 (February): 1–26. doi:10.3389/fphar.2018.00108.

Zhang, Qing Wen, Li Gen Lin, and Wen Cai Ye. 2018. "Techniques for Extraction and Isolation of Natural Products: A Comprehensive Review." *Chinese Medicine (United Kingdom)* 13 (1). BioMed Central: 1–26. doi:10.1186/s13020-018-0177-x.

Zhang, Yan-qiong, Yan Shen, Ming-mei Liao, Xia Mao, Gu-jie Mi, Chen You, Qiu-yan Guo, et al. 2019. "Galactosylated Chitosan Triptolide Nanoparticles for Overcoming Hepatocellular Carcinoma: Enhanced Therapeutic Efficacy , Low Toxicity , and Validated Network Regulatory Mechanisms." *Nanomedicine: Nanotechnology, Biology, and Medicine* 15 (1). Elsevier Inc.: 86–97. doi:10.1016/j.nano.2018.09.002.

Zhao, Zhaoyan, Yuchen Xiao, Lingqing Xu, Ye Liu, Guanmin Jiang, Wei Wang, Bin Li, et al. 2021. "Glycyrrhizic Acid Nanoparticles as Antiviral and Anti-Inflammatory Agents for COVID-19 Treatment." *ACS Applied Materials and Interfaces* 13 (18): 20995–21006. doi:10.1021/acsami.1c02755.

Zheng, Xiao, Fei Wu, Xiao Lin, Lan Shen, and Yi Feng. 2018. "Developments in Drug Delivery of Bioactive Alkaloids Derived from Traditional Chinese Medicine." *Drug Delivery* 25 (1). Informa Healthcare USA, Inc: 398–416. doi:10.1080/10717544.2018.1431980.

Zhou, Fei, Ting Yu, Ronghui Du, Guohui Fan, Ying Liu, Zhibo Liu, Jie Xiang, et al. 2020. "Clinical Course and Risk Factors for Mortality of Adult Inpatients with COVID-19 in Wuhan , China : A Retrospective Cohort Study." *The Lancet* 395 (10229). Elsevier Ltd: 1054–1062. doi:10.1016/S0140-6736(20)30566-3.

Zwenger, S., and C. Basu. 2008. "Plant Terpenoids: Applications and Future Potentials." *Biotechnology and Molecular Biology Reviews* 3 (February): 1–7.

4 Medicinal Plant-Based Lignin and Its Role in Nanoparticles Synthesis and Applications

Limenew Abate Worku, Mesfin Getachew Tadesse, Archana Bachheti, Azamal Husen, and Rakesh Kumar Bachheti

CONTENTS

DOI: 10.1201/9781003213727-4

4.1 INTRODUCTION

In plant biomass, lignocellulose refers to plant dry matter; without compromising global food security, it possesses a larger capacity than fossil material as an alternative means to generate second-generation sourced materials and chemicals as well as biofuels (Zoghlami and Paes, 2019). Lignocellulose is a complex organic molecule formed by the cell wall in many different plants. It possesses three important components: hemicellulose and cellulose as carbohydrate polymers and lignin aromatic polymer (Figure 4.1) (Koch et al., 2020). Next to cellulose, lignin is the most renewable compound; it is the most available biomacromolecule globally (Gordobil et al., 2018). In plants, it is useful for the transportation of nutrition and water by making the cell wall hydrophobic and has a significant role in giving stiffness to the cell wall and mechanical strength to the wood.

Moreover, lignin protects the plants against microbial degradation (Saratale et al., 2019; Bachheti et al., 2021a; Bachheti et al., 2020). Because of its chemical structure complexity and more stable properties, lignin is considered a useful biomolecule to generate fuels in the cosmetics, paints, paper, chemicals, concrete, and lubricants industries (Maija-Liisa et al., 2018). It is widely subjugated to prepare various biomaterials and broadly examined for tissue regeneration, drug delivery, antimicrobial, and antioxidant properties (Figueiredo et al., 2017). Lignin uses covalent bond linkage to hemicellulose and cellulose and forms a lignin-carbohydrate complex (LCC). Each year, the paper and pulp industries produce between 50 and 70 million tonnes of lignin as industrial waste (Zeng et al., 2014).

Sinapyl, coniferyl and p-coumaryl alcohol are the raw material and the main phenylpropanoid building block (monomeric unit) for the preparation of lignin polymers through oxidative coupling reaction (Lourenco and Pereira, 2018). Through oxidative polymerization, each monolignol produces syringyl (S), p-hydroxyphenyl (H), and guaiacyl (G) residues (Novo et al., 2009). Cereal, sugarcane bagasse, pulp, and wood are the main sources of lignin. Raw lignin contents are 15–25% in grasses, 24–33% in softwoods, and 19–28% in hardwoods. The lignin functional groups like methoxyl, carbonyl, hydroxyl, carboxyl, etc., connecting to aliphatic or aromatic moieties, with different proportions and amounts, produce various structures and compositions of lignin (Kai et al., 2016). In coupling reactions, different linkages such as α -O-4, β -β ', 5–5′, 4-O-5, β -1, β -O-4, and

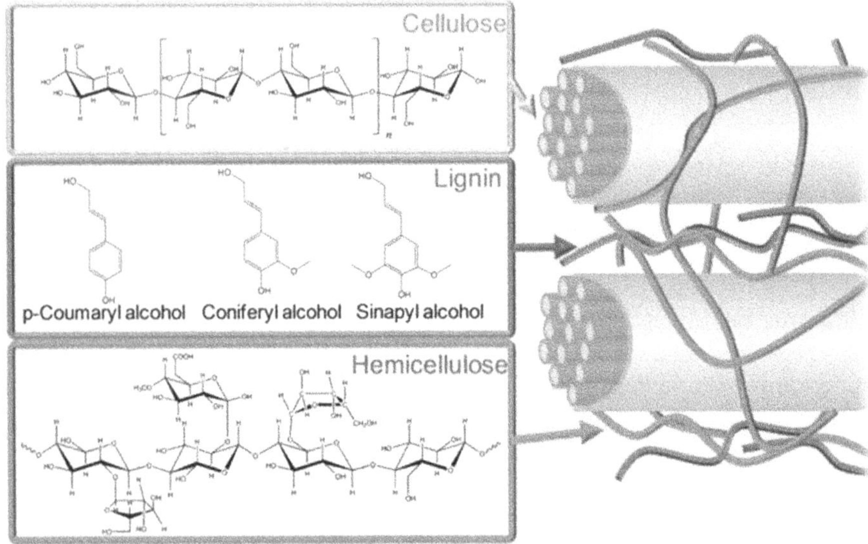

FIGURE 4.1 Lignocellulosic biomass from plant cell wall (adapted from Alonso et al., 2012).

β -5 are involved in the biosynthesis of lignin (Lu et al., 2017). Various types of lignins, for instance, soda, sulphite, organosolv, and kraft lignin, are formed through different degradation processes (Upton and Kasko, 2016).

Due to its many sources or separation processes, lignin has low miscibility, low solubility, and varied structure. Also, it possesses a complex structure, random microstructure, and broad molecular weight distributions (Gao and Fatehi, 2019). These properties make lignin difficult to perform well in different applications (Myint et al., 2016).

Nanotechnology enhances the application and performance of lignin nanomaterial (Khan et al., 2019). The material in nanoscale has unique, new, and superior chemical and physical behaviour than its bulk structure because of the presence of a larger surface area to volume ratio (Khan et al., 2019; Bachheti et al., 2021b). A green nanoparticles (NPs) synthesis method is currently a rapidly developing field; plant-mediated preparation of NPs receives more attention from researchers because it is green in nature, environmentally friendly, non-toxic, and cost effective (Aadila et al., 2019; Roopan, 2017; Bachheti, et al., 2020a, 2020b; Bachheti et al., 2019a; Husen et al., 2019). The stability of NPs in plants is due to the formation of stable bonds (Mohamad et al., 2014; Painuli et al., 2020; Bachheti et al., 2019b). The synthesis of lignin nanoparticles (LNPs) is the best strategy to enhance lignin's chemical and physical characteristics because they have a larger surface area compared to original lignin. The presence of many functional groups like phenolic hydroxyl, aliphatic hydroxyl, and thiols also plays a significant role in reducing agents for the conversion into NPs and permitting modification of its surface (Roopan, 2017; Chauhan, 2018). LNPs enhance lignin's performance and application by alleviating the problems described previously (Gao and Fatehi, 2019). Various applications of LNPs have been reported, for example in antibacterial, emulsion stabilizer, antioxidant, free radical scavenger, polymer industry, sunscreen, anticancer therapy, and controlled release drug delivery contexts (Siddiqui et al., 2020). However, structural heterogeneity and complexity, type of solvent, variability of the biomass sources, and pretreatment procedures are major challenges for LNP application (Tang et al., 2020). Most lignin-based NP preparation methods involve costly or environmentally harmful solvents like dimethylformamide (DMF), tetrahydrofuran (THF), dimethyl sulfide (DMS) (Tang et al., 2020). As a result, low-cost and environmentally friendly solvents like ethanol and water must be utilized to replace DMF, dimethyl sulfoxide (DMSO), and THF in the preparation of LNPs, and create preferable preparation methods. This chapter aims to provide updated information on the physical, mechanical, and chemical techniques involved in preparing different LNPs and their important industrial applications.

4.2 STRUCTURE OF LIGNIN

4.2.1 Chemical Structure

Lignin has an important function in woody plants by providing strength and rigidity to the cell walls. It regulates fluid flow and protects against biochemical stress by preventing enzymes from degrading other components (Boerjan et al., 2003). It typically contributes 15–40% of the dry matter in woody plants (Lin and Dence, 2012). The phenyl propane units in its chemical structure come from three monolignols such as sinapyl, coniferyl, and p-coumaryl alcohols. Syringyl (S) is derived from sinapyl alcohol, guaiacyl (G) obtained from coniferyl alcohol and p-hydroxyphenyl (H) comes from coumaryl alcohol (Figure 4.2). The content and composition of lignin are influenced by sources of the plant and the environment in which it is present; for instance, in softwood lignin, G units are major components with a small number of H units. In hardwood lignin of H units are major components with a small number of G (Gao and Fatehi, 2019). The phenyl propane units vary by the nature and quantity of substitutes in aromatic rings in the lignin molecule. The ratio of these phenyl propene subunits makes it have different types of lignin (Gao and Fatehi, 2019). The

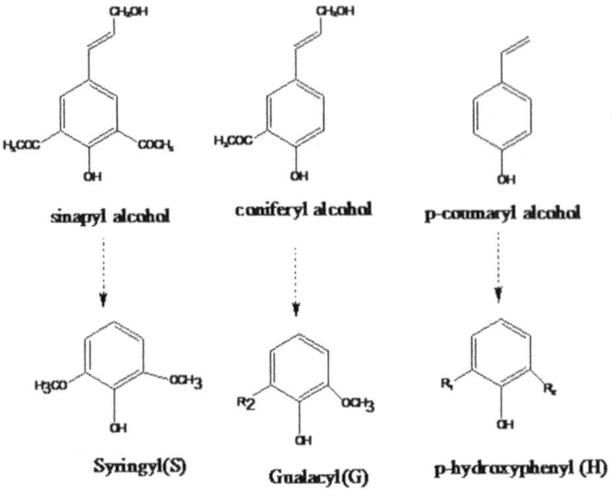

FIGURE 4.2 Types of monolignol precursors for lignin syntheses.

TABLE 4.1
Percentage of Monomers in Lignin

	Abundance (%)		
	p-Coumaryl	Conifery	Sinapyl
Softwood	<5	<95	Small amount
Hardwood	0–8	25–50	46–75
Grasses	5–53	33–80	20–54

content of the monomers in lignin varies based on the type of plant species (Table 4.1) (Yinghuai et al., 2013).

4.2.2 Linkages between Monolignols

Oxidation of the three monolignols can be used to form lignin through enzymatic polymerization (Demuner et al., 2019). Monolignols are randomly joined by carbon–carbon or ether bonds at various places throughout the lignification process, resulting in a variety of interunit connections in multiple lignin substructures (Zhu et al., 2020). β-O-4′ ether linkage, which covers half of the whole interunit linkages, is the well-known linkage in lignin (Figueiredo et al., 2018). During the lignin fragmentation, the linkage β-O-4′ is a well-known reactive type of linkage in depolymerization; the variation in the content of β-O-4′ linkage is the main character in lignin studies (Galkin and Samec, 2016). Other vital covalent bonds like β–β resinol, β-O-4-aryl ether, β-1-(1,2-diarylpropane), 4-O-5-diarylether, 5–5-biphenyl, and β-5-phenylcoumaran are well known. In different plants and woods, the proportions of these linkages vary significantly (Azadi et al., 2013). Via the coupling reactions involved in lignin biosynthesis, various types of linkages, either of carbon–oxygen or carbon–carbon type, are formed in different quantities. Some of them include phenylcoumaran (β-β′), pino/resinol (β-5), aryglycerol-α-ether dimer (α-O-4), diphenylethane (β-1), biphenyl/dibenzodioxocin (5–5′), aryglycerol-β-ether dimer (β-O-4), siaryl ether (4-O-5), and spirodienon (Lu et al., 2017)

TABLE 4.2
The Content of a Functional Group in Every 100 Propane Units

Functional group	Hardwood lignin/100 units	Softwood lignin/100 units
Methoxyl	140–160	90–95
Aliphatic hydroxyl	110–115	115–120
Phenolic hydroxyl	10–20	20–30
Carbonyl	15	20

4.2.3 Functional Groups

The phenolic hydroxyl, methoxy, carboxyl, and carbonyl functional groups are the most common functional groups present in hard and softwood lignin. The phenolic hydroxyl and methoxy groups have the most significant impact on the reaction activity of lignin (Adler and Hernestam, 1955). Strong electronegativity has been seen in the oxygen atoms of methoxy and phenolic hydroxyl groups. With the electron cloud of the benzene ring, the unshared p electron pairs can form a p-conjugated system. Because a benzene ring is offset by the p electron cloud of oxygen atoms, additional electron clouds are formed for the ortho- and para-sites of the benzene ring, resulting in increased chemical reactivity (El Mansouri and Salvadó, 2007). The content of a functional group for every 100 propane units in hard and softwood lignin is listed in Table 4.2 (Alén, 2000).

4.3 TYPE AND EXTRACTION OF LIGNIN

Depending on the sulphur content of lignin, most commercially available lignin can be classified as sulphur-free and sulphur-containing lignin. The odour of sulphur-containing lignin is an unpleasant smell, is less eco-friendly, and has less potential application than sulphur-free lignin (Bajwaa et al., 2019). It is nearly impossible to separate lignin in its natural, unadulterated state. It is important to use drastic procedures to separate lignin from cellulose and cell walls, as these methods modify the original structure of the lignin (Radotic and Micic, 2016).

4.3.1 Sulphur-Containing Lignin

After the extraction process, sulphur in the lignin is used to form sulphur-containing lignin. Hydrolyzed lignin, sulphite lignin, and kraft lignin are commonly known as sulphur-containing lignin (Bajwaa et al., 2019). Most of the time they are obtained from the paper and pulp industries.

4.3.1.1 Kraft Lignin

Kraft lignin (KL) is considered one of the most important sources of technical lignin. It is primarily used on an industrial scale (Li and Takkellapati, 2019). It is estimated that 70–75% of the hydroxyl groups become sulphonated, which highly modifies the lignin structure. Due to highly polar organic solvents, it is soluble in a solution of alkali and basic (Lange et al., 2013). Compared to lignosulphonate, Klignin is purer, contains inorganic impurities, has smaller residues of carbohydrate, and has less sulphur (Sameni et al., 2014). KL, like lignosulphonate lignin, is primarily used to power paper mills (Chakar and Ragauskas, 2004). In terms of structure, it has a more branched carbon–carbon skeleton, and the number of hydroxide (OH) groups present in the phenolic compound is also higher and has β-aryl bond breakage (Demuner et al., 2019). KL contains 0.5–3.0% ash and 3–7% sulphur; thus, it is important in resins, binders, carbon fibres, pesticides, and fertilizers (Bajwaa et al., 2019).

4.3.1.2 Lignosulphonate Lignin

Lignosulphonate is sulphonated lignin; it is removed from the wood of the plant using sulphite pulping (Lange et al., 2013). The sulphite pulping method uses a similar process to kraft pulping; however, it uses a medium of acidic solution for its process (Espinoza-Acosta et al., 2018). They are water-soluble lignin-based products currently used as dispersant polymers for various applications such as drug delivery (Inwood et al., 2018). Various products like adhesive and cement additives, and a component of feed, colloidal suspension, stabilizer, detergents, particleboard, binders, and dispersants can be formed from sulphite lignin (L'udmila et al., 2015).

4.3.2 Sulphur-Free Lignin

Sulphur-free lignins are an emerging class of lignin compounds with a small macromolecular size. These lignins have a structure that is similar to that of native lignins. They have intriguing features that make them a promising source of low-molar-mass phenols and aromatic chemicals. Sulphur-free lignins are split into two categories, soda and organosolv lignins (Mandlekar et al., 2018).

4.3.2.1 Organosolv Lignin

Organosolv lignin is derived as a separate process stream after isolation of components via organic solvent treatment (Lange et al., 2013). The most well-known property of organosolv lignin is its great solubility. It can be soluble in organic solvents and alkaline solutions. However, due to its hydrophobic behaviours, it does not become soluble in water (L'udmila et al., 2015; Espinoza-Acosta et al., 2018). Because of the dissolution effect of solvents, it has a comparatively homogeneous molecular distribution. However, it contains condensed structures (Grossman and Wilfred, 2019), with approximately 1.7% ash content (L'udmila et al., 2015). In biomaterial synthesis, organosolv lignin showed an attractive result because of it is sulphur-free, has a less modified structure, and has high purity (Asawaworarit et al., 2019).

4.3.2.2 Soda Lignin

Soda lignin is obtained through soda anthraquinone or the soda pulping method (Figueiredo et al., 2018). This isolation method has been carried out since the 19th century from non-woody species (Tribot et al., 2019). Anthraquinone is also used as a catalyst to dissolve the lignin and simultaneously weaken carbohydrate degradation (Galkin and Samec, 2016). In soda pulping, the breaking of ether–aryl bonding of lignin causes the loss of primary aliphatic OH and the production of small quantities of hydroxyls in phenolic compounds (Tejado et al., 2007). Soda lignin has 0.7–2.3% of ash content, which is comparatively higher compared to organosolv lignin (Mandlekar et al., 2018). In the soda-based cooking process, annual plants such as flax and straw are vital (Rodríguez et al., 2010). Soda lignin is essential for polymer synthesis, phenolic resin production, dispersion production, and animal nutrition (Bajwaa et al., 2019).

4.4 LIGNIN NANOPARTICLES SYNTHESIS

Because of their appealing features, such as increased surface area, improved aqueous dispersion stability, and specific geometries (solid/hollow particle, multilayer, and fibrous architectures), LNPs have attracted a lot of attention (Gao and Fatehi, 2019). The transformation of plant-derived lignin into LNPs is both sustainable and cost effective when compared to synthetic lignin (Iravani and Varma, 2020). Different solvents (water, ethylene glycol, and castor oil) are important for the formation of LNPs. These solvents affect the size and morphologies of LNPs. For instance, it was shown that about 15–20 nm LNPs with a spherical shape were observed in ethylene glycol and castor oil solvent (Iravani and Varma, 2020a; Rahman et al., 2018). Using THF solvent, spherical LNPs with a mean diameter of 200–500 nm were observed by Lievonen et al. (2016). Hu and

Hsieh (2015) synthesized silver nanoparticle – alkali lignin – cellulose fibrous membranes (AgNPs–AL-Cells) using alkali lignin and cellulose. Previously prepared cellulose (Cell) fibrous membranes were dipped in 15 wt% alkali lignin using 1 wt% aqueous solution of NaOH for 30 min at 80°C to prepare NPs. The size of synthesized NPs increases with time and becomes more polydispersed in size distribution. Transmission electron microscopy (TEM) analysis showed uniformly distributed and spherical AgNPs on fibre surfaces.

There is a change in UV-absorption wavelength during the conversion of raw lignin into LNPs; it increases to 280 nm due to reducing the energy level of the excited electrons by π-stacking and decreasing their surface energy due to the development of spherical forms (Schneider et al., 2021). In LNP preparation, to increase the hydrophobicity, stability, shape, and particle size, lignin should be chemically modified (Tian et al., 2017; Qian et al., 2014). For instance, Qian et al. (2014) acetylated alkali lignin to form LNPs, because lignin contains many phenolic hydroxy groups. This functional group prevents the synthesis of dense packing sphere-shaped and increased hydrogen bonding between water molecules and the lignin and π–π interaction.

The type of lignin, extraction process, and the mechanism of LNP production all influence the formation of these NPs. Spherical, clusters, aggregates, hollow sphere, nonspherical, cuboidal, irregular form, and quasi-spheres are some of the diverse forms of LNPs (Schneider et al., 2021) (Table 4.3).

4.4.1 Methods for Lignin Nanoparticles Synthesis

Nanomaterials are materials in nanoscale; usually, they range in size from 1 to 100 nm. Nanomaterials have considerably different properties because of the presence of a larger surface area compared to larger material of the same composition (Khan et al., 2019). Compared to bulk material, the properties of polymer and lignin at the nanoscale are more attractive. Because of their appealing properties, LNPs have gained significant interest from many researchers. The high surface area properties of nanomaterial enhanced the stability in aqueous dispersion (Gao and Fatehi, 2019). Many different methods are used to synthesize LNPs (Frangville et al., 2012; Duval and Lawoko, 2014). Some of the methods that have been used to prepare LNPs are described below.

4.4.1.1 Ultrasonication Lignin Nanoparticle Preparation Methods

Ultrasonication is one of the most used LNP preparation procedures for reducing particle size to the nanoscale scale. Even though huge size distributions are one of its drawbacks, it is an essential strategy due to its ease and comfort (Tang et al., 2020). Gilca et al. (2015) made LNPs from wheat straw and sardana grass lignin by sonicating the aqueous lignin suspensions for 60 min and then drying them under moderate conditions. The results showed that the resulting NPs had a consistent form, with the size of particles ranging between 10 and 50 nm. Furthermore, the ultrasonication energy used would not result in substantial structural or compositional changes in NPs. Zhou et al. (2020) used ultrasonication to make alkali lignin/polydopamine-based and alkali lignin nanocapsules. By combining acid precipitation and ultrasonication Agustin et al. (2019) also developed a green and quick technique of manufacturing LNPs that is free from sulphur straight from an alkaline pulping fluid. The combined technique produced spherical LNPs with a hierarchical nanostructure and a considerably negative surface charge within 5 min of sonication.

Wang et al. (2018) discovered that varied ultrasonic intensities were used to produce NPs of diverse sizes and shapes. LNPs formed without ultrasonic treatment were irregular in shape and dispersed widely, with only a small number of regular large size NPs. The conclusion was verified by TEM pictures, which revealed that most of the particles were irregular in shape and poorly distributed. More regular and well-separated LNPs were prepared with a low-intensity ultrasound (100 W) treatment. An excellent spherical shape and suitable diameter distribution were attained with a further increase in ultrasonic intensity by increasing ultrasound to 200 W. In general, the duration of ultrasonic intensities reaches a specific point in chemical reactions and affects the shape and

TABLE 4.3
Lignin Shape, Size, and Characterization and Their Significance

Lignin sources	Size (nm)	Shape	Characterization techniques	Roles (significances)	References
Alkali lignin	5–100	Spherical	UV-Vis, EDX, SEM, TGA, TEM, and DSC	Synthesis on electrospun cellulose membranes of fibrous	Hu and Hsieh (2015)
Tall oil fatty acid conjugated kraft lignin	300–400	Spherical	TEM and SEM	Antimicrobial and bioactive, papers, hygiene products, medical textiles, hand sheets, self-adhesive tapes	Setala et al. (2020)
Steam exploded rice straw lignin	15–20	Spherical	TEM, DLS, and SEM	It is used as anticorrosive nanofillers via epoxy coatings. It is environmentally friendly and use a facile approach	Rahman et al. (2018)
Elephant grass (*Pennisetum purpureum*) acid-alkali treatment for lignin isolation	Average 55	Spherical	TEM, FESEM, DLS, STEM, and UV-Vis	Antioxidant activities	Trevisan and Rezende (2020)
Commercial kraft lignin	Average 38	Spherical	FESEM, HRTEM, BET, UV-Vis, TG/DTA, DSC, XRD, XPS, and ATR-FTIR,	Health and drug delivery systems and cosmetics,	Myint et al. (2016)
Soft wood kraft lignin	200	Spherical	AFM, TEM, and DLS	Cosmetics, food, medicine, and different sectors from chemical industry	Mattinena et al. (2018)
Softwood lignin	200–500	Spherical	TEM and zeta potential	Antimicrobial materials, wound healing, Pickering emulsions, Coatings, glue, drug delivery, and composites.	Lievonen et al. (2016)
Alkali lignin	Average 13	Spherical	DLS, SEM, TEM, UV-Vis, and TG	For making medicine and packaging UV-shielding films	Ju et al. (2019)
Pinus densiflora tree	11–14	Spherical morphology	UV-Vis, FT-IR, TEM, XPS, and XRD	Reducing agent and supporting material	Han et al. (2020)
Acacia wood	10–50	Spherical	UV–Vis, XRD and TEM	Packaging material, food, biosensor, textile, drug delivery biomedical, drug delivery, and textiles.	Aadila et al. (2019)
Acacia wood	10–50	Spherical	UV-Vis, TEM, AFM, and XRD	In various industrial and research applications. In breast cancer MCF-7 was used as cytotoxic effect and melanoma A375 cell lines.	Aadila et al. (2016)
Biorefinery lignin produced from the solid residue	>100	Spherical	TEM, SEM, DLS, and XRD	Sustainable and green method; very much stable particles. The NPs have, both lignin and carbohydrates. Because of their bicomponent it can be used for exclusive applications	Sadeghifar et al. (2019)
Wheat straw (Herbaceous biomass)	70–90	Uniform and spherical	SEM and FSEM	In high yield well-dispersed and highly pure lignin nanoparticle were prepared	Lou et al. (2019)

(*Continued*)

TABLE 4.3 (CONTINUED)
Lignin Shape, Size, and Characterization and Their Significance

Lignin sources	Size (nm)	Shape	Characterization techniques	Roles (significances)	References
Kraft and organosolv	45–250	Spherical	TEM, UV-Vis, and Zata potential	Sustainable and green delivery nanovesicles for pesticide, fungicide, antimicrobial antimicrobials, fungicide pedrugs. Sustainable and scalable method.	Richter et al. (2016)
Natural lignocelluloses for source of rice straw	Mean 100	Spherical	SEM, TEM, HRTE, and FTIR	Highly stable lignin nanoparticle and monodisperse functional groups with large potentials for application of biomedical.	Si et al. (2018)
Waste obtained from kraft pulping	200–500	Spherical	TEM and zeta potential	The process usually does not need chemical modification and it is a simple process. via adsorption of oppositely charged polyelectrolyte it can be surface modifiable: potentially for scale-up	Lievonen et al. (2016)
Commercial kraft lignin	Mean 38	Uniform, quasi-spherical	UV-Vis, TG/DTA, DSC, XRD, XPS, and ATR-FTIR	Environmentally friendly and facile method. Uniformly thermal degradation activity. relatively accelerated solubility, dispersion stability, higher UV absorbing	Myint et al. (2016)
Alkali lignin	200–800	Irregular shape	SEM, TEM, UV-Vis, and NMR	Relatively green and Facile method. The NPs without dialysis by centrifugation can be separated and rapidly produced. the applied solvent could be reused, reduced cost amenable and recycled to industrial scale-up production	Wang et al. (2019)
Alkaline lignin, pretreated lignin, enzymatic lignin, hydrolysis lignin, kraft lignin	80–230	Quasi-spherical	SEM, TEM, and UV-Vis	By employing hydrotropic chemistry and the synergistic dissociation of the entrapped p-toluene sulfonate, intrinsic phenolic hydroxyl, and carboxylic acid moieties of the lignin nanoparticle, undesirable factors such as processing pH and lignin species were removed.	Chen et al. (2018)
Lignin from giant reeds	10–50	Irregular shaped	TGA, DTG, FTIR, and FETEM	The nanocomposite used as hydrogel this could be possibly used in wound dressing, food packaging and drug delivery	Ingtipi and Moholkar (2019)
Lignin from coir by sulphite pulping	11–50	Spherical	TEM and UV-Vis,	Fluorescent in nature. Without any physical damage, it can support cell division and minimize photo quenching. It can be prepared with low energy-intensive and less chemical and the synthesis route is eco-friendly.	Pillai et al. (2018)
Organosolv lignin extracted from rice husks	Average 225	Spherical	FTIR and SEM,	In cosmetics and sunscreens, natural UV-blocking additive	Lee et al. (2020)
Dioxane lignin and alkali lignin	80–104	Spherical	TEM, FTIR, UV-Vis, and DSC	Food, pharmaceutical and cosmetic industries	Yearlaa and Padmasree (2015)

(Continued)

TABLE 4.3 (CONTINUED)
Lignin Shape, Size, and Characterization and Their Significance

Lignin sources	Size (nm)	Shape	Characterization techniques	Roles (significances)	References
Colloidal lignin	320–360	Spherical	AFM and TEM,	Gene transfer and drug delivery carrier, food additives, adhesives for tissue engineering and wound sealing	Henn and Mattinen (2019)
Kraft lignin (KL)	40–60	Irregular shape	FESEM, FTIR, TGA, DSC, and UV-Vis	Reduced water sensitivity and improved UV resistance, thermal stability, tensile strength and modulus, important in nanocomposite film production.	Yang et al. (2015)
Enzymatic hydrolysis lignin (EHL)	419–566	Hollow sphere	XPS, NMR, SEM, FTIR, and TEM	It shows greater applicability and chemical activity than kraft lignin or lignosulfonate and has good solubility in THF.	Xiong et al. (2017)
Kraft lignin (KL)	195	Spherical	Zeta potential, UV-Vis, TGA, SEM, DLS, DSC, and TEM,	Antioxidant activity, high UV shielding efficiency, biocompatibility, production of nanocomposite film	Tian et al. (2017)
Lignosulphonate (LS)	278–375	Square-shaped	SEM, FTIR, and UV-Vis	Antioxidants enhanced the scavenging of free radical activities	Ge et al. (2014)
Alkaline lignin (AL) from corn cob	130	Uniform sphere	SEM, TEM, zeta Potential, and UV-Vis	Used as drug carrier	Dai et al. (2017)
Low-sulfonated lignin (IAT)	100-200	Spherical	TEM and DLS	As emulsion and foam stabilizers and, they are used as sorbent for ions of heavy metal and drug delivery vehicles, other environmental pollutants. Method for the preparation of lignin nanoparticle is cheap and simple	Frangville et al. (2012)
Acidolysis lignin	40–60		TGA, DSC, FESEM, and DSC	When the NPs content is greater than 3 wt%, induced higher degradation rate.	Yang et al. (2015)
Low sulphonated light brown fine powder lignin (lignin protobind-1000)	50–250	Shape	DLS, TEM, EDX, and XRD.	Lignin nanoparticle derived from chemically modified lignin applicable for natural filler for polymer nanocomposite at lower temperature	Gupta et al. (2014)
AL separated from wheat pulping black liquor	Average 80	Sphere	DLS, AFM, TEM, FTIR, and XPS	Applications in technologies and various nanoscience for example, microencapsulation of pesticides and drug delivery systems	Qian et al. (2014a)
Kraft lignin, alkali lignin, organosolv lignin	~30–2 μm	Spherical	SEM, TEM, DLS, and zeta potential	As surfactant-free emulsification	Ago et al. (2016)
Softwood kraft black liquor	>100	Irregular in shape	SEM, TGA, DSC, DTG, and 13P NMR	No chemical modification and structure increase thermal stability	Nair et al. (2014)

size of LNPs. The ultrasonic energy may cause the lignin to distribute uniformly. High-intensity ultrasonic treatment appears to improve the surface shape and homogeneity of LNPs while also reducing particle size

4.4.1.2 Crosslinking/Polymerization

The preparation of nano or microcapsules took place in water/oil interface by miniemulsion or micro crosslinking/polymerization, which permits hydrophobic and hydrophilic components encapsulation in lignin which possess amphiphilic properties (Antonietti and Landfester, 2002; Pavel, 2004). Tortora et al. (2014) created oil-filled KL microcapsules utilizing a procedure in which oil in water emulsion was made initially, followed by an ultrasonic-assisted process in which high cross-linking lignin was produced at the water/oil interface. The interaction of hydrophobic lignin within the oil phase causes this reaction, followed by covalent cross-linking of the resulting spherical microsystems using high-intensity ultrasound. Additional optical microscopies revealed that the generated lignin microparticles had a consistently spherical morphology, which added to this justification.

The work of Yiamsawas et al. (2014) presents hollow lignin nanocontainer synthesis in an aqueous core. Using toluene diisocyanate as a crosslinking agent and a selective polyaddition reaction the NP diameters formed were 150–200 nm. Li et al. (2015) made porous lignosulphonate spheres by cross-linking lignosulphonate and sodium alginate with epichlorohydrin, then dropped the spheres into $CaCl_2$ solutions for gelation and solidification. According to the findings, the lignosulphonate spheres produced exhibited an obvious porous structure with a large pore volume and high porosity. Chen et al. (2016) used crosslinking/polymerization procedures to make LNPs. Their research used lignin as a precursor monomer to create an environmentally responsive nanocapsule. Initially, an allyl group was etherified onto a water-soluble lignosulphonate. Then, using surfactant and a costablizer, an ultrasonic system was used to create a stable oil/water miniemulsion system. This miniemulsion system included a crosslinker in the organic phase and a modified amphiphilic lignosulphonate macromonomer at the water/oil interface. A free radical-induced thiolene reaction was used to crosslink two compounds that were in touch at the interface.

4.4.1.3 Self-Assembly Method

Self-assembly is a process in which an ordered or organized structure is formed due to noncovalent interactions, including hydrophobic, electrostatic, hydrogen-bonding, and van der Waals interactions between molecules, in the absence of any external directions. Making NPs via self-assembly is a frequently used method (Tang et al., 2020). To make homogenous lignin-based colloidal spheres, Qian et al. (2014a) employed a self-assembly approach to construct homogeneous lignin-based colloidal spheres. The alkali lignin was converted into acetylated lignin and then dissolved in THF after acetylation. The acetylated lignin molecules began to join to synthesize colloidal spheres through the hydrophobic contact as water was gradually added to the ACL/THF mixtures. The experiment's result confirmed a formation of colloidal spheres with a hydrodynamic radius of 110 nm. Self-assembly of lignin molecules mediated by van der Waals and interactions between the aromatic rings of lignin produces NPs when the solvent environment in the mixture is modified (Gao and Fatehi, 2019).

Lignin cannot self-assemble into homogeneous particles in selective solvents unless it is hydrophobically modified and completely dissolved in a solvent like THF (Qian et al., 2017). Despite being practically water-insoluble, lignin and ACL samples could be dissolved in THF to generate a solution. On the other hand, lignin could only dissolve at a low concentration in THF (2.0 mg/mL), whereas ACL could completely dissolve in THF to prepare a higher dosage. The solubility of THF for the hydrophobic skeletons of ACL decreased when 30% water was applied to the lignin solution; however, the lignin molecules began to combine slowly. The LNP solution became turbid when 70% water was added, apparently due to the emulsion of LNPs (Wang et al., 2019).

4.4.1.4 Acid Precipitation Methods

The pH changing method is another LNP preparation technique; it has similarities with a method of solvent shifting. In the pH shifting method, pH is altered from basic to acidic or acidic to basic. The shifting of pH highly depends on the physicochemical characteristics of the substrate used for the formation of LNPs. Lignin changes its miscibility or solubility when the value of pH is also changed (Beisl et al., 2017). Two standard ways of LNP precipitation were introduced by Frangville et al. (2012). The first standard method is founded on changing the solvent and pH at the same time and the second method is focused on the change of pH from basic to acidic in an aqueous medium. Both techniques are used to synthesize NPs with significantly various stability and properties. In a water medium, low-sulphonated lignin cannot dissolve under an acidic and neutral environment (Bobleter, 1994). On the other hand, at a higher pH value (pH >10) in water and ethylene glycol medium the lignin is significantly soluble.

In this method, two steps have been used for LNP preparation. In the first, ethylene glycol is used as a solvent, followed by HCl(aq) and step-by-step addition of water for precipitation. The second method at pH of 12 uses water with sodium hydroxide as a solvent for lignin. By adding HNO_3, the pH value was lowered, and precipitation occurred (Frangville et al., 2012). Using acid precipitation methods, Gupta et al. (2014) effectively demonstrate the synthesis of LNPs from lignin. The lignin monomer was dissolved in ethylene glycol, and then hydrochloric acid was added. The product formed was recovered by precipitation at a pH of 2 and subjected to NP analysis.

4.4.1.5 Electrospinning Method

The electrospinning method is another technique used for electrostatic fibre formation. It is a well-known choice for forming nonwoven fabrics, fibre arrays, and continuous threads with fibre diameters of less than 1 μm (Rutledge and Fridrikh, 2007). It is a well-known, viable, versatile method for producing ultrathin fibres Xue et al. (2019). In this process, a polymer solution in an appropriate solvent is pumped via a very small nozzle with 100 μm inner diameters. It works as counter electrode and electrode simultaneously and an electric field in the range of 100–500 kV/m is used. In laboratory setups, the distance between the nozzle and counter electrode in which fibres are collected is in the range of 10–25 cm (Greiner and Wendorff, 2007).

Technical lignins possess branching and heterogeneous structures, and during biomass delignification, instead of forming continuous fibres, they are converted to low molecular weight particles. On the other hand, technical lignins can use electrospinning methods to generate fibres by dampening instabilities (Dallmeyer et al., 2010) (Table 4.4).

4.4.2 Factors Affecting the Formation of Lignin Nanoparticles

The process of lignin extraction, the raw material, and, most importantly, the method of NP preparation all affect the size of LNPs. The size of the NPs to be manufactured, and hence the process used will be determined by their intended use. For example, for cancer treatment, for drug delivery applications, the size of LNPs in the range of 70–250 nm is preferable (Gaumet et al., 2008; Figueiredo et al., 2017). However, NPs utilized in hybrid nanocomposites, such as Pickering emulsions (Qian et al., 2014), fibres (Kai et al., 2015), films (Yang et al., 2016), and hydrogels (Chen et al., 2019), do not need to be as small because they're used in larger-scale applications and interact with larger-scale components.

Light exposure, ionic strength, pH, temperature, concentration, and pressure flow rate all affect the storage and stability of LNPs. LNPs are stable at low temperatures (4°C) or at room temperatures when the pH range is 5–11 and the ionic strength is less than 0.01 (Schneider et al., 2021). According to Myint et al. (2016), increasing the concentration and temperature of lignin and decreasing the solution and pressure flow rate increases the degree of particle coalescence/aggregation and the size

TABLE 4.4
Shape, Size, and Method of Extraction of Lignin Nanoparticles from Different Literature Sources (Adapted from Beisl et al., 2017)

Type of nanoparticle	Method of extraction	Size range (nm)
Solid	Solvent shifting	41–2700
	PH shifting	48–375
	Crosslinking polymerization	15–800
	Mechanical methods	20–100
	Ice segregation	~80
	Aerosol process	31–7817
	CO_2 antisolvent	38–74
Hollow	Solvent shifting	100–566
	Crosslinking polymerization	100–1100
	Ice segregation	~130
Fibre	Ice segregation	<100
	Electrospinning	61–3000

distribution. Zhao et al. (2017) also indicated the effect of formation of LNPs via π–π intermolecular hydrogen bonding and π–π interactions. This intermolecular force makes lignin difficult to solubilize in solvents. In the lignin structure, the presence of the aliphatic hydroxyl groups decreases the solubility of lignin because of the formation of strong hydrogen bonds with solvent molecules. Ju et al. (2019), investigated the synthesis of LNPs using various solvents such as DMF, DMSO, CH_3OH, and THF. DMF, DMSO, CH_3OH, and THF produced mean particle sizes of 850, 400, 450, and 850 nm, respectively. From this information, DMSO seems a suitable solvent for lignin. In another synthesized lignin NPs, the result indicated that DMSO is considered a good solvent and lignin is highly soluble in it. However, it is slightly soluble in deionized water at a pH of 6.68 and practically insoluble in dilute HCl(aq) at a pH of 5.3 (Siddiqui et al., 2020). The number of bio nanocomposites of LNPs that were loaded on wheat gluten (WG) films affects the value of UV-Vis light transmittance. In the spectrum of visible light at 550nm, the UV-Vis spectroscopy exhibited that the transmittance of light in LNP loadings in 3 wt% of WG, LNP loadings is 1 wt%. of WG, and neat WG nanocomposites are 56%, 72%, and 89% respectively. This indicates that the presence of LNPs decreases the amount of light transmittance (Yang et al., 2015b). The effect of amount and concentration on the size of LNPs was also shown by Gupta et al. (2014). Their finding indicates that the variation in the amount of HCl(aq) depends on the size of LNPs. The lowest sizes and excellent stability of NPs were observed at 4.8% vol. of added 0.25 M HCl, at a higher percentage of 24% vol. of 0.25 M HCl, much larger and polydisperse NPs were observed. At a higher concentration of HCl, the particle size of lignin increased. The smallest particle size of LNPs was produced at the lowest concentration of HCl. By combining gelatin with LNPs, Yin et al. (2018) constructed the LNPs-gelatin complex. The complex showed better properties of flocculation than LNPs. The competence of flocculation is highly affected by concentration and pH against *Escherichia coli* strains and *Staphylococcus aureus*. The findings of Lievonen et al. (2016) also demonstrated the solvent's effect on LNP formation. Their findings revealed that utilizing ethylene glycol as a solvent resulted in particles with a less uniform round shape than those made using THF. This is due to ethylene glycol being more polar than THF, and this difference in polarity could be one of the reasons for the differences in properties.

4.5 APPLICATION OF LIGNIN NANOPARTICLES

In different industrial processes in the last decade, LNPs received great attention because of their biotechnological potential (Chauhan, 2018). Since LNPs are green and renewable, they are also promising candidates for different biomedical applications, such as for antioxidant, antimicrobial, cancer therapy, drug delivery, diagnosis, etc. (Iravani and Varma, 2020). They also have applications in coatings, biomedical science, biocomposites, agriculture, and many other fields (Mishra and Ekielski, 2019).

4.5.1 Antibacterial Agent

LNPs have antibacterial activities against different bacteria strains (Barapatre et al., 2016). For example, Cavallo et al. (2021) investigated the antibacterial properties of polylactic acid (PLA) and 3 wt% of LNPs PLA nanocomposite films against *Escherichia coli* and *Micrococcu luteus*. The results showed that PLA/3LNPs inhibited microorganism development in both bacterium strains when compared to clean PLA. Bacteria strains such as *Pseudomonas aeruginosa*, *Escherichia coli*, *Klebsiella pneumonia*, *Staphyloccocus haemolyticus*, *Micrococcus flavus*, *Bacillus licheniformis*, *Corynebacterium xerosis*, and *Staphyloccocus aureus* inhibited their growth by using LNPs, which were obtained from linen fabrics using ultrasonic treatment (Zimniewska et al., 2008). Richter et al. (2015) also demonstrated the effect of LNPs on different bacteria strains. LNP inculcated with Ag^+ and coated with a cationic polyelectrolyte layer encourages the adhesion of the particles to bacterial cell membranes and, together with Ag^+, can kill a broad spectrum of bacteria. It has its advantage over metallic NPs because of its higher antimicrobial activity and green. The antimicrobial activities of AgNPs and LNPs were examined by Richter et al. (2015). The result obtained explained the currently used AgNPs and the antimicrobial properties of lignin-core NPs. Ag^+ ions are released in both NPs and cause the bacteria to die. After utilization Ag^+ does not present in the silver-infused lignin particle. Therefore, it remained as biodegradable lignin and kept the environment from hazards. However, in AgNPs a silver ion is present after utilization. It is hazardous to the environment.

4.5.2 Antioxidant Agent

A range of endogenous systems and exposure to various disease states or physiochemical circumstances produce free radicals like reactive nitrogen and oxygen species in human bodies (Lobo et al., 2010). Lignin can slow down oxidation reactions by decreasing oxygen radicals. The phenolic structure on the lignin structure possesses antioxidant effects. The hydroxide group scavenges free radicals, which contain reactive oxygen (Dizhbite et al., 2004). The antioxidant behaviour of nanoscale lignin biomaterial can be affected by smaller particle size, lower molecular weight, less quantity of carboxyl group, and a higher quantity of Ph-OH (Ge et al., 2014). A higher free radical scavenging activity is exhibited when the size of the LNP is smaller (Frangville et al., 2012). For instance, based on the finding of (Yearlaa and Padmasree, 2015), alkali LNPs, exhibit more antioxidant activity than their parent polymers. In the research work of Trevisan and Rezende (2020), LNPs from elephant grass of *Pennisetum purpureum* showed higher antioxidant activities than lignin in solution. According to their result, LNPs have higher relative strength index (RSI) values and lower EC_{50} concentrations than the lignin solution. The antioxidant properties of LNPs are also demonstrated in the study by Zhang et al. (2019), based on their result, as the size of LNPs increases, the IC_{50} value also increases. This indicated that as the size LNPs increase the antioxidant capacity of LNPs decrease.

Several researchers have employed PVA/lignin composite films to investigate lignin's antioxidant properties and UV shielding capabilities (Xiong et al., 2018). Sadeghifar et al. (2015), discovered that the antioxidant capacity of polyethylene/lignin blends is proportional to their thermal-oxidative stability. Tian et al. (2017) also found that LNPs boosted PVA's thermal stability due to its antioxidant properties. PVA degrades by removing side hydroxyl groups and chain scission processes, which produce a large number of free radicals. These radicals can be neutralized by LNPs, and retard thermal degradation (Tian et al., 2017). Zhang et al. (2019) indicated that in the derivative thermogravimetry (DTG) curve, pure PVA film showed peaks at 432, 340, and 258°C. However, thermal stability improvement is observed with the addition of CC95 LNPs. The deterioration peak of pure PVA film increased from 340°C to around 348°C.

4.5.3 UV Absorbents

In plants, the effect of sunshine radiation can be prevented by lignin, thus it is possible to make natural macromolecular sunscreen to protect human skin from UV radiation. It contains UV-absorbing functional groups. Methoxyl and hydroxyl groups of auxochromes enhance the UV-absorbance and expand the spectrum of aromatic rings of lignin (Qian et al., 2017). However, lignin as a UV absorbent is greatly hindered because of its poor compatibility behaviour. Thus, LNP synthesis can improve the application of UV absorbance lignin. LNPs have significant UV protectant activity and improve their dispersion and compatibility properties compared to original lignin (Yearlaa and Padmasree, 2015). UV transmittance decreases as LNPs increase (Zhang et al., 2019a). By merging LNPs with pure skin cream (Qian et al., 2017), it synthesizes lignin-based sunscreens. Yearlaa and Padmasree (2015) also showed the UV protectant potential of dioxane lignin nanoparticles (DLNP) and alkali lignin nanoparticles (ALNP) by controlling the rates of survival of *Escherichia coli* upon UV-induced mortality. In shielding *Escherichia coli* from UV-irradiation-induced mortality, both NPs were more effective than dioxane lignin and alkali lignin. These properties possess the LNPs used as a cosmetic product in the cosmetic industry (Varma and Iravani, 2020). The UV-radiation absorbance properties of lignin nanospheres were synthesized by Xiong et al. (2018) and used to produce transparent nanocomposite films with UV absorbing capacity using PVA as the polymer matrix; the result confirmed that the LNP exhibited good candidates for medicine bottles and application of food packaging. In the work of Trevisan and Rezende (2020), LNPs from elephant grass (*Pennisetum purpureum*) showed high radical scavenging ability and can have higher UV-Vis light blocking capability.

In the work of Yang et al. (2015) excellent UV barrier properties occurred in the presence of LNPs. It decreases the light transmittance of the tested material. This result explains that these nanocomposite materials could be important for agricultural bags and food packaging. Alkali LNPs with dioxane LNPs, obtained using the nanoprecipitation method, shows a good application in food and pharmaceutical industries. Because of UV absorption properties they also exhibited excellent cosmetic products (Varma and Iravani, 2020). Zhang et al. (2019) prepared a film using LNPs as a composite. The result showed that with increasing lignin concentration, the films become darker. This UV-Vis transmittance data (as the amount of LNPs increases, the transmittance of UV light decreases) showed that the composite films could selectively block UV light.

4.5.4 Drug Delivery

Drug delivery is the method or process of delivering a pharmacological substance to achieve a therapeutic effect in humans or animals (Tiwari et al., 2012). Because of ease of surface modification and their stability, a polymeric NP formed from synthetic and natural polymers has higher consideration for drug delivery systems (Singh and Lillard, 2009). LNPs are one of the biopolymers used as a higher potential candidate for drug delivery application. Many research findings have explained the performance of lignin-based NPs in controlling drug release, which is critical in medicine. NPs formed from lignin are inexpensive, stable, biodegradable, and non-toxic. These four primary characteristics demonstrate their utility as effective medication delivery systems in human diseases (Iravani and Varma, 2020). LNPs have hydrophobic properties which allow them to be loaded with hydrophobic drugs and other surface-active materials. To facilitate the loading of hydrophilic drugs, polymers that are sensitive in pH are added to LNPs (Richter et al., 2016). In anticancer treatment, to enlarge the drug's effectiveness, LNPs have great importance. Water-soluble or poorly water-soluble anticancer drugs loaded on LNPs increase the growth of inhibitory and anticancer effects in various cancer cell lines (Figueiredo et al., 2017). In the past, toluene diisocyanate and cross-linked LNPs by poly(ethylene glycol) diglycidyl ether have been utilized to form nanodrug carriers (Gao and Fatehi, 2019). The insoluble anticancer medication resveratrol (RSV) and Fe_3O_4 magnetic NPs were successfully loaded into the lignin-based magnetic NPs drug delivery system. In animal and cytological tests, this magnetic bioactive molecule resveratrol (RSV) loaded LNPs showed enhanced tumour reduction, good anticancer effect, drug accumulation, in vitro RSV stability and release, and lower adverse effects effect than free drugs (Dai et al., 2017). Furthermore, hollow shaped LNPs can be effectively utilized as polymer nanocapsules that have a polymeric shell and a liquid core to encapsulate different guest molecules for the release and delivery of a drug (Gao and Fatehi, 2019).

Chen et al. (2018) formed a LNP as a drug carrier with a load capacity of 37.4 wt% of the tested drug on LNPs. They sustained biocompatibility and drug-releasing potential in vivo and in vitro analysis. They were loading low water-soluble drugs, for instance, in a sustainable manner for sorafenib and benzazulene (BZL) in the pH range of 7.4 and 5.5 and increasing their release profiles (Chauhan, 2018). This indicated those LNPs are promising applicants for the delivery of the drug. Due to the superparamagnetic properties of Fe_3O_4LNP, it is promising for the treatment and diagnosis of cancer (Figueiredo et al., 2017). A lignin-based targeted polymer NP platform, folic acid-polyethylene glycol-alkali lignin conjugate, was constructed in another work for the self-assembling delivery of anticancer drugs (hydroxycamptothecin, HCPT). The outcome revealed a high level of medication loading efficiency (Liu et al., 2018). Iravani and Varma (2020) prepared a lignin-based targeted polymeric NP platform, using self-assembly methods. Folic acid–polyethyleneglycol–alkaline lignin conjugates were formed for delivery of HCPT. The result showed that about 150 nm-sized lignin-based NPs were produced and possess more significant drug loading efficiency, improved cellular uptake, a longer time for blood circulation, and outstanding biocompatibility.

4.5.5 Nanocomposite

LNPs have also been used in nanocomposites and polymer matrices as reinforcing agents. The resulting copolymers have better biocompatibility, thermal, and mechanical properties than raw polymers (Chauhan, 2018). Del Saz-Orozco et al. (2012) explained that 8.5 wt% LNPs made from softwood lignosulphonates increased the compressive modulus and compressive strength of phenolic foams by 28 and 74%, respectively, compared to foams without reinforcement. The tensile strength (σ) of *polymer* polyvinyl alcohol increased from 45.7 to 51.4 when 3 wt% LNP was incorporated into the polymer. Nevárez et al. (2011) used LNPs to study the effects of KL, hydrolytic, and organosolv lignin in non-acetylated and acetylated forms to study the mechanical properties of cellulose triacetate (CAT) membranes. According to the findings, adding 1 wt% lignin increased the mechanical properties significantly. To synthesize hybrid nanocomposites, LNPs can improve the interaction of amorphous lignin with other materials. Biological applications and Pickering stabilization can benefit from LNPs with more spherical morphologies, as they are effective transporters of hydrophobic compounds in target cells (Sipponen et al., 2017; Österberg et al., 2020). Jiang et al. (2013) reported the use of LNPs. According to their finding, 110–380 nm-sized complex of nano lignin poly (diallyldimethylammonium chloride) (PDADMAC) improved a natural rubber matrix's mechanical and thermal behaviour. In the work of Gupta et al. (2014), the result obtained from analysis of differential scanning calorimetry (DSC) reported that the presence of LNPs increases the value of crystalline temperature (Tc), glass transition temperature (Tg), and melting temperature (Tm). The analysis of TGA LNPs also indicated enhancement in thermal stability. The glass transition temperature (Tg) value of LNPs at 153.7°C, which is slightly increased as compared to the Tg (151.5°C) of lignin. The Tc of LNPs was found to be 142.4°C, which was higher than the Tc of lignin (134.5°C). The *Tm* of lignin and LNPs was observed around 80.6°C and 87.7°C, respectively. Chollet et al. (2019) show the potential of LNPs used to raise the flame-retardant characteristics of polylactide (PLA). The degradation of PLA during the melting process is greatly hindered when phosphorylated modified LNPs are added as a flame retardant.

4.6 CONCLUSION AND FUTURE PERSPECTIVES

Lignin is a low-cost, recyclable natural resource with a functional group that contains various phenolic hydroxyl groups responsible for various applications. Currently, between 50 and 70 million tonnes of lignin by-products are formed each year in the pulp and paper industry. In natural biomass, it contains 15–25% in grasses, 24–33% in softwoods, and 19–28% in hardwoods. However, due to very low solubility and wide distribution of molecular weight, the complexity and a random microstructure of the lignin limit the application of lignin. Synthesis and fabrications of LNPs have alleviated this problem by enhancing the performance and application of lignin. It can be easily scaled up for commercial applications. Different types of LNPs are obtained from different sources such as kraft, lignosulfonates, organosolv, soda lignin, etc. They are promising prospects for a variety of biological applications such as cancer therapy, drug delivery and diagnosis, antimicrobial activities, antioxidant activities, nanocomposite, flame retardant, and UV absorbents, and nanosized coating. However, the solvent used for the fabrication process is a challenge for applying LNPs. The heterogeneity and complexity of the structure of lignin is also another problem to prepare defined regular shaped and high yield LNPs. The variability of the biomass sources and pretreatment procedures are an additional difficulty for LNP application and synthesis. The use of green and safe solvents like water, ethanol, and acetone should be selected and used to synthesize LNPs. It is vital to design innovative methods for producing time-saving, cost-effective, affordable, and well-organized LNPs under environmentally friendly conditions with the desired characteristics, morphologies, and structures acceptable for various applications.

REFERENCES

Aadila, K., A. Barapatrea, A. Meena, H. Jhaa. 2016. Hydrogen peroxide sensing and cytotoxicity activity of Acacia ligninstabilized silver NPs. *International Journal of Biological Macromolecules*, 82, 39–47.

Aadila, K., N. Pandey, S. Mussattoc, H. Jha. 2019. Green synthesis of silver NPs using acacia lignin, their cytotoxicity, catalytic, metal ion sensing capability and antibacterial activity. *Journal of Environmental Chemical Engineering*. https://doi.org/10.1016/j.jece.2019.103296.

Adler, E., S. Hernestam. 1955. Estimation of phenolic hydroxyl groups in lignin. *Association Acta Chemica Scandinavica*, 9(2), 319–334.

Ago, M., S. Huan, M. Borghei, J. Raula, E. Kauppinen, O. Rojas. 2016. High-throughput synthesis of lignin particles (~30 nm to ~2 μm) via aerosol flow reactor: Size fractionation and utilization in pickering emulsions. *ACS Applied Materials and Interfaces*, 8, 23302–23310.

Agustin, M., P. Penttila, M. Lahtinen, K. Mikkonen. 2019. Rapid and direct preparation of LNPsfrom alkaline pulping liquor by mild ultrasonication. *ACS Sustainable Chemistry and Engineering*, 7(24), 19925–19934.

Alén, R. 2000. Structure and chemical composition of wood. *Forest Products Chemistry*, 3, 11–57.

Alonso, D. M., S. G. Wettstein, J. A. Dumesic. 2012. Bimetallic catalysts for upgrading of biomass to fuels and chemicals. *Chemical Society Reviews*, 41(24), 8075–8098.

Antonietti, M., K. Landfester. 2002. Polyreactions in miniemulsions. *Progress in Polymer Science*, 27(4), 689–757.

Asawaworarit, P., P. Daorattanachai, W. Laosiripojana, C. Sakdaronnarong, A. Shotipruk, N. Laosiripojana. 2019. Catalytic depolymerization of organosolv lignin from bagasse by carbonaceous solid acids derived from hydrothermal of lignocellulosic compounds. *Chemical Engineering Journal*, 356, 461–471.

Azadi, P., O. Inderwildi, R. Farnood, D. King. 2013. Liquid fuels, hydrogen and chemicals from lignin: A critical review. *Renewable and Sustainable Energy Reviews*, 21, 506–523.

Bachheti, A., A. Sharma, R. K. Bachheti, A. Husen, V. K. Mishra. 2019a. Plant-mediated synthesis of copper oxide nanoparticles and their biological applications. In *Nanomaterials and Plant Potential.* Cham: Springer, pp. 221–237. https://doi.org/10.1007/978-3-030-05569-1_8.

Bachheti, A., R. K. Bachheti, L. Abate, Azamal Husen. 2021a. Current status of Aloe-based nanoparticle fabrication, characterization and their application in some cutting-edge areas. *South African Journal of Botany.* https://doi.org/10.1016/j.sajb.2021.08.021.

Bachheti, R. K., A. Fikadu, Archana Bachheti, Azamal Husen. 2020. Biogenic fabrication of nanomaterials from flower-based chemical compounds, characterization and their various applications: A review. *Saudi Journal of Biological Sciences*, 27(10), 2551–2562.

Bachheti, R. K., A. Sharma, A. Bachheti, A.Husen, G. M. Shanka, D. P. Pandey. 2020b. Nanomaterials from various forest tree species and their biomedical applications. In Husen A., Jawaid M. (eds) *Nanomaterials for Agriculture and Forestry Applications.* Elsevier, pp. 81–106. https://doi.org/10.1016/B978-0-12-817852-2.00004-4.

Bachheti, R. K., L. Abate, A. Bachheti, A. Madhusudhan, A. Husen. 2021. Algae-, fungi-, and yeast-mediated biological synthesis of nanoparticles and their various biomedical applications. In *Handbook of Greener Synthesis of Nanomaterials and Compounds.* Elsevier, pp. 701–734. https://doi.org/10.1016/B978-0-12-821938-6.00022-0.

Bachheti, R. K., R. Konwarh, V. Gupta, A. Husen, Archana Joshi. 2019. Green synthesis of iron oxide nanoparticles: Cutting edge technology and multifaceted applications. In *Nanomaterials and Plant Potential.* Cham: Springer, pp. 239–259. https://doi.org/10.1007/978-3-030-05569-1_9.

Bachheti, R. K., Y. Godebo, A.Bachheti, M. O. Yassin, Azamal Husen. 2020a. Root-based fabrication of metal/metal-oxide nanomaterials and their various applications. In Husen A., Jawaid M. (eds) *Nanomaterials for Agriculture and Forestry Applications.* Elsevier, pp. 135–166. https://doi.org/10.1016/B978-0-12-817852-2.00006-8.

Bajwaa, D., G. Pourhashemb, A. Ullahb, S. Bajwac. 2019. A concise review of current lignin production, applications, products and their environmental impact. *Industrial Crops and Products.* https://doi.org/10.1016/j.indcrop.2019.111526.

Barapatre, A., K. R. Aadil, H. Jha. 2016. Synergistic antibacterial and antibiofilm activity of silver NPs biosynthesized by lignin-degrading fungus. *Bioresources and Bioprocessing*, 3, 1–13.

Beisl, S., A. Miltner, A. Friedl. 2017. Lignin from micro- to nanosize: Production methods. *International Journal of Molecular Sciences*, 18(6), 1–31.

Bobleter, O. 1994. Hydrothermal degradation of polymers derived from plants. *Progress in Polymer Science*, 19(5), 797–841.

Boerjan, W., J. Ralph, M. Baucher. 2003. Lignin biosynthesis. *Annual Review of Plant Biology*, 54, 519–546.

Cavallo, E., X. He, F. Luzi, F. Dominici, P. Cerrutti, C. Bernal, M. Foresti, L. Torre, D. Puglia. 2021. UV protective, antioxidant, antibacterial and compostable polylactic acid composites containing pristine and chemically modified lignin NPs. *Molecules*, 26, 1–20.

Chakar, F., A. Ragauskas. 2004. Review of current and future softwood kraft lignin process chemistry. *Industrial Crops and Products*, 20(2), 131–141.

Chauhan, P. 2018. Lignin NPs: Eco-friendly and versatile tool for new era. *Bioresource Technology Reports.* https://doi.org/10.1016/j.biteb.2019.100374.

Chen, L., X. Zhou, Y. Shi, B. Gao, J. Wu, T. Kirk, J. Xu, W. Xue. 2018. Green synthesis of lignin nanoparticle in aqueous hydrotropic solution toward broadening the window for its processing and application. *Chemical Engineering Journal*, 346, 217–225.

Chen, N., L. A. Dempere, Z. Tong. 2016. Synthesis of pH-responsive lignin-based nanocapsules for controlled release of hydrophobic molecules. *ACS Sustainable Chemistry Engineering in Life Sciences*, 4(10), 5204–5211.

Chen, Y., K. Zheng, L. Niu, Y. Zhang, Y. Liu, C. Wang, F. Chu. 2019. Highly mechanical properties nanocomposite hydrogels with biorenewable lignin NPs. *International Journal of Biological Macromolecules*, 128, 414–420.

Chollet, B., J.-M. Lopez-Cuesta, F. Laoutid, L. Ferry. 2019. LNPsas a promising way for enhancing lignin Flame Retardant Effect in polylactide. *Materials Letters*, 12, 1–19.

Dai, L., R. Liu, L.-Q. Hu, Z. F. Zou, C. L. Si. 2017. Lignin nanoparticle as a novel green carrier for the efficient delivery of resveratrol. *ACS Sustainable Chemistry and Engineering*, 5(9), 8241–8249.

Dallmeyer, I., F. Ko, J. Kadla. 2010. Electrospinning of technical lignins for the production of fibrous Networks. *Journal of Wood Chemistry and Technology*, 30(4), 315–329.

Del Saz-Orozco, B., M. Oliet, M. Alonso, E. Rojo, F. Rodríguez. 2012. Formulation optimization of unreinforced and lignin nanoparticle-reinforced phenolic foams using an analysis of variance approach. *Composites Science and Technology*, 72(6), 667–674.

Demuner, I., J. Colodette, A. Demuner, C. Jardim. 2019. Kraft lignins for added value. *Bioresources*, 14(3), 7543–7581.

Dizhbite, T., G. Telysheva , V. Jurkjane, U. Viesturs. 2004. Characterization of the radical scavenging activity of lignins––Natural antioxidants. *Bioresource Technology*, 95(3), 309–317.

Duval, A., M. Lawoko. 2014. A review on lignin-based polymeric, micro-and nanostructured materials. *Reactive and Functional Polymers*, 85, 78–96.

El Mansouri, N. E., J. Salvadó. 2007. Analytical methods for determining functional groups in various technical lignins. *Industrial Crops and Products*, 26(2), 116–124.

Espinoza-Acosta, J., P. Torres-Chávez, J. Olmedo-Martínez, A. Vega-Rios, S. Flores-Gallardo, E. Zaragoza-Contreras. 2018. Lignin in storage and renewable energy applications: A review. *Journal of Energy Chemistry*, 27(5), 1422–1438.

Figueiredo, P., K. Lintinen, A. Kiriazis, V. Hynninen, Z. Liu, T. Bauleth-Ramos, A. Rahikkala, A. Correia, T. Kohout, B. Sarmento. 2017. In vitro evaluation of biodegradable lignin-based NPs for drug delivery and enhanced antiproliferation effect in cancer cells. *Biomaterials*, 121, 97–108.

Figueiredo, P., K. Lintinen, J. Hirvonen, M. Kostiainen, H. Santos. 2018. Properties and chemical modifications of lignin: Towards lignin-based nanomaterials for biomedical applications. *Progress in Materials Science*, 93, 233–269.

Frangville, C., M. Rutkevicius, A. Richter, O. Velev, S. Stoyanov, V. Paunov. 2012. Fabrication of environmentally biodegradable lignin NPs. *ChemPhysChem*, 13(18), 4235–4243.

Galkin, M., J. Samec. 2016. Lignin valorization through catalytic lignocellulose fractionation: A fundamental platform for the future biorefinery. *ChemSusChem*, 9(13), 1544–1558.

Gao, W., P. Fatehi. 2019. Lignin for polymer and nanoparticle production: Current status and challenges. *Canadian Journal of Chemical Engineering*, 97(11), 2827–2842.

Gaumet, M., A. Vargas, R. Gurny, F. Delie. 2008.NPs for drug delivery: The need for precision in reporting particle size parameters. *European Journal of Pharmaceutics and Biopharmaceutics*, 69(1), 1–9.

Ge, Y., Q. Wei, Z. Li. 2014. Preparation of evaluation of the free radical scavenging activities of nanoscale lignin biomaterial. *bioresource.com*, 9, 6699–6706.

Gilca, I. A., V. I. Popa, C. Crestini. 2015. Obtaining lignin nanoparticles by sonication. *Ultrasonics Sonochemistry* 23, 369–375.

Gordobil, O., R. Herrera, M. Yahyaoui, S. Ilk, M. Kaya, J. Labidi. 2018. Potential use of kraft and organosolv lignins as a natural additive for healthcare products. *RSC Advances*, 6(43), 24525–24533.

Greiner, A., J. Wendorff. 2007. Electrospinning: A fascinating method for the preparation of ultrathin fibers. *Angewandte Chemie*, 46(30), 5670–5703.

Grossman, A., V. Wilfred. 2019. Lignin-based polymers and nanomaterials. *Current Opinion in Biotechnology*, 56, 112–120.

Gupta, A., S. Mohanty, S. Nayak. 2014. Synthesis, characterization and application of LNPs(LNPs). *Materials Focus*, 3(6), 444–454.

Han, J. K., A. Madhusudhan, R. Bandi, C. W. Park, J. C. Kim, Y. K. Lee, S. H. Lee, J. M. Won. 2020. Green synthesis of AgNPs. *BioResources*, 15(2), 2119–2132.

Henn, A., M. Mattinen. 2019. Chemo-enzymatically prepared LNPsfor value-added applications. *World Journal of Microbiology and Biotechnology*, 35(125), 1–9.

Hu, S., Y. L. Hsieh. 2015. Synthesis of surface bound silver NPs on cellulose fibers using lignin as multifunctional agent. *Carbohydrate Polymers*, 131, 134–141.

Husen, A., Q. I. Rahman, M. Iqbal, M. O. Yassin, R. K. Bachheti. 2019. Plant-mediated fabrication of gold nanoparticles and their applications. In *Nanomaterials and Plant Potential*. Cham: Springer, pp. 71–110. https://doi.org/10.1007/978-3-030-05569-1_3.

Ingtipi, K., V. Moholkar. 2019. Sonochemically synthesized LNPsand its application in the development of nanocomposite hydrogel. *Materials Today: Proceedings*, 17, 362–370.

Inwood, J., L. Pakzad, P. Fatehi. 2018. production of sulfur containing kraft lignin products. *BioResources*, 13(1), 53–70.

Iravani, S., R. Varma. 2020. Greener synthesis of LNPsand their applications. *Green Chemistry*. https://doi.org/10.1039/C9GC02835H.

Iravani, S., R. S. Varma. 2020. Greener synthesis of LNPsand their applications. *Green Chemistry*, 22(3), 612–636.

Jiang, C., H. He, H. Jiang, L. Ma, D. Jia. 2013. Nano-lignin filled natural rubber composites: Preparation and characterization. *Express Polymer Letters*, 7(5), 480–493.

Ju, T., Z. Zhang, Y. Li, X. Miaoa, J. Ji. 2019. Continuous production of LNPsusing a microchannel reactor and its application in UV-shielding films. *RSC Advances*, 9(43), 24915–24921.

Kai, D., M. Tan, P. Chee, Y. Chua, Y. Yap, X. Loh. 2016. Towards lignin-based functional materials in a sustainable world. *Green Chemistry*, 18(5), 1175–1200.

Kai, D., S. Jiang, Z. W. Low, X. J. Loh. 2015. Engineering highly stretchable lignin-based electrospun nanofibers for potential biomedical applications. *Journal of Materials Chemistry B*, 3(30), 6194–6204.

Khan, I., K. Saeed, I. Khan. 2019. NPs: Properties, applications and toxicities. *Arabian Journal of Chemistry*, 12(7), 908–931.

Koch, D., M. Paul, S. Beisl, A. Friedl, B. Mihalyi. 2020. Life cycle assessment of a lignin nanoparticle biorefinery: Decision support for its process development. *Journal of Cleaner Production*. https://doi.org/10.1016/j.jclepro.2019.118760.

Lange, H., S. Decina, C. Crestini. 2013. Oxidative upgrade of lignin – Recent routes reviewed. *European Polymer Journal*, 49(6), 1151–1173.

Lee, S., E. Yoo, S. Lee, K. Won. 2020. Preparation and application of light-colored LNPsfor broad-spectrum sunscreens. *Polymers*, 12(3), 1–14.

Li, T., S. Takkellapati. 2019. The current and emerging sources of technical lignins and their applications. *Biofuel Bioprod Biorefin*. https://doi.org/10.1002/bbb.1913.

Li, Z., Y. Ge, L. Wan. 2015. Fabrication of a green porous lignin-based sphere for the removal of lead ions from aqueous media. *Journal of Hazardous Materials*, 285, 77–83.

Lievonen, M., J. Valle-Delgado, M. L. Mattinen, E. L. Hult, K. Lintinen, M. Kostiainen, A. Paananen, G. Szilvay, H. Setala, M. Osterberg. 2016. A simple process for lignin nanoparticle preparation. *Green Chemistry*. https://doi.org/10.1039/c5gc01436k.

Lin, S. Y., C. W. Dence. 2012. *Methods in Lignin Chemistry*. New York: Springer Science & Business Media.

Liu, K., D. Zheng, H. Lei, J. Liu, J. Lei, L. Wang, X. Ma & Engineering. 2018. Development of novel lignin-based targeted polymeric nanoparticle platform for efficient delivery of anticancer drugs. *ACS Biomaterials Science*, 4, 1730–1737.

Lobo, V., A. Patil, A. Phatak, N. Chandra. 2010. Free radicals, antioxidants and functional foods: Impact on human health. *Pharmacognosy Reviews*, 4(8), 118–126.

Lou, R., R. Ma, K. T. Lin, A. Ahamed, X. Zhang. 2019. Facile extraction of wheat straw by Deep Eutectic Solvent (DES) to produce lignin NPs. *ACS Sustainable Chemistry and Engineering*, 7(12), 10248–10256.

Lourenco, A., H. Pereira. 2018. Compositional variability of lignin in biomass. *Lignin – Trends and Applications*. http://doi.org/10.5772/intechopen.71208.

Lu, Y., H. Hu, F. Xie, X. Wei, X. Fan. 2017. Structural characterization of lignin and its degradation products with spectroscopic methods. *Journal of Spectroscopy*. https://doi.org/10.1155/2017/8951658.

Ľudmila, H., J. Michal, Š. Andrea, H. Aleš. 2015. Lignin, potential products and their market value. Technical Report, CRDLR US Army Chemical Research and Development Laboratories, 60, 973–986.

Maija-Liisa, M., J. Delgadoa, T. Leskinen, T. Anttila, G. Riviere, M. Sipponen, A. Paananen, K. Lintinen, M. Kostiainen, M. Österberg. 2018. Enzymatically and chemically oxidized LNPsfor biomaterial applications. *Enzyme and Microbial Technology*, 111, 48–56.

Mandlekar, N., A. Cayla, F. Rault, S. Giraud, E. Salaün, G. Malucelli, J.-P. Guan. 2018. An overview on the use of lignin and its derivatives in fire retardant polymer systems. *Lignin-Trends Applications*.

Mattinena, M., J. Valle-Delgadoa, T. Leskinena, T. Anttilaa, G. Rivierea, M. Sipponena, A. Paananenb, K. Lintinenc, M. Kostiainenc, M. Osterberg. 2018. Enzymatically and chemically oxidized LNPsfor biomaterial applications. *Enzyme and Microbial Technology*, 111, 48–56.

Mishra, P., A. Ekielski. 2019. A simple method to synthesize lignin NPs. *Colloids and Interfaces*, 3, 1–6.

Mohamad, N., N. Arham, J. Jai, A. Hadi. 2014. Plant extract as reducing agent in synthesis of metallic NPs: A review. *Advanced in Materials Research*, 832, 350–355.

Myint, A., H. Lee, B. Seo, W. S. Son, J. Yoon, T. Yoon, H. Park, J. Yu, J. Yoona, Y. W. Lee. 2016. One-pot synthesis of environmentally friendly LNPswith compressed liquid carbon dioxide as an an solvent. *Green Chemistry*, 18(7), 2129–2146.

Nair, S., S. Sharma, Y. Pu, Q. Sun, S. Pan, J. Zhu, Y. Deng, A. Ragauskas. 2014. Production of first and second-generation biofuels: A comprehensive review. *Renewable and Sustainable Energy Reviews*, 14, 578–597.

Nevárez, L. A. M., L. B. Casarrubias, A. Celzard, V. Fierro, V. T. Muñoz, A. C. Davila, J. R. T. Lubian, G. G. Sánchez. 2011. Biopolymer-based nanocomposites: Effect of lignin acetylation in cellulose triacetate films. *Science and Technology of Advanced Materials*, 12(4), 1–18.

Novo, U., R. L. V. Gomez, F. Pomar, M. Bernal, A. Paradela, J. Albar, B. Ros. 2009. The presence of sinapyl lignin in Ginkgo biloba cell cultures changes our views of the evolution of lignin biosynthesis. *Physiologia Plantarum*, 135(2), 196–213.

Österberg, M., M. H. Sipponen, B. D. Mattos, O. J. Rojas. 2020. Spherical lignin particles: A review on their sustainability and applications. *Green Chemistry*, 22(9), 2712–2733.

Painuli, S., P. Semwal, A. Bacheti, R. K. Bachheti, A. Husen. 2020. Nanomaterials from nonwood forest products and their applications. In Husen A., Jawaid M. (eds) *Nanomaterials for Agriculture and Forestry Applications*. Elsevier, pp. 15–40. https://doi.org/10.1016/B978-0-12-817852-2.00002-0.

Pavel, F. M. 2004. Microemulsion polymerization. *Journal of Dispersion Science and Technology*, 25(1), 1–16.

Pillai, M., K. Karpagam, R. Begam, R. Selvakumar, A. Bhattacharyya. 2018. Green synthesis of lignin based fluorescent nanocolorants for live-cell imaging. *Materials Letters*, 212, 78–81.

Qian, Y., Y. Deng, X. Qiu, H. Li, D. Yang. 2014a. Formation of uniform colloidal spheres from lignin, a renewable resource recovered from pulping spent liquor. *Green Chemistry*, 16(4), 2156–2163.

Qian, Y., Q. Zhang, X. Qiu, S. Zhu. 2014b. CO_2-responsive diethylaminoethyl-modified LNPsand their application as surfactants for CO_2/N_2-switchable Pickering emulsions. *Green Chemistry*, 6, 4963–4968.

Qian, Y., X. Zhong, Y. Li, X. Qiu. 2017. Fabrication of uniform lignin colloidal spheres for developing natural broad-spectrum sunscreens with high sun protection factor. *Industrial Crops and Products*, 101, 54–60.

Radotic, K., M. Micic. 2016. Methods for extraction and purification of lignin and cellulose from plant tissues. In *Sample Preparation Techniques for Soil, Plant, and Animal Samples*. Springer, pp. 365–376.

Rahman, O., S. Shi, J. ding, D. Wang, S. Ahmad, H. Yu. 2018. Lignin NPs: Synthesis, characterization and their corrosion protection performance. *New Journal of Chemistry*. https://doi.org/10.1039/C7NJ04103A.

Richter, A., B. Bharti, H. Armstrong, J. Brown, D. Plemmons, V. Paunov, S. Stoyanov, O. Velev. 2016. Synthesis and characterization of biodegradable LigninNPs with tunable surface properties. *Langmuir*, 32(25), 6468–6477.

Richter, A., J. Brown, B. Bharti, A. Wang, S. Gangwal, K. Houck, E. Hubal, V. Paunov, S. Stoyanov, S. Velev. 2015. An environmentally benign antimicrobialnanoparticle based on a silver-infused lignin core. *Nature Nanotechnology*, 10(9), 817–823.

Rodríguez, A., R. Sánchez, A. Requejo, A. Ferrer. 2010. Feasibility of rice straw as a raw material for the production of soda cellulose pulp. *Journal of Cleaner Production*, 18(10–11), 1084–1091.

Roopa, S. 2017. An overview of natural renewable bio-polymer lignin towards nano and biotechnological applications. *International Journal of Biological Macromolecules*. http://doi.org/10.1016/j.ijbiomac.2017.05.103.

Rutledge, G., S. Fridrikh. 2007. Formation of fibers by electrospinning. *Advanced Drug Delivery Reviews*, 59(14), 1384–1391.

Sadeghifar, H., D. S. Argyropoulos and Engineering. 2015. Correlations of the antioxidant properties of softwood kraft lignin fractions with the thermal stability of its blends with polyethylene. *ACS Sustainable Chemistry*, 3, 349–356.

Sadeghifar, H., R. Venditti, J. Pawlak, J. Jur. 2019. Bi-component carbohydrate and lignin nanoparticle production from Bio-refinery lignin: A rapid and green method. *BioResources*, 14(3), 6179–6185.

Sameni, J., S. Krigstin, D. dos Santos Rosa, A. Leao, M. Sain. 2014. Thermal characteristics of lignin residue from industrial processes. *BioResources*, 9(1), 725–737.

Saratale , R., G. Saratale, G. Ghodake, S. Choc, A. Kadama, G. Kumar , B. Jeon, D. Pant, A. Bhatnagar, H. Shin. 2019. Wheat straw extracted lignin in silver NPs synthesis: Expanding its prophecy towards antineoplastic potency and hydrogen peroxide sensing ability. *International Journal of Biological Macromolecules*, 128, 391–400.

Schneider, W., A. Dillon, M. Camassola. 2021. LNPsenter the scene: A promising versatile green tool for multiple applications. *Biotechnology Advances*, 47, 1–23.

Setala, H., H.-L. Alakomi, A. Paananen, G. Szilvay, M. Kellock, M. Lievonen, V. Liljestrom, E. L. Hult, K. Lintinen, M. Osterberg, M. Kostiainen. 2020. LNPsmodified with tall oil fatty acid for cellulose functionalization. *Cellulose*, 27(1), 273–284.

Si, M., J. Zhang, Y. He, Z. Yang, X. Yan, M. Liu, S. Zhuo, S. Wang, X. Min, C. Gao, L. Chai, Y. Shi. 2018. Synchronous and rapid preparation of LNPsand carbon quantum dots from natural lignocellulose. *Green Chemistry*. https://doi.org/10.1039/C8GC00744F.

Siddiqui, L., J. Bag, D. Seetha, A. Mittal, H. Mishra Leekha, M. Mishra, A. Verma, P. Mishra, A. Ekielski, Z. Iqbal, S. Talegaonkar. 2020. Assessing the potential of LNPsas drug carrier: Synthesis, cytotoxicity and genotoxicity studies. *International Journal of Biological Macromolecules*. https://doi.org/10.1016/j.ijbiomac.2020.02.311.

Singh, R., J. M. Lillard. 2009. Nanoparticle-based targeted drug delivery. *Experimental and Molecular Pathology*, 86(3), 215–223.

Sipponen, M. H., M. Smyth, T. Leskinen, L. S. Johansson, M. Österberg. 2017. All-lignin approach to prepare cationic colloidal lignin particles: Stabilization of durable Pickering emulsions. *Green Chemistry*, 19(24), 5831–5840.

Tang, Q., Y. Qian, D. Yang, X. Qiu, Y. Qin, M. J. P. Zhou. 2020. Lignin-Based NPs: A review on their preparations and applications. Polymers, 12, 2471.

Tejado, A., C. Pena, J. Labidi, J. Echeverria, I. Mondragon. 2007. Physico-chemical characterization of lignins from different sources for use in phenol-formaldehyde resin synthesis. *Bioresource Technology*, 98(8), 1655–1663.

Tian, D., J. Hu, J. Bao, R. Chandra, J. Saddler, C. Lu. 2017. Lignin valorization: LNPsas high-value bioadditive for multifunctional nanocomposites. *Biotechnology for Biofuels*, 10, 1–11.

Tiwari, G., R. Tiwari, B. Sriwastawa, L. Bhati, S. Pandey, P. Pandey, S. K. Bannerjee. 2012. Drug delivery systems: An updated review. *International Journal of Pharmaceutical Investigation*, 2(1), 2.

Tortora, M., F. Cavalieri, P. Mosesso, F. Ciaffardini, F. Melone, C. Crestini. 2014. Ultrasound driven assembly of lignin into microcapsules for storage and delivery of hydrophobic molecules. *Biomacromolecules*, 15(5), 1634–1643.

Trevisan, H., C. Rezende. 2020. Pure, stable and highly antioxidant LNPsfrom elephant grass. *Industrial Crops and Products*. https://doi.org/10.1016/j.indcrop.2020.112105.

Tribot, A., G. Amer, M. Alio, H. de Baynast, C. Delattre, A. Pons, J. Mathias, J. M. Callois, C. Vial, P. Michaud, C. Dussap. 2019. Wood-lignin: Supply, extraction processes and use as bio-based material. *European Polymer Journal*, 112, 228–240.

Upton, B., A. Kasko. 2016. Strategies for the conversion of lignin to high-value polymeric materials: Review and perspective. *Chemical Reviews*, 116(4), 2275–2306.

Varma, R., S. Iravani. 2020. Greener synthesis of LNPsand their applications. *Green Chemistry*. https://doi.org/10.1039/C9GC02835H.

Wang, B., D. Sun, H. M. Wang, T. Q. Yuan, R. C. Sun. 2018. Green and facile preparation of regular LNPswith high yield and their natural broad-spectrum sunscreens. *ACS Sustainable Chemistry and Engineering*, 7, 2658–2666.

Wang, B., D. Sun, H.-M. Wang, T. Q. Yuan, R. C. Sun. 2019. Green and facile preparation of regular LNPswith high yield and their natural broad-spectrum sunscreens. *ACS Sustainable Chemistry and Engineering*, 7(2), 2658–2666.

Xiong, F., Y. Han, S. Wang, G. Li, T. Qin, Y. Chen, F. Chu. 2017. Preparation and formation mechanism of renewable lignin hollow nanospheres with a single hole by self-assembly. *ACS Sustainable Chemistry and Engineering*, 5(3), 2273–2281.

Xiong, F., Y. Wu, G. Li, Y. Han, F. Chu. 2018. Transparent nanocomposite films of lignin nanospheres and poly (vinyl alcohol) for UV-absorbing. *Industrial and Engineering Chemistry Research*, 57(4), 1207–1212.

Xue, J., T. Wu, Y. Dai, Y. Xia. 2019. Electrospinning and electrospun nanofibers: Methods, materials, and applications. *Chemical Reviews*, 119(8), 5298–5415.

Yang, W., E. Fortunati, F. Dominici, I. Kenny, D. Puglia. 2015. Effect of processing conditions and lignin content on thermal, mechanical and degradative behavior of lignin NPs/ polylactic (acid) bionanocomposites prepared by melt extrusion and solvent casting. *European Polymer Journal*, 71, 126–139.

Yang, W., J. Kenny, D. Puglia. 2015. Structure and properties of biodegradable wheat glutenbionanocomposites containing lignin NPs. *Industrial Crops and Products*, 74, 348–356.

Yang, W., J. Owczarek, E. Fortunati, M. Kozanecki, A. Mazzaglia, G. Balestra, J. Kenny, L. Torre, D. Puglia. 2016. Antioxidant and antibacterial LNPsin polyvinyl alcohol/chitosan films for active packaging. *Industrial Crops and Products*, 94, 800–811.

Yearlaa, S., K. Padmasree. 2015. Preparation and characterisation of lignin NPs: Evaluation of their potential as antioxidants and UV protectants. *Journal of Experimental Nanoscience*. http://doi.org/10.1080/17458080.2015.1055842.

Yiamsawas, D., G. Baier, E. Thines, K. Landfester, F. R. Wurm. 2014. Biodegradable lignin nanocontainers. *RSC Advances*, 4(23), 11661–11663.

Yin, H., L. Liua, X. Wanga, T. Wanga, Y. Zhoua, B. Liua, Y. Shana, L. Wangb, X. Lüa. 2018. A novel flocculant prepared by lignin NPs-gelatin complex from switchgrass for the capture of Staphylococcus aureus and Escherichia coli. *Colloids and Surfaces A*, 545, 51–59.

Yinghuai, Z., K. T. Yuanting, N. S. Hosmane. 2013. Applications of ionic liquids in lignin chemistry. In *Ionic Liquids—New Aspects for the Future*. London, UK: IntechOpen, pp. 315–346.

Zeng, J., Z. Tong, L. Wang, J. Zhu, L. Ingram. 2014. Isolation and structural characterization of sugarcane bagasse lignin after dilute phosphoric acid plus steam explosion pretreatment and its effect on cellulose hydrolysis. *Bioresource Technology*, 154, 274–281.

Zhang, X., M. Yang, Q. Yuan, G. Cheng. 2019. Controlled preparation of corncob LNPsand their size-dependent antioxidant properties: Toward high value utilization of lignin. *ACS Sustainable Chemistry Engineering in Life Sciences*, 7, 17166–17174.

Zhao, W., L. Xiao, G. Song, R. Sun, L. He, S. Singh, B. Simmons, G. Cheng. 2017. from lignin subunits to aggregate: Insight into lignin solubilization. *Green Chemistry*. https://doi.org/10.1039/C7GC00944E.

Zhou, Y., Y. Qian, J. Wang, X. Qiu, H. Zeng. 2020. Bioinspired lignin-polydopamine nanocapsules with strong bioadhesion for long-acting and high-performance natural sunscreens. *Biomacromolecules*, 21(8), 3231–3241.

Zhu, J., C. Yan, X. Zhang, C. Yang, M. Jiang, X. Zhang. 2020. A sustainable platform of lignin: From bioresources to materials and their applications in rechargeable batteries and supercapacitors. *Progress in Energy and Combustion Science*. https://doi.org/10.1016/j.pecs.2019.100788.

Zimniewska, M., R. Kozłowski, J. Batog. 2008. Nanolignin modified linen fabric as a multifunctional product. *Molecular Crystals and Liquid Crystals*, 484, 409.

Zoghlami, A., G. Paes. 2019. Lignocellulosic biomass: Understanding recalcitrance and predicting hydrolysis. *Frontiers in Chemistry*, 7, 1–11.

5 Medicinal Plant-Based Alkaloids

A Suitable Precursor for Nanoparticle Synthesis and Their Various Applications

Shubhangee Agarwal, Anuj Chauhan, Jigisha Anand, Limenew Abate, and Nishant Rai

CONTENTS

DOI: 10.1201/9781003213727-5

ABBREVIATION

ABTS	2′-azino-bis(3-ethylbenzothiazoline-6-sulfonic acid
AFM	Atomic Force Microscopy
DLS	Dynamic Light Scattering
DPPH	1, 1-Diphenyl-2-Picryl Hydrazyl
EGCG	Epigallocatechin Gallate
FE-SEM	Field Emission Scanning Electron Microscopy
FRAP	Ferric Reducing Antioxidant Power Assay
H_2O_2	Hydrogen peroxide
FTIR	Fourier Transform Infrared
PCS	Photon Correlation Spectroscopy
PDI	Polydispersity Index
ROS	Reactive Oxidative Species
SPR	Surface Plasmon Resonance
TEM	Transmission Electron Microscope
TEM	Transmission Electron Microscopy
TIA	Terpenoid Indole Alkaloids.

5.1 INTRODUCTION

Alkaloids are natural compounds, i.e., organic compounds containing amino or amido groups or nitrogen atoms are present in their structures. Since nitrogen atoms are alkaline, they provide alkalinity to these compounds (Robert 2001; Cushnie et al. 2014; Qiu et al. 2014). In preclinical and clinical tests, some of the identified alkaloid compounds have shown to be particularly effective. For instance, indoles alkaloids obtained from plant sources have been reported to have anticancer, antibacterial, antiviral, antimalarial, antifungal, and anti-inflammatory properties (Abate et al. 2021; Omar et al. 2021; Abate et al. 2022; Asfaw et al. 2022).

Nanoparticles (NPs) are particles ranging in size from 1 to 100 nm. They have unique properties, i.e., higher surface area to volume ratio, optical properties, uniformity, fictionalization, and quantum confinement. NPs are becoming useful in various fields such as the chemical, food, electronics, and healthcare industries (Yasin et al. 2014; Contado et al. 2015; Ibrahim et al. 2016; Husen et al. 2019; Painuli et al.2020; Bachheti et al. 2021a,b). Various approaches are used to synthesize NPs, including physical, chemical, and biological approaches. Chemical methods such as the electrochemical method (Sangeetha et al. 2015), precipitation method, the sonochemical method (Safarifard et al. 2012), sol-gel method (Pandiyarajan et al. 2013; Yahia et al. 2016), the hydrothermal approach (Hasan et al. 2008), chemical bath deposition (Jiang et al. 2015), chemical reduction (Karthik and Geetha 2013), and chemical vapor deposition (Dang et al. 2011; Ealia et al. 2017; Cele 2020) are very important methods for the synthesis of NPs. The synthetic harsh chemicals sodium borohydride (Liu et al. 2012), hypophosphite (Zhu et al. 2004), and hydrazine (Su et al. 2007) are used as reducing agents in a chemical approach that leads to the adsorption of harsh chemicals on the surface of the synthesized NPs and eventually, increase their toxicity (Iravani et al. 2011). On the other hand, the physical methods such as plasma, pulsed laser, gamma radiation, and mechanical milling normally require high energy and are more time-consuming to scale up than green synthesis (Kumar et al. 2014; Khan et al. 2019).

In recent years, the green chemistry method has become a significant focus of researchers for the synthesis of NPs. Green synthesis of NPs is more effective than physical and chemical synthesis and is a clean, economical, and ecologically sound approach. It discards the use of harsh, poisonous, and costly chemicals and uses biological entities like microbes, yeast, fungi, algae, and plants (Mandal et al. 2006; Kumar et al. 2014; Chaudhary et al. 2020; Chauhan et al. 2022; Bachheti et al. 2019a; Bachheti et al. 2019b; Bachheti et al., 2020).

Plant-based NPs are now an interesting area of research due to their different applications. Unlike biosynthesis of microbes-based NPs, the plant-based NPs doesn't involves complex protocols (Malik et al. 2014). Furthermore, no gene mutation occurs in plants, and plants are broadly allocated and easily procurable compared to microbes (Sastry et al. 2003; Kavitha et al. 2013). Hence, many such features have made plant-mediated NP synthesis advantageous and promising compared to other methods (Fahmy et al. 2011; Kumar et al. 2014). However, the plant-mediated mechanism for the synthesis of NPs is still unclear and ambiguous, although it has been reported through several studies that biomolecules in the plant extract such as flavonoids, alkaloids, glycosides, protein, and phenolics play a significant role in the reduction of metal ions to their respective NPs and used as capping and stabilizing agents for the biosynthesized NPs (Krishnaraj et al. 2010; Behravan et al. 2019).

5.2 MEDICINALLY IMPORTANT ALKALOIDS AND THEIR STRUCTURE

Important alkaloids isolated from plants are pyrrole, indole, isoquinoline, quinoline, matrine, berberine, sanguinarine, plamatine, etc. as shown in Figure 5.1).

5.3 PLANT-BASED ALKALOIDS, NANOPARTICLES SYNTHESIS, AND CHARACTERIZATION

5.3.1 Gold Nanoparticles

5.3.1.1 *Rauwolfia serpentina*

The perennial evergreen shrub, *Rauwolfia serpentina* belongs to the Apocynaceae family, also known as sarpagandha or Indian snakeroot (Deshmukh et al. 2012; Alshahrani et al. 2021). The plant stem is smooth with very few to no hairs. The leaves are a lance shape and have tapered ends with long petioles, and the fruit is purple in colour and spherical (Biradar et al. 2016). The roots and leaves of this plant contain various types of alkaloids (Deshmukh et al. 2012). The content of alkaloids present in the bark is approximately 90%, and in roots, it is approximately 1.7–3%. The ajmaline and reserpine alkaloids are the most important components of *Rauwolfia serpentina*. Other alkaloids are ajmalinine, isoajmaline, rawolfinine, rauwolscine, renoxycine, rescinamine, reserpilline, reserpine, sarpagine, serpentine, serpentine, and tetraphyllicine (Alshahrani et al. 2021).

Gold nanoparticles (AuNPs) play an essential role in nanoscience, i.e., nanomedicine and nanotechnology. Introducing novel technology such as nano enhances the ability to harness NPs and shows better pharmacological activity. According to a study, it has been reported that inorganic salts can be reduced to NPs by plant extract and stabilized by capping them (Das et al. 2017; Usman et al. 2019). They show pharmacological activity against antibacterial and anti-inflammatory agents and are highly used by pharmaceutical companies (Al-Rashed et al. 2021). The productiveness and empathy of the drug are boosted so it can be used to treat various diseases (Liang et al. 2008; Baker et al. 2021). Gold salt (HAuC14) is reduced by alkaloids present in *Rauwolfia serpentine* aqueous leaf extract) into AuNPs, and then they are encapsulated to prevent the AuNPs from aggregation and provide stability to the R-AuNPs; hence, R-AuNPs are synthesized, which show antibacterial, antioxidant, and anticancer activity (Alshahrani et al. 2021).

5.3.1.1.1 Characterization

The surface plasmon resonance (SPR) technique detects an abnormal process in noble metal NPs. The quality of intense electromagnetic fields is imparted onto the NP's surface, resulting in scattering and absorption (Baker et al. 2021). Hence, UV-visible spectroscopy confirms the synthesis of R-AuNPs. It detects the absorption peak of the R-AuNPs at 528 nm, which coincides with the SPR band of the R-AuNPs. The change in colour from light yellow to ruby red at 528 nm confirms the synthesis of R-AuNPs. A transmission electron microscope (TEM) gives the exact shape, size, and

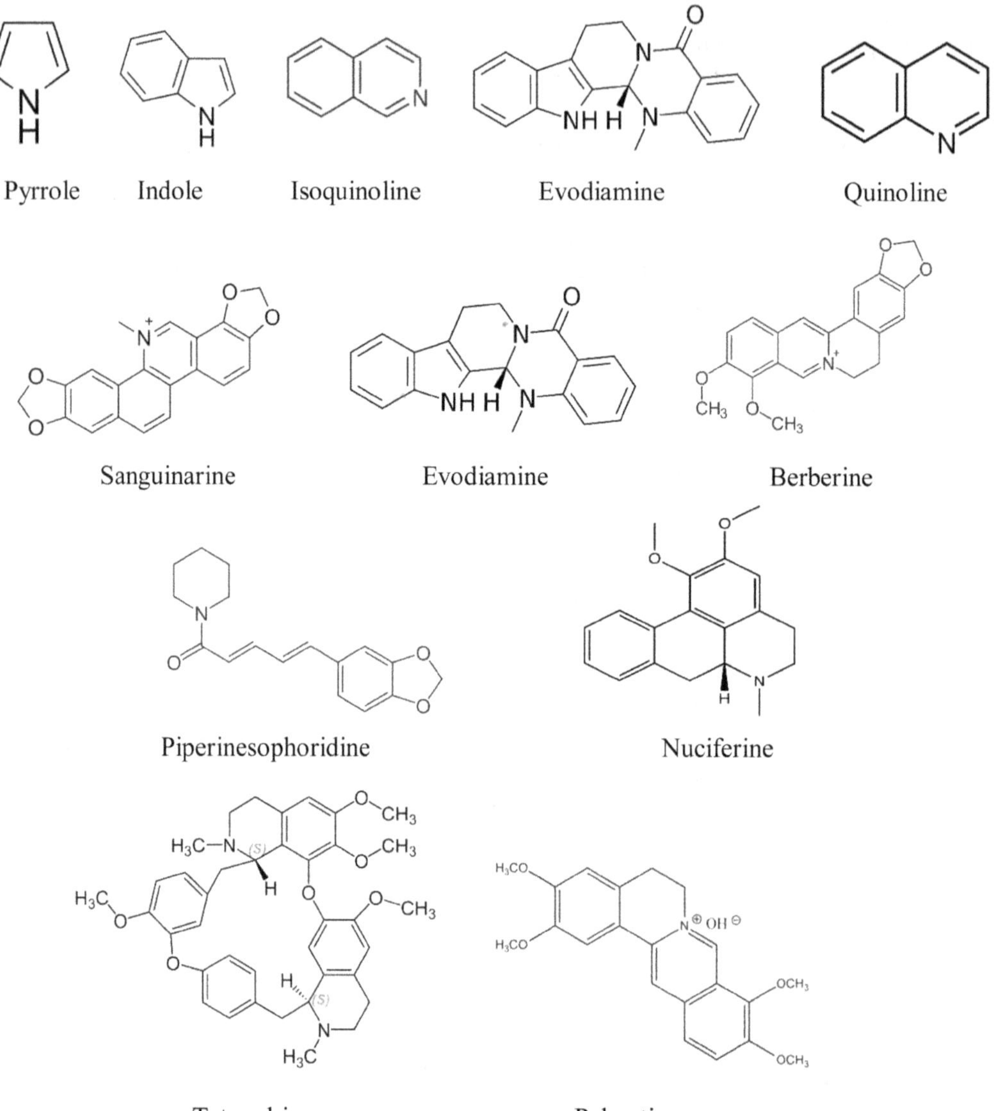

FIGURE 5.1 Structure of major alkaloids isolated from medicinal plants.

2-D morphology of R-AuNPs. The R-AuNPs are spherical, with a size of 17 nm, and are monodispersed. The particle size distribution profile and the average particle size of R-AuNPs can be determined using the dynamic light scattering (DLS) technique. Homogeneous size distribution is indicated by the polydispersity index (PDI), which is 0.215 and an average particle size of R-AuNPs is 68 nm.

5.3.1.2.1 Characterization

Atomic force microscopy (AFM) analysis is used for the surface morphology of the synthesized AuNPs. The TEM shows that the AuNPs are spherical and the size ranges between 5 and 50 nm with an average size of about 25 nm. The absorption peak is at 523 nm in UV-Vis spectrophotometer due to the SPR, which indicates the formation of AuNPs. To identify the possible biomolecules responsible for the reduction of the Au^{3+} ions and capping of the reduced Au^0, FTIR is performed.

The strong FTIR bands seen at 3,464 cm^{-1} correspond to the primary amines' N–H stretching vibration, while the band at 2,066 cm^{-1} corresponds to the C–N stretching of possible R–N=C=S bonds. Furthermore, the study revealed the band at 1,372 and 1,109 cm^{-1} of C–N and –C–O–C stretching modes, respectively, thus confirming the presence of tannin, terpenoid, saponins, flavonoids, glycosides, and polyphenolic compounds responsible for reducing Au^{3+} ions (Gopinath et al. 2014).

5.3.1.2 *Terminalia arjuna*

Terminalia arjuna is an evergreen tree that is commonly known as *Arjunindia*. The tree belongs to the Combretaceae family and is 20–30 m high (Kumar and Maulik 2013). It is found in the Indo-sub-Himalayan regions like Uttar Pradesh, South Bihar, Madhya Pradesh, Delhi, and Deccan. This plant is cultivated near ponds and rivers. The plant is also wildly spread in the forests of Sri Lanka, Burma, and Mauritius (Jain et al. 2009). It contains branches that spread horizontally and has a buttressed trunk. The branches are dropped downwards. The presence of a single-layered epidermis is confirmed by the plant's grey bark, which is smooth and has hair-like projections and a few scattered lenticels. The leaves are conical oblong or elliptic, 10–15 cm long, and 4–7 cm wide (Thakur et al. 2021).

The stem bark of *Terminalia arjuna* is rich in triterpenes like arjunic acid, arjunolic acid, arjungenin, arjunetin, tannins, and glycosides. It contains huge amounts of flavonoids such as quercetin, kaempferol, luteolin, and pelargonidin and also contains ellagic acids, phytosterols, and minerals like calcium, magnesium, zinc, and copper (Ali et al. 2003; Cheng et al. 2002; Singh et al. 2002). Phytoconstituents for different chemical compounds are reported in the plant's stem bark, root bark, leaves, and fruits (Jain et al. 2009).

AuNPs are synthesized using *Terminalia arjuna* fruit extract. Compared to the chemical method, the green synthesis method is very effective. It is advantageous for drug delivery applications. It is biocompatible and can be used for large-scale production. The addition of *Terminalia arjuna* fruit extract with $HAuCl_4$ solution within 15 min indicates the synthesis of AuNPs. A change in the colour of the solution from yellow to dark red indicates the formation of AuNPs (Gopinath et al. 2014).

5.3.2 Silver Nanoparticles

5.3.2.1 *Solanum tuberosum*

Solanum tuberosum is a member of the Solanaceae or nightshade family, a family of flowering plants that also includes eggplant and tomato. Its starchy tubers, called potatoes, are one of the world's most commonly grown and essential food crops. In India, the potato is said to be a national vegetable and is considered the kitchen king as it is part of every Indian delicacy. Steroidal alkaloids, also known as glycoalkaloids, are secondary metabolites in numerous solanaceous plants, including potatoes. These nitrogenous compounds, which have an oligosaccharide chain connected to the C-3 position of the nitrogenous alkaloid backbone are generated via the triterpene pathway and are considered an essential component of the plant's chemical weaponry against insects, viruses, bacteria, and fungi. The most common glycoalkaloids present in potatoes are α-solanine and α-chaconine. A green approach is used to make silver nanoparticles (AgNPs) from potato alkaloids, which are then tested as biocontrol fungicides (Almadiy and Nehaah 2018).

5.3.2.1.1 Characterization

Extracts are prepared from the leaves of *Solanum tuberosum* and bioactive compounds like glycoalkaloids are identified and purified using spectroscopic instrumentation, including UV, IR, and NMR. TLC analysis of compounds reveals the presence of α-chaconine and α-solanine. AgNPs are synthesized by the reduction of silver nitrate solution using potato alkaloids. The colour of the resulting solution changes from colourless to greenish brown over time, indicating the formation of AgNPs. The size of NPs is between 39.5 and 80.3 nm. At 25& 143232#xB0;C, photon correlation spectroscopy (PCS) determines the particle size distribution of AgNPs. Green synthesis techniques for nanomaterials are becoming more popular, allowing for greater shape and size control, crystal development, and nanomaterial stabilization (Almaidy et al. 2018a).

5.3.2.2 *Capsicum annuum*

The plant *Capsicum annuum* is native to southern North America, the Caribbean, and northern South America. Capsaicinoids, an alkaloid, are isolated from the fruits of capsicum species. The capsaicinoids found in the chili pepper are dihydrocapsaicin followed by nordihydrocapsaicin, homodihydrocapsaicin, and homocapsaicin and the primary capsaicinoid present in chili pepper is capsaicin. The two most effective and major capsaicinoids present in chili pepper fruit are capsaicin and dihydrocapsaicin. The only difference in their molecular structure is the acyl group saturation (Kobata et al. 1998; Reyes-Escogido et al. 2011). The pungency i.e., spicy flavour of chili pepper fruit, is because of capsaicinoids, the compound known to be present in capsicum. Capsaicin (trans-8-methyl-N-vanillyl-6- nonenamide) is an alkaloid with the molecular formula $C_{18}H_{27}NO_3$ which is crystalline, lipophilic, colourless, and odourless. Its molar mass is 305.40 g/mol, and is fat-soluble, alcohol-soluble, and oil-soluble (Reyes-Escogido et al. 2011). The NPs, which are then formed with capsaicin alkaloids are the AgNPs. Silver nitrate reduces to AgNPs as capsaicin which acts as a bioreductant (Jin et al. 2001; Amruthraj et al. 2015).

5.3.2.2.1 Characterization

The shape and size of the synthesized AgNPs can be seen by TEM after 24 h, which is nearly spherical with sizes ranging from 20 to 30 nm (Amruthraj et al. 2015). UV-Vis spectroscopy, at 0 h, on mixing silver nitrate solution with capsaicin solution, no change takes place, the only peak of capsaicin is observed at 220 nm. With the increase in time, there is a decrease in the capsaicin peak at 220 nm and peak increases at 458 nm, and silver nitrate is reduced to AgNPs, which changes the colour of the solution to pale grey.

5.3.2.3 *Chenopodium album*

The *Chenopodium album* family is the Chenopodiaceae family. This weedy plant is used for various pharmacological activities. The plant is an erect herb about 3.5 m high and 4,700 m altitude, which is polymorphous and mealy white. It is used as an ingredient in paratha and they are cooked as a vegetable. Their tender shoots are eaten raw in a salad or with curd (Singh et al. 2007; Pandey and Gupta 2014). The plant is rich in cinnamic acid amide (Cutillo et al. 2003), alkaloid chinoalbicin (Cutillo et al. 2004), apocarotenoid (DellaGreca et al. 2004), xyloside (DellaGreca et al. 2005), phenols and lignans (Cutillo et al. 2006). Biosynthesis of AgNPs is carried out at room temperature using *Chenopodium album* plant extracts (Anandalakshmi 2021).

5.3.2.3.1 Characterization

The reaction of silver nitrate and leaf extract shows the change in yellow colour solution to yellowish brown, confirming AgNP formation. FTIR analysis of AgNPs shows the peaks of the aqueous *Chenopodium album* leaf extract are in the range of 684–3,412 cm^{-1}. The FTIR analysis showed peaks at 684, 1,006, 1,629, 2,924, and 3,412 cm^{-1} corresponding to the C–H bending vibration of alkynes, C–O stretching vibration of carboxylic acid, N–H bending vibration of amines, C–H stretching vibration of alkanes, and O–H stretching vibration of carboxylic acids respectively. X-ray diffraction (XRD) analysis shows the average crystalline size of Ag nanoparticles as 28 nm. SEM determines the shape. The particle's shape is like spherical beads in the form of clusters (Anandalakshmi 2021).

5.3.2.4 *Gloriosa superba*

Gloriosa superba is a perennial creeper belonging to the Liliaceae family and is often known as Malabar glory, native to Africa and South-East Asia. *Gloriosa superba* is known for its properties to treat various diseases such as skin diseases and wounds and is also used as an antidote for snake bites. As well as all these properties, it also has antimicrobial, antitumour, and antioxidant activity. *Gloriosa superba* is rich in phytochemicals like flavonoids, alkaloids, phenols, fatty acids, and proteins (Altuntas et al. 2005; Jothi and Jabamalar 2019).

Muthukrishnan et al. (2018) mentioned the process of NP synthesis of AgNPs from *Gloriosa superba* and its various characterization techniques. Methanolic leaf extract of *Gloriosa superba* was used for the synthesis of AgNPs. The FTIR value reveals a reduction and capping of the silver ion. The particle size of around 20–69 nm, spherical in form, and highly polydisperse characteristics were revealed. The high-resolution TEM investigation revealed spherical-shaped and crystalline AgNPs. It is found that the edges of the NPs were lighter and the centre was thick, indicating that alkaloids (colchicines and their derivatives) in leaf extract capped the AUNPs, resulting in Ag^+ ions being reduced to Ag^0. The FTIR signal of methanolic *Gloriosa superba* leaf extract ranges between 3,595.31 and 3,414 nm. The significant peaks were detected at 2,924 corresponding to –CH2– group, peak found at 1,627 reduced to 1,635 (AgNP) attributed to —CO– groups, and peak at 1,442 reduced to 1,415 (AgNP) corresponding to -NH2- groups of colchicine and colchicine derivatives. The presence of these functional groups suggested that they are responsible for capping and stability, reduction of silver ions, and formation of the NPs. The zeta particle was tested for the particle size distribution of NP DLS and the stability of NPs.

5.3.3 Selenium Nanoparticles

5.3.3.1 *Trigonella foenum-graecum*

Fenugreek (*Trigonella foenum-graecum*), an annual Leguminosae herb belonging to the Fabaceae family, is used as a spice and herb. Fenugreek is also known as Greek hay. The aroma of the seeds is strong and tastes bitter. Fenugreek is cultivated in Southern Europe, the Mediterranean region, and Western Asia. As these seeds are aromatic, they are grown from Western Europe to China for their aroma. The plant is grown as a semi-arid crop all over the world (Altuntas et al. 2005). The polysaccharide galactomannan is prominent in fenugreek seeds. They're also a good source of saponins like diosgenin, yamogenin, and gitogeninetc. Other bioactive constituents of fenugreek include flavonoids, essential oils, and alkaloids such as choline and trigonelline. Proteins, amino acids, carbohydrates, alkaloids, flavonoids, cardiac glycosides, and saponins are present in the aqueous extract of fenugreek seeds (Ramamurthy et al. 2013).

Selenium nanoparticles (SeNPs) synthesized from plants like *Vitis vinifera*, *Capsicum annum*, and *Trigonella foenum-graecum* are less harmful than selenium made chemically. SeNPs have risen to prominence in the modern era, with various uses in the food, textile, and pharmaceutical industries. In some research articles, it has been shown that SeNPs are active against human breast cancer cells and activate intrinsic apoptosis signalling in cancer cells (Faghfuri et al. 2021), whereas another study highlighted the potential of SeNPs in upregulation of miR-16 and inducing antitumour activity in prostate cancer (Liao et al. 2020).

5.3.3.1.1 Characterization

The absorbance of SeNPs is measured using a UV-Vis spectrophotometer between 200 and 400 nm, and the colour shifts from colourless to ruby red. SEM shows that the particle size ranges between 50 and 150 nm. The XRD pattern indicates that it is nanocrystalline (Ramamurthy et al. 2013). The reduction of SeNPs is due to the presence of a reducing group.

5.3.4 Zinc Nanoparticles

5.3.4.1 *Aquilegia pubiflora*

Aquilegia pubiflora is a perennial herb with medicinal value and is most commonly found in the Himalayan region of India, northern Pakistan, and Afghanistan. This herbaceous plant is also known as hairy flowered columbine or Himalayan columbine but locally it is known by the name thandi buti, or domba (Dhar and Samant 1993). The *Aquilegia* is a genus that belongs to the Ranunculaceae family, which consists of approximately 60 different species of plant and shows various medicinal properties against many diseases all over the world, mainly in South Asia.

It has been reported that *Aquilegia pubiflora* shows various important pharmacological and medicinal activities and also works as astringent, dyspepsia, cardiotonic, antiasthmatic, antipyretic, stimulant, and antijaundice. Many diseases like eye disorders, snakebites, inflammation, toothaches, and nervous system disorders have also been treated with the dried roots of this plant (Hazrat et al. 2011). Furthermore, *Aquilegia pubiflora* uses its pharmacological properties in many health ailments such as influenza, skin burns, wound healing, jaundice, gynecology, circulatory, and cardiovascular disease (Jan et al. 2020; Jan et al. 2021a).

Various phytochemicals in this plant are ferulic acid, β-sitosterol, apigenin, aquilegiolide, magnoflorine, berberine, caffeic acid, and p-coumaric acid, genkwanin, glochidionolactone A, and resorcylic acid (Mushtaq et al. 2016; Jan et al. 2020).

The leaf extract of *Aquilegia pubiflora* has been used as a reducing and stabilizing agent for the biosynthesis of zinc nanoparticles (ZnONPs) (Jan et al. 2021a). Phytochemicals present in *Aquilegia pubiflora* leaf extract act as reducing and capping agents for the synthesis of NPs.

5.3.4.1.1 Characterization

The high-performance liquid chromatography (HPLC) technique quantifies the fundamental phytochemicals that cause reduction and capping during the synthesis of NPs. A total of eight constituents are observed and quantified, which include four hydroxycinnamic acid derivatives (sinapic acid, ferulic acid, chlorogenic acid, and p-coumaric acid) and four flavonoids (orientin, vitexin, isoorientin, and isovitexin). The presence of pure and crystalline NPs indicates the synthesis of ZnONPs by the pattern of XRD which shows the peaks at high diffraction between the range of 002 and 212. The FTIR spectra of the synthesized NPs are estimated in the spectral range of 400–4,000 cm^{-1}. The absorbance peak observed at 3,100 cm^{-1} corresponds to the O–H stretching mode of the hydroxyl group, 1,459 cm^{-1} corresponds to (–NH) vibration stretch indicating amine, 1,028 cm^{-1} and 1,384 cm^{-1} representing O–H and C–N stretching vibrations of aromatic amines in biomolecules. Thus, the analysis confirms the role of amines and carboxyl groups in forming ZnONPs. SEM is used to determine the shape and size of the synthesized ZnONPs, the NPs appear to be spherical, and by using the ImageJ program, the average size of the NPs observed is 34.23 nm (Suresh et al. 2015; Khan et al. 2018). Raman spectroscopy observed the peaks of the synthesized NPs between 435 and 572 cm^{-1} by indicating NP synthesis (Diallo et al. 2015; Jan et al. 2020).

5.3.4.2 *Cayratia pedata*

Cayratia pedata belongs to the Vitaceae family which is commonly known as 'birdfoot grapevine' (Khare 2007). This plant is commonly known as kattuppirandai, ainthilai kodi (5-pedata) in Tamil, goalilatain Hindi, godhapadi in Sanskrit, and velutta sori valli in Malayalam. This species is native to Thiashola and the Korakundah range and scrambles over the hedges and trees (Sharmila et al. 2018). This medicinal plant is seen in the southern part of India. It grows mainly in tropical forests, has a cylindrical stem, and is a woody climber. This plant in nature is an indigenous endangered species.

This plant's leaf treats many diseases such as ulcers, inflammation, and scabies. The plant contains many phytoconstituents such as alkaloids, tannins, phenolic compounds, flavonoids, and terpenoids (Stanley et al. 2012). Chlorophyll, amino acids, proteins, and common sugars are the primary constituents, while flavonoids, essential oils, tannins, terpenoids, alkaloids, and phenolic compounds are the secondary constituents (Edeoga et al. 2005; Krishnaiah et al. 2007). The crude plant extract acts as an antimicrobial (Nayak and Lazar 2014), antiulcer (Karthik et al. 2010), anti-inflammatory (Rajendran et al. 2011a), anti-arthritic (Selvarani and Bai 2014), antidiarrhoeal (Karthik et al. 2011), antioxidant (Rice-Evans et al. 1997), and antinociceptive agent (Rajendran et al. 2011b).

Phytochemicals are natural antioxidants and toxin-free; therefore, they act as both reducing and stabilizing agents (Singh et al. 2018). The metal precursors are converted to metal NPs by the phytoconstituents present in the plant extract, which act as a reducing agent. To synthesize ZnONPs, zinc nitrate was used as a precursor (Jayachandran et al. 2021).

5.3.4.2.1 Characterization

Field emission scanning electron microscopy (FE-SEM) is used to determine the shape and size of the synthesized NPs. The peak at 320 nm is observed by a UV-visible spectrophotometer, which is specific for ZnONPs and indicates the formation of ZnONPs. The absorption peak is observed in the wavelength range of 310–360 nm for ZnONPs (Ghamsari et al. 2016). The formation of NPs is confirmed by the morphology of the surface, which is observed in the agglomerated form. XRD observes the crystalline shape of ZnO nanoparticles. The average crystalline size of the ZnONPs is 52.24 nm. Energy-dispersive X-ray spectroscopy (EDX) observes the high signal for zinc and oxygen, which indicates the presence of zinc in the oxide form, indicating the formation of ZnONPs. The interaction of phenolic compounds, alkynes, terpenoids, and flavonoids results in the formation of ZnONPs, as revealed in FTIR spectra of the synthesized ZnONPs estimated to be in the range of 400–4,000 cm^{-1}. The peak, observed at 3,275 cm^{-1} corresponds to O–H stretching of phenolic compounds, and the peak observed at 1,624 cm^{-1} is attributed to the presence of alkene groups. The band at 1,313 cm^{-1} corresponds to C–N stretching bonds of the amines, while the C–O stretching of esters and carboxylic functional groups at 491 cm^{-1} and 435 cm^{-1} is attributed to the presence of Zn–O stretching bands (Jayachandran et al. 2021).

5.3.5 Iron Nanoparticles

5.3.5.1 *Euphorbia milii*

As an ornamental plant, *Euphorbia milii* is most commonly cultivated in India and is native to Madagascar, a flowering plant belonging to the Euphorbiaceae family. It has also been reported that this plant has many pharmacological activities (Yadav et al. 2006). This plant has latex which uses its embryofoetal toxicity to control molluscs and is also used as a traditional medicine against many diseases like liver fluke, and schistosomiasis in sheep, cattle, and even humans (Schall et al. 2001; Yadav et al. 2006). Several alkaloids present in this plant are b-sitosterol, euphol, euphorbol, and euphorbolhexacosanoate which show medicinal properties and a potent antileukaemic macrolide-lasiodiplodin (Yadav et al. 2006). The green synthesis of iron nanoparticles (FeNPs) is eco-friendly and cost-effective; thus, used for large-scale production of FeNPs. These FeNPs can be used in the treatment of effluent and also in the remediation process of the environment. This synthesis method of NPs is simple, easy, cost effective, and biological, which would be useful in environmental, biotechnological, and biomedical applications (Shah et al. 2014).

Iron has potent magnetic and catalytic properties, showing a great deal at the nanoscale (Huber 2005). As a NP, iron has been neglected many times as it is highly reactive, which makes the study of FeNPs difficult and inappropriate for practical applications. The plant shows powerful pharmacological activities due to the primary and secondary metabolic products present in plants: flavonoids, flavones, isoflavones, isothiocyanates, carotenoids, and polyphenols, and acts as an important precursor for the synthesis of metallic NPs (Park et al. 2011). Various secondary metabolites such as terpenoids, flavonoids, and tannins are present in the aerial part of the plant (Qaisar et al. 2012). The reduction of ferric ions into nano forms is due to the presence of phytochemicals in the plant parts. This bioreduction of the ferric ions leads to the synthesis of FeNPs (Shah et al. 2014).

5.3.5.1.1 Characterization

The average size of FeNPs has been determined using a particle size analyzer. It is reported that the average size of FeNPs was between 13 and 21 nm. The change in colour from yellowish-brown to dark blue, which looks like blackish blue, indicates the formation of FeNPs. Colour change is characterized by UV-Vis spectroscopy which indicates the formation of FeNPs (Song and Kim 2009). The absorption peak of the synthesized FeNPs is observed to be in the 190–202 nm range. The shape of the FeNPs is spherical when determined using TEM (Shah et al. 2014).

5.3.5.2 *Camellia sinensis*

For more than 5,000 years, the tea plant, i.e., *Camellia sinensis* or *Thea sinensis* has been cultivated, especially in the tropical and temperate regions of Asian countries such as China, Japan, India, and Thailand, and South America, and in African countries (Yu et al. 1986; Cao et al. 2013; Chen et al. 2016; Naveed et al. 2018).

Green tea is rich in polyphenols (flavonoids and catechins) and contains various organic groups, so it acts as a reducing agent. It is used in the synthesis of iron oxide NPs. The major catechin is the epigallocatechin gallate (EGCG). The standard potential of EGCG is 0.57 V, which takes part in the reduction process and hence reduces the ferric ion to iron (standard potential of iron −0.036 V). Two steps are involved in reducing ferric ion to iron; first, an O–H bond breaks by adding a precursor and a complex is formed and then forms a partial bond with a metal ion. Then partial bond breaks and electrons transfer, reducing the ferric ions to FeNPs, and hence, is oxidized to ortho-quinone (Gottimukkala 2017).

5.3.5.2.1 Characterization

The FTIR analysis ranges between 1,020 and 3,419 cm^{-1}, confirming the formation of FeNPs. The FTIR printing showed vibrations stretching at 1,632 cm^{-1} for C=C and 3,452 cm^{-1} for O–H. 2,926 cm^{-1} and 1,383 cm^{-1} correspond to C–H and C–N adsorption bands which confirm the presence of polyphenols in the solution of NPs and indicate their involvement in reducing FeNPs. It is also reported that the polyphenols present in the green tea extract act as reducing and capping agents. The average diameter of FeNPs was found to be about 116 nm by SEM analysis (Gottimukkala 2017). XRD shows the crystalline nature of FeNPs with face-centred cubic geometry and a spherical shape (Sangode et al. 2021) (Figure 5.2).

5.4 APPLICATIONS

Plants are rich in bioactive compounds like vitamins, resins, and secondary metabolites such as flavonoids, tannins, terpenes, phenolic compounds, and alkaloids. Most plant parts, bark, roots, leaves, and flowers are used for the green synthesis of NPs. These parts of plants are rich in alkaloids (berberine, berberine, jatrorrhizine, columbamine, berberubine, oxicanthine, palmatine) which are used for the synthesis of NPs. They have various pharmacological uses like antimicrobial, antiemetic, antipyretic, antipruritic, and cholagogue properties, and medical properties such as anticancer,

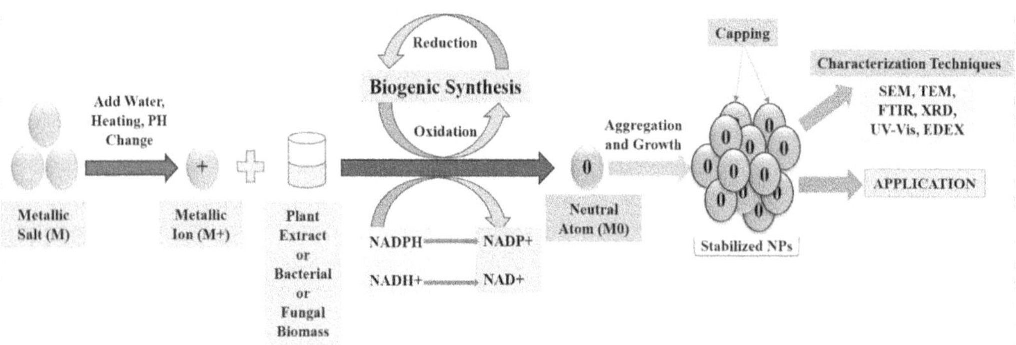

FIGURE 5.2 Synthesis, characterization, and application of plant-based nanoparticles.

anti-inflammatory, and antioxidant activities (Hadaruga et al. 2010). Some of the other applications of alkaloids are discussed below (Table 5.1).

5.4.1 Anticancer Activities

Plant-based indole alkaloids (Table 5.2) have a lot of medicinal properties, and the indole nucleus is thought to play a big role (Omar et al. 2021).

Nanotechnology is a novel approach to treating different diseases of the central nervous system, particularly gliomas. Capsaicin is one of the alkaloid molecules present in chili peppers that have outstanding biological activity. Several reports have shown that capsaicin exerts significant antitumoural effects. According to Benavidez et al. (2021) capsaicin seems to have a methoxy phenyl group that is extremely reactive. Capsaicin develops unique chemical interactions with the Ca^{2+} channel through hydrophobic attractions, implying dynamic conformational shifts.

As various alkaloids are present in *R. serpentina*, it plays a very important role in the pharmaceutical industry and is used to treat diseases like cardiovascular diseases, high blood pressure, hypertension, arrhythmia, various psychiatric diseases, mental disorders, breast cancer, human promyelocytic leukaemia. Previous *in vitro* study has revealed the antineoplastic potential of the alkaloid-rich *R. serpentina* leaf extract encapsulated AuNPs. The study showed anticancer activity against the human cervical cancer (HeLa) cell line with IC50 88.3 μg/mL Alshahrani et al. 2021).

> Among the biomedical properties of *Perilla frutescens* leaf extract. synthesized AgNPs investigated include antibacterial, antioxidant and anticancer activities. The plant extract-based AgNPs showed significant apoptotic effect against human colon cancer (COLO205) and prostate adenocarcinoma (LNCaP) and resulted into cell shrinkage, membrane blebbing, chromatin condensation, DNA fragmentation, and formation of apoptotic bodies
>
> **(Reddy et al. 2021).**

5.4.2 Antimicrobial Activities

Different botanical components of medicinal and aromatic plants and trees, such as flowers, stems, roots, leaves, fruits, roots, and bark, provide a rich source of natural compounds with antibacterial, antifungal, and antiviral properties. Plant parts contain natural compounds like alkaloids, flavonoids, and other secondary metabolites which have been exploited for their various biological activities. One major activity is antimicrobial property against a wide variety of microorganisms. Many research works documented the plant metabolites' bioactivity, in the form of NPs for their antimicrobial properties (Salem et al. 2021).

AgNPs synthesized from *P. frutescens* leaf extract showed antibacterial effect against *Escherichia coli* (14.4 mm), *Bacillus subtilis* (10.6 mm), and *Staphylococcus aureus* (10.3 mm) (Reddy et al. 2021). AgNPs from flower extract of *Handelia trichophylla* (Yazdi et al. 2019), marigold flower (Padalia et al. 2015), *Tragopogon collinus* (Seifipour et al. 2020) *C. tinctorius* (Aboutorabi et al. 2018), FeNPs from *Blumea eriantha* (Chavan et al. 2020) have depicted antibacterial activity in different *in vitro* studies.

The studies have also concluded the possible mechanism of antibacterial action pertaining to the ultrasmall size of NPs permitting easy penetration of the AgNPs across the cell membrane, cell wall pore formation, and the presence of antibacterial flavonoids, and polyphenols of the *P. frutescens* leaf extract in the encapsulated NPs (Reddy et al. 2021; Nguyen et al. 2022).

Furthermore, the efficacy of AgNPs synthesized from alkaloids and flavonoids from *Cassia alata*, *Euphorbia hirta*, *Hespesia populnea*, and *Wrightia tinctoria* have been studied against *Bacillus subtilis* and *Pseudomonas aeruginosa*. The alkaloids, and flavonoids of AgNPs were inhibitory and shown to disrupt the bacterial cell membrane and inhibit their motility (Raji et al. 2019).

TABLE 5.1
Plant-Based Alkaloids Metal/Metal Oxide Nanoparticles Synthesis, Particle Size, Shapes, and Their Application

Plants	Family	NP	Size (nm)	Shape	Application	Key reference
Alternanthera tenella	Amaranthaceae	Ag	48	Spherical	Anticancer	Sathishkumar et al. (2016)
Andrographis paniculate	Acanthaceae	Ag	67–88	Spherical	Hepatocurative activity	Suriyakalaa et al. (2013)
Angelicae pubescentis	Apiaceae	Ag, Au	12.8	Quasi spherical		Markus et al. (2017)
Aquilegia pubiflora	Ranunculaceae	ZnO	34.23	Spherical	Antiasthmatic	Hazrat et al. (2011)
Benzoin gum	Hammamelidaceae	Ag	12–38	Spherical	Antibiofilm, Anticancer	Du et al. (2016)
Camellia sinensis	Theaceae	Fe	116	Spherical	Bioremediation	Fahmy et al. (2018)
Capsicum annuum	Solanaceae	Ag	20–30	Spherical	Anti-inflammatory	Reyes-Escogido et al. (2011)
Chenopodium album	Chenopodiaceae	Ag	28	Spherical	Antibacterial	Anandalakshmi (2021)
Conocarpus lancifolius		Ag	75	Spherical	Antibacterial	Raheema et al. (2020)
Euphorbia milii	Euphorbiaceae	Fe	13–21	Spherical	Bioremediation	Shah et al. (2014)
Piper nigrum	Piperaceae	Ag	20–50	Spherical	Biosensing devices, water purification, and nanoelectronics,	Shukla et al. (2010)
Mentha pulegium	Lamiaceae	Ag	5–10	Anisotropic	Anticancer	Kelkawi et al. (2017)
Peganum harmala	Nitrariaceae	Ag	22–66	Spherical	Insecticidal	Almadiy et al. (2018)
Peganum harmala	Nitrariaceae	PLGA (chitosan coated)	2–202	Spherical	Antibacterial and wound healing	Azzazy et al. (2021)
Peganum harmala	Nitrariaceae	Pt, Pd	20–35	Spherical	Antineoplastic	Fahmy et al. (2021)
Rauwolfia serpentina	Apocynaceae	Au	17	Spherical	Anticancer	Alshahrani et al. (2021)
Terminalia arjuna	Combretaceae	Au	25	Spherical	Antioxidant	Thakur et al. (2021)

TABLE 5.2
Alkaloids with Anticancer Activities

Alkaloids	Plant	Mode of action	Reference
Berberine	*Rhizoma coptidis*	Inhibitor of several enzymes N-acetyltransferase (NAT) COX_2, Telomarase	Sun et al. (2009)
Evodiamine	*Evodia rutaecarpa*	Inducescaspase-dependent and caspase-independent apoptosis, downregulates Bcl-2 expression	Lee et al. (2006)
Matrine	*Sophora* plants	Apoptosis	Jiang et al. (2007)
Piperine	*Piper nigrum* and *Piper longum*	Inhibitor of NF-κB, c-Fos, cAMP	Pradeep et al. (2004)
Sanguinarine	*Sanguinaria canadensis* L. and *Chelidonium majus*	Inhibitor of mitogen-activated protein kinase phosphatase 1 (MKP-1)	Vogt et al. (2005)
Tetrandrine	*Stephania tetrandra*	activation of p38 mitogen-activated protein kinase (p38 MAPK)	McCubrey et al. (2012)
Nitidine chloride	*Zanthoxylum nitidum*	Arrest G2/M phase	Dan-dan et al. (2019)

5.4.3 Antioxidant Activities

Plant antioxidant activity and biosynthesis are often boosted in response to stress conditions to minimize oxidative damage caused by increased ROS generation. Surprisingly, plant stress tolerance is defined by the ability to battle oxidative stress by increasing antioxidant enzyme activity and the production of antioxidant metabolites. Antioxidants found in plant-derived substances and products have a wide range of medicinal benefits, according to several studies. These antioxidants aid in the prevention and treatment of non-infectious chronic diseases such as cardiovascular, inflammatory, and neurodegenerative disorders, metabolic syndrome, and cancer. Antioxidants can be classified into different categories, but low molecular weight substances like terpenoids, alkaloids, and most notably, polyphenols have sparked much attention among antioxidants. A recent study by Reddy et al. (2021) estimated the antioxidant activity of *P. frutescens* leaf extract AgNPs, revealed its high antioxidant properties. The study showed 68.02% free radical scavenging activity against 1,1-diphenyl-2-picryl hydrazyl (DPPH) radicals with an IC_{50} concentration of 54.52 μg/mL (Reddy et al. 2021).

Ageing can be triggered by many neurodegenerative disorders, which result from the synthesis of uncontrolled free radicals and can be regulated by exogenous antioxidants. Antioxidant activity of AgNPs biosynthesized from leaf extracts of *Chenopodium murale* species has been investigated by DPPH radical scavenging, 2,2′-azino-bis(3-ethylbenzothiazoline-6-sulfonic) acid (ABTS), ferric reducing antioxidant power assay (FRAP), hydrogen peroxide (H_2O_2) assay, etc. Studies have noticed a positive correlation between the antioxidant activity of plant extract encapsulated NPs and the content of alkaloids, signifying that the compounds induced the radical scavenging potentiality. However, the mechanism of scavenging action by stimulating cytokines (interleukin [IL] -2, IL-4, IL-12, IFN-γ, and tumour necrosis factor-alpha [TNF-γ]) remains to be elucidated (Jain et al. 2011; Jayakumar and Murugan 2016).

5.4.4 Anti-Inflammatory Activity

Aquilegia pubiflora contains flavonoids such as orientin, isoorientin, vitexin, and isovitexin. The ZnONPs synthesized from aqueous leaf extract of *A. pubiflora* exerts anti-inflammatory stress in several ways like cyclooxygenase inhibition with selective activity on COX-1 vs COX-2, phospholipase

A2, and lipoxygenases (enzymes developing eicosanoids), and reducing the concentration of leukotrienes and prostanoids which are actively involved in developing inflammation (Rathee et al. 2009; Jan et al. 2021b). *A. pubiflora*-mediated ZnONPs showed efficient inhibition of enzymes, secreted phospholipases A2 and 15-lipooxygenase involved in inflammatory processes (Jan et al. 2021b). The anti-inflammatory activity of biosynthesized AgNPs from aqueous leaf extract of *Terminalia catappa* and *T. mellueri* was revealed using a carrageenan-induced hind paw oedema model. The study demonstrated 92.13–95.7% inhibition of oedema exhibited by *T. catappa*, *T. mellueri*, and *T. bellerica* AgNPs comparable to the anti-inflammatory effect (92.1%) of standard drug indomethacin (El-Rafie and Hamed 2014).

5.4.5 Environmental Remediation

5.4.5.1 Catalytic Activities

Herbicides, insecticides, pesticides, and synthetic agricultural dyes are made from 4-nitrophenol and its derivatives, which can harm the ecosystem as common natural pollutants in wastewater. 4-nitrophenol is an important problem for the environment because of its toxicity and inhibiting activities. As a result, lowering these contaminants is critically necessary. $NaBH_4$ as a reductant and metal catalysts such as AgNPs, AuNPs, copper oxide nanoparticles (CuONPs), and palladium nanoparticles (PdNPs) are the most reliable and efficient technique used to lower the 4-nitrophenol (Singh et al. 2018). Because of the high rate of surface adsorption and high surface area to volume ratio, metal NPs have excellent catalytic potential. By enhancing the adsorption of reactants on their surface, metallic NPs can help to accelerate the reactions and lower the activation energy barriers. In a study by Yuan et al. (2017), AgNPs are synthesized using stem extract of *Chenopodium aristatum* as a reducing and capping agent. The AgNPs synthesized have good antibacterial action against *Escherichia coli* and *Staphylococcus aureus*, as well as strong catalytic activity in the breakdown of 4-nitrophenol.

5.4.5.2 Removal of Pollutant Dye

Organic dyes are extremely important since they are in such high demand in the paper, textile, plastic, leather, culinary, printing, and pharmaceutical industries. Approximately 60% of dyes are utilized in the textile industry during the pigmentation process for many materials. These manufacturing plants' pollutants are the most significant contributors to environmental contamination. Cationic and anionic dyes are two types of organic pollutants commonly utilized in such manufacturing facilities. They lead to unwanted turbidity in the water, which reduces sunlight penetration, leading to photochemical synthesis resistance and biological attacks on aquatic and marine biodiversity. So, to counter such problems, metal oxides such as ZnO, TiO, SnO, and CuO are preferred for the photolytic activity of chemically synthetic dyes. The advantages of these nano photocatalysts (e.g., ZnO and TiO_2 NPs) are due to their high surface area to mass ratio, which enhances organic pollutant adsorption. The large number of reactive surface sites available on the NP surfaces raises the surface energy of the NPs. As a result, at very low NP concentrations, the pollutant removal rate increases. This leads to a much lower requirement for nanocatalysts than bulk material to treat polluted water. Plant-based metal oxide NPs are proven very effective as photocatalytic agents to treat polluted water with harmful dyes.

5.4.5.3 Heavy Metal Sensing

Heavy metals (such as Pb, Ni, Hg, Fe, Cd, Cu, Cr, Zn, Co, Pb, and Mn) are well-known air, soil, and water contaminants. Mining waste, car emissions, natural gas, paper, plastic, coal, and dye industries contribute to heavy metal contamination. As a result, identifying hazardous metals in the biological and aquatic environment has become critical for effective remediation (Singh et al. 2018). Due to their variable size and distance-dependent optical characteristics, studies show that metallic NPs have been favoured for detecting heavy metal ions in polluted water systems. Simplicity, cost-effectiveness, and high sensitivity at sub-ppm levels are all advantages of utilizing metal NPs as colorimetric sensors for heavy metal ions in environmental systems/samples. AgNPs are synthesized from different plant extracts such as neem leaves, neem barks, pepper seeds, green tea, mango leaves, etc. for use as colorimetric sensors for the detection of heavy metal ions like Hg, Cr, Cd, Ca, and Zn (Cd^{2+}, Cr^{3+}, Hg^{2+}, Ca^{2+}, and Zn^{2+}) in water (Singh et al. 2018).

The calorimeter sensor uses NPs prepared from plant extracts. Neem extract exhibits selective sensing for Hg and Cd, similarly AgNPs prepared from pepper seed extract and green tea extract (GT-AgNPs) showed selective sensing properties for Hg^{2+}, Pb^{2+}, and Zn^{2+} ions. Interestingly, these green synthesized AgNPs selectively detected the presence of harmful metal ions in aqueous solutions throughout a wide pH range (2.0–11), which is a highly desirable characteristic from the standpoint of several sources of water contamination (Singh et al. 2018).

5.5 CONCLUSION AND FUTURE PROSPECTS

This study described numerous plant-based alkaloids and addressed their respective pharmacological and environmental relevance. Plant-based NP synthesis is an ecologically sound and safer approach to exploring the plant bioactive compounds for their broad-spectrum applications in food, medicine, textile, and environmental remediation, etc. Some alkaloids such as ajmalinine, rauwolscine, renoxycine, rescinamine, and reserpilline and flavonoids like quercetin, kaempferol, luteolin, and pelargonidin are used as a reductant for the preparation of gold, silver, iron, and zinc NPs. Alkaloids obtained from plant and other bioactive compounds exhibit anticancer, antimicrobial, antioxidant, anti-inflammatory, and catalytic activities as well as in mediation of different metal and metal oxide-based NPs. Although the medicinal value of these substances has been established, there have been preliminary studies to determine their efficacy and safety in various situations. Only a few reports have been published on plant-based alkaloids and their synthesis of metal-based NPs. Because of a lack of comprehensive pharmacological data and research studies on these plant-based alkaloids and other bioactive compounds from prior investigations, new clinical studies utilizing well-developed methodologies are still required.

REFERENCES

Abate, L., A. Bachheti, R. K. Bachheti & A. Husen. 2021. Antibacterial properties of medicinal plants: Recent trends, progress, and challenges. In Husen A. (ed.) *Traditional Herbal Therapy for the Human Immune System*. Boca Ration, FL: CRC Press, pp. 13–54.

Abate, L., M. G. Tadesse, A. Bachheti & R. K. Bachheti. 2022. Traditional and phytochemical bases of herbs, shrubs, climbers, and trees from Ethiopia for their anticancer response. *BioMed Research International* 2022 (1589877):1–27.

Aboutorabi, S.N., M. Nasiriboroumand & P. Mohammadi. 2018. Biosynthesis of silver nanoparticles using safflower flower: Structural characterization, and its antibacterial activity on applied wool fabric. *Journal of Inorganic and Organometallic Polymers and Materials* 28:2525–2532.

Ali, A., S.T. Abdullah, H. Hamid, M. Ali & M.S. Alam. 2003. Two new pentacyclic triterpenoid glycosides from the bark of *Terminalia arjuna*. *Indian Journal of Chemistry* 42B:2905–2908.

Almadiy, A.A. & G.E. Nenaah. 2018. Ecofriendly synthesis of silver nanoparticles using potato steroidal alkaloids and their activity against phytopathogenic fungi. *Brazilian Archives of Biology and Technology* 61:1–14.

Almadiy, A.A., G.E. Nenaah & D.M. Shawer. 2018. Facile synthesis of silver nanoparticles using harmala alkaloids and their insecticidal and growth inhibitory activities against the khapra beetle. *Journal of Pest Science* 91:727–737.

Al-Rashed, S., A. Baker, S.S. Ahmad, A. Syed, A.H. Bahkali, A.M. Elgorban & M.S. Khan. 2021. Vincamine, a safe natural alkaloid, represents a novel anticancer agent. *Bioorganic Chemistry* 107:1–9.

Alshahrani, M.Y., Z. Rafim, N.M. Alabdallahm, A. Shoaibm, I. Ahmad, M. Asiri, G.S. Zaman, S. Wahab, M. Saeed & S. Khan. 2021. A comparative antibacterial, antioxidant, and antineoplastic potential of *Rauwolfia serpentina* (L.) leaf extract with its biologically synthesized gold nanoparticles (R-AuNPs). *Plants (Basel)* 10(11):2278.

Altuntas, E., O.E. Ozgoz & F. Taser. 2005. Some physical properties of fenugreek (*Trigonella foenum-graceum* L.) seeds. *Journal of Food Engineering* 71:37–43.

Amruthraj, N.J., J.P.P. Raj & A. Lebel. 2015. Capsaicin-capped silver nanoparticles: Its kinetics, characterization and biocompatibility assay. *Applied Nanoscience* 5:403–409.

Anandalakshmi, K. 2021. Green synthesis, characterization and antibacterial activity of silver nanoparticles using *Chenopodium album* leaf extract. *Indian Journal of Pure & Applied Physics* 59:456–461.

Asfaw, T.B., T.B. Esho, A. Bachheti, R.K. Bachheti, D.P. Pandey & A. Husen. 2022. Exploring important herbs, shrubs, and trees for their traditional knowledge, chemical derivatives, and potential benefits. In Husen A. (ed.) *Herbs, Shrubs, and Trees of Potential Medicinal Benefits*. Boca Raton, FL: CRC Press, pp. 1–26.

Azzazy, H.M.E.S., S.A. Fahmy, N.K. Mahdy, M.R. Meselhy & U. Bakowsky. 2021. Chitosan- coated PLGA nanoparticles loaded with *Peganum harmala* alkaloids with promising antibacterial and wound healing activities. *Nanomaterials* 11(2438):1–16.

Bachheti, A., A. Sharma, R.K. Bachheti, A. Husen & V.K. Mishra. 2019a. Plant-mediated synthesis of copper oxide nanoparticles and their biological applications. In *Nanomaterials and Plant Potential*. Cham: Springer, pp. 221–237.

Bachheti, A., R.K. Bachheti, L. Abate & A. Husen. 2021. Current status of Aloe-based nanoparticle fabrication, characterization and their application in some cutting-edge areas. *South African Journal of Botany* (in Press). https://doi.org/10.1016/j.sajb.2021.08.021.

Bachheti, R.K., A. Sharma, A. Bachheti, A. Husen, G.M. Shanka & D.P. Pandey. 2020. Nanomaterials from various forest tree species and their biomedical applications. In Husen A., Jawaid M. (eds) *Nanomaterials for Agriculture and Forestry Applications*. Amsterdam, Netherlands: Elsevier, pp. 81–106.

Bachheti, R.K., L. Abate, A. Bachheti, A. Madhusudhan & A. Husen. 2021. Algae-, fungi-, and yeast-mediated biological synthesis of nanoparticles and their various biomedical applications. In Kharisov B., Kharissova O. (eds) *Handbook of Greener Synthesis of Nanomaterials and Compounds*. Amsterdam, Netherlands: Elsevier, pp. 701–734.

Bachheti, R.K., R. Konwarh, V. Gupta, A. Husen & A. Joshi. 2019b. Green synthesis of iron oxide nanoparticles: Cutting edge technology and multifaceted applications. In *Nanomaterials and Plant Potential*. Cham: Springer, pp. 239–259.

Baker, A., I. Wahid, B.M. Hassan, S.S. Alotaibi, M. Khalid, I. Uddin, J.J. Dong & M.S. Khan. 2021. Silk cocoon-derived protein bioinspired gold nanoparticles as a formidable anticancer agent. *Journal of Biomedical Nanotechnology* 17:615–626.

Behravan, M., A.H. Panahib, A. Naghizadeh, M. Ziaeed, R. Mahdavi & A. Mirzapour. 2019. Facile green synthesis of silver nanoparticles using *Berberis vulgaris* leaf and root aqueous extract and its antibacterial activity. *International Journal of Biological Macromolecule* 124:148–154.

Biradar, N., I. Hazar & V. Chandy. 2016. Current insight to the uses of *Rauwolfia*: A review. Research & Reviews A Journal of Pharmacognosy 3:1–4.

Cao, H. 2013. Polysaccharides from Chinese tea: Recent advance on bioactivity and function. *International Journal of Biological Macromolecules* 62:76–79.

Cele, T. 2020. Preparation of nanoparticles. In *Engineered Nanomaterials-Health and Safety*. London, UK: IntechOpen.

Chaudhary, R., K. Nawaz, A.K. Khan, C. Hano, B.H. Abbasi & S. Anjum. 2020. An overview of the algae-mediated biosynthesis of nanoparticles and their biomedical applications. *Biomolecules* 10:1–36.

Chauhan, A., J. Anand & N. Rai. 2022. Biogenic synthesis: A sustainable approach for nanoparticles synthesis mediated by fungi. *Inorganic and Nano-Metal Chemistry*. https://doi.org/10.1080/24701556.2021.2025078.

Chavan, R.R., S.D. Bhinge & M.A. Bhutkar. Characterization, antioxidant, antimicrobial and cytotoxic activities of green synthesized silver and iron nanoparticles using alcoholic Blumea eriantha DC plant extract. *Materials Today Communications* 24:101320.

Chen, G.J., Q.X. Yuan, M. Saeeduddin, S.Y. Ou, X.X. Zeng & H. Ye. 2016. Recent advances in tea polysaccharides: Extraction, purification, physicochemical characterization and bioactivities. *Carbohydrate Polymers* 153:663–678.

Cheng, H.Y., C.C. Lin & T.C. Lin. 2002. Antiherpes simplex virus type-2 activity of casuarinin from the bark of *Terminalia arjuna* Linn. *Antiviral Research* 55:447–455.

Contado, C. 2015. Nanomaterials in consumer products: A challenging analytical problem. *Frontiers in Chemistry* 3:1–20.

Cushnie, T.P., B. Cushnie & A.J. Lamb. 2014. Alkaloids: An overview of their antibacterial, antibiotic enhancing and anti-virulence activities. *International Journal of Anti-Microbial Agents* 44:377–386.

Cutillo, F., B. D'Abrosca, M. Dellagreca, C.D. Marino, A. Golino, L. Previtera & A. Zarrellia. 2003. Cinnamic acid amides from *Chenopodium album*: Effects on seeds germination and plant growth. *Phytochemistry* 64:1381–1387.

Cutillo, F., B. D'Abrosca, M. DellaGreca & A. Zarrelli. 2004. Chenoalbicin, a novel cinnamic acid amide alkaloid from *Chenopodium album. Chemistry and Biodiversity* 1:1579–1583.

Cutillo, F., M. DellaGreca, M. Gionti, L. Previtera & A. Zarrelli. 2006. Phenols and lignans from *Chenopodium album. Phytochemical Analysis* 17:344–349.

Dan-dan, X., F. Zhen-bo, L. Ze-feng, Y. Qin, L. Li-min, F. Hao-xuan, H. Rong-Quan, W. Hua-Yu, D. Yi-wu, C. Gang & L. Dian-Zhong. 2019. High throughput circRNA sequencing analysis reveals novel insights into the mechanism of nitidine chloride against hepatocellular carcinoma. *Cell Death & Disease* 10. https://doi.org/10.1038/s41419-019-1890-9.

Dang, T.M.D., T.T.T. Le, E. Fribourg-Blanc & M.C. Dang. 2011. Synthesis and optical properties of copper nanoparticles prepared by a chemical reduction method. *Advances in Natural Sciences: Nanoscience and Nanotechnology* 2:1–7.

Das, R.K., V.L. Pachapur, L. Lonappan, M. Naghdi, R. Pulicharla, S. Maiti, M. Cledon, L.M.A. Dalila, S.J. Sarma & S.K. Brar. 2017. Biological synthesis of metallic nanoparticles: Plants, animals and microbial aspects. *Nanotechnology for Environmental Engineering* 2:1–21.

DellaGreca, M., C. DiMarino, A. Zarrelli & B. D'Abrosca. 2004. Isolation and phytotoxicity of apocarotenoids from *Chenopodium album. Journal of Natural Products* 67:1492–1495.

DellaGreca, M., L. Previtera & A. Zarrelli. 2005. A new xyloside from *Chenopodium album. Journal of Pharmacognosy and Phytochemistry* 19:87–90.

Deshmukh, S.R., D.S. Ashrit & B.A. Patil. 2012. Extraction and evaluation of indole alkaloids from *Rauwolfia serpentina* for their antimicrobial and antiproliferative activities. *International Journal of Pharmacy and Pharmaceutical Sciences* 4:329–334.

Dhar, U. & S. Samant. 1993. Endemic plant diversity in the Indian Himalaya I. *Ranunculaceae* and *Paeoniaceae. Journal of Biogeography* 20:659–668.

Diallo, B.D., N.E. Park & M. Maaza. 2015. Green synthesis of ZnO nanoparticles by *Aspalathus linearis*: Structural & optical properties. *Journal of Alloys and Compounds* 646:425–430.

Du, J., H. Singh & T.H. Yi. 2016. Antibacterial, anti-biofilm and anticancer potentials of green synthesized silver nanoparticles using *benzoin gum* (*Styrax benzoin*) extract. *Bioprocess and Biosystems Engineering* 39:1923–1931.

Ealia, S.A.M. & M.P. Saravanakumar. 2017. A review on the classification, characterization, synthesis of nanoparticles and their application. *IOP Conf. Series: Materials Science and Engineering* 263:1–15.

Edeoga, H.O., D.E. Okwu & B.O. Mbaebie. 2005. Phytochemical constituents of some Nigerian medicinal plants. *African Journal of Biotechnology* 4:685–688.

El-Rafie, H.M. & M.A. Hamed. 2014. Antioxidant and anti-inflammatory activities of silver nanoparticles biosynthesized from aqueous leaves extracts of four *Terminalia* species. *Advances in Natural Sciences: Nanoscience and Nanotechnology* 5:035008.

Faghfuri, E., R. Ajideh, F. Shahverdi, M. Hosseini, F. Mavandadnejad, M.H. Yazdi & A.R. Shahverdi. 2021. Fabrication of calcium sulfate coated selenium nanoparticles and corresponding *In-Vitro* cytotoxicity effects against 4T1 breast cancer cell line. *Avicenna Journal of Medical Biotechnology* 13(4):201–206.

Fahmy, H.M., F.M. Mohamed, M.H. Marzouq, A.B. Mustafa, A.M. Alsoudi, O.A. Ali, M.A. Mohamed & F.A. Mahmoud. 2018. Review of green methods of iron nanoparticles synthesis and applications. *BioNanoScience* 8:491–503.

Fahmy, S.A., I.M. Fawzy, B.M. Saleh, M.Y. Issa, M.Y.U. Bakowsky & H.M.E.S. Azzazy. 2021. Green synthesis of platinum and palladium nanoparticles using *Peganum harmala* L. seed alkaloids: Biological and computational studies. *Nanomaterials* 11(4):965.

Fahmy, T.Y.A. & F. Mobarak. 2011. Green nanotechnology: A short cut to beneficiation of natural fibers. *International Journal of Biological Macromolecules* 48:134–136.

Ghamsari, M.S., S. Alamdari, W. Han & H. Park. 2016. Impact of nanostructured thin ZnO film in ultraviolet protection. *International Journal of Nanomedicine* 12:207–216.

Gopinath, K., S. Gowri, V. Karthika & A. Arumugam. 2014. Green synthesis of gold nanoparticles from fruit extract of *Terminalia arjuna*, for the enhanced seed germination activity of *Gloriosa superba. Journal of Nanostructure* 4:115.

Gottimukkala, K.S.V. 2017. Green synthesis of iron nanoparticles using green tea leaves extract. *Journal of Nanomedicine and Biotherapeutic Discovery* 7(1):1–4.

Hadaruga, D.I., N.G. Hadaruga, G.N. Bandur, A. Rivis, C. Costescu, V.L. Ordodi& A. Ardelean. 2010. *Berberis vulgaris* extract/β cyclodextrin nanoparticles synthesis and characterization. *Revista de Chimie (Bucharest)* 61:669–675.

Hasan, S.S., S. Singh, R.Y. Parikh, M.S. Dharne, M.S. Patole, B. Prasad & Y.S. Shouche. 2008. Bacterial synthesis of copper/copper oxide nanoparticles. *Journal of Nanoscience and Nanotechnology* 8:3191–3196.

Hazrat, A., M. Nisar, J. Shah & S. Ahmad. 2011. Ethnobotanical study of some elite plants belonging to Dir, Kohistan valley, Khyber Pukhtunkhwa, Pakistan. *Pakistan Journal of Botany* 43:787–795.

Huber, D.L. 2005. Synthesis, properties, and applications of iron nanoparticles. *Small* 1:482–501.

Husen, A., Q.I. Rahman, M. Iqbal, M.O. Yassin & R.K. Bachheti. 2019. Plant-mediated fabrication of gold nanoparticles and their applications. In *Nanomaterials and Plant Potential*. Cham: Springer, pp. 71–110.

Ibrahim, S., T. Charinpanitkul, E. Kobatake & M. Sriyudthsak. 2016. Nanowires nickel oxide and nanospherical manganese oxide synthesized via low temperature hydrothermal technique for hydrogen peroxide sensor. *Journal of Chemistry* 2016:1–6.

Iravani, S. 2011. Green synthesis of metal nanoparticles using plants. *Green Chemistry* 13:2638–2650.

Jain, R., A. Sharma, S. Gupta, I.P. Sarethy & R. Gabrani. 2011. *Solanum nigrum*: Current perspectives on therapeutic properties. *Alternative Medicine Review* 16:78–85.

Jain, S., P.P. Yadav, V. Gill, N. Vasudeva & N. Singla. 2009. *Terminalia arjuna* a sacred medicinal plant: Phytochemical and pharmacological profile. *Phytochemistry Reviews* 8:491–502.

Jan, H., H. Usman, M. Shah, G. Zaman, S. Mushtaq, S. Drouet, C. Hano & B.H. Abbasi. 2021b. Phytochemical analysis and versatile *in vitro* evaluation of antimicrobial, cytotoxic and enzyme inhibition potential of different extracts of traditionally used *Aquilegia pubiflora* Wall. Ex Royle. *BMC Complementary Medicine and Therapies* 21:1–19.

Jan, H., M. Shah, A. Andleeb, S. Faisal, A. Khattak, M. Rizwan, S. Drouet, C. Hano & B.H. Abbasi. 2021a. Plant-based synthesis of Zinc Oxide Nanoparticles (ZnO-NPS) using aqueous leaf extract of *Aquilegia pubiflora*: Their antiproliferative activity against HepG2 cells inducing reactive oxygen species and other *in vitro* properties. *Oxidative Medicine and Cellular Longevity* 2021:4786227.

Jan, H., M. Shah, H. Usman, M.A. Khan, M. Zia, C. Hano & B.H. Abbasi. 2020. Biogenic synthesis and characterization of antimicrobial and anti-parasitic zinc oxide (ZnO) nanoparticles using aqueous extracts of the Himalayan columbine (*Aquilegia pubiflora*). *Frontiers in Materials* 7.

Jayachandran, A., T.R. Aswathy & A.S. Nair. 2021. Green synthesis and characterization of zinc oxide nanoparticles using *Cayratiapedata* leaf extract. *Biochemistry and Biophysics Reports* 26:1–8.

Jayakumar, K. & K. Murugan. 2016. Solanum alkaloids and their pharmaceutical roles: A review. *Journal of Analytical and Pharmaceutical Research* 3:1–12.

Jiang, H., C. Hou, S. Zhang, H. Xie, W. Zhou, Q. Jin, X. Cheng, R. Qian & X. Zhang. 2007. Matrine upregulates the cell cycle protein E2F-1 and triggers apoptosis via the mitochondrial pathway in K562 cells. *European Journal of Pharmacology* 559:98–108.

Jiang, T., Y. Wang, D. Meng & M. Yu. 2015. Facile synthesis and photocatalytic performance of self-assembly CuO microspheres. *Superlattices and Microstructures* 85:1–6.

Jin, R, Y.W. Cao, C.A. Mirkin, K.L. Kelly, G.C. Schatz & J.G. Zheng. 2001. Photoinduced conversion of silver nanospheres to nanoprisms. *Science* 294:1901–1903.

Jothi, U. & A. Jebamalar. 2019. Study on estimation and antioxidant activity of *Gloriosa superba* L. whole plant extract. *International Journal of Scientific Research in Biological Sciences* 6(3):50–55.

Karthik, A.D. & K. Geetha. 2013. Synthesis of copper precursor, copper and its oxide nanoparticles by green chemical reduction method and its antimicrobial activity. *Journal of Applied Pharmaceutical Science* 3:16–21.

Karthik, P., P. Amudha & J. Srikanth. 2010. Study on phytochemical profile and antiulcerogenic effect of *Cayratiapedata* Lam in albino wistar rats. *Journal of Pharmacology and Pharmacotherapeutics* 2:1017–1029.

Karthik, P., R.N. Kumar & P. Amudha. 2011. Anti-diarrheal activity of the chloroform extract of *Cayratiapedata* Lam in albino wistar rats. *Journal of Pharmacology* 2:69–75.

Kavitha, K.S., B. Syed, D. Rakshith, H.U. Kavitha, H.C.Y. Rao, B.P. Harini & S. Satish. 2013. Plants as green source towards synthesis of nanoparticles. *International Research Journal of Biological Sciences* 2:66–76.

Kelkawi, A.A.H., A.A. Kajani & A.K. Bordbar. 2017. Green synthesis of silver nanoparticles using *Mentha pulegium* and investigation of their antibacterial, antifungal and anticancer activity. *IET Nanobiotechnology* 11:370–76.

Khan, I., K. Saeed & I. Khan. 2019. Nanoparticles: Properties, applications and toxicities. *Arabian Journal of Chemistry* 12:908–931.

Khan, S.A., F. Noreen, S. Kanwal, A. Iqbal & G. Hussain. 2018. Green synthesis of ZnO and Cu-doped ZnO nanoparticles from leaf extracts of *Abutilon indicum*, *Clerodendrum infortunatum*, *Clerodendrum inerme* and investigation of their biological and photocatalytic activities. *Materials Science and Engineering* 82:46–59.

Khare, C.P. 2007.*Indian Medicinal Plants- an Illustrated Dictionary.* 1st Indian Reprint Springer Pvt. Ltd., New Delhi, India.

Kobata, K., M. Kawamura, M. Toyoshima, Y. Tamura, S. Ogawa & T. Watanabe. 1998. Lipase catalyzed synthesis of capsaicin analogs by amidation of vanillylamine with fatty acid derivatives. *Biotechnology Letters* 20:451–454.

Krishnaiah, D., R. Sarbatly & A. Bono. 2007. Phytochemical antioxidants for health and medicine a move towards nature. *Biotechnology and Molecular Biology Reviews* 2:97–104.

Krishnaraj, C., E.G. Jagan, S. Rajasekar, P. Selvakumar, P.T. Kalaichelvan & N. Mohan. 2010. Synthesis of silver nanoparticles using *Acalypha indica* leaf extracts and its antibacterial activity against water borne pathogens. *Colloids and Surfaces B Biointerfaces* 76:50–56.

Kumar, P.P.N.V., S.V.N. Pammi, P. Kollu, K.V.V. Satyanarayana & U. Shameem. 2014.Green synthesis and characterization of silver nanoparticles using *Boerhaaviadiffusa* plant extract and their antibacterial activity. *Industrial Crops and Products* 52:562–566.

Kumar, S. & S.K. Maulik. 2013. Effect of *Terminalia arjuna* on cardiac hypertrophy. In *Bioactive Food as Dietary Interventions for Cardiovascular Disease.* Amsterdam, Netherlands: Academic Press, pp. 673–680.

Lee, T.J., E.J. Kim, S. Kim, E.M. Jung, J.W. Park, S.H. Jeong, S.E. Park, Y.H. Yoo & T.K. Kwon. 2006. Caspase-dependent and caspase-independent apoptosis induced by evodiamine in human leukemic U937 cells. *Molecular Cancer Therapeutics* 5:2398–2407.

Liang, X.J., C. Chen, Y. Zhao, L. Jia & P.C. Wang. 2008. Biopharmaceutics and therapeutic potential of engineered nanomaterials. *Current Drug Metabolism* 9:697–709.

Liao, G., J. Tang, D. Wang, H. Zuo, Q. Zhang, Y. Liu & H. Xiong. 2020. Selenium nanoparticles (SeNPs) have potent antitumor activity against prostate cancer cells through the upregulation of miR-16. *World Journal of Surgical Oncology* 18(1):81.

Liu, Q.M., D.B. Zhou, Y. Yamamoto, R. Ichino & M. Okido. 2012. Preparation of Cu nanoparticles with $NaBH_4$ by aqueous reduction method. *Transactions of Nonferrous Metals Society of China* 22:117–123.

Malik, P., R. Shankar, V. Malik, N. Sharma & T.K. Mukherjee. 2014. Green chemistry based benign routes for nanoparticle synthesis. *Journal of Nanoparticles* 2014:1–14.

Mandal, D., M.E. Bolander, D. Mukhopadhyay, G. Sarkar & P. Mukherjee. 2006. The use of microorganisms for the formation of metal nanoparticles and their application. *Applied Microbiology and Biotechnology* 69:485–492.

Markus, J., D. Wang, Y.J. Kim, S. Ahn, R. Mathiyalagan, C. Wang & D.C. Yang. 2017. Biosynthesis, characterization, and bioactivities evaluation of silver and gold nanoparticles mediated by the roots of Chinese herbal *Angelica pubescens* Maxim. *Nanoscale Resesach Letter* 12:1–12.

Martínez-Benavidez, E., I. Higuera-Ciapara, S.E. Herrera-Rodríguez, O.Y. Lugo-Melchor, F.M. Goycoolea & F.J.G. Jazo. 2021. Capsaicin nanoparticles as therapeutic agents against Gliomas. *Nano Biomedicine and Engineering* 13:433–445.

McCubrey, J.A., L.S. Steelman, S.L. Abrams, N. Misaghian, W. Chappell, J. Basecke, F. Nicoletti, M. Libra, G. Ligresti, F. Stivala, D. Maksimovic-Ivanic, S. Mijatovic, G. Montalto, M. Cervello, P. Laidler , A. Bonati , C. Evangelisti, L. Cocco & A.M. Martelli. 2012. Targeting the cancer initiating cell: The ultimate target for cancer therapy. *Current Pharmaceutical Design* 18:1784–1795.

Mushtaq, S., M.A. Aga, P.H. Qazi, M.N. Ali, A.M. Shah, S.A. Lone, A. Shah, A. Hussain, F. Rasool, H. Dar, Z.H. Shah & S.H. Lone. 2016. Isolation, characterization and HPLC quantification of compounds from *Aquilegia fragrans Benth*: Their *in vitro* antibacterial activities against bovine mastitis pathogens. *Journal of Ethnopharmacology* 178:9–12.

Muthukrishnan, S., B. Vellingiri & G. Murugesan. 2018. Anticancer effects of silver nanoparticles encapsulated by *Gloriosa superba* (L.) leaf extracts in DLA tumor cells. *Future Journal of Pharmaceutical Sciences* 4:206–214.

Naveed, M., J. BiBi, A. A. Kamboh, I. Suheryani, I. Kakar, S.A. Fazlani, X.F. Fang, S.A. Kalhoro, L. Yunjuan, M.U. Kakar, M.E. Abd El-Hack, A.E. Noreldin, S. Zhixiang, C. LiXia & Z. XiaoHui. 2018. Pharmacological values and therapeutic properties of black tea (*Camellia sinensis*): A comprehensive overview. *Biomedicine & Pharmacotherapy* 100:521–531.

Nayak, K. & J. Lazar. 2014. Antimicrobial efficacy of leaf extract of *Cayratiapedata* Lam, Vitaceae. *International Journal of ChemTech Research* 6:5721–5725.

Nguyen, N.T.T., L.M. Nguyen, T.T.T. Nguyen, T.T. Nguyen, D.T.C. Nguyen & T.V. Tran. 2022. Formation, antimicrobial activity, and biomedical performance of plant-based nanoparticles: A review. *Environmental Chemistry Letters* 25:1–41.

Omar, F., A.M. Tareq, A.M. Alqahtani, K. Dhama, K.M.A. Sayeed, T.B. Emran & J. Simal-Gandara. 2021. Plant-based indole alkaloids: A comprehensive overview from a pharmacological perspective. *Molecules* 26(8):2297.

Padalia, H., P. Moteriya & S. Chanda. 2015. Green synthesis of silver nanoparticles from marigold flower and its synergistic antimicrobial potential. *Arabian Journal of Chemistry* 8:732–741.

Painuli, S., P. Semwal, A. Bachheti, R.K. Bachheti & A. Husen. 2020. Nanomaterials from nonwood forest products and their applications. In Husen A., Jawaid M. (eds) *Nanomaterials for Agriculture and Forestry Applications*. Amsterdam, Netherlands: Elsevier, pp. 15–40.

Pandey, S. & R.K. Gupta. 2014. Screening of nutritional, phytochemical, antioxidant and antibacterial activity of *Chenopodium album* (Bathua). *Journal of Pharmacognosy and Phytochemistry* 3:1–9.

Pandiyarajan, T., R. Udayabhaskar, S. Vignesh, R.A. James & B. Karthikeyan. 2013. Synthesis and concentration dependent antibacterial activities of CuO nanoflakes. *Material Science and Engineering. C* 33:2020–2024.

Park, Y., Y.N. Hong, A. Weyers, Y.S. Kim & R.J. Linhardt. 2011. Polysaccharides and phytochemicals: A natural reservoir for the green synthesis of gold and silver nanoparticles. *IET Nanobiotechnol* 5:69–78.

Pradeep, C.R. & G. Kuttan. 2004. Piperine is a potent inhibitor of nuclear factor-κB (NF-κB), c-Fos, CREB, ATF-2 and proinflammatory cytokine gene expression in B16F-10 melanoma cells. *International Immunopharmacology* 4(14):1795–1803.

Qaisar, M., S.N. Gilani & S. Farooq. 2012. Preliminary comparative phytochemical screening of *Euphorbia* species. *American-Eurasian Journal of Agricultural & Environmental Sciences* 12:1056–1060.

Qiu, S., H. Sun, A.H. Zhang, H.Y. Xu, G.L. Yan, Y. Han & X.J. Wang. 2014. Natural alkaloids: Basic aspects, biological roles, and future perspectives. *Chinese Journal of Natural Medicines* 12:401–406.

Raheema, R.H. & R.M. Shoker. 2020. Phytochemical screening and antibacterial activity of silver nanoparticles, phenols, and alkaloids extracts of *Conocarpuslancifolius*. *European Asian Journal of BioSciences* 14:4829–4835.

Rajendran, V., S. Indumathy & V. Gopal. 2011b. Anti-nociceptive activity of *Cayratiapedata* in experimental animal models. *Journal of Pharmaceutical Research* 4:852–853.

Rajendran, V., V. Rathinambal & V. Gopal. 2011a. A preliminary study on anti-inflammatory activity of *Cayratiapedata* leaves on Wister albino rats. *Der Pharmacia Lettre* 3:433–437.

Raji, P., A.V. Samrot, D. Keerthana & S. Karishma. 2019. Antibacterial activity of alkaloids, flavonoids, saponins and tannins mediated green synthesised silver nanoparticles against *Pseudomonas aeruginosa* and *Bacillus subtilis*. *Journal of Cluster Science* 30:881–895.

Ramamurthy, C.H., K.S. Sampath, P.A. Kumar, M.S. Kumar, V. Sujatha, K. Premkumar & C. Thirunavukkarasu. 2013. Green synthesis and characterization of selenium nanoparticles and its augmented cytotoxicity with doxorubicin on cancer cells. *Bioprocess and Biosystems Engineering* 36:1131–1139.

Rathee, P., H. Chaudhary, S. Rathee, D. Rathee, V. Kumar & K. Kohli. 2009. Mechanism of action of flavonoids as anti-inflammatory agents: A review. *Inflammation & Allergy Drug Targets* 8:229–235.

Reddy, N.V., H. Li, T. Hou, M.S. Bethu, Z. Ren & Z. Zhang. 2021. Phytosynthesis of silver nanoparticles using *Perilla frutescens* leaf extract: Characterization and evaluation of antibacterial, antioxidant, and anticancer activities. *International Journal of Nanomedicine* 16:15–29.

Reyes-Escogido, M.D.L., E.G. Gonzalez-Mondragon & E. Vazquez-Tzompantzi. 2011. Chemical and pharmacological aspects of capsaicin. *Molecules* 16:1253–1270.

Rice-Evans, C., N. Miller & G. Paganga. 1997. Antioxidant properties of phenolic compounds. *Trends in Plant Sciences* 2:152–159.

Robert, A.M. 2001. *Encyclopedia of Physical Science and Technology – Alkaloids*, 3rd edition. New York: Academic Press.

Safarifard, V. & A. Morsali. 2012. Sonochemical syntheses of a nano-sized copper (II) supramolecule as a precursor for the synthesis of copper (II) oxide nanoparticles. *Ultrasonics Sonochemistry* 19:823–829.

Salem, S.S. & A. Fouda. 2021. Green synthesis of metallic nanoparticles and their prospective biotechnological applications: An overview. *Biological Trace Element Research* 199:344–370.

Sangeetha, S., G.P. Kalaignan & J.T. Anthuvan. 2015. Pulse electrodeposition of self-lubricating Ni–W/PTFE nanocomposite coatings on mild steel surface. *Applied Surface Sciences* 359:412–419.

Sangode, C.M., S.A. Mahant, P.C. Tidke, M.J. Umekar & R.T. Lohiya. 2021. Green synthesized of novel iron nanoparticles as promising antimicrobial agent: A review. *GSC Biological and Pharmaceutical Sciences* 15:117–127.

Sastry, M., A. Ahmad, M.I. Khan & R. Kumar. 2003. Biosynthesis of metal nanoparticles using fungi and actinomycete.*Current Science* 85:162–170.

Sathishkumar, P., K. Vennila, R. Jayakumar, A.R.M. Yussof, T. Hadibarat & T. Palvannan. 2016. Phytosynthesis of silver nanoparticles using *Alternanthera tenella* leaf extract: An effective inhibitor for the migration of human breast adenocarcinoma (MCF-7) cells. *Bioprocess and Biosystems Engineering* 39:651–659.

Schall, V.T., M.C. Vasconcellos, R.S. Rocha, C.P. Souza & N.M. Mendes. 2001. The control of the scistosome-transmitting snails *Biomphalaria glabrata* by the plant molluscicide *Euphorbia splendense* var. hislopii (syn milii Des. Moul): A longitudinal field study in an endemicarea in Brazil. *Acta Tropica* 79:165–170.

Seifipour, R., M. Nozari & L. Pishkar. 2020. Green synthesis of silver nanoparticles using tragopogon collinus leaf extract and study of their antibacterial effects. *Journal of Inorganic and Organometallic Polymers and Materials* 30:2926–2936.

Selvarani, K. & G.V.S. Bai. 2014. Anti-arthritic activity of *Cayratiapedata* leaf extract in Freund's adjuvant induced arthritic rats. *International Journal of Research in Pharmaceutical Sciences* 4:55–59.

Shah, S., S. Dasgupta, M. Chakraborty, R. Vadakkekara & M. Hajoori. 2014. Green synthesis of iron nanoparticles using plant extracts. *International Journal of Biological & Pharmaceutical Research* 5:549–552.

Sharmila, S., K. Kalaichelvi & S.M. Dhivya. 2018. Pharmacognostical and phytochemical analysis of *cayratiapedata* var. glabra – A Vitaceae member. *International Journal of Pharmaceutical Sciences and Research* 9:218–226.

Shukla, V.K., R.P. Singh & A.C. Pandey. 2010. Black pepper assisted biomimetic synthesis of silver nanoparticles. *Journal of alloys and compounds* 507:L13–L16.

Singh, D.V., R.K. Verma, S.C. Singh & M.M. Gupta. 2002. RP-LC determination of oleane derivatives in *Terminalia arjuna. Journal of Pharmaceutical and Biomedical Analysis* 28:447–52.

Singh, J., T. Dutta, K.H. Kim, M. Rawat, P. Samddar & P. Kumar. 2018. Green synthesis of metals and their oxide nanoparticles: Applications for environmental remediation. *Journal of Nanobiotechnology* 16:1–24.

Singh, L., N. Yadav, A.R. Kumar, A.K. Gupta, J. Chacko, K. Parvin & U. Tripathi. 2007. Preparation of value-added products from dehydrated bathua leaves (*Chenopodium album* L.). *Natural Product Radiance* 6:6–10.

Song, J.Y. & B.S. Kim. 2009. Rapid biological synthesis of silver nanoparticles using plant leaf extracts. *Bioprocess and Biosystems Engineering* 32:79–84.

Stanley, A.L., V.A. Ramani & A. Ramachandran. 2012. Phytochemical screening and GC-MS studies on the ethanolic extract of *Cayratia pedate. International Journal of Pharmacy and Phytopharmacology Research* 1:112–116.

Su, X., J. Zhao, H. Bala, Y. Zhu, Y. Gao, S. Ma & Z. Wang. 2007. Fast synthesis of stable cubic copper nanocages in the aqueous phase. *Journal of Physical Chemistry C* 111:14689–14693.

Sun, Y., K. Xun, Y. Wang & X. Chen. 2009. A systematic review of the anticancer properties of berberine, a natural product from Chinese herbs. *Anticancer Drugs* 20:757–769.

Suresh, D., P.C. Nethravathi, Udayabhanu, H. Rajanaika, H. Nagabhushana & S.C. Sharma. 2015. Green synthesis of multifunctional zinc oxide (ZnO) nanoparticles using *Cassia fistula* plant extract and their photodegradative, antioxidant and antibacterial activities. *Materials Science in Semiconductor Processing* 31:446–454.

Suriyakalaa, U., J.J. Antony, S. Suganya, D. Siva, R. Sukirtha, S. Kamalakkannan, P.B.T. Pichiah & S. Achiaman. 2013. Hepatocurative activity of biosynthesized silver nanoparticles fabricated using *Andrographis paniculate. Colloids and Surfaces B* 102:189–194.

Thakur, S., H. Kaurav & G. Chaudhary. 2021. *Terminalia arjuna*: A potential ayurvedic cardio tonic. *International Journal for Research in Applied Sciences and Biotechnology* 8(2):227–236.

Usman, A.I., A.A. Aziz & O.A. Noqta. 2019. Application of green synthesis of gold nanoparticles: A review. *Jurnal Teknologi* 81:171–182.

Vogt, A., A. Tamewitz, J. Skoko, R.P. Sikorski, K.A. Giuliano & J.S. Lazo. 2005. *Journal of Biological Chemistry* 280:19078–19086.

Yadav, S.C., M. Pande & M.V. Jagannadham. 2006. Highly stable glycosylated serine protease from the medicinal plant *Euphorbia milii. Phytochemistry* 67:1414–1426.

Yahia, I.S., A.A.M. Farag, S. El-Faify, F. Yakuphanoglu & A.A. Al-Ghamdi. 2016. Synthesis, optical constants, optical dispersion parameters of CuO nanorods. *Optik* 127:1429–1433.

Yasin, S.M.M., S. Ibrahim & M.R. Johan. 2014. Effect of zirconium oxide nanofiller and dibutyl phthalate plasticizer on ionic conductivity and optical properties of solid polymer electrolyte. *Scientific World Journal* 2014:1–8.

Yazdi, M.E.T., M.S. Amiri & H.A. Hosseini. 2019. Plant-based synthesis of silver nanoparticles in Handelia trichophylla and their biological activities. *Bulletin of Materials Science* 42:155.

Yu, F.L. 1986. Discussion on the originating place and the originating center of tea plant. *Journal of Tea Science* 6:1–8.

Yuan, C., C. Huo, B. Gui, P. Liu, & C. Zhang. 2017. Green synthesis of silver nanoparticles using chenopodium aristatum L. stem extract and their catalytic/antibacterial activities. *J Clust Sci* 28:1319–1333.

Zhu, H.T. & Y.S. Lin. 2004. A novel one-step chemical method for the preparation of copper nanofluids. *Journal of Colloid and Interface Science* 277:100–103.

6 Sustainable Synthesis of Nanoparticles Using Saponin-Rich Plants and Its Pharmaceutical Applications

Kamalanathan Pouthika, Gunabalan Madhumitha, and Selvaraj Mohana Roopan

CONTENTS

6.1 INTRODUCTION

Nanoscience has been the only way to discover the unique assets of materials (Sahu et al. 2021). By customizing the behaviour of metals at incredibly small scale, nanotechnology seems to have a significant impact on the advancement of research into materials science (Shreema et al. 2021). Nanoparticles (NPs) are particles with an inherent structure of less than 100 nm. Because of their distinct physical and chemical abilities, NPs are used constantly (Jaswal and Gupta 2021). NPs can be made in various ways, including chemical, physical, and biological procedures; however, biological approaches are more environmentally friendly than other methods. Sustainable biotic approaches for producing NPs have sparked attention in recent years because the formulation is environmentally benign, easy, cost-effective, non-toxic, and preferred to other procedures (Kalaiselvi et al. 2015; Bachheti et al. 2019, 2019a; Bachheti 2020; Bachheti 2021). The physical method involves ion

DOI: 10.1201/9781003213727-6

implantation (Stepanov et al. 2013), sputter deposition (Suzuki et al. 2015), laser ablation (Piriyawong et al. 2012), high energy ball milling (Cheng et al. 2018), and electric arc deposition methods. The chemical method involves the reduction method (Khan et al. 2011), sol-gel method (Hasnidawani et al. 2016), and precipitation method (Phiwdang et al. 2013). The biological method involves a micro-organism-based approach (Jayaseelan et al. 2013) and a green synthesis method (Kumar et al. 2014; Elango et al. 2015). The primary advantage of using plant extracts to produce metal/metal oxide NPs is their versatility, stability, and lack of toxicity, all of which aid in reducing metal ions and stabilizing metal NPs. (Zhang et al. 2006; Anandalakshmi et al. 2016; Moodley et al. 2018; Husen et al. 2019; Aritonang et al. 2019; Bachheti et al. 2020a, 2020b; Bachheti et al. 2021a).

Saponins are a natural/organic surfactant that can replace commercial surfactants since aglycones possess different forms of lipophilicity and sugar chains to contain significant hydrophilicity (Liao et al. 2021). Saponins comprise polycyclic organic compounds with many structures obtained in plants and some aquatic species. Saponins have several qualities, including bitterness and sweetness, haemolytic capabilities, emulsification, and pharmacological and antimicrobial capabilities (Geng et al. 2021). This chapter clearly explains the synthesis of saponin-mediated NPs using medicinal plants and their application in various fields.

6.2 SAPONINS

Saponins are non-volatile, surface-active compounds found in nature, especially in plant parts such as roots, leaves, bark, seeds, and so on (Tschesche et al. 1969; Hostettmann et al. 2005). Saponin is derived from the Latin word 'sapo' (soap), and it refers to saponins' surfactant properties and potential to create foam (Oleszek et al. 2002). Naturally, saponins are amphiphilic structures with one or more hydrophilic sugar groups and a lipophilic steroid/triterpenic portion (Figure 6.1). Therefore,

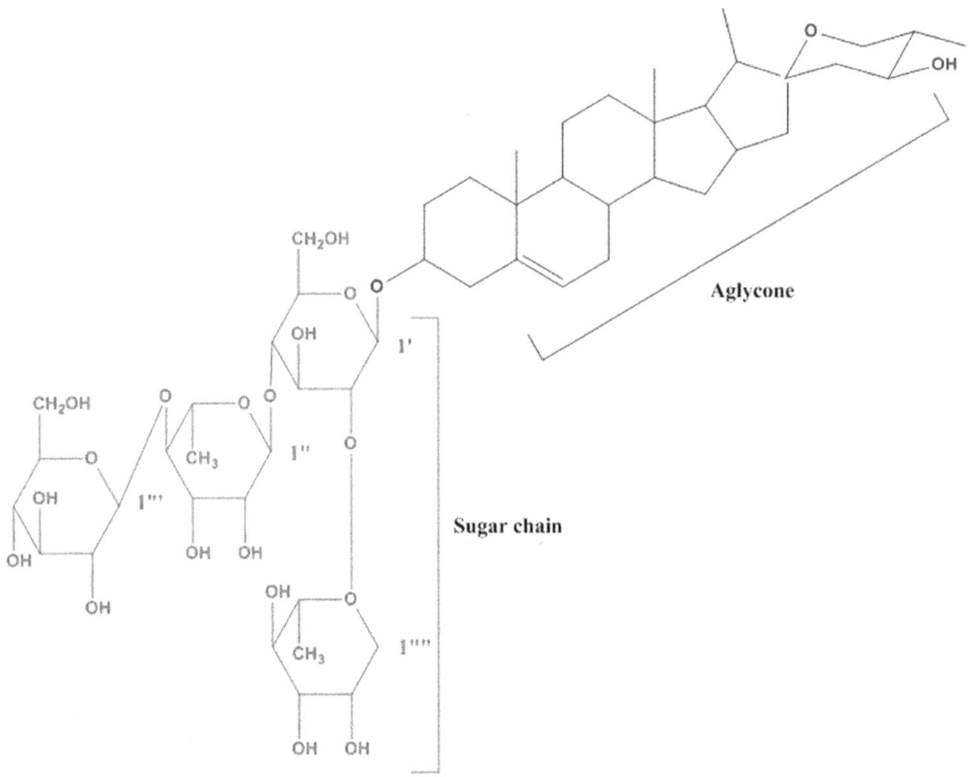

FIGURE 6.1 Structure of saponin (Stepanov et al. 2013).

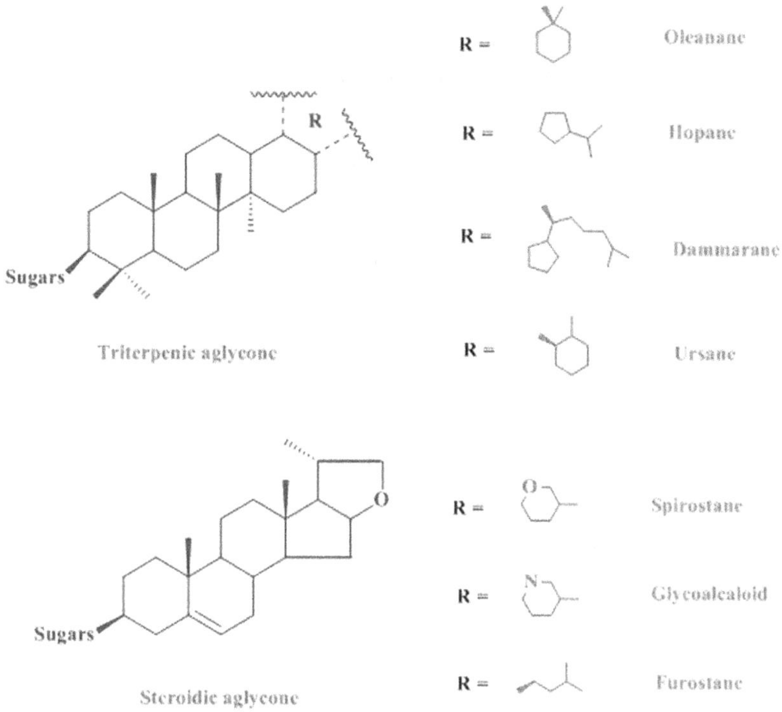

FIGURE 6.2 Different possible structures of saponin aglycones (Chaieb 2010).

saponins are categorized as steroidal and triterpenoid saponins based on the chemical character of the aglycone (also known as sapogenin) (Figure 6.2). Plant steroidal saponins comprise compounds with 27 carbon atoms in their core structures, such as spirostan. Aglycones with 30 carbon atoms or their derivatives make up the majority of triterpenoid saponins. Pentacyclic oleanans and tetracyclic dammarans are the most familiar core structures (Vincken et al. 2007). Saponins in the triterpenoid glycosides category are the familiar saponins found in plants.

Saponins are commonly employed in the food, pharmaceutical, and cosmetic industries due to their ability to function as a 'biosurfactant'. Saponins' ability to minimize interfacial energy between different phases (hydrophobic–hydrophilic) allows them to stabilize emulsions (Balakrishnan et al. 2006). Saponins can help to stabilize nanosuspensions and nanoemulsions, as these are biphasic systems of very small droplets (less than 100 nm). They have a variety of fascinating pharmacological and medicinal properties, such as a reduction in haemolysis (de Ven et al. 2010) and an improvement in antigen-specific immune responses (Cao et al. 2010). Saponins' self-aggregating properties are a promising research field that could develop entirely new nano-objects and NPs. They work well as a reducing agent in the biosynthesis of NPs.

6.2.1 Saponin in Plant Extract – Laboratory Test

Different saponin tests, as detailed below, can be used to assess and confirm the presence of saponins in the laboratory.

6.2.1.1 Typical Foam Test

Crude plant powder (3 g) was mixed with 300 ml of warm purified water. The aqueous extricate was cooled at room temperature, stirred well, filtered, and stored at 4°C in the refrigerator for 24 h. In a test tube, 5 ml of the plant extract was taken and 5 ml of distilled water (DW) was added to it

TABLE 6.1
Some Plants Species Rich in Saponin

Triterpenoid saponin	Steroid saponin
Poaceae	**Poaceae**
Avena strigose	*Avena sativa*
Avena sativa	*Avena strigosa*
	Panicum coloratum
	Panicum virgatum
Chenopodiaceae	**Solanaceae**
Beta vulgaris	*Solanum lycopersicum*
Chenopodium quinoa	*Capsicum frutescens*
	Solanum tuberosum
Leguminosae	**Alliaceae**
Glycine max	*Allium schoenoprasum*
Pisum sativum	*Allium nutans*
Medicago truncatula	*Allium sativum*
Medicago sativa	*Allium cepa*
Phaseolus vulgaris	*Alium porrum*
Theaceae	
Camellia sinensi	

for dilution. The resulting mixture was vigorously shaken for 2 min. The existence of saponins was indicated by the occurrence of foam that lasts for at least 15 min. Alternatively, once olive oil was introduced, an emulsion formed.

6.2.1.2 Wet Foam Test

After diluting the test solution with water and vigorously shaking it for 1–2 min, a constant foam appeared at the top of the sample test tube.

6.2.1.3 Dry Foam Test

About 0.5 g of dry plant residue was taken, mixed with 5 ml DW, shaken, and warmed in a water bath in a test tube. Three drops of olive oil were added to the froth and shaken vigorously. The formation of emulsion indicated the occurrence of saponins.

6.2.1.4 Foam Test for Fresh Samples

A 2 g plant sample (leaves) was added to 20 ml DW with a mixer, filtered, and the filtrate reduced to half of the original capacity by evaporation in a water bath, then transferred into a test tube. Three drops of olive oil were added to the froth and vigorously shaken. The existence of saponins was indicated by the formation of emulsion (El Aziz et al. 2019) (Table 6.1).

6.2.2 Saponin Extraction Techniques

The solubility of a solute from plant materials in a solvent is the basis for conventional extraction. As a result, it also consumes a large amount of solvent to remove the desired solute, which is often assisted by heating and mechanical stirring or shaking. Less toxic chemical synthesis, safer materials, energy conservation, and pollution reduction are all part of the green extraction process. Various saponin extraction techniques are shown in Figure 6.3.

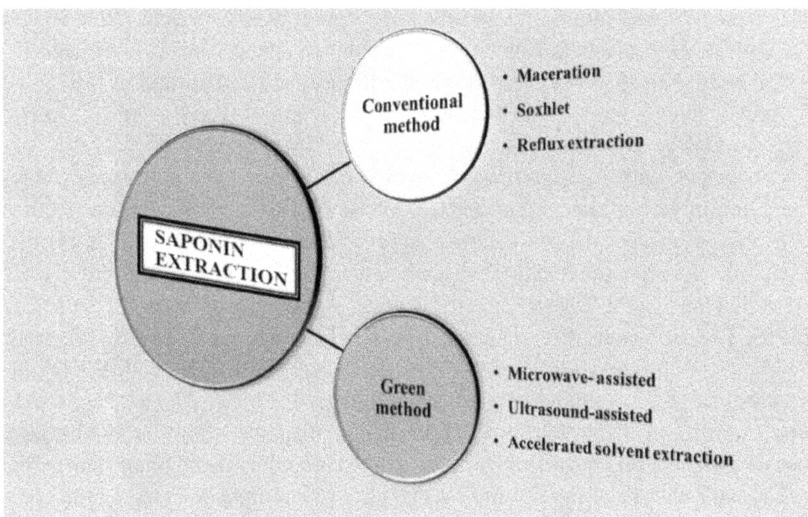

FIGURE 6.3 Various saponin extraction techniques.

6.3 SYNTHESIS OF SAPONIN-MEDIATED NANOPARTICLES

As mentioned earlier, saponins play an important role in synthesizing NPs due to their ability to perform as biosurfactant. Paramesh et al. used a saponin extracted from *Simarouba glauca* seed meal to produce silver nanoparticles (AgNPs). Initially, the seed's bark was removed, stripped, and defatted. Methanol was used to extract a saponin-rich extract from defatted sample. A rotary evaporator was used to concentrate the extract, which was then dried and processed at 4°C for later use. The AgNP was produced using a green method with saponin extract. UV-visible spectroscopy (UV-vis) displayed the absorption band at 424 nm, confirming the formation of AgNPs. A slight shift in the IR band indicates saponin adsorption on the AgNPs' surface, and the saponins' efficiency in stabilizing and capping the NPs. The NPs were discovered to have a diameter of 4.61 nm and a spherical shape (Paramesh et al. 2021).

While processing *Chenopodium quinoa*, Segura et al., discovered significant quantities of saponin-rich husk by-products. Sixteen saponin analogues were identified by liquid chromatography-mass spectroscopy (LC-MS),which occupies 60% weight in quinoa husk extract. The saponins obtained are triterpene-built, with the aglycon groups comprising β-amyrin covalently connected to double oligosaccharide chains at the C_3 and C_{28} sites, permitting them to be classified as budesonide. Under mild conditions, the quinoa husk extract obtained acted as a reducing/capping agent for silver ions, resulting in stable AgNPs. The particles are irregular, but some have faceted outlines (truncated triangles and hexagons). The reported AgNPs have a mean diameter of around 19 nm, and the particles are mostly between 5 and 50 nm in size.

The Fourier-transform infrared spectroscopy (FTIR) spectrum confirmed that oleanane triterpenoid saponin in quinoa extract was responsible for the synthesis of NPs (Segura et al. 2020).

Isolation of saponin from *Trianthema decandra* leaves was carried out by maceration extraction. Preparative high performance liquid chromatography (HPLC) purified the attained saponin extract. Geethalakshmi and Sarada used the green method to make AgNPs and gold nanoparticles (AuNPs) from the saponin extract. In the presence of saponin, silver nitrate and chloroauric acid were hastily reduced. The generation of AgNPs and AuNPs was confirmed by the colouration from yellow to orange and yellow to dark ruby. These NPs were characterized by UV-vis, FTIR, scanning electron microscopy (SEM), and energy dispersive X-ray spectroscopy (EDAX). The UV-vis spectra showed a strong absorption peak at 440–580 nm for AuNPs and 260–380 nm for AgNPs. The FTIR absorption peak at

1637.76 cm^{-1} is typical of gold atoms and the silver absorption bands are detected at 1637.06 cm^{-1} and 1120.01 cm^{-1}. AuNPs were cubic, hexagonal, and circular in shape with 37.7–79.9 nm in size, while AgNPs were spherical with 17.9–59.6 nm in size (Geethalakshmi and Sarada 2013).

AgNPs obtained from saponin-rich leaf extract of *Ocimum tenuiflorum* and their antifungal potential were studied by Nguyen et al. (2020). Using a foam test, the occurrence of saponin in leaf extract was sensed, and the saponin content was quantified. The leaf extract was discovered to be high in saponin in phytochemical tests. UV-vis, FTIR, transmission electrom microscope (TEM), and EDAX were used to characterize AgNPs. The colour shift of leaf extract from light brown to reddish-brown after the reaction with silver nitrate indicated the formation of AgNPs. The peak located at 434 nm in the UV-vis spectra reflects AgNPs absorption due to surface plasmon resonance (SPR). The vibrations of O–H stretching, C–H stretching, C=C or C=O stretching, O–H bending, and C–O stretching were allocated to the peaks at 3407, 2923, 1634, 1384, and 1076 cm^{-1}. AgNPs were encapsulated by phytochemical constituents of *O. tenuiflorum* leaf extract, particularly saponin, according to these findings. TEM images displayed that the synthesized NPs were 5–61 nm (the majority were 5 nm) in size with oval, triangular, hexagonal, and cylindrical morphologies. As a result, the saponins (Figure 6.4) acted as capping/stabilizing agents, preventing the silver from growing in an anisotropic manner.

Berezin et al. (2013) demonstrated using plant saponin NPs as a delivery system for a flu vaccine. Gg6 and Ah6 saponins were obtained from *G. glabra* and *A. hippocastanum*, for NP synthesis. A 95% ethanol extraction was used to extract crude saponins from roots and seeds, further purified by HPLC fractionation. The dialysis method was used to make NPs by combining purified saponins, separated NA and HA antigens, and lipids/fatty acid in a non-ionic cleansing agent. Electron microscopy was used to determine the morphology of the synthesized NPs. NPs appeared as exact virus-like structures with a size of 60 nm, which is about half the size of the flu virus.

Sharma et al. (2018) developed saponin-incorporated AgNPs from the seeds of *Madhuca longifolia,* which exposed hopeful anti-inflammatory bioefficacy. Seed powder was mixed with aqueous ethanol and placed in a microwave oven, followed by ultrasonication. Saponin was extracted from the resulting extract by treating it with diethyl ether. Chromatographical studies revealed the presence of four saponins, namely Mi-saponin A, Mi-saponin B, madhucoside A and, madhucoside B in the seed extract. AgNPs were produced by utilizing the secluded saponin (Figure 6.5). A colour change from pale yellow to dark brown confirmed the formation of NPs. Saponin is composed of a triterpenoid or steroid aglycone (sapogenin) that is glycosidically linked to sugar moieties. A glycosidic linkage's oxygen atom with unpaired electrons effectively reduced Ag^+ to Ag^0 ions, establishing saponin-fabricated AgNPs.

Furthermore, –OH groups present in the sugar portion and sapogenin can aid in the reduction of Ag^+ ions. This medicated coating on synthesized AgNPs improves medical efficiency by protecting them from aggregation and holding them in a nano state. As a result of its high synergetic reduction potential, saponin acts as a reducing and capping agent, resulting in the formation of stable AgNPs. FTIR studies showed the occurrence of saponin on the fabricated AgNPs' surface, indicated by the peaks representing the saponin structure (–C--H–, –C--O–C–, –OH, and –C=C–). As a result, these peaks in the FTIR spectra suggest saponin's dual function as a reducing and stabilizing agent. SEM analysis revealed that the formed NPs had a virtually spherical morphology, ranging from 20 to 48 nm.

One-step sustainable synthesis of *Memecylon umbellatum* L. extract-regulated AgNPs and AuNPs were reported by Arunachalam et al. in 2013. Phytochemical studies displayed that *M. umbellatum* leaf extract was very rich in saponin and they act as the best starting material for the production of these NPs. The eco-friendly synthesized NPs were described by FTIR, UV-vis, TEM, and EDAX studies. UV-vis spectra displayed an absorption band at 440 nm for AgNPs and 540 nm for AuNPs. Plant extracts comprising saponins and phenolic compounds attached to NPs *via* free amine groups in proteins. Saponins, quinones, and phenolic compounds from *M. umbellatum* effectively stabilized the AgNPs by silver ion bioreduction. TEM analysis showed that AgNPs were spherical shapes with sizes ranging from 15 to 20 nm in diameter. The AuNPs were triangular,

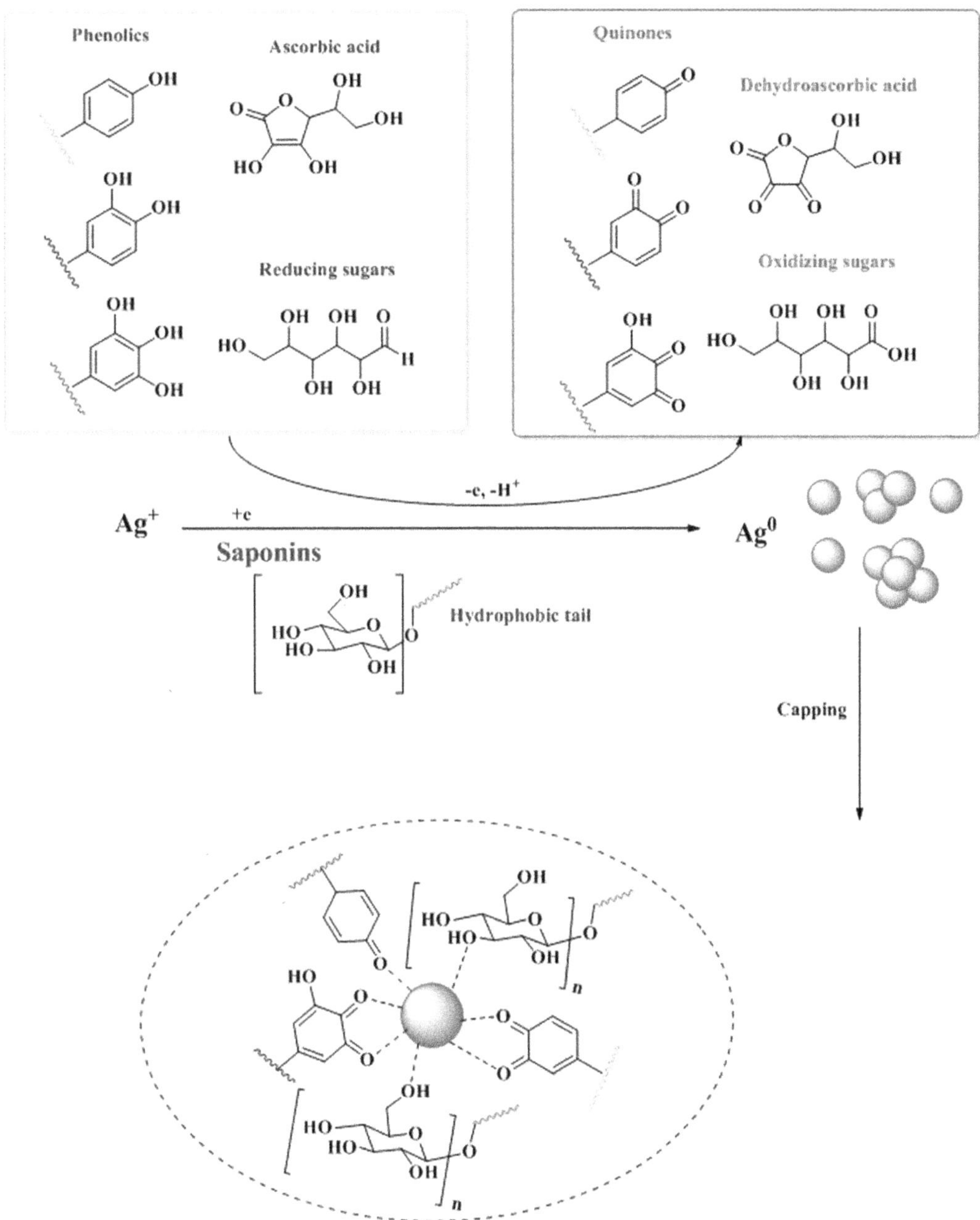

FIGURE 6.4 Role of saponin in nanoparticle formation.

hexagonal, and spherical, and 15–20 nm in size. The EDAX spectrum demonstrated that both the manufactured NPs are crystalline in nature. The FTIR spectrum exhibited the shift in the peak of AgNPs and AuNPs compared to the plant extract. Proteins in the saponins, phenolic substance, and quinones from *M. umbellatum* may be attached to AgNPs and AuNPs through free amine groups or cysteine residues, thereby stabilizing the NPs produced by the surface-bound proteins (Arunachalam et al. 2013).

Chandrasekhar and Vinay illustrated the formation of AgNPs using yellow-coloured blooms of *Turnera ulmifolia* and *Argemone mexicana*, as well as the analysis of their antibacterial and

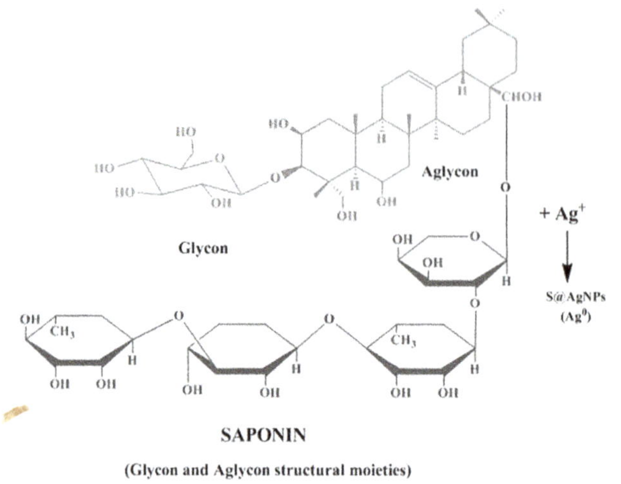

FIGURE 6.5 Saponin present in *Madhuca longifolia* seeds.

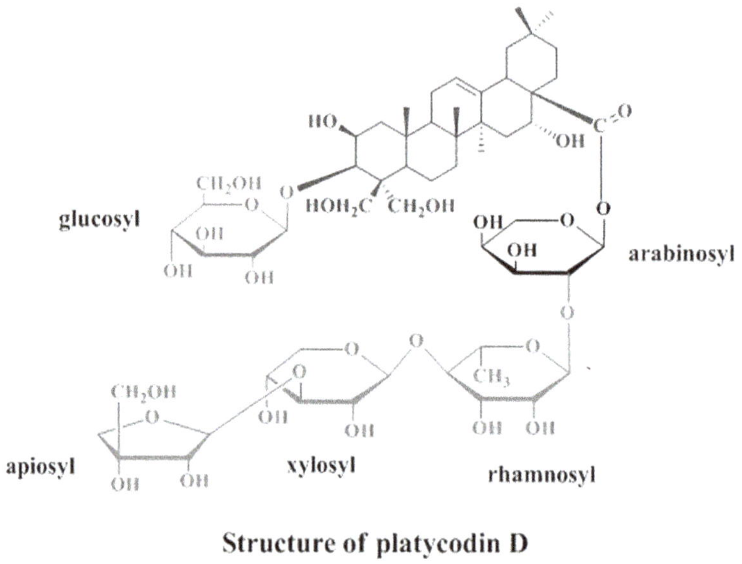

FIGURE 6.6 Structure of platycodin D.

antioxidant function (Chandrasekhar et al. 2017). According to the results of the phytochemical inspection, *T. ulmifolia* bloom extract was high in saponins and flavonoids, while *A. mexicana* was high in alkaloids, saponins, flavonoids, and phenolic constituents. Owing to the formation of AgNPs, the mixture's golden yellow colour faded after 24 h and turned dark brown. The synthesized NPs were examined using UV-vis, FTIR, X-ray diffraction (XRD), SEM, and EDAX. UV-vis displayed the highest absorbance peak for *A. mexicana* at 398 nm and *T. ulmifolia* at 423 nm, indicating the formation of AgNPs. FTIR spectra of AgNPs showed that phytochemicals such as saponins, alkaloids, flavonoids, etc., were actively involved in AgNP biosynthesis. The particles are crystalline, as per XRD analysis. According to SEM analysis, the particle size is spherical, with a size of 29.34 nm for *A. mexicana* AgNPs. *T. mexicana* AgNPs were found to be 32.42 nm in size.

Choi et al. (2018) developed AuNPs and AgNPs in a green manner using saponins from *Platycodon grandiflorum*. Enzymatic conversion of a *Platycodon grandiflorum* root extract resulted in the enrichment of main triterpenoidal saponin Platycodin D (PD) (Figure 6.6). The PD-AuNPs had a wine-purple tint and an SPR band of 536 nm. At 427 nm, the SPR band of PD-AgNPs, which had a yellow tint, was detected by UV spectra. HR-TEM revealed spherical NPs with a mean size of 14.94 ± 2.14 nm for PD-AuNPs and 18.40 ± 3.20 nm for PD-AgNPs. There was no aggregation of AuNPs, implying that the saponin also served as a capping/stabilizing agent. Curvature-dependent progression was used to improve atomic force microscopy (AFM) scans and the calculation of NP sizes. The enhanced AFM results determined that the sizes were 29.93 nm for PD-AgNPs and 19.14 nm for PD-AuNPs. Strong diffraction patterns from XRD studies established structures for both NPs. FTIR peaks at C–O, and C–H vibrations appeared at 1035 cm^{-1} confirming the existence of triterpene aglycone and sugars to form glycosides. The FTIR outcomes show the aromatic C–H, C–O, C=C groups, and AgNPs and AuNPs.

Ganguly and his co-workers (2018) showed that a natural saponin stabilized nanocatalyst (AgNPs-rGO) can be used as a dye-degradation catalyst (Ganguly et al. 2018). Using *Bacopa monniera* (Brahmi) leaf extract, AgNP-adorned reduced graphene oxide (r-GO) is fabricated using the solvothermal process. Brahmi leaf extract acts as a suitable reducing agent and colloid stabilizer since it consists of two water-soluble saponins bacoside A1 and bacoside A3. UV spectra showed a high band at 420 nm that resembles the AgNPs' formation. According to TEM images, the particles were grown successfully as spherical particles in the range of 4–30 nm on the r-GO surface. Ag (0) NPs (nanodots) with dimensions of less than 8 nm were also visible in several areas. The same thing happened in HR-TEM with lattice fringe. This finding supports AgNPs' spherical crystalline form supporting XRD results.

Karthik et al. (2016) investigated the ability of *Quillaja saponaria* (Figure 6.7) to produce solid lipid NPs in the presence of stearic acid (SA) as a lipid carrier and imatinib mesylate (IM) as a model compound. *Quillaja saponaria* was used as a probable emulsifier to prepare solid lipid NPs. Hot homogenization and ultrasonication were used to synthesize IMSLNs. IMSLNs were found to be neutrally charged, quasi-spherical in shape, and measuring 143.5–641.9 nm.

Rao and Paria in 2015 used *Aegle marmelos* L. and plant saponins (reetha, acacia) (Figure 6.8) to make AgNPs and AuNPs. The Taguchi methodology improved pH, temperature, reactant ratio, reaction medium, colour, and light wavelength. Table 6.2 presents the optimal conditions for AgNP and AuNP formulation.

The synthesized NPs were defined using UV-vis, AFM, FTIR, TEM, EDAX, zeta potential, and XRD. By refining the parameters, the Taguchi method employed even smaller spherical AgNPs of ~13 nm and non-spherical AuNPs of ~23 nm in size. UV-vis spectra revealed maximum bands at

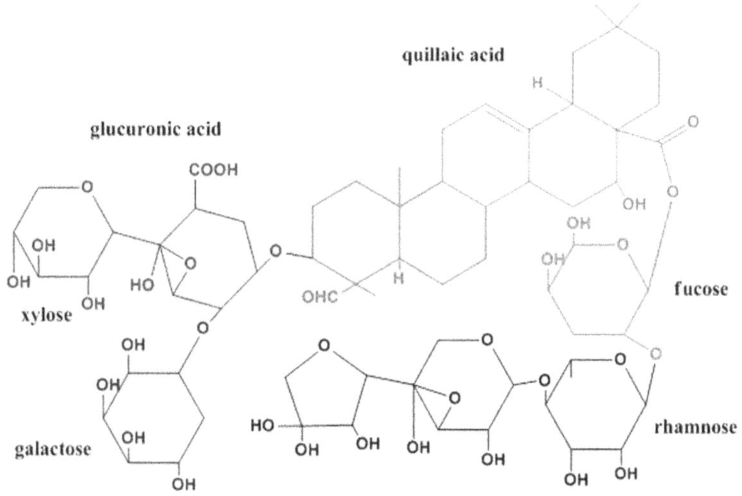

FIGURE 6.7 Structure of *Quillaja* saponin.

$$RCHO\,(aq) + OH^- \longrightarrow RCOO^- + H_2O + e^- \Longrightarrow 1$$

$$2\,AgNO_3 + 2e^- \longrightarrow 2\,Ag + O_2 + 2NO_2 \Longrightarrow 2$$

$$HAuCl_4 + 3e^- \longrightarrow 2\,Au + 4Cl^- + H^+ \Longrightarrow 3$$

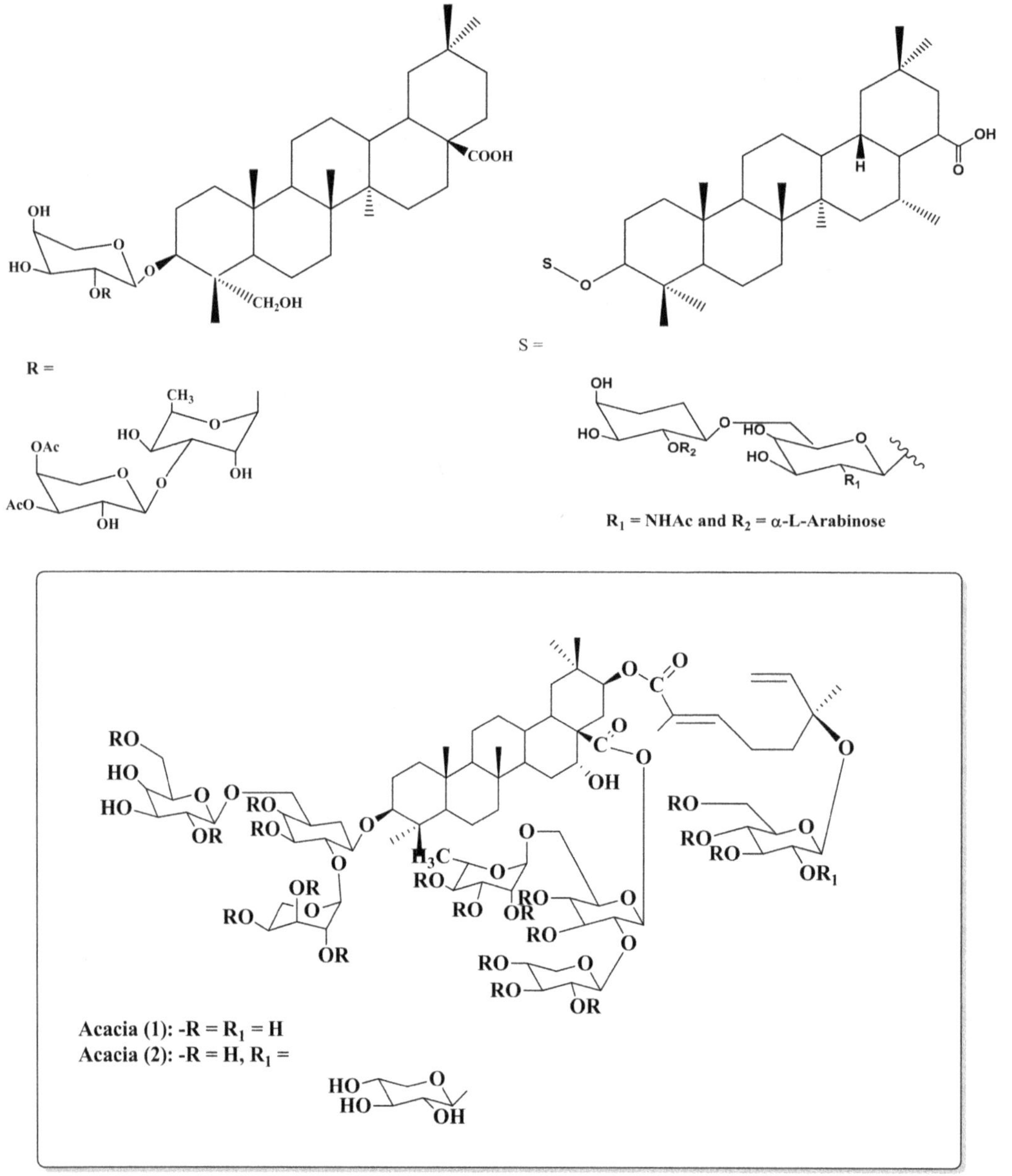

FIGURE 6.8 Structure of three plant saponin reetha, acacia.

TABLE 6.2
Optimal Conditions for Silver and Gold Nanoparticle Formulation

Nanoparticle	Reactant ratio	pH	Medium	Temperature	Light
AgNPs	1 mM/0.3%	7.0	Acacia	55°C	white
AuNPs	1 mM/1.0%	8.5	Reetha	40°C	dark

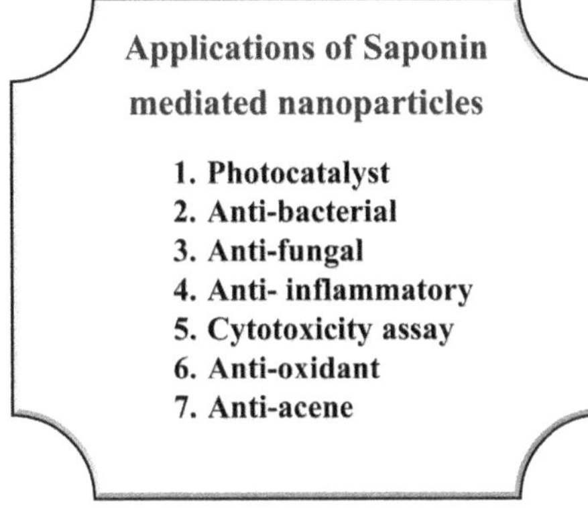

FIGURE 6.9 Application of saponin nanoparticles.

476 and 543 nm for AgNPs and AuNPs, respectively. The AgNPs are spherical with an average size of ~13.83 ± 4.89 nm, and the AuNPs are cuboids spherical, smooth-edged cubes, and pentagons, with an average size of ~23 nm, according to TEM data. The XRD results indicated that the NPs are naturally crystalline, with silver and gold face-centred cubic (fcc) structures. Because of the capping of saponins on the particle's surface, zeta potential studies exhibited negative zeta potential values for both synthesized NPs. FTIR results confirmed that the existence of the glycone moiety (aldehyde group) from the natural saponins is crucial in the reduction of NPs. The mechanisms are described below.

The steroidal saponin from the *Chenopodium album* leaf was extracted by using ethyl acetate as solvent (Srivastava et al. 2020). The saponin obtained was characterized using FTIR, thin layer chromatography (TLC), and HPLC. Srivastava et al. synthesized the AgNPs from the extracted saponin using the green method. The resulting AgNPs were studied by FTIR spectroscopy, UV spectrophotometer, zeta sizer, SEM, and zeta potential. According to UV-vis analysis, the overall absorbance of the AgNPs was 421 nm. Saponins were found to be well capped by saponins in FTIR tests. According to an SEM investigation, the produced AgNPs had a size of 30–40 nm.

Raji et al. (2019), evaluated saponin-mediated AgNPs fabricated from *Wrightia tinctoria*, *Cassia alata*, *Thespesia populnea*, and *Euphorbia hirta* leaf extract. The synthesized NPs were examined using UV-vis, SEM, AFM, and XRD. The size of the NPs was found to be 17–30 nm. According to XRD results, the NPs were found to be crystalline in nature with a fcc structure (Figure 6.9 and Table 6.3).

6.4 APPLICATIONS

6.4.1 Dye Degradation

The degradation of three pollutants, Congo red, methyl orange, 2-chloro-4-nitrophenol, and methylene blue, was tested using AgNPs from *Simarouba glauca* deoiled cake as a photocatalyst

TABLE 6.3
Saponin-Mediated Green Synthesis of Metal Nanoparticles

	Name of the plant	Plant part	Saponin	Nanoparticles	Size	Reference
1	*Simarouba glauca*	Oilseed meal	Triterpenoid aglycone	AgNPs	4.61 nm	Paramesh et al. (2021)
2	*Chenopodium quinoa*	Quinoa husk	Oleanane triterpenoid saponin	AgNPs	19 nm	Segura et al. (2020)
3.	*Trianthema decandra*		Not mentioned	AgNPs, AuNPs	AgNPs: 17.9–59.6 nm ,AuNPs: 37.7–79.9 nm	Geethalakshmi et al. (2013)
4.	*Ocimum tenuiflorum*	Leaf	Not mentioned	AgNPs	5–61 nm	Nguyen et al. (2020)
5	*Glycyrrhiza glabra* and *Aesculus hippocastanum*	Roots and seeds	Gg6 and Ah6 saponins	NPs incorporated with flu virus antigens, lipids, and saponins Gg6 and Ah6	60 nm	Berezin et al. (2013)
6.	*Madhuca longifolia*	Seeds	Mi-saponin A, Mi-saponin B, Madhucoside A, and Madhucoside B	AgNPs	48 nm	Sharma et al. (2018)
7.	*Memecylon umbellatum*	Leaf	Not mentioned	AgNPs,AuNPs	15–20 nm	Arunachalam et al. (2013)
8.	*Argemone mexicana* and *Turnera ulmifolia*	Blooms	Not mentioned	AgNPs	32–42 nm	Chandrasekhar et al. (2017)
9.	*Platycodon grandiflorum*	Root	Triterpenoidal platycodon saponin	AgNPs, AuNPs	AuNPs : 14.94 ± 2.14 nm AgNPs : 18.40 ± 3.20 nm	Choi et al. (2018)
10.	*Bacopa monniera*	Leaf	Bacoside A1 and bacoside A3	AgNPs	4–30 nm	Ganguly et al. (2018)
11.	*Quillaja*	Bark	*Quillaja* saponin	Solid lipid NPs	143.5–641.9 nm	Karthik et al. (2016)
12.	*Aegle marmelos*	Leaf	Reetha, acacia obtained from saponin	AgNPs, AuNPs	AgNPs: ~13 nm, AuNPs: ~23 nm	Rao et al. (2015)
13.	*Chenopodium album*	Leaf	Steroidal saponin	AgNPs	30–40 nm	Srivastava et al. (2020)
14.	*Wrightia tinctoria*, *Cassia alata*, *Thespesia populnea*, and *Euphorbia hirta*	Leaf	Not mentioned	AgNPs	17–30 nm	Raji et al. (2019)

(Paramesh et al. 2021). The test was carried out with $NaBH_4$ in water at 27°C. At a 1 mg/ml concentration, the catalyst is efficient in dye reduction within the time scale of 27 to 80 min. Also, the catalyst is quickly recovered and recyclable for up to five cycles.

The AuNPs manufactured from *Platycodon* saponin were tested for their catalytic performance by reducing 4-nitrophenol in the attendance of $NaBH_4$ (Choi et al. 2018). At 0.04 mM of AuNPs, the 4-nitrophenol gets reduced into 4-aminophenol within 720 s in surplus $NaBH_4$. Plant secondary substances have desirable properties, which, when combined with the intrinsic behaviours of NPs, also have synergistic abilities, which are responsible for the reduction process.

AgNPs synthesized from natural saponin were encapsulated r-GO and examined its reduction efficiency against the organic dye methylene blue and Congo red (Ganguly et al. 2018). A measure of 5 mg of r-GO-Ag composite was mixed with two dyes separately in the presence of $NaBH_4$ and examined in UV-vis spectrometer. The typical peaks detected for Congo red and methylene blue were 493 nm and 663 nm. These peaks had entirely vanished within 45 s and 5 min. The total elimination of the appropriate peaks indicates that the equipped catalyst has fully reduced/degraded the dyes. Metallic NPs with different scales might interfere with many organic dye molecules. The dye molecules were previously physisorbed over the catalyst's surface due to the large surface area of AgNPs. Later, the dye molecules gross electrons and reduce to colourless molecules. As a result, the electron relay mechanism suggests that AgNPs play the most important role in catalytic action.

6.4.2 Pharmacological Activities

6.4.2.1 Antibacterial Activity

The disc diffusion method was used to test the antimicrobial performance of AgNPs and AuNPs synthesized from *Trianthema decandra* L. extract (Geethalakshmi and Sarada 2013). The test was carried out on eight different bacteria. Out of eight, AuNPs showed an inhibition zone of about 11.5 ± 0.5 mm for *Yersinia enterocolitica*. This was attributed to the small size of AuNPs, which allowed them to easily enter the bacterial cells and become caught by the AuNPs' biofilm. Consequently, the constituent emitted by cells causes the cell wall to distort, resulting in morphological changes and cell demise. AgNPs displayed an inhibition zone of about 20.3 ± 0.5 mm for *Staphylococcus aureus*. The activity of AgNPs was due to their larger surface area and ability to attach to bacterial cells and disrupt their respiratory mechanisms. The antimicrobial performance of *M. umbellatum* leaf extract was carried out by the disc diffusion method (Arunachalam et al. 2013). The inhibition zone for *Streptococcus pyogenes* appeared to be 14.5 mm, 10.3 mm for *Proteus mirabilis*, 10 mm for *Pseudomonas aeruginosa*, 9.7 mm for *Proteus vulgaris*, 9.4 mm for *Citrobacter freundii*, 7.4 mm for *Escherichia coli*, and 6.9 mm for *Enterococcus faecalis*. The disc diffusion technique was used to estimate the antimicrobial assets of AgNPs attained from the blooms extract of *T. ulmifolia* and *A. mexicana* against the microbes, namely *P. aeruginosa*, *S. aureus*, *K. aerogenes*, and *E. coli* (Chandrasekhar and Vinay 2017). The AgNPs synthesized from *T. ulmifolia* showed high antibacterial properties (inhibition rate 21 to 27 mm) compared to *A. Mexicana* (inhibition rate 20 to 24 mm).

The antibacterial assessment of the AgNP-adorned r-GO nanocomposite was executed against the bacteria *E. coli* (Ganguly et al. 2018). The inhibition zone was found to be 1.8 cm. The colloidal distribution of r-GO-AgNPs aids in effective bacterial cell wall interaction. The capturing incident was observed once the AgNPs within r-GO sheets interfered with the bacterial membrane, accompanied by the origination of Ag^+ ions, resulting in the production of reactive oxygen species (ROS). The ability of ROS to kill the bacterial lung tissue, which is lethal to bacteria, is incredible. The r-GO layers can serve as a strong reinforcement for AgNPs and have the ability to stabilize nanometals. Antibacterial properties against two bacterial cultures comprising *P. aeruginosa* and *B. subtilis* were studied using the prepared AgNPs (Raji et al. 2019). The activity was carried out by the agar well diffusion approach. The AgNPs showed good activity against both bacteria.

6.4.3 Antifungal Activities

AgNPs and AuNPs synthesized from *Trianthema decandra* L. extract were utilized to study the antifungal activity against the fungus *Candida albicans* (Geethalakshmi and Sarada 2013). Saponin-mediated AgNPs were prepared from the *Ocimum tenuiflorum* (OT) leaf extract and tested its antifungal activities against the fungus *A. niger*, *A. flavus*, and *F. oxysporum* by the agar method (Nguyen et al. 2020). The inhibition zone was 10 ± 0.3 mm for AuNPs and 10 ± 0.5 mm for AgNPs. The leaf extract alone does not show any activity. But in the case of OT-AgNPs, the inhibition rate was increased with an increase in dosage rate. Compared to *A. flavus* and *F. oxysporum*, *A. niger* exhibited good antifungal action by decreasing the growth rate. When the NP dosage was increased from 10 to 50 ppm, it decreased the spreading proficiency of *A. niger* to 1.5-fold. This was because that OT-AgNPs with a smaller size of about 100 nm, allow them effortlessly to permit into the fungal layers and hinder fungal DNAase.

6.4.4 Anti-Inflammatory Assay

Saponin-mediated AgNPs were utilized to estimate *in vivo* acute toxicity in Swiss albino mice (male) by oral direction (Sharma et al. 2018). The designed AgNPs at 1.5 mg/kg/bw concentration displayed inhibition of about 70.99%. This was due to the microbes and tissue damage persuaded by biphasic activity. Due to their smaller scale, high surface area, use of the bioactive principle, biocompatibility, remarkable optical assets related to SPR, and sensible stability, they easily pierce the tissues. Silver's anti-inflammatory capabilities contribute to the improvement in inflammatory qualities.

6.4.5 Antioxidant Activity

The *in vitro* antioxidant approach by 2,2-diphenyl-1-picrylhydrazyl assay (DPPH) has shown that green mediated AgNPs attained from *A. mexicana* ensured better action with 87.06% on 500 μg ml^{-1} than the AgNPs attained from *T. ulmifolia* with 84.62% on 500 μg ml^{-1} (Chandrasekhar and Vinay 2017). The solid lipid NPs produced from the *Quillaja* saponin were used to investigate the cytotoxicity study (Karthik et al. 2016). Imatinib mesylate (IM)-SLN with IM concentrations 1.0, 2.5, 5.0, 7.5, and 10 μM showed majority cytotoxic toward MCF7 cells. This is due to lipid NPs' high cell absorption and synergism response around *Quillaja* saponin and the IM present in the IMSLN. The higher cytotoxic prospects for MCF7 cells were due to *Quillaja* saponin's presence.

6.4.6 Antiacne Activity

The AgNPs were obtained from *Chenopodium album* L. extract were employed to describe the antiacne action opposing the acne-triggering bacteria (Rao et al. 2015). The disc diffusion technique achieved this activity. The results revealed that the saponin-rich plant extricate had a more significant inhibitory effect at 14 mm than the NP produced, which had a suppression zone of about 11 mm. This activity was due to the existence of saponin in the plant extract attributable to its competence to instigate microbial cell leakage of enzymes and proteins.

6.4.7 Vaccine

HA and NA antigens of the flu virus were encapsulated into the fabricated NPs (Berezin et al. 2013). This appreciably enhanced IgA, IgM, and IgG antibodies and the development of IL-2 and IFN gamma after vaccination. In contrast to immunization with an entire virus-denatured vaccine or vaccine blended with alum hydroxide adjunct, mucosa immunization with NPs comprising HA+NA antigens and Ah6 or Gg6 saponins provided chicks with a substantially superior shield to exposure to the virulent H7N1 avian flu virus.

6.5 CONCLUSION

This chapter assessed NPs made using saponin-rich medicinal herbs, which had good qualities and might be used as photocatalysts and other pharmaceutical purposes. The foam test can determine whether or not the plants contain saponin. Because of its amphipathic nature, saponin found in plants plays a vital role in forming metal NPs. Its function as a biosurfactant aids in tuning NP size and shape, which has a significant practical benefit. As a result, more study on saponin-mediated NPs is needed, as they will be highly valuable in the future.

REFERENCES

Anandalakshmi, K., J. Venugobal & V. Ramasamy. 2016. Characterization of silver nanoparticles by green synthesis method using *Pedalium murex* leaf extract and their antibacterial activity. *Applied Nanoscience* 63(3): 399–408.

Aritonang, H. F., H. Koleangan & A. D. Wuntu. 2019. Synthesis of silver nanoparticles using aqueous extract of medicinal plants *Impatiens balsamina* and *Lantana camara* fresh leaves and analysis of antimicrobial activity. *International Journal of Microbiology* 2019: 1–8.

Arunachalam, K. D., S. K. Annamalai & S. Hari. 2013. One-step green synthesis and characterization of leaf extract-mediated biocompatible silver and gold nanoparticles from *Memecylon umbellatum. International Journal of Nanomedicine* 8: 1307.

Bachheti, A., A. Sharma, R. K. Bachheti, A. Husen & V. K. Mishra. 2019a. Plant-mediated synthesis of copper oxide nanoparticles and their biological applications. In *Nanomaterials and Plant Potential.* Cham: Springer, pp. 221–237. https://doi.org/10.1007/978-3-030-05569-1_8.

Bachheti, A., R. K. Bachheti, L. Abate & A. Husen. 2021a. Current status of Aloe-based nanoparticle fabrication, characterization and their application in some cutting-edge areas. *South African Journal of Botany.* https://doi.org/10.1016/j.sajb.2021.08.021.

Bachheti, R. K., A. Fikadu, A. Bachheti & A. Husen. 2020. Biogenic fabrication of nanomaterials from flower-based chemical compounds, characterization and their various applications: A review. *Saudi Journal of Biological Sciences* 27(10): 2551–2562.

Bachheti, R. K., A. Sharma, A. Bachheti, A.Husen, G. M. Shanka & D. P. Pandey. 2020b. Nanomaterials from various forest tree species and their biomedical applications. In Husen A., Jawaid M. (eds) *Nanomaterials for Agriculture and Forestry Applications.* Elsevier, pp. 81–106. https://doi.org/10.1016/B978-0-12-817852-2.00004-4.

Bachheti, R. K., L. Abate, A. Bachheti, A. Madhusudhan & A. Husen. 2021. Algae-, fungi-, and yeast-mediated biological synthesis of nanoparticles and their various biomedical applications. In *Handbook of Greener Synthesis of Nanomaterials and Compounds.* Elsevier, pp. 701–734. https://doi.org/10.1016/B978-0-12-821938-6.00022-0.

Bachheti, R. K., R. Konwarh, V. Gupta, A. Husen & A. Joshi. 2019. Green synthesis of iron oxide nanoparticles: Cutting edge technology and multifaceted applications. In *Nanomaterials and Plant Potential.* Cham: Springer, pp. 239–259. https://doi.org/10.1007/978-3-030-05569-1_9.

Bachheti, R. K., Y. Godebo, A.Bachheti, M. O. Yassin & A. Husen. 2020a. Root-based fabrication of metal/metal-oxide nanomaterials and their various applications. In Husen A., Jawaid M. (eds) *Nanomaterials for Agriculture and Forestry Applications.* Elsevier, pp. 135–166. https://doi.org/10.1016/B978-0-12-817852-2.00006-8.

Balakrishnan, S., S. Varughese & A. P. Deshpande. 2006. Micellar characterisation of saponin from *Sapindus mukorossi. Tenside, Surfactants, Detergents* 435(5): 262–268.

Berezin, V. E., A. P. Bogoyavlenskiy, A. S. Turmagambetova, P. G. Alexuk, I. A. Zaitceva, E. S. Omirtaeva & N. S. Sokolova. 2013. Nanoparticles from plant saponins as delivery system for mucosal influenza vaccine. *American Journal of Infectious Diseases* 11: 1–4.

Cao, F., W. Ouyang & Y. Wang. 2010. Preparation of O/W ginseng saponins-based nanoemulsion and its amplified immune response. *Zhongguo Zhong Yao Za Zhi= Zhongguo ZhongYao Zazhi= China Journal of Chinese Materia Medica* 354: 439–443.

Chaieb, I. 2010. Saponins as insecticides: A review. *Tunisian Journal of Plant Protection* 51: 39–50.

Chandrasekhar, N. & S. P. Vinay. 2017. Yellow-colored blooms of *Argemone mexicana* and *Turnera ulmifolia* mediated synthesis of silver nanoparticles and study of their antibacterial and antioxidant activity. *Applied Nanoscience* 78(8): 851–861.

Chen, D., B. Yang, Y. Jiang & Y. Z. Zhang. 2018. Synthesis of Mn_3O_4 nanoparticles for catalytic application via ultrasound-assisted ball milling. *Chemistry Select* 314: 3904–3908.

Choi, Y., S. Kang, S. H. Cha, H. S. Kim, K. Song, Y. J. Lee & Y. Park. 2018. Platycodon saponins from *Platycodi Radix, Platycodon grandiflorum* for the green synthesis of gold and silver nanoparticles. *Nanoscale Research Letters* 131: 1–10.

de Ven, H. V., L. Van Dyck, W. Weyenberg, L. Maes & A. Ludwig. 2010. Nanosuspensions of chemically modified saponins: Reduction of hemolytic side effects and potential tool in drug targeting strategy. *Journal of Controlled Release: Official Journal of the Controlled Release Society* 1481(1): e122–123.

El Aziz, M. M. A., A. S. Ashour & A. S. G. Melad. 2019. A review on saponins from medicinal plants: Chemistry, isolation, and determination. *Journal of Nanomedicine Research* 81: 282–288.

Elango, G., S. M. Kumaran, S. S. Kumar, S. Muthuraja & S. M. Roopan. 2015. Green synthesis of SnO_2 nanoparticles and its photocatalytic activity of phenolsulfonphthalein dye. *Spectrochimica Acta Part A: Molecular and Biomolecular Spectroscopy* 145: 176–180.

Ganguly, S., S. Mondal, P. Das, P. Bhawal, T. kanti Das, M. Bose & N. C. Das. 2018. Natural saponin stabilized nano-catalyst as efficient dye-degradation catalyst. *Nano-Structures and Nano-Objects* 16: 86–95.

Geethalakshmi, R. & D. V. L. Sarada. 2013. Characterization and antimicrobial activity of gold and silver nanoparticles synthesized using saponin isolated from *Trianthema decandra* L. *Industrial Crops and Products* 51: 107–115.

Geng, P., P. Chen, L. Z. Lin, J. Sun, P. Harrington & J. M. Harnly. 2021. Classification of structural characteristics facilitate identifying steroidal saponins in Alliums using ultra-high-performance liquid chromatography high-resolution mass spectrometry. *Journal of Food Composition and Analysis* 102: 1–13.

Hasnidawani, J. N., H. N. Azlina, H. Norita, N. N. Bonnia, S. Ratim & E. S. Ali. 2016. Synthesis of ZnO nanostructures using sol-gel method. *Procedia Chemistry* 19: 211–216.

Hostettmann, K. & A. Marston. 2005. *Saponins*. Cambridge UK: Cambridge University Press.

Husen, A., Q. I. Rahman, M. Iqbal, M. O. Yassin & R. K. Bachheti. 2019. Plant-mediated fabrication of gold nanoparticles and their applications. In *Nanomaterials and Plant Potential*. Cham: Springer, pp. 71–110. https://doi.org/10.1007/978-3-030-05569-1_3.

Jaswal, T. & J. Gupta. 2021. A review on the toxicity of silver nanoparticles on human health. *Materials Today: Proceedings*. https://doi.org/10.1016/j.matpr.2021.04.266.

Jayaseelan, C., A. A. Rahuman, S. M. Roopan, A. V. Kirthi, J. Venkatesan, S. K. Kim & C. Siva. 2013. Biological approach to synthesize TiO_2 nanoparticles using *Aeromonas hydrophila* and its antibacterial activity. *Spectrochimica Acta Part A: Molecular and Biomolecular Spectroscopy* 107: 82–89.

Kalaiselvi, A., S. M. Roopan, G. Madhumitha, C. Ramalingam & G. Elango. 2015. Synthesis and characterization of palladium nanoparticles using *Catharanthus roseus* leaf extract and its application in the photocatalytic degradation. *Spectrochimica Acta Part A: Molecular and Biomolecular Spectroscopy* 135: 116–119.

Karthik, S., C. V. Raghavan, G. Marslin, H. Rahman, D. Selvaraj, K. Balakumar & G. Franklin. 2016. Quillaja saponin: A prospective emulsifier for the preparation of solid lipid nanoparticles. *Colloids and Surfaces B: Biointerfaces* 147: 274–280.

Khan, Z., S. A. Al-Thabaiti, A. Y. Obaid & A. Al-Youbi. 2011. Preparation and characterization of silver nanoparticles by chemical reduction method. *Colloids and Surfaces B: Biointerfaces* 822(2): 513–517.

Kumar, D. A., V. Palanichamy & S. M. Roopan. 2014. Green synthesis of silver nanoparticles using *Alternanthera dentata* leaf extract at room temperature and their antimicrobial activity. *Spectrochimica Acta Part A: Molecular and Biomolecular Spectroscopy* 127: 168–171.

Liao, Y., Z. Li, Z. Qing, M. Sheng, Q. Qu, Y. Shi & X. Shi. 2021. Saponin surfactants used in drug delivery systems: A new application for natural medicine components. *International Journal of Pharmaceutics* 603, 1–14.

Moodley, J. S., S. B. N. Krishna, K. Pillay & P. Govender. 2018. Green synthesis of silver nanoparticles from *Moringa oleifera* leaf extracts and its antimicrobial potential. *Advances in Natural Sciences: Nanoscience and Nanotechnology* 91: 015011.

Nguyen, D. H., T. N. N. Vo, T. N. T. T. Le, D. P. N. Thi & T. T. H. Thi. 2020. Evaluation of saponin-rich/poor leaf extract-mediated silver nanoparticles and their antifungal capacity. *Green Processing and Synthesis* 91(1): 429–439.

Oleszek, W. A. 2002. Chromatographic determination of plant saponins. *Journal of Chromatography A* 9671(1): 147–162.

Paramesh, C. C., G. Halligudra, V. Gangaraju, J. B. Sriramoju, M. Shastri, D. Rangappa & P. D. Shivaramu. 2021. Silver nanoparticles synthesized using saponin extract of *Simarouba glauca* oil seed meal as effective, recoverable and reusable catalyst for reduction of organic dyes. *Results in Surfaces and Interfaces* 3: 100005.

Phiwdang, K., S. Suphankij, W. Mekprasart & W. Pecharapa. 2013. Synthesis of CuO nanoparticles by precipitation method using different precursors. *Energy Procedia* 34: 740–745.

Piriyawong, V., V. Thongpool, P. Asanithi & P. Limsuwan. 2012. Preparation and characterization of alumina nanoparticles in deionized water using laser ablation technique. *Journal of Nanomaterials* 2: 1–6.

Raji, P., A. V. Samrot, D. Keerthana & S. Karishma. 2019. Antibacterial activity of alkaloids, flavonoids, saponins, and tannins mediated green synthesised silver nanoparticles against *Pseudomonas aeruginosa* and Bacillus subtilis. *Journal of Cluster Science* 304(4): 881–895.

Rao, K. J. & S. Paria. 2015. Aegle marmelos leaf extract and plant surfactants mediated green synthesis of Au and Ag nanoparticles by optimizing process parameters using Taguchi method. *ACS Sustainable Chemistry and Engineering* 33(3): 483–491.

Sahu, T., Y. K. Ratre, S. Chauhan, L. V. K. S. Bhaskar, M. P. Nair & H. K. Verma. 2021. Nanotechnology based drug delivery system: Current strategies and emerging therapeutic potential for medical science. *Journal of Drug Delivery Science and Technology* 63: 102487.

Segura, R., G. Vásquez, E. Colson, P. Gerbaux, C. Frischmon, A. Nesic & G. Cabrera-Barjas. 2020. Phytostimulant properties of highly stable silver nanoparticles obtained with saponin extract from Chenopodium quinoa. *Journal of the Science of Food and Agriculture* 10013: 4987–4994.

Sharma, M., S. Yadav, M. Srivastava, N. Ganesh & S. Srivastava. 2018. Promising anti-inflammatory bio-efficacy of saponin loaded silver nanoparticles prepared from the plant *Madhuca longifolia. Asian Journal of Nanosciences and Materials* 14: 244–261.

Shreema, K., R. Mathammal, V. Kalaiselvi, S. Vijayakumar, K. Selvakumar & K. Senthil. 2021. Green synthesis of silver doped zinc oxide nanoparticles using fresh leaf extract *Morinda citrifolia* and its antioxidant potential. *Materials Today: Proceedings* 47: 2126–2131.

Srivastava, N., M. Choudhary, G. Singhal & S. S. Bhagyawant. 2020. SEM studies of saponin silver nanoparticles isolated from leaves of *Chenopodium album* L. for in vitro anti-acne activity. *Proceedings of the National Academy of Sciences, India Section B: (Biological Sciences)* 902(2): 333–341.

Stepanov, A. L., M. F. Galyautdinov, A. B. Evlyukhin, V. I. Nuzhdin, V. F. Valeev, Y. N. Osin & B. N. Chichkov. 2013. Synthesis of periodic plasmonic microstructures with copper nanoparticles in silica glass by low-energy ion implantation. *Applied Physics A* 111(1): 261–264.

Suzuki, S., Y. Tomita, S. Kuwabata & T. Torimoto. 2015. Synthesis of alloy Au-Cu nanoparticles with the L1 0 structure in an ionic liquid using sputter deposition. *Dalton Transactions* 449: 4186–4194.

Tschesche, R., M. Tauscher, H. W. Fehlhaber & G. Wulff. 1969. Steroidsaponine mit mehr als einer Zuckerkette, IV. Avenacosid A, ein bisdesmosidisches Steroidsaponin aus *Avena sativa. Chemische Berichte* 1026(6): 2072–2082.

Vincken, J. P., L. Heng, A. de Groot & H. Gruppen. 2007. Saponins, classification and occurrence in the plant kingdom. *Phytochemistry* 683(3): 275–297.

Zhang, W., X. Qiao & J. Chen. 2006. Synthesis and characterization of silver nanoparticles in AOT microemulsion system. *Chemical Physics* 3303(3): 495–500.

7 Synthesis, Characterization, and Application of Nanoparticles from Medicinal Plant-Based Carotenoids

S. Saranyadevi, Lekshmi Gangadhar, Siva Sankar Sana, Methaq Hadi Lafta, and Aseel Abdulabbas Kadem

CONTENTS

7.1 INTRODUCTION

Nanotechnology (NT) is an area of science that deals with small objects and the study of materials on the nanoscale, usually between 10 and 100 nm. NT is a rapidly growing discipline that entails the synthesis, characterization, and development of a diverse range of nanomaterials (NMs) that play critical roles in everyday life by delivering beneficial goods that improve industrial output, agriculture, telecommunications, and healthcare (Dubchak et al. 2010; Mubarak Ali et al. 2013; He

DOI: 10.1201/9781003213727-7

et al. 2019; Beshah et al. 2020; Bachheti et al. 2020; 2020a). In terms of physicochemical characteristics, nanoparticles (NPs) differ significantly from corresponding bulk components (Thakkar et al. 2010; Bachheti 2021, 2021a). The huge surface area to volume proportion of these materials causes significant changes in catalytic and heat activity, melting point, conductance, mechanical characteristics, and optical absorption. Because of these characteristics, NPs can be used in practically any sector (Schmid 2011; Govindaraju et al. 2008; Painuli et al. 2020; Bachheti 2020b, 2021a). NPs play important roles in biodiagnostic and optical biosensing, nanophotonics, imaging, and therapy of various disorders that influence human health. Because of their small size and wide surface area, silver nanoparticles (AgNPs) may interrelate efficiently with microorganism surfaces and thus be utilized as antibacterial agents (Durán et al. 2016). AgNPs integrated with cotton fibre and produced with *Fusarium keratoplasticum* demonstrated considerable antimicrobial activity toward pathogens (Mohamed et al. 2017). A study by Bansal et al. proved that the fungus *Fusarium oxysporum* can be challenged with aqueous ZrF_6^{2-} anions to create zirconia NPs by the extracellular protein-hydrolysis. The use of anionic complexes leads to the ability to work at room temperature. Nanocrystalline zirconia synthesis of metal anions were hydrolyzed by cationic proteins with a molecular weight of 24–28 kDa. Identical to silicate in nature, these NPs were found to be responsible for opening up the production of large-scale biological synthesis and is an exciting prospect for oxide materials that are technologically significant (Bansal et al. 2004).

Moreover, a study by Alagarsamy et al. exhibited a green synthesis method that offers an alternative, environmentally benign approach for synthesizing zirconium oxide (ZrO_2) NPs that are effective at removing methylene blue. More fed-batch and column studies are required to make S-ZrON a dynamic adsorbent for real-time effluents (Alagarsamy et al. 2021). Iron oxide (FeO) NPs are employed in various ways in resonance imaging and cancer diagnosis (Park et al. 2008; Bachheti et al. 2019). Because of their lower toxicity, gold nanoparticles (AuNPs) are used in various sectors (Parial et al. 2012). AuNPs also exhibit a strong photothermal conductivity. They are also good at photoacoustic activity and are used in cancer photothermal treatment. Biogenic copper oxide (CuO) NPs, zinc oxide (ZnO) NPs, and selenium oxide (SeO_2) NPs are reported for their antitumour activity (Hassan et al. 2019; Bachheti et al. 2019a; Fouda et al. 2018; Salem et al. 2020).

The physicochemical methods employed to make NPs are frequently inefficient, costly, and release hazardous by-products that endanger ecosystems (Gour and Jain 2019; Khanna et al. 2019). Biogenic NP production has emerged as a viable alternative to chemical and physical production to overcome these constraints (Sharma et al. 2016). Metal predecessors have been significantly lowered to their equivalent NPs by diatoms, mushrooms, algae, plants, fungus, bacterial, acinetobacter, lichen, cyanobacteria, and microalgae (Asmathunisha and Kathiresan 2013). Biological samples have converted raw material to nanoforms via intracellular and extracellular green synthesis processes (Dahoumane et al. 2012). These nanostructures can be created by adding biological substances such as flavanones, amides, catalysts, proteins, colours, carbohydrates, phenolics, saponins, and alkaloids which act as reducing and stabilization agents. (Dahoumane et al. 2012; Husen et al. 2019). The target physical property of NPs is a large surface area to volume ratio, which leads to their wide application and ability to survive harsh environments (Dahoumane et al. 2016). Different abiotic variables, such as pH, temperature (T), the type of the microbes, their metabolic action, and relationships with heavy metals, can influence the form and volume of NPs synthesized by microbes (Lengke et al. 2007; Lengke et al. 2006; Hamouda et al. 2019).

7.2 CAROTENOIDS

Carotenoids are lipophilic secondary metabolites that play important roles in floras as well as being nutritionally important (Etoh et al. 2000). They have received much attention not only because of their potential antioxidant properties but also because they are abundant in many fruits and vegetables (such as oranges, tomatoes, and flowers, which are an essential part of the human diet (Meléndez-Martnez et al. 2014). Carotenoids have been revealed as essential factors in preventing ageing disorders, such as tumours, cardiac disease, cataracts, and age-related macular degeneration, over decades of study (Bowen et al. 2015).

7.3 STRUCTURE AND PRODUCTION

Carotenoids' fundamental structural characteristic is their structure of conjugated systems, also known as the 'polyene chain'. The colourless carotenoids phytoene and phytofluene are unusual in the carotenoid group because their networks are substantially smaller than most carotenoids (Figure 7.1). Such evident disparities in the characteristics and activities of other carotenoids are predicted to affect the properties and activities of such colourless carotenoids. This generates attention at several levels, particularly in health and cosmetics development (Meléndez-Martínez et al. 2018).

7.4 APPLICATIONS OF CAROTENOIDS

Because of their colour qualities, carotenoids have long been employed in food and animal feed. Natural carotenoids are employed to enhance the colour of fish, which improves the level of satisfaction among consumers. For instance, carotenoids are added to the fish diet to give farmed salmon colour. Carotenoids' nutraceutical qualities are of interest to the food sector as well. Numerous scientific studies have proved the medical benefits of carotenoids, and for this reason their use is quickly increasing. Furthermore, carotenoids have been considered high-value chemicals that could help make microalgae-based biofuel production more cost-effective (Pulz and Gross 2004). β-carotene, the most well-known of the carotenoids, is a vitamin A producer, and numerous other carotenoids are believed to be as well. Carotenoids have antioxidant qualities, and extensive research has proven their health advantages. Carotenoids are supposed to lessen the risk of degenerative diseases and cancer, particularly in older people (Albanes et al. 1996; Poggetti et al. 2016).

Algae are becoming more popular for green synthesis since they are a high group of secondary metabolites, proteins, peptides, and dyes that can be used as nanobiofactories (Patel et al. 2015). They are also excellent targets for biological NP synthesis because of their quick growth rate, ease of harvesting, and cost-effective scale-up (Chandran et al. 2011). Algae seem to be the most basic species on Earth, occupying many environments and serving as the primary photosynthetic organisms (Fawcett et al. 2017). Algae have the capacity to hyper-accumulate metals and convert them to NPs, making them perfect for green production (Dhillon et al. 2012). Various metallic and metal oxide NPs are generated from diverse algae families, including blue-green algae (Cyanophyceae), brown algae (Phaeophyceae), green algae (Chlorophyceae), and red algae (Rhodophyceae) (Sahoo et al. 2014; Agnihotri et al. 2014).

Algae have also long been used as food, feed, flavours, medications, fertilizer, and bioremediation substances in both commercial and industrial settings. The usage of algae for green synthesis

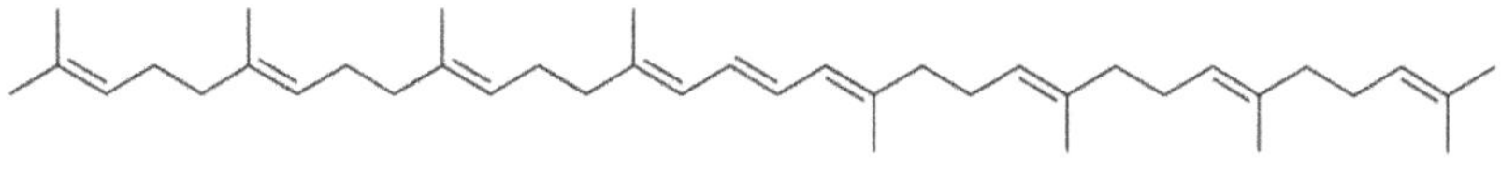

Phytoene

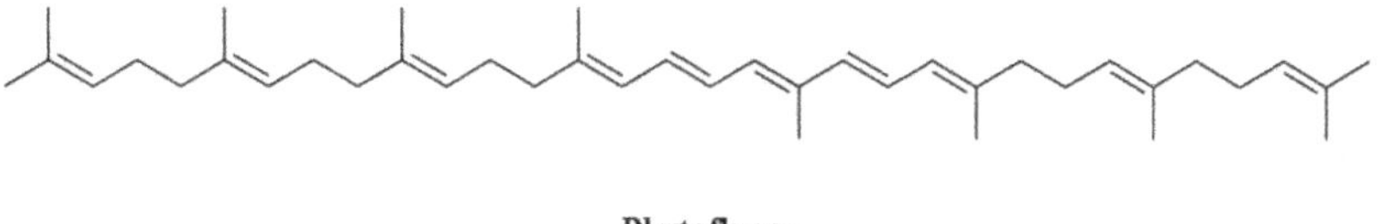

Phytofluene

FIGURE 7.1 Phytoene and phytofluene chemical structures.

is justified since it requires no external decreasing or capping agents, has a minimal energy input, the cost of synthesis is minimal, and is a normal reaction. Nevertheless, the utilization of algae for NP synthesis, termed physio-NT, is now in its early stages. As a result, the promise of algae-based NP production, the intricate processes in their biosynthesis, and significant industrial or biomedical uses of algae-linked NPs are critically reviewed in this chapter. Eventually, the biogenic production of metal and metal oxide NPs utilizing marine algae and seagrasses is summarized in this study.

7.5 ALGAE-ASSOCIATED SYNTHESIS OF THE NANOPARTICLE MECHANISM

Algae were recognized for their ability to hyper-accumulate heavy metal ions and restructure them into more pliable forms. Algae have now been proposed as mouse models for producing many NMs, particularly metallic NPs, because of these appealing characteristics (Fawcett et al. 2017). Antioxidants, pigment, minerals, polysaccharides, lipids, proteins, unsaturated fatty acids, and other phytochemicals are among the biological substances found in algae that aid in the reduction of metal-based ions charged to zero-valent condition. Biological deduction is a three-step procedure. The production of NPs begins when a preliminary metal solution is digested with algal extract. The activation phase of bioreduction includes metal ion reductions and nucleation owing to enzymes released within algal cells, as evidenced by an alteration in the colour change. Nucleated metal elements merge with one another throughout the development phase, generating NPs of various sizes and forms that are metastable. In the last termination phase, NPs attain their final form. The physical features of NPs are controlled by factors like temperature, pH, time, static circumstances, substrate concentration, and agitation (Prasad et al. 2016). We have explicitly examined the process responsible for the biosynthesis of AuNPs by algae, which are nearly identical to that biosynthetic pathway of other NPs.

Dependent on where the NPs are generated, the biological production of NPs from algae may be intra- or extracellular. An intracellular approach was a dosage-based procedure in that NP production occurs within the algal cell. Reducing agents are nicotinamide adenine dinucleotide phosphate (NADPH)/NADPH-dependent reductase, which is emitted during metabolic methods like nitrogen fixation, photosynthesis, and respiratory (Sharma et al. 2016; Dahoumane et al. 2014). Chloroauric acid (CA) was incubated with *Ulva intestinalis* and *Rhizoclonium fontinale* algae at 20°C for 72 hours to produce AuNPs. A noticeable colour change from purple to green suggested the production of AuNPs throughout the thallus.

Additionally, no colour change was seen when the gold metal solution was incubated with feedstock, indicating that no intracellular enzymes or metabolites were involved in the biological reduction mechanism. In another study, the chlorophyll of *Klebsormidium flaccidum* contained in silica gel dispersion changed from green to purple, indicating the cells' ability to decrease gold precursors (CA). The existence of condensed Au precursor salt even by NADPH-dependent reductase enzyme or NADPH was further verified by transmission electron microscope (TEM) study, which revealed dark circles inside the thylakoid membrane, suggesting the existence of lowered Au precursor salt by both the NADPH-dependent reductase enzyme or NADPH (Sicard et al. 2010). Senapati et al. (2012) revealed the intracellular production of AuNPs in *Tetraselmis kochinensis* by an algal cell membrane. Extracellular production was not detected using UV-visible spectroscopy. The AuNPs were found in higher concentrations in the cell wall than in the cytoplasm, most probably attributable to the concentration of bioactive components involved in biological reduction.

Metal ions were applied to the top of algal cells in the external mode of production, and metabolites like receptors, lipids, non-protein RNA, DNA, ions, carotenoids, and enzymes decrease them just at the surface (Vijayan et al. 2014). Although the extracellular form of production (Figure 7.2) is more efficient because NPs were readily purified, it does necessitate some necessary pre-treatments such as washing and mixing algal material (Dahoumane et al. 2016). The size, form, and aggregation of NPs are influenced by physio-chemical variables such as pH, T, and the type and starting

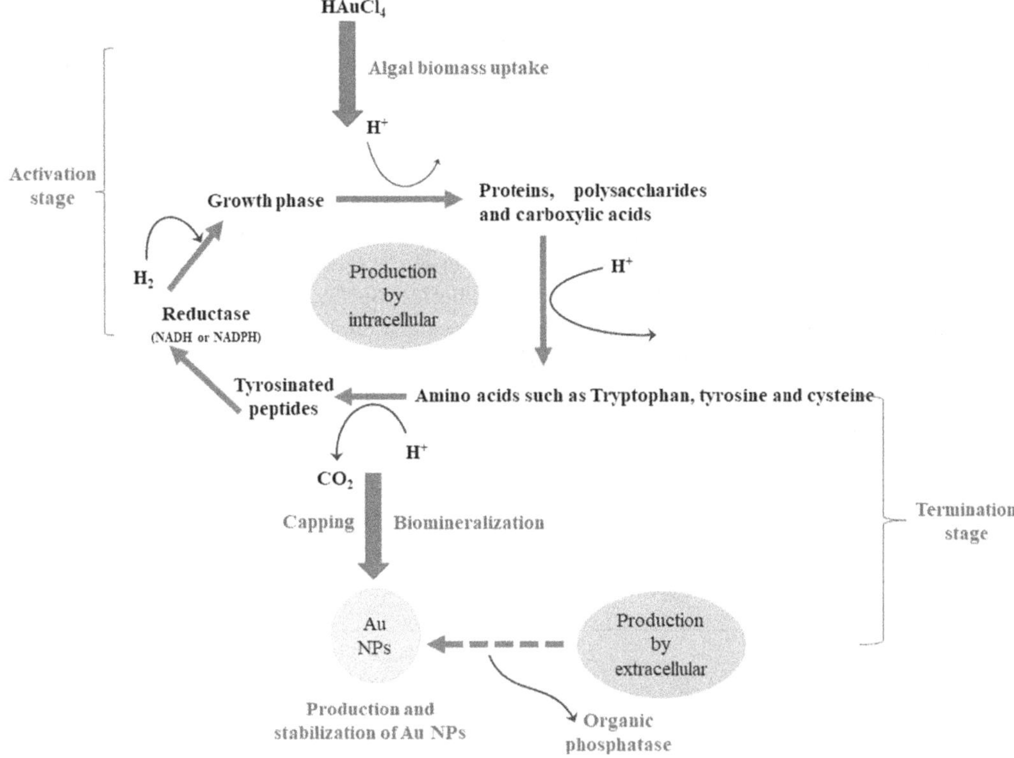

FIGURE 7.2 Schematic design of mechanism intricated in extra- and intra-cellular production of algae-mediated silver nanoparticles.

concentrations of metal and substrates (Oza et al. 2012). Increased pH reduces the reducing power of functional groups, which inhibits NP aggregation (Parial and Pal 2015; Parial et al. 2012). By interacting only with amine assemblies of surface-bound proteins and their excess amino acids, basic pH aids in the capping and stability of NPs (Namvar et al. 2015). The presence of a surface plasmon signal at 530 nm, which indicates the participation of proteins, enzymes, and macromolecules in algae-associated NP production, was validated by extracellular production of AuNPs generated by *S. platensis* at various doses of CA (Kalabegishvili et al. 2012).

7.6 GREEN SYNTHESIS OF ALGAE-BASED NANOPARTICLES

Brown, red, and green algae are a varied collection of photoautotrophic, eukaryotic, marine, uni/multicellular organisms that have been classed based on the colouring they emit (Sahayaraj et al. 2012; Prasad et al. 2013). Algae are often employed to produce numerous metallic and metal oxide NPs since they cultivate quickly, are easy to manage, and their volume development is ten times larger than that of larger plants on average. The various algal strains have been investigated for the greener synthesis of various kinds of NPs. The research for the production of NPs from several kinds of algae is carefully examined in this section.

7.7 BROWN ALGAE NANOPARTICLES

Brown algae pertain to the Fucales order as well as the Sargassaceae group. Sterols like cholesterols, sulphated polysaccharides, and functional assemblies including glucuronic acid, alginic acid, and vinyl-based derivatives, that act as reducing and capping substances for forming NPs, are the most abundant constituents of Fucales (Kumar et al. 2011). As shown in Table 7.1, various varieties of brown algae were used to manufacture metallic (Ag and Au) and metal oxide (ZnO and titanium oxide (TiO_2) NPs. AgNPs, AuNPs, and CuNPs are among the most commonly produced NPs from brown algae (Azizi et al. 2014; Rajeshkumar et al. 2012; Liu et al. 2005; Ghodake and Lee 2011; Rajeshkumar et al. 2013). Among the many metallic NPs, the production of AgNPs from various algae strains accounts for more than half of the published data. That is because AgNPs have better physicochemical properties than their bulk counterparts, making them particularly valuable in a variety of industries, including jewellery, paints, textiles, dental metals, medication delivery, and wound repair (Mohanpuria et al. 2008; Vaidyanathan et al. 2010; Maneerung et al. 2008). Several taxa have been discussed in the literature to biosynthesize AgNPs from brown algae, including *Gelidiella acerosa*, *Sargassum polycystum*, *Padina pavonica*, and *Cystophora moniliformis*, etc., (Azizi et al. 2014; Rajeshkumar et al. 2012; Liu et al. 2005; Ghodake and Lee 2011; Rajeshkumar et al. 2013). Extracellularly produced spherical AgNPs (96 nm) from *T. conoides* were found to have significant antimicrobial effects on *S. aureus*. In one study, there was epidermis, *E. coli*, and *C. albicans*. (Rajeshkumar et al. 2012). Turbinaria species have been known to reduce agents of prelude Ag salts used during the production of AgNPs (Khalil et al. 2014; Kumar-Krishnan et al. 2015). The natural conjugates, free hydroxyl, and carbonyl assemblies of Turbinaria species have already been stated to act as reducing agents of prelude Ag salts employed in producing AgNPs.

TABLE 7.1
Diverse Kinds of Algae-Mediation Production of Metallic Nanoparticles

S. No.	Algae	Type of algae	NPs	Location	Refs.
1	*Turbinaria conoides*	Brown	Ag	Extracellular	Rajeshkumar et al. (2012)
2	*Fucus vesiculosus*	Brown	Au	Extracellular	Mata et al. (2009)
3	*Sargassum muticum*	Brown	Ag	Extracellular	Madhiyazhagan et al. (2015)
4	*Sargassum muticum*	Brown	ZnO	Extracellular	Mata et al. (2009)
5	*Ecklonia cava*	Brown	Au	Extracellular	Venkatesan et al. (2014)
6	*Sargassum myriocystum*	Brown	Au	Extracellular	Dhas et al. (2012)
7	*Sargassum tenerrimum*	Brown	Au	Extracellular	Ramakrishna et al. (2016)
8	*Sargassum wightiigrevilli*	Brown	Ag	Extracellular	Alagarsamy et al. (2021)
9	*Gracilaria edulis*	Red	Ag	Extracellular	González-Ballesteros et al. (2019)
10	*Kappaphycus alvarezii*	Red	Au	Extracellular	Rajasulochana et al. (2012)
11	*Acanthophora spicifera*	Red	Ag	Extracellular	Kumar et al. (2012)
12	*Pterocladia capillacae*	Red	Ag	Extracellular	[92]
13	*Desmarestia menziesii*	Red	Ag/Au	Extracellular	González-Ballesteros et al. (2018)
14	*Gracilaria sp.*	Red	Ag/Au	Extracellular	Al-Naamani et al. (2017)
15	*Chlorella vulgaris*	Micro green	Au	Extracellular	Wei and Qian (2008)
16	*Pithophora oedogonia*	Micro green	Ag	Extracellular	Lengke et al. (2006)
17	*Plectonema boryanum*	Micro green	Ag	Extracellular	Khanna et al. (2019)
18	*Tetraselmis suecica*	Micro green	Au	Extracellular	Shakibaie et al. (2010)
19	*Chlamydomonas reinhardtii*	Micro green	Ag	Extracellular	Barwal et al. (2011)
20	*Euglena intermedia*	Micro green	Ag	Extracellular	Li et al. (2015)
21	*Rhizoclonium fontinale*	Macro green	Au	Intracellular	Dhanalakshmi et al. (2012)
22	*Ulva reticulata*	Macro green	Ag	Extracellular	Kannan et al. (2013)
23	*Ulva intestinalis*	Macro green	Ag/Au	Extracellular	González-Ballesteros et al. (2019)
24	*Gracilaria edulis*	Macro green	Ag and ZnO	Extracellular	Priyadharshini et al. (2014)
25	*Spirogyra varians*	Macro green	Ag	Extracellular	Salari et al. (2016)

Brown algae have been found to biosynthesize metal oxide NPs, like ZnONPs and TiO_2NPs, in combination with metallic NPs (Sirelkhatim et al. 2015). ZnONPs were synthesized from *S. muticum* and zinc acetate solution algal powder. The mixture was stirred continuously for hours until NPs were generated. The ZnONPs were hexagonal and varied in size from 35 to 57 nm, with bioactive functional groups such as sulphate, amines, hydroxyl, and carbonyl capped on top (Azizi et al. 2014; Azizi et al. 2017).

7.8 RED ALGAE NANOPARTICLES

Red algae, members of the Rhodophyta family, are primarily consumed as food in various nations due to their distinct flavour and high vital vitamins and proteins (Yoon et al. 2006). Such vitamins and proteins may be the best candidates for reducing and stabilizing NP production in algae. Nevertheless, due to self-aggregation, sluggish crystal development, and constancy difficulties, the production of NPs from red seaweed algae has always been in the growth processes (Singaravelu et al. 2007; Ramakritinan et al. 2020). *Porphyra vietnamensis* is one of the most prominent red algae strains that has been frequently used for the fabrication of various types of NPs. It is due to the presence of a powerful reducing agent, such as sulphated polysaccharides that comprise anionic disaccharides units.

Compared to the physio-chemical technique, red algae-mediated AgNPs were effective, eco-friendly, time consuming, and cost effective (Kumar et al. 2012). The size and form of NPs are critical considerations in medical applications. AgNPs produced from several red algae strains are mostly spherical in shape and range in size from 20 to 60 nm (Pugazhendhi et al. 2018). These extra-cellular-generated AgNPs demonstrate anti-microfouling action, which is of considerable interest in the healthcare environment (Pugazhendhi et al. 2018; Vadlapudi and Amanchy 2017). Installing microfouling in internal NP synthesis takes several processes; however, removing microfouling in extracellular production is a tiny step. Installing microfouling in internal NP synthesis takes several steps, whereas microfouling reduction in extracellular production is a one-step method (Nabikhan et al. 2010). Another red alga, *Gelidium amansii*, has been implicated in manufacturing AgNPs and reducing microfouling using the 96-well technique (Kumar et al. 2013; Selvam and Sivakumar 2015).

In contrast to AgNPs, only a few investigations have been conducted on the red algae-associated production of AuNPs. The production of AuNPs in *Lemanea fluviatilis*, a marine red alga, is being studied using CA as a preparatory salt. It resulted in 5.9 nm face-centred cubic and poly-dispersed crystallized AuNPs, as revealed by TEM (Singh et al. 2015; Murugesan et al. 2017). *Corallina officinalis* was also used in the extracellular production of spherical AuNPs using reducing agents such as OH–, phenol, and carbonyl functional group in another study (El-Kassas and El-Sheekh 2014). There are also *L. flaviatilis*, *C. flaviatilis*, and many other red algae plants that have been reported to produce AuNPs, including *Kappaphycus alvarezii*, and so on (Castro et al. 2013). Using varied molar proportions (1:1, 1:3, and 3:1) of $AgNO_3$ and $HAuC_{14}$, the red algae strain *Gracilaria edulis* successfully biosynthesized bimetallic Ag/Au NPs (Al-Naamani et al. 2017). These manufactured bimetallic NPs have shown significant cytotoxic activity in human breast cancer cells.

7.9 GREEN ALGAE NANOPARTICLES

Micro and macro green algae are the two different green algae categories based on their environment (Kim et al. 2011). Microgreen algae are unicellular and primarily found in fresh water. In contrast, macro green algae are multiple aquatic plant-like species (Deglint et al. 2019). Green algae production of diverse mono-, bi-metallic, and metal oxide NPs is now extensively used (Singh et al. 2013). This chapter presents the production of NPs from both kinds of green algae as micro- and macro-mediated production.

7.10 GREEN MICROALGAE SYNTHESIS

Cladophorales algae are microgreen algae or green microalgae extensively studied for industrial, medical, and biological applications. They contain several essential parts like alkaloids, phenols, flavonoids, sugars, and functional assemblies that can act as both reducing and stabilizing substances in micro-mediated NP production (Yousefzadi et al. 2014). Amongst many monometallic NPs, AgNPs are the NPs most often created in vitro from numerous microgreen algae species. For the production of AgNPs, over 20 species of green microalgae have been used thus far. When studied using spectroscopic and microscopic methods like FT-IR (Fourier transform infrared spectroscopy), DLS (dynamic light scattering), XRD (X-ray diffraction), EDX (energy-dispersive X-ray) spectroscopy and SEM (scanning electron microscopy), AgNPs produced from various species reveal fascinating and diverse physicochemical features (Khanna et al. 2019; Sinha et al. 2015; Vigneshwaran et al. 2007; Lengke et al. 2006; Jena et al. 2014). Almost all extracellular environments utilized microgreen algae types, like oedogonia, *Chlorococcum humicola*, and *Chlorella vulgaris*. Comparable to AgNPs, much evidence on green microalgae-associated production of AuNPs has been developed recently. *Pithophora crispa* from high altitudes were among the most extensively used microalgae to manufacture AuNPs by decreasing CA precursor salt using intra- and extracellular receptors and peptides (Xie et al. 2007). Proteins, peptides, cyclic molecules, and carboxylic acids are the most commonly described primary metabolites important in preparing metallic NPs from green microalgae (Castro et al. 2013; Kannan et al. 2013; Beganskien et al. 2004).

Microgreen algae, in addition to AgNPs and AuNPs, have been employed to make semiconductor NPs. Synthesizing silicon NPs from microgreen algae has been attempted numerous times. Silicon nanoparticles (SiNPs) are semiconductors employed as bioindicators in numerous industrial wastes to detect harmful chemicals. They also play a critical part in the environmental cycle, as they contribute to creating oxygen and recycling nutrients. *C. edulis*, green algae, is used for this (vulgaris was considered), silicon alkaloids were employed as a silicon precursor, and combined with algae extracts. Peptides and proteins in C were used to hydrolyze and poly-condensate silicon alkaloids, resulting in the SiNP extract of vulgaris (Wei and Qian 2008). Aside from SiNPs, biogenesis of other metallic, bimetallic, metal oxide, and semiconductor NPs is underway, with a series of studies and experimentation in the early stages.

7.11 GREEN MACROALGAE SYNTHESIS

Because they contain various valuable chemicals essential for NP reduction and capping, green macroalgae were also known as biological factories for producing metallic NPs (Singh et al. 2013; Priyadharshini et al. 2014). These chemicals have anticancer, antiviral, antimicrobial, and cytotoxic effects on microorganisms and are involved in forming NPs. (Salari et al. 2016). *Ulva fasciata* is amongst the most beneficial green macroalgae species. It was used to make nanosized Ag colloids, which were then treated to cotton fabric in the existence and exclusion of citric acid to see how effective they were against bacteria (El-Rafie et al. 2013). *Gracilaria edulis* (high in amide, carboxylic, and nitro substances were employed to make spherical AgNPs and octahedral ZnONPs in that other study (Priyadharshini et al. 2014; Madhiyazhagan et al. 2017). Some other significant species of seaweed are green macroalgae and *Chaetomorpha linum*. *Gracilaria edulis* was also utilized to synthesize AgNPs (Priyadharshini et al. 2014).

AuNPs have attracted researchers in current years due to their possible utility in cancer care as a tailored medication delivery system. Due to their non-reproducibility at the ideal size and form, AuNP production is always difficult, but green macroalgae overcome this problem (Priyadharshini et al. 2014; Dhanalakshmi et al. 2012; Parial and Pal 2014). Green macroalgae species like *Prasiola crispa* and *Rhizoclonium fontinale* have also been found to produce AuNPs in addition to AgNPs (Dhanalakshmi et al. 2012; Parial and Pal 2014) (Table 7.1).

7.12 PLANT-BASED SYNTHESIS OF NANOPARTICLES

Green chemistry is a novel arena that promotes using a set of standards to reduce the use and development of potentially hazardous wastes (Varma et al. 2012). As a consequence, green techniques reduce the environmental impact of industrial labour. Researchers are working to create alternatives to the costly processes and toxic materials that can be discovered when employing classic physicochemical synthetic routes (Nogueira et al. 2020; Hekmati et al. 2020; Khalaj et al. 2020; Yadi et al. 2018). The use of environmentally friendly solvents and chemicals, low-energy processes, and non-toxic molecules like DNA, proteins, enzymes, carbohydrates, and plant extracts, make it possible to synthesize biocompatible metal-based NPs through plummeting metal ions in aqueous substances (Hou et al. 2020; Fasciotti 2017). Noble metal NPs such as AuNPs and AgNPs have sparked interest because of their unique physicochemical and biological features (Dauthal and Mukhopadhyay 2016). These metal NPs offer appealing properties like strong electrical and thermal conductivity, chemical solidity, higher catalytic activity, and, most notably for farming, antibacterial action towards a wide range of microbes. Their nanosized substances, adjustable shape, and surface morphology all play a role in these features (Sreeprasad and Pradeep 2013). There were two types of manufacturing approaches for generating AuNPs and AgNPs, as per Jamklande et al. (2019), relying on the precursor material for their manufacture. To begin, when the feedstock is larger than nanoscale, the top-down approach is used, requiring the particles to be broken down through grinding, milling, lithography methods, or thermal ablation. This particle size reduction method consumes a lot of energy and may cause surface defects in NPs, which can significantly impact their physicochemical properties (Yadi et al. 2018; Jamklande et al. 2019; Narayanan and Sakthivel 2011). The bottom-up route, also called 'self-assembly strategy', uses chemical and biological mechanisms to produce NPs by atoms growing in nucleation centres. The biological synthesis of AuNPs and AgNPs were a low-cost bottom-up technique that produces a large number of NPs in a short period. The NPs produced using this green method have fewer flaws and more uniform chemical characteristics (Jamklande et al. 2019; Gour and Jain 2019; Golinska et al. 2014; Rafique et al. 2017; Hussain et al. 2016).

Medicinal plants are becoming more popular as an environmentally and economically beneficial substitute for synthesizing AuNPs and AgNPs utilizing a variety of approaches, like employing $AgNO_3$ at room T in a matter of minutes to hours (Gour and Jain 2019; Sani and Tatiana 2017). Furthermore, compared to microbes or fungi, this synthesis technique is quicker (Rafique et al. 2017). Leaves, roots, twigs, skins, fruits, and essential oils are all examples of floral parts or products that can be extracted. Polyphenols, flavonoids, carbohydrates, enzymes, and receptors were abundant. These phytochemicals are isolated and used directly in the extracellular production of metallic NPs as reducing and stabilizing substances, substituting potentially dangerous compounds such as sodium borohydride ($NaBH_4$) (Dauthal and Mukhopadhyay 2016). Owing to the excessive diversity of phytoconstituents found in the extracts, the particular mode of action underlying this phenomenon has yet to be understood. Though polyphenols and receptors were thought to be the principal reducing agents, the diverse phytochemicals function in concert (Dauthal and Mukhopadhyay 2016). Generally, this strategy can be cost effective and appropriate for larger-scale manufacturing methods (Bhattarai et al. 2018).

Furthermore, these innovative AuNP and AgNP synthesis methods offer the benefit of creating vast numbers of NPs free of pollution and with better-defined size and morphology. In terms of form, NPs made through plant extract-mediated synthesis have a more energetic advantageous spherical shape that provides the required reactivity for various uses, especially in agriculture activities (Arias Ortiz et al. 2019; Noah et al. 2019). Four investigations (Gopinath et al. 2012; Kumar et al. 2014; Nakkala et al. 2014; Nakkala et al. 2015) have described the production of spherical AgNPs using extracts from *Tribulus terrestris* fruit, *Acorus calamus* roots, and the complete flora of *Boerhaavia diffusa* species. Other research has created NPs from different medicinal plants, with changes in form anticipated when different shapes within the same plant are used in the extraction procedure (Muthukumar et al. 2020; Pereira et al. 2020). Nevertheless, Rajakumar et al. found that using *Eclipta prostrata* leaves to synthesize AuNPs resulted in triangular, pentagonal, and hexagonal forms (Rajakumar and Abdul Rahuman 2011). One downside of these approaches is that the nature of the raw resources restricts the circumstances in which they may be employed, which can affect the creation of NPs. As a result, well-defined requirements for T, pH, metallic solution mixture, and reaction time must be provided (Arias Ortiz and Palma Holguín 2019).

Arreche et al. investigated two commercial types of yerba mate to manufacture aqueous-based extracts for producing AgNPs using $AgNO_3$ at room T (Figure 7.3). Spherical, hexagonal, and triangular NPs were produced by average particle dimensions of 50 nm and a surface plasmon peak of 460 nm. Antibacterial activities were further tested towards *E. coli* and *S. aureus* bacteria. The therapy brands 1 and 2 yielded 7.66 and 17.66 gm/L of *E. coli*, respectively. The minimal inhibitory concentrations needed for *E. coli* are listed below.

On the other hand, the values for *S. aureus* are quite different. The concentrations of *S. aureus* in the therapy brands 1 and 2 were 23.25 and 50.60 gm/L, respectively. According to the findings, polyphenols included in yerba mate floral extract function as a reduction agent and NP stabilizer (Arreche et al. 2020).

Furthermore, Sasidharan et al. employed the pericarp of *Myristica fragans* fruit extract for environmentally friendly AgNP production. The aqueous fruit extract of the flora served as a reducing and stabilizing substance in this method, and the resulting AgNPs had good catalytic and antimicrobial activity (Sasidharan et al. 2020). Alkhalaf et al. investigated the impact of green production of AgNPs from a *Nigella sativa* floral extract, which produced NPs with antioxidant capacity (Alkhalaf et al. 2020).

On the other hand, various attempts have implemented AuNPs using a green method. Kesarla et al. developed a green process that included an aqueous-based extract of a powdered form of *Terminalia bellirica* dry fruit pericarp that served as a decreasing and stabilizing agent. One gram of the powder was added to 100 mL of deionized water, heated and maintained at 90°C for 1 hour, then cooled and filtered through a 0.2 mm cellulose nitrate filter membrane. The newly produced extract was then mixed with 2 mL of $HAuCl_4$ 1 M and stirred vigorously. The synthesis was fast, requiring less than 10 seconds to produce the NPs, as seen by the colour shift from yellow to reddish

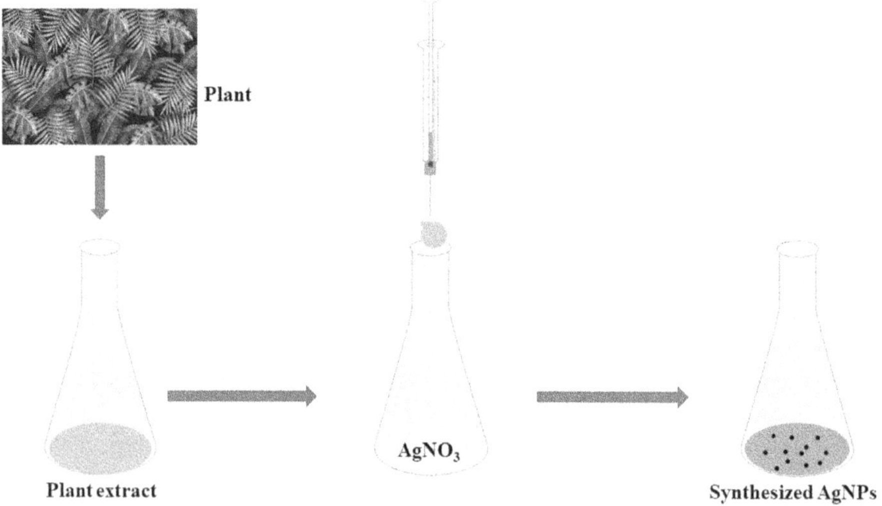

FIGURE 7.3 Synthesis of silver nanoparticles by employing natural extracts from waste substances.

pink. The large quantities of polyphenols in T explain the quick decrease and stabilization. UV-Vis spectroscopy at 530 nm was also used to verify the production of AuNPs (Kesarla et al. 2014).

Both NPs have been synthesized and described in numerous studies. Sk et al. made them from the aqueous extract of *Malva verticillata* leaves. AuNPs showed excellent catalytic action in the hydride transfer reductions of aromatic nitro Schiff bases, whereas AgNPs showed promising antimicrobial effects (Sk et al. 2020). The antioxidant capacity and production of AuNPs and AgNPs were attributed to phenolic chemicals in turmeric extract and blackberry (Sk et al. 2020).

7.13 BIOMEDICAL USES OF ALGAE-MEDIATED NANOPARTICLES

Algae species can avoid the need for hazardous chemicals in the reduction and stabilization of NPs since they comprise naturally produced biomolecules with low or no toxicity, making them ideal for various biomedical uses (González-Ballesteros et al. 2019; Bhattacharya et al. 2019). Even though they do not require any exterior capping/reducing agents during the creation of NPs, they are generally biocompatible and devoid of harmful compounds entrapped on their surfaces, and consequently have lower toxicity than chemically manufactured NPs. Several uses of algae-mediated NPs, primarily biomedical, are addressed here.

7.13.1 Antimicrobial Activity

The antimicrobial properties of NPs produced from algae have been tested against various microbial species. AgNPs made from the brown algae *Padina tetrastromatica* effectively delayed the growth of *P. aeuroginosa* and other *Bacillus* spp. (Rajeshkumar et al. 2012). An additional study found that steady and colloidal-shaped AgNPs made from the aqueous extract of the green marine algae *Caulerpa serrulata* had remarkable antibacterial activity towards *Shigella* sp., *S. aureus*, at lower doses. *S. aureus*, *S. typhi*, *E. coli*, and *P. aeruginosa*.

The maximum inhibition zone of 21 mm of AgNPs were obtained against *E. coli*, but towards *S. typhimurium*, the narrowest zone of inhibition was 10 mm at 50 L AgNPs. (Aboelfetoh et al. 2017). AgNPs made from *Pithophora oedogonia* aqueous extract have also shown antimicrobial effects on *E. coli*, *S. aureus*, *Micrococcus luteus*, *B. subtilis*, *P. aeruginosa*, and so on. Antibiotics are widely used to treat microbial illnesses, which has resulted in the creation of multi-drug-resistant bacterial species. A major worldwide health concern is finding a safe and effective therapy for drug-resistant species of bacteria. As a result, there has been a trend toward using NPs as a substitute for antimicrobial activity, which is effective and superior in terms of bactericidal activity. Because NPs disrupt the cellular membrane and generate reactive oxygen species (ROS), they have broad spectrum antibacterial activity against both Gram +ve and −ve bacteria (Wang et al. 2017). For *P. aeruginosa*, the zone of inhibition (17.2 mm) was assessed, indicating that AgNPs have outstanding antimicrobial activity towards highly resilient Gram −ive rods.

Additionally, spherical AuNPs made from a protein extract of the blue-green alga *S. platensis* development was severely slowed by *S. aureus* and *B. subtilis* (Suganya et al. 2015). The antimicrobial property of AuNPs made from *Ecklonia cava* and *Nitzschia* has been investigated compared to *E. coli*, *S. aureus*, *P. aeruginosa*, and *B. subtilis*, etc. (Venkatesan et al. 2014; Borase et al. 2017). When compared to the tetracycline antibiotic standard, AuNPs produced from *Stoechospermum marginatum* showed higher antimicrobial activity toward *Enterobacter faecalis* (Rajathi et al. 2012). In that other study, AgNPs and AuNPs produced by *Neodesmus pupukensis* were examined for antimicrobial activities against various bacterium types. *Pseudomonas* sp. (43 mm); *E. coli* (24.5 mm); *K. pneumoniae* (27 mm); *S. marcescens* (39 mm), while AuNPs were exclusively active against *Pseudomonas* sp. (27.5 mm) as well as *S. marcescens* (28.5 mm) (Omomowo et al. 2020). These results suggest that algae-mediated NPs could be used as antimicrobial agents.

7.13.2 Antifungal Activity

Due to the scarcity of antifungal medications and the emergence of antifungal drug resistance, fungal pathogens have become a rising threat to public health. There is a tremendous motivation to produce new antifungal medicines that are both strong and effective. Because of their high fungicidal action, NPs can be a new therapy option for fungal infections (Mallmann et al. 2015). So far, AgNPs are the most powerful antifungal drug created using the green approach. AgNPs were synthesized from *Sargassum longifolium* and tested for antifungal activity at different dosages against a variety of pathogenic fungal species, namely *A. fumigatus*, *Fusarium* sp., and *C. albicans*. AgNPs effectively suppressed the development of each fungal stress in a dosage-based way, according to the findings (Rajeshkumar et al. 2014). AgNPs made from the green algae *Ulva latica* and the red algae *Hypnea musciformis* has been shown to prevent the progression of *A. niger*, *C. albicans*, and *C. parapsilosis* fungi (Dhavale et al. 2019). AgNPs were produced using aqueous extract from the red seaweed *Gelidiella acerosa* and evaluated for antifungal effectiveness towards *Fusarium dimerum*, *Mucor indicus*, *Humicola insolens*, and *Trichoderma reesei* in another work. AgNPs showed significant antifungal activity compared to traditional antifungal drugs (Vivek et al. 2011).

7.13.3 Antineoplastic Activity

The usage of NPs for tumour therapy and targeted therapies of anticancer medications is among the most active fields of NT study (Cheung et al. 2015). Several studies on the anticancerous properties of algae-associated NPs were developed recently. *Sargassum vulgare*-derived AgNPs (10 nm) were found to have anticancer action against HeLa cells and human myeloblastic leukaemia cells HL60 research (LewisOscar et al. 2016). Photothermal agents were Ag nanotriangles coated with algal-derived chitosan polymers (Chit-AgNPs) directed at the NCI-H460 non-small human lung cancer cell line (KJP 2017). Furthermore, in vitro cytotoxicity of *S. muticum*-linked AgNPs towards the MCF7 breast tumour cells has been demonstrated. The MCF7 cell line was exposed to different doses of AgNPs from 3 to 50 g/mL for roughly 48 hours, with the most significant viability percentage of 100.36% reported at the 12.5 g/mL concentration. These AgNPs have produced ROS intracellularly, causing tumour cells to apoptosis and subsequently die (Supraja et al. 2016). In a separate investigation, using the MTT assay, AgNPs generated by myriocystum were tested for cytotoxicity towards the HeLa cell line at diverse concentrations varying from 0, 2, 4, 8, 16, 32, 64, 128, 256, and 512 g/mL. The AgNP-treated HeLa cell line exhibited 50% inhibiting and apoptosis actions, with overall cytotoxic abilities increasing as the concentration of AgNPs in the media (Balaraman et al. 2020). Likewise, the in vitro toxicity of algal-linked AgNPs compared to the MCF-7 tumour model was investigated at different concentrations (0–100 g/mL) for 24, 36, and 48 hours. AgNPs were found to have a maximal inhibitory concentration of 20 g/mL towards breast cancer cells that showed nuclear fragmentation, apoptotic, and death of cells, indicating their antitumour activity (Gopu et al. 2020).

Algae-derived AuNPs have also been shown to have potent anticancer properties against various cell types. *Acanthophora spicifera*-mediated AuNPs were found to have potent anticancer properties in the colorectal adenocarcinoma HT-29 cell line in one investigation. AuNPs at doses of 1.88, 3.75, 7.5, 15, and 30 g/mL were applied and maintained for around 24 hours before being examined using the MTT test. In tumour cell lines, a maximal inhibitory dosage of 21.86 g/mL was found, which caused apoptotic, loss of morphology, and cell shrinkage (Babu et al. 2020). Another study found that *Chaetomorpha linum*-mediated AuNPs have antitumour activity in vitro towards the HCT-116 colon tumour cells. After incubating with all these NPs, dosage-dependent lethal impacts of AuNPs were observed in colon cells. A sequence of apoptotic screenings was examined, leading to the release of apoptotic caspase 3 and 9 and a decrease in antiapoptotic proteins, including Bcl-xl and Bcl-2, proving that AuNPs generated by algae were effective antitumour agents (Acharya et al. 2021). Likewise, compared to the antitumour cytokine tumour necrosis factor-alpha, AuNPs covered by polyethylene glycol can kill most tumour cells (Cai et al. 2008; van Horssen et al. 2006). This research suggests that algal-derived NPs have anticancer properties.

7.13.4 Uses of Microalgae-Mediated Nanoparticles

Scenedesmus quadricauda and *Selenastrum capricornutum* have both been found to accumulate and degrade polycyclic aromatic hydrocarbons in recent research. Traditionally, algal systems have been used as a tertiary wastewater treating method. They have also been mentioned as a possible subsequent therapy option. Algae can be used in treatment procedures to remove coliform microbes, reduce chemical and biochemical oxygen demand, and remove N, P, and heavy metals (Rao et al. 2011). The healing process, antifungal, antitumour, and antimicrobial activities are among the medical use of algal NPs. Metallic NPs (Au, Ag, and CuO) made from *Bifurcaria bifurcate*, *Galaxaura elongate*, *Sargassum plagiophyllum*, and *Chlorococcum humicola*, etc. show antimicrobial action towards Gram +ve and −ve microbes, according to research. Diatoms engross carbon dioxide from the atmosphere and release oxygen, raising dissolved oxygen (DO) levels in marine bodies. It has the potential to boost biological diversity, marine life, and water quality. Algal NPs, which may penetrate different polysaccharide (EPS) and cellular membranes, could be employed as antibiofilm agents for multidrug-resistant microbes in the future. Additionally, they can also be used in nanocomposites and biosensing uses.

7.14 APPLICATIONS OF PLANT-MEDIATED NANOPARTICLES

7.14.1 Agriculture

The production of NPs was exciting because of the exclusive features that might be used in composite fibres, biosensors, cryogenic superconducting substances, cosmetics, and electrical machinery (Zhang et al. 2020). The production of AuNPs and AgNPs from floral extracts, and more from waste materials, is an important topic to support sustainability in agro-industrial processes, due to global warming and the loss of natural resources. Because plants represent the foundation of this green synthesis, the NPs produced can be employed in various agroindustry activities, from the soil used to the food chain, thanks to their lower toxicity (Cox et al. 2017; Dasgupta et al. 2015; Awad et al. 2019). The Food and Agriculture Organization (FAO) and the World Health Organization (WHO) jointly declared nanotechnological food and agricultural uses in June 2009, covering a wide range of topics such as nanostructured additives, nanosized biofortification, packaged food, nanocoating, and -filtration (Takeuchi et al. 2014). NPs can also operate as ‘magic bullets’, delivering nutrition materials to particular plant locations or structures, like advantageous genes and organic molecules, to boost productivity. As a result, NPs are smart nanodelivery schemes for farming management focusing on crop nutrition (Marchiol et al. 2014).

Food packaging is one of the most common indirect uses depending on the antibacterial action of NPs (Marchiol et al. 2014).

Many studies on the direct use of AuNPs and AgNPs in farming have concentrated on seed sprouting, root extension, and plant reactions to metal NPs, like cellular oxidative damage or cytotoxicity (Cox et al. 2017; Ribeiro et al. 2019). Metal NPs can also be used to produce nanofertilizers and nanopesticides (Vijayaraghavan and Ashokkumar 2017). The above-mentioned uses have been extensively tackled in the agroindustry in a huge range of goods comprising NPs of these metals at atom sizes ranging from 100–250 nm, improving their water solubility and action (Prasad et al. 2017).

7.14.2 Antimicrobial Uses

Green synthesized metal NPs may be employed as antioxidants, biological sensors, and heavy metal detectors (Chavan et al. 2020; Akjouj and Mir 2020; Teodoro et al. 2019). Their unique physicochemical properties, such as their ability to bind biological molecules, higher specific surface area to volume ratio, higher surface reactivity, ease of synthesizing and characterizing, lower cytotoxicity, and ability to enhance the expression of genes for redox reactions, enable them to be used as antibacterial agents against floral disorder pathogens and others that can cause foodborne disorders (Giljohann et al. 2010; Murphy et al. 2008; Bhattacharya and Mukherjee 2008; Pardhi et al. 2020). A wide range of floral extracts employed to make AuNPs and AgNPs have been prepared and their antibacterial action towards microbes (Ali et al. 2019; Nishanthi et al. 2019; Taghavizadeh Yazdi et al. 2019). These NPs, on the other hand, have almost no antibacterial action when made using standard procedures (Mishra et al. 2011). The difference could be due to its synergistic impact on AuNPs or AgNPs and extracts containing significant amounts of steroids, carbohydrates, and flavonoids, which function as ion reducers and cover agents, helping NP stability (Yugay et al. 2020). The antibacterial usage is mainly due to their ultra-small dimensions and shape (250 times smaller than microbes). This allows an electrostatic interaction among the Au or Ag from the NPs and the −ve charge on the cell wall of microbes, ensuing in deformations of membrane power functions like porosity, osmoregulation, electron transport, and respiration, resulting in the development of antibiotic resistance. Smaller NPs will have a larger bactericidal impact through binding with the microbe's cellular membrane's huge surface area, which depends on surface accessibility (Islam et al. 2019; Rashmi et al. 2020; Nadeem et al. 2017).

Nonetheless, the specific antibacterial mode for AuNPs and AgNPs is unknown, and numerous theories have been offered to explain it. First and foremost, NPs have a high predilection for interacting with sulfhydryl and phosphorus families on the cell wall, resulting in significant destruction and releasing microbial cell elements (Ahmad et al. 2016). Another theory claims that NPs can enter the microbial cellular membrane and ascribe to NADH dehydrogenases, causing a large number of ROS, causing adenosine triphosphate (ATP) to be depleted, and disrupting the respiratory chain. These radicals can interrelate with proteins, sulphur, phosphorus-containing cell components and DNA, causing them to be destroyed (Figure 7.4) (Ahmad et al. 2016; Cui et al. 2012). Furthermore, since DNA lacks its repetition capacity and proteins become inactive after interfacing with these ions, the discharge of Au or Ag ions by NPs may assist to their antibacterial effects. This biocidal action is thought to be dimension and dosage dependent, with greater NP concentrations interacting more with cytoplasmic structures and microbial nucleic acids (Chauhan et al. 2013; Manivasagan et al. 2013).

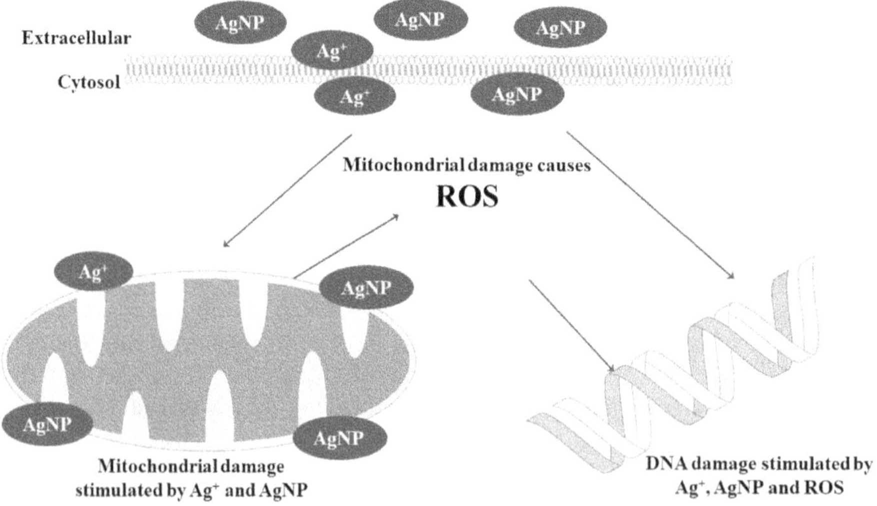

FIGURE 7.4 Silver nanoparticles' antibacterial mode of action via microbial cellular membrane diffusion, disruption of the respiratory chain, and DNA impairment through reactive oxygen species.

Furthermore, antibacterial action differs depending on the nature of the bacterial cell wall. Gram −ve microbes are more resistant to AuNPs and AgNPs than Gram +ve microbes. Gram +ve microbes have a thick coating of peptidoglycan, which is made up of linear polysaccharide chains cross-linked through shorter peptides and produces a hard, rigid system for NPs to enter, whereas Gram −ve bacteria have a thinner layer that indicates a pierceable configuration (Muthuvel et al. 2014; Kaviya et al. 2011). Nonetheless, Nirmala et al. and Padalia et al. observed that surface modification by aminoglycoside antibiotics increased the antibacterial activity of NPs, widening their activity range to Gram +ve microbes (Nirmala Grace and Pandian 2007; Padalia et al. 2015). Wang et al. (2016) reported that adding gentamicin to AgNPs increased their bactericidal toxicity towards the foodborne microbes *S. aureus, E. coli*, and gentamicin resistant. Because the medication stimulates NP breakdown, raises Ag ion concentration, and develops bacteria inhibition and mortality, it was effective against *E. coli* (Wang et al. 2016). It is worth noting that this could be an issue for agricultural workers who use antibiotics. Even though various study groups were increasing the responsiveness of the delinquent and asking people to stop using antibiotics in fields other than medicine, this strategy can be highly beneficial and reduce drug resistance difficulties when used in a controlled atmosphere and under strict regulations (Chang et al. 2018).

Most AuNPs have a spherical shape, according to many scientific studies wherein the NPs have demonstrated potential antibacterial activity. Despite this, rod-shaped, triangular, and cubic NPs have been discovered in the combination (Ramesh et al. 2019). Thangamani and Bhuvaneshwari (2019) used *Simarouba glauca* leaf extract to make AuNPs. The dimensions and form of NPs were responsive to leaf broth concentration; atoms were likely to shrink in dimensions as leaf broth concentration increased, while diverse morphologies, including a combination of prism and spherical-like particles, were generated. They also tested the antimicrobial action of the compounds towards Gram +ve and Gram −ve microbes. *S. aureus* performed better in the antibacterial assay, *S. mutans*, *B. subtilis*, etc. (Thangamani and Bhuvaneshwari 2019).

7.14.3 Use in Water Treatment

Water quality is important for expanding agricultural productivity, but wastewater action is also important because agricultural production uses around 70% of the world's renewable water resources (Omara et al. 2019). Heavy metals, fertilizers, and pesticides employed in agriculture have lowered the availability of fresh water for drinking and crop irrigation (Fernández-Luqueño et al. 2018). NT can be essential in determining the quality and purity of a longer-term productivity scheme. Noble metals are the most appealing and cost-effective NMs for ecological sustainability and water clean-up operations. Plant extract-mediated green-produced AuNPs and AgNPs can be used for water monitoring, purifying, drinking water, and agricultural wastewater because of their higher surface area to volume proportion, chemical solidity, and increased catalytic action features (Graily-Moradi et al. 2020; Palit and Hussain 2020).

Furthermore, by integrating these NPs into sensors for the direct diagnosis of harmful chemicals like pesticides and heavy metals, research organizations have benefited from their various properties, such as their higher reactivity for recognizing harmful chemicals (Dobrucka 2020; Vinci and Rapa 2019). Adsorption, photocatalytic deprivation, and nanofiltration approaches utilizing NPs can all be used to detoxify water and wastewater (Sheng and Liu 2011). Using AuNPs and AgNPs, multiple writers have documented the mechanism of pesticide mineralization in water, including chlorpyrifos, malathion, and atrazine (Manimegalai et al. 2011). Pesticides are extracted via adsorption onto NPs that hold them on their surface and interact with them for lengthy periods until the complex precipitates. As a result, these NPs are a feasible, practical, and cost-effective method of eliminating pesticides from drinking water or irrigation water (Manimegalai et al. 2011; Saifuddin et al. 2011). Furthermore, due to their strong adsorption capability, AuNPs and AgNPs were considered to be an intriguing strategy for removal of heavy metals in water (Sushma and Richa 2015).

7.15 CONCLUSION

Plant and algae-based extracts are being used to synthesize metal NPs, which are gaining popularity owing to their excellent efficiency over standard physicochemical approaches. Green synthesized AuNPs and AgNPs have prospective uses in agriculture and agroindustry, particularly as antibacterial agents for certain microbes, where effectiveness has been experimentally demonstrated. Furthermore, algae are unquestionably good candidates for the green synthesis of NPs because they are rich in secondary metabolites that work as reducing and capping agents. In order to determine and establish the involvement of specific biomolecules responsible for the reduction and capping of NPs throughout the algae-mediated biosynthetic pathways. Algae can be utilized in wastewater for various purposes, including biological oxygen demand (BOD) reduction, nitrogen and phosphorus elimination, coliform suppression, and heavy metal reduction. As a result, microalgal waste management via biological and physicochemical processes could be an appealing complement to conventional biological treatment. Microalgae have previously been shown to accumulate lipids and carotenoids in intracellular quantities. Algae were unquestionably excellent fits for the green production of NPs since they are rich in secondary metabolites, which act as reducing and stabilizing agents. This may be investigated in the future. A systematic formulation of algal-based NPs might be carried out, which would assist in refining application areas for algal-mediated NPs. In this chapter, we have summarized the green-synthesized metallic NPs from flora and algae, their mode of synthesis, and their uses in various fields.

REFERENCES

Aboelfetoh, E.F., R.A. El-Shenody & M.M. Ghobara. 2017. Eco-friendly synthesis of silver nanoparticles using green algae (*Caulerpa serrulata*): Reaction optimization, catalytic and antibacterial activities. *Environmental Monitoring and Assessment* 189(7): 1–15.

Acharya, D., S. Satapathy, P. Somu, U.K. Parida & G. Mishra. 2021. Apoptotic effect and anticancer activity of biosynthesized silver nanoparticles from marine algae *Chaetomorpha linum* extract against human colon cancer cell HCT-116. *Biological Trace Element Research* 199(5): 1812–1822.

Agnihotri, S., S. Mukherji & S. Mukherji. 2014. Size-controlled silver nanoparticles synthesized over the range 5–100 nm using the same protocol and their antibacterial efficacy. *RSC Advances* 4(8): 3974–3983.

Ahmad, A., F. Syed, M. Imran, A.U. Khan, K. Tahir, Z.U.H. Khan & Q. Yuan. 2016. Phytosynthesis and antileishmanial activity of gold nanoparticles by *Maytenus royleanus*. *Journal of Food Biochemistry* 40(4): 420–427.

Akjouj, A. & A. Mir. 2020. Design of silver nanoparticles with graphene coatings layers used for LSPR biosensor applications. *Vacuum* 180: 109497.

Alagarsamy, A., S. Chandrasekaran & A. Manikandan. 2021. Green synthesis and characterization studies of biogenic zirconium oxide (ZrO_2) nanoparticles for adsorptive removal of methylene blue dye. *Journal of Molecular Structure* 11: 131275.

Albanes, D., O.P. Heinonen, P.R. Taylor, J. Virtamo, B.K. Edwards, M. Rautalahti, A.M. Hartman, J. Palmgren, L.S. Freedman, J. Haapakoski, M.J. Barrett, P. Pietinen, N. Malila, E. Tala, K. Liippo, E.R. Salomaa, J.A. Tangrea, L. Teppo, F.B. Askin, E. Taskinen, Y. Erozan, P. Greenwald & J.K. Huttunen. 1996. α-Tocopherol and β-carotene supplements and lung cancer incidence in the Alpha-Tocopherol, beta-carotene Cancer Prevention Study: Effects of base-line characteristics and study compliance. *JNCI: Journal of the National Cancer Institute* 88(21): 1560–1570.

Ali, J., N. Ali, L. Wang, H. Waseem & G. Pan. 2019. Revisiting the mechanistic pathways for bacterial mediated synthesis of noble metal nanoparticles. *Journal of Microbiological Methods* 159: 18–25.

Alkhalaf, M.I., R.H. Hussein & A. Hamza. 2020. Green synthesis of silver nanoparticles by Nigella sativa extract alleviates diabetic neuropathy through anti-inflammatory and antioxidant effects. *Saudi Journal of Biological Sciences* 27(9): 2410–2419.

Al-Naamani, L., S. Dobretsov, J. Dutta & J.G. Burgess. 2017. Chitosan-zinc oxide nanocomposite coatings for the prevention of marine biofouling. *Chemosphere* 168: 408–417.

Arias Ortiz, J.D. & M.I. Palma Holguín. 2019. Elaboración de un Compuesto Antimicrobial con Nanopartículas de Plata Sintetizadas a Partir del Extracto de Hojas de Romero (*Rosmarinus officinalis*), Para ser Aplicado en Frutas Frescas. Ph.D. Thesis, Facultad de Ingeniería Química, Universidad de Guayaquil: Guayaquil, Ecuador.

Arreche, R.A., G. Montes de Oca-Vásquez, P. Vázquez & J.R. Vega-Baudrit. 2020. Synthesis of silver nanoparticles using extracts from yerba mate (*Ilex paraguariensis*) wastes. *Waste and Biomass Valorization* 11(1): 245–253.

Asmathunisha, N. & K. Kathiresan. 2013. A review on biosynthesis of nanoparticles by marine organisms. *Colloids and Surfaces, Part B: Biointerfaces* 103: 283–287.

Awad, M.A., N.E. Eisa, P. Virk, A.A. Hendi, K.M.O.O. Ortashi, A.S.A. Mahgoub, M.A. Elobeid & F.Z. Eissa. 2019. Green synthesis of gold nanoparticles: Preparation, characterization, cytotoxicity, and antibacterial activities. *Materials Letters* 256: 126608.

Azizi, S., M.B. Ahmad, F. Namvar & R. Mohamad. 2014. Green biosynthesis and characterization of zinc oxide nanoparticles using brown marine macroalga *Sargassum muticum* aqueous extract. *Materials Letters* 116: 275–277.

Azizi, S., M. Mahdavi Shahri & R. Mohamad. 2017. Green synthesis of zinc oxide nanoparticles for enhanced adsorption of lead ions from aqueous solutions: Equilibrium, kinetic and thermodynamic studies. *Molecules* 22(6): 831.

Babu, B., S. Palanisamy, M. Vinosha, R. Anjali, P. Kumar, B. Pandi, M. Tabarsa, S. You & N.M. Prabhu. 2020. Bioengineered gold nanoparticles from marine seaweed *Acanthophora spicifera* for pharmaceutical uses: Antioxidant, antibacterial, and anticancer activities. *Bioprocess and Biosystems Engineering* 43(12): 2231–2242.

Bachheti, R.K., A. Fikadu, A. Bachheti & A. Husen. 2020. Biogenic fabrication of nanomaterials from flower-based chemical compounds, characterization and their various applications: A review. *Saudi Journal of Biological Sciences* 27(10): 2551–2562.

Bachheti, R.K., Y. Godebo, A. Bachheti, M.O. Yassin & A. Husen. 2020a. Root-based fabrication of metal/metal-oxide nanomaterials and their various applications. In Husen, A. & Jawaid, M., Eds. *Nanomaterials for Agriculture and Forestry Applications*. Elsevier, pp. 135–166. https://doi.org/10.1016/B978-0-12-817852-2.00006-8.

Bachheti, R.K., L. Abate, A. Bachheti, A. Madhusudhan & A. Husen. 2021. Algae-, fungi-, and yeast-mediated biological synthesis of nanoparticles and their various biomedical applications. In *Handbook of Greener Synthesis of Nanomaterials and Compounds*. Elsevier, pp. 701–734. https://doi.org/10.1016/B978-0-12-821938-6.00022-0.

Bachheti, A., R.K. Bachheti, L. Abate & A. Husen. 2021a. Current status of Aloe-based nanoparticle fabrication, characterization and their application in some cutting-edge areas. *South African Journal of Botany*. https://doi.org/10.1016/j.sajb.2021.08.021.

Bachheti, R.K., A. Sharma, A. Bachheti, A.Husen, G.M. Shanka & D.P. Pandey. 2020b. Nanomaterials from various forest tree species and their biomedical applications. In Husen, A. & Jawaid, M., Eds. *Nanomaterials for Agriculture and Forestry Applications*. Elsevier, pp. 81–106. https://doi.org/10.1016/B978-0-12-817852-2.00004-4.

Bachheti, R.K., R. Konwarh, V. Gupta, A. Husen & A. Joshi. 2019. Green synthesis of iron oxide nanoparticles: Cutting edge technology and multifaceted applications. In *Nanomaterials and Plant Potential*. Springer, Cham, pp. 239–259. https://doi.org/10.1007/978-3-030-05569-1_9.

Bachheti, A., A. Sharma, R.K. Bachheti, A. Husen & V.K. Mishra. 2019a. Plant-mediated synthesis of copper oxide nanoparticles and their biological applications. In *Nanomaterials and Plant Potential*. Springer, Cham, pp. 221–237. https://doi.org/10.1007/978-3-030-05569-1_8.

Balaraman, P., B. Balasubramanian, D. Kaliannan, M. Durai, H. Kamyab, S. Park, S. Chelliapan, C.T. Lee, V. Maluventhen & A. Maruthupandian. 2020. Phyco-synthesis of silver nanoparticles mediated from marine algae *Sargassum myriocystum* and its potential biological and environmental applications. *Waste and Biomass Valorization* 11(10): 5255–5271.

Bansal, V., D. Rautaray, A. Ahmad & M. Sastry. 2004. Biosynthesis of zirconia nanoparticles using the fungus *Fusarium oxysporum*. *Journal of Materials Chemistry* 14(22): 3303–3305.

Barwal, I., P. Ranjan, S. Kateriya & S.C. Yadav. 2011. Cellular oxido-reductive proteins of *Chlamydomonas reinhardtii* control the biosynthesis of silver nanoparticles. *Journal of Nanobiotechnology* 9(1): 56.

Beganskienė, A., V. Sirutkaitis, M. Kurtinaitienė, R. Juškėnas & A. Kareiva. 2004. FTIR, TEM and NMR investigations of stöber silica nanoparticles. *Material Science (Medžiagotyra)* 10: 287–290.

Beshah, F., Y. Hunde, M. Getachew, R.K. Bachheti, A. Husen & A. Bachheti. 2020. Ethnopharmacological, phytochemistry and other potential applications of *Dodonaea* genus: A comprehensive review. *Current Research in Biotechnology* 2: 103–119.

Bhattacharya, P., S. Swarnakar, S. Ghosh, S. Majumdar & S. Banerjee. 2019. Disinfection of drinking water via algae mediated green synthesized copper oxide nanoparticles and its toxicity evaluation. *Journal of Environmental Chemical Engineering* 7(1): 102867.

Bhattacharya, R. & P. Mukherjee. 2008. Biological properties of "naked" metal nanoparticles. *Advanced Drug Delivery Reviews* 60(11): 1289–1306.

Bhattarai, B., Y. Zaker & T.P. Bigioni. 2018. Green synthesis of gold and silver nanoparticles: Challenges and opportunities. *Current Research in Green and Sustainable Chemistry* 12: 91–100.

Borase, H.P., C.D. Patil, R.K. Suryawanshi, S.H. Koli, B.V. Mohite, G. Benelli & S.V. Patil. 2017. Mechanistic approach for fabrication of gold nanoparticles by Nitzschia diatom and their antibacterial activity. *Bioprocess and Biosystems Engineering* 40(10): 1437–1446.

Bowen, P.E., M. Stacewicz-Sapuntzakis & V. Diwadkar-Navsariwala. 2015. Carotenoids in human nutrition. In Chen, C., Ed. *Pigments in Fruits and Vegetables*, Springer, New York, pp. 31–67.

Cai, W., T. Gao, H. Hong & J. Sun. 2008. Applications of gold nanoparticles in cancer nanotechnology. *Nanotechnological Science Applications* 1: 17.

Castro, L., M.L. Blázquez, J.A. Muñoz, F. González & A. Ballester. 2013. Biological synthesis of metallic nanoparticles using algae. *IET Nanobiotechnology* 7(3): 109–116.

Chandran, P.R., M. Naseer, N. Udupa & N. Sandhyarani. 2011. Size controlled synthesis of biocompatible gold nanoparticles and their activity in the oxidation of NADH. *Nanotechnology* 23(1): 015602.

Chang, Q., W. Wang, G. Regev-Yochay, M. Lipsitch & W.P. Hanage. 2018. Antibiotics in agriculture and the risk to human health: How worried should we be? *Evolutionary Applications* 8(3): 240–247.

Chauhan, R., A. Kumar & J.A. Abraham. 2013. A Biological approach to the synthesis of silver nanoparticles with Streptomyces sp JAR1 and its antimicrobial activity. *Science Pharmaceutical* 81(2): 607–624.

Chavan, R.R., S.D. Bhinge, M.A. Bhutkar, D.S. Randive, G.H. Wadkar, S.S. Todkar & M.N. Urade. 2020. Characterization, antioxidant, antimicrobial and cytotoxic activities of green synthesized silver and iron nanoparticles using alcoholic *Blumea eriantha* DC plant extract. *Materials Today Communications* 24: 101320.

Cheung, R.C.F., T.B. Ng, J.H. Wong & W.Y. Chan. 2015. Chitosan: An update on potential biomedical and pharmaceutical applications. *Marine Drugs* 13(8): 5156–5186.

Cox, A., P. Venkatachalam, S. Sahi & N. Sharma. 2017. Reprint of: Silver and titanium dioxide nanoparticle toxicity in plants: A review of current research. *Plant Physiology and Biochemistry: PPB* 110: 33–49.

Cui, Y., Y. Zhao, Y. Tian, W. Zhang, X. Lü & X. Jiang. 2012. The molecular mechanism of action of bactericidal gold nanoparticles on Escherichia coli. *Biomaterials* 33(7): 2327–2333.

Dahoumane, S.A., C. Djediat, C. Yéprémian, A. Couté, F. Fiévet, T. Coradin & R. Brayner. 2012. Species selection for the design of gold nanobioreactor by photosynthetic organisms. *Journal of Nanoparticle Research* 14(6): 883.

Dahoumane, S.A., C. Yéprémian, C. Djédiat, A. Couté, F. Fiévet, T. Coradin & R. Brayner. 2014. A global approach of the mechanism involved in the biosynthesis of gold colloids using micro-algae. *Journal of Nanoparticle Research* 16(10): 2607.

Dahoumane, S.A., E.K. Wujcik & C. Jeffryes. 2016. Noble metal, oxide and chalcogenide-based nanomaterials from scalable phototrophic culture systems. *Enzyme and Microbial Technology* 95: 13–27.

Dasgupta, N., S. Ranjan, D. Mundekkad, C. Ramalingam, R. Shanker & A. Kumar. 2015. Nanotechnology in agro-food: From field to plate. *International Food Research Journal* 69: 381–400.

Dauthal, P. & M. Mukhopadhyay. 2016. Noble metal nanoparticles: Plant-mediated synthesis, mechanistic aspects of synthesis, and applications. *Industrial and Engineering Chemistry Research* 55(36): 9557–9577.

Deglint, J.L., J.L. Jin & C.A. Wong. 2019. Investigating the automatic classification of algae using the spectral and morphological characteristics via deep residual learning. In Karray, F., Campilho, A. & Yu, A., Eds. *Proceedings of the International Conference on Image Analysis and Recognition*. Springer, Cham, Switzerland, pp. 269–280.

Dhanalakshmi, P., R. Azeez, R. Rekha, S. Poonkodi & T. Nallamuthu. 2012. Synthesis of silver nanoparticles using green and brown seaweeds. *Phykos* 42: 39–45.

Dhas, T.S., V.G. Kumar, L.S. Abraham, V. Karthick & K. Govindaraju. 2012. *Sargassum myriocystum* mediated biosynthesis of gold nanoparticles. *Spectrochimica Acta, Part A: Molecular and Biomolecular Spectroscopy* 99: 97–101.

Dhavale, R., S. Jadhav & G. Sibi. 2019. Microalgae mediated silver nanoparticles (Ag-NPs) synthesis and their biological activities. *Journal of Critical Reviews* 7: 2020.

Dhillon, G.S., S.K. Brar, S. Kaur & M. Verma. 2012. Green approach for nanoparticle biosynthesis by fungi: Current trends and applications. *Critical Reviews in Biotechnology* 32(1): 49–73.

Dobrucka, R. 2020. Metal nanoparticles in nanosensors for food quality assurance. *Log Forum* 16(2): 271–278.

Dubchak, S., A. Ogar, J. Mietelski & K. Turnau. 2010. Influence of silver and titanium nanoparticles on arbuscular mycorrhiza colonization and accumulation of radiocaesium in *Helianthus annuus*. *Spanish Journal of Agricultural Research* 8(1): 103–108.

Durán, N., M. Durán, M.B. De Jesus, A.B. Seabra, W.J. Fávaro & G. Nakazato. 2016. Silver nanoparticles: A new view on mechanistic aspects on antimicrobial activity. *Nanomedicine: Nanotechnology, Biology, and Medicine* 12(3): 789–799.

El-Kassas, H.Y. & M.M. El-Sheekh. 2014. Cytotoxic activity of biosynthesized gold nanoparticles with an extract of the red seaweed *Corallin officinalis* on the MCF-7 human breast cancer cell line. *Asian Pacific Journal of Cancer Prevention: APJCP* 15(10): 4311–4317.

El-Rafie, H., M. El-Rafie & M. Zahran. 2013. Green synthesis of silver nanoparticles using polysaccharides extracted from marine macro algae. *Carbohydrate Polymers* 96(2): 403–410.

Etoh, H., Y. Utsunomiya, A. Komori, Y. Murakami, S. Oshima & T. Inakuma. 2000. Carotenoids in human blood plasma after ingesting paprika juice. *Bioscience, Biotechnology, and Biochemistry* 64(5): 1096–1098.

Fasciotti, M. 2017. Perspectives for the use of biotechnology in green chemistry applied to biopolymers, fuels and organic synthesis: From concepts to a critical point of view. *Sustainable Chemistry and Pharmacy* 6: 82–89.

Fawcett, D., J.J. Verduin, M. Shah, S.B. Sharma & G.E.J. Poinern. 2017. A review of current research into the biogenic synthesis of metal and metal oxide nanoparticles via marine algae and seagrasses. *Journal of Nanoscience* 1(1): 1–15.

Fernández-Luqueño, F., G. Medina-Pérez, F. López-Valdez, R. Gutiérrez-Ramírez, R.G. Campos-Montiel, E. Vázquez-Núñez, S. Loera-Serna, I. Almaraz-Buendía, O.E. Del Razo-Rodríguez & A. Madariaga-Navarrete. 2018. Use of Agro nano biotechnology in the Agro-Food industry to preserve environmental health and improve the welfare of farmers. In López-Valdez, F. & Fernández-Luqueño, F., Eds. *Agricultural Nanobiotechnology: Modern Agriculture for a Sustainable Future*. Springer International Publishing, Cham, Switzerland, pp. 3–16.

Fouda, A., E. Saad, S.S. Salem & T.L. Shaheen. 2018. In-vitro cytotoxicity, antibacterial, and UV protection properties of the biosynthesized zinc oxide nanoparticles for medical textile applications. *Microbial Pathogenesis* 125: 252–261.

Ghodake, G. & D.S. Lee. 2011. Biological synthesis of gold nanoparticles using the aqueous extract of the brown algae *Laminaria japonica*. *Journal of Nanoelectronics and Optoelectronics* 6(3): 268–271.

Giljohann, D.A., D.S. Seferos, W.L. Daniel, M.D. Massich, P.C. Patel & C.A. Mirkin. 2010. Gold nanoparticles for biology and medicine. *Angewandte Chemie - International Edition* 49(19): 3280–3294.

Golinska, P., M. Wypij, A.P. Ingle, I. Gupta, H. Dahm & M. Rai. 2014. Biogenic synthesis of metal nanoparticles from actinomycetes: Biomedical applications and cytotoxicity. *Applied Microbiology and Biotechnology* 98(19): 8083–8097.

González-Ballesteros, N., J. González-Rodríguez, M. Rodríguez-Argüelles & M. Lastra. 2018. New application of two Antarctic macroalgae *Palmaria decipiens* and *Desmarestia menziesii* in the synthesis of gold and silver nanoparticles. *Polar Science* 15: 49–54.

González-Ballesteros, N., L. Diego-González, M. Lastra-Valdor, M. Rodríguez-Argüelles, M. Grimaldi, A. Cavazza, F. Bigi & R. Simón-Vázquez. 2019. Immunostimulant and biocompatible gold and silver nanoparticles synthesized using the *Ulva intestinalis* L. aqueous extract. *Journal of Materials Chemistry B* 7(30): 4677–4691.

Gopinath, V., D. MubarakAli, S. Priyadarshini, N.M. Priyadharsshini, N. Thajuddin & P. Velusamy. 2012. Biosynthesis of silver nanoparticles from *Tribulus terrestris* and its antimicrobial activity: A novel biological approach. *Colloids and Surfaces, Part B: Biointerfaces* 96: 69–74.

Gopu, M., P. Kumar, T. Selvankumar, B. Senthilkumar, C. Sudhakar, M. Govarthanan, R.S. Kumar & K. Selvam. 2020. Green biomimetic silver nanoparticles utilizing the red algae *Amphiroa rigida* and its potent antibacterial, cytotoxicity and larvicidal efficiency. *Bioprocess and Biosystems Engineering* 44(2): 217–223.

Gour, A. & N.K. Jain. 2019. Advances in green synthesis of nanoparticles. *Artificial Cells, Nanomedicine and Biotechnology* 47(1): 844–851.

Govindaraju, K., S.K. Basha, V.G. Kumar & G. Singaravelu. 2008. Silver, gold and bimetallic nanoparticles production using single-cell protein (*Spirulina platensis*) Geitler. *Journal of Materials Science* 43(15): 5115–5122.

Graily-Moradi, F., M. Mallak & M. Ghorbanpour. 2020. Biogenic synthesis of gold nanoparticles and their potential application in agriculture. In Ghorbanpour, M., Bhargava, P., Varma, A., & Choudhary, D.K. (Eds.) *Biogenic Nano-Particles and Their Use in Agro-Ecosystems*. Springer, Singapore, pp. 187–204.

Hamouda, R.A., M.H. Hussein, R.A. Abo-Elmagd & S.S. Bawazir. 2019. Synthesis and biological characterization of silver nanoparticles derived from the cyanobacterium *Oscillatoria limnetica. Scientific Reports* 9(1): 13071.

Hassan, S.E.D., A. Fouda & A.A. Radwan. 2019. Endophytic actinomycetes Streptomyces spp mediated biosynthesis of copper oxide nanoparticles as a promising tool for biotechnological applications. *JBIC, Journal of Biological Inorganic Chemistry* 24(3): 377–393.

Hekmati, M., S. Hasanirad, A. Khaledi & D. Esmaeili. 2020. Green synthesis of silver nanoparticles using extracts of Allium rotundum l, *Falcaria vulgaris* Bernh, and *Ferulago angulate* Boiss, and their antimicrobial effects in vitro. *Gene Reports* 19: 100589.

He, X., H. Deng & H.M. Hwang. 2019. The current application of nanotechnology in food and agriculture. *Journal of Food and Drug Analysis* 27(1): 1–21.

Hou, D. & D. O'Connor. 2020. Chapter 1: Green and sustainable remediation: Concepts, principles, and pertaining research. In Hou, D., Ed. *Sustainable Remediation of Contaminated Soil and Groundwater.* Butterworth-Heinemann, Waltham, MA, pp. 1–17.

Hussain, I., N.B. Singh, A. Singh, H. Singh & S.C. Singh. 2016. Green synthesis of nanoparticles and its potential application. *Biotechnology Letters* 38(4): 545–560.

Husen, A., Q.I. Rahman, M. Iqbal, M.O. Yassin & R.K. Bachheti. 2019. Plant-mediated fabrication of gold nanoparticles and their applications. In *Nanomaterials and Plant Potential.* Springer, Cham, pp. 71–110. https://doi.org/10.1007/978-3-030-05569-1_3.

Islam, N., K. Jalil, M. Shahid, A. Rauf, N. Muhammad, A. Khan, M.R. Shah & M.A. Khan. 2019. Green synthesis and biological activities of gold nanoparticles functionalized with Salix alba. *Arabian Journal of Chemistry* 12(8): 2914–2925.

Jamkhande, P.G., N.W. Ghule, A.H. Bamer & M.G. Kalaskar. 2019. Metal nanoparticles synthesis: An overview on methods of preparation, advantages and disadvantages, and applications. *Journal of Drug Delivery Science and Technology* 53: 101174.

Jena, J., N. Pradhan, R.R. Nayak, B.P. Dash, L.B. Sukla, P.K. Panda & B.K. Mishra. 2014. Microalga Scenedesmus sp.: A potential low-cost green machine for silver nanoparticle synthesis. *Journal of Microbiology and Biotechnology* 24(4): 522–533.

Kalabegishvili, T.L., E.I. Kirkesali, A.N. Rcheulishvili, E.N. Ginturi, I.G. Murusidze, D.T. Pataraya, M.A. Gurielidze, G.I. Tsertsvadze, V.N. Gabunia & L.G. Lomidze. 2012. Synthesis of gold nanoparticles by some strains of Arthrobacter genera. *Materials Science and Engineering. Part A, Structural Materials: Properties, Microstructure and Processing* 2: 164–173.

Kannan, R., W. Stirk & J. Van Staden. 2013. Synthesis of silver nanoparticles using the seaweed *Codium capitatum* PC Silva (Chlorophyceae). *South African Journal of Botany* 86: 1–4.

Kaviya, S., J. Santhanalakshmi, B. Viswanathan, J. Muthumary & K. Srinivasan. 2011. Biosynthesis of silver nanoparticles using *Citrus sinensis* peel extract and its antibacterial activity. *Spectrochimica Acta. Part A: Molecular and Biomolecular Spectroscopy* 79(3): 594–598.

Kesarla, M.K., B.K. Mandal & P.R. Bandapalli. 2014. Gold nanoparticles by *Terminalia bellirica* aqueous extract—A rapid green method. *Journal of Experimental Nanoscience* 9(8): 825–830.

Khalaj, M., M. Kamali, M.E.V. Costa & I. Capela. 2020. Green synthesis of nanomaterials—A scientometric assessment. *Journal of Cleaner Production* 267: 122036.

Khalil, M.M., E.H. Ismail, K.Z. El-Baghdady & D. Mohamed. 2014. Green synthesis of silver nanoparticles using olive leaf extract and its antibacterial activity. *Arabian Journal of Chemistry* 7(6): 1131–1139.

Khanna, P., A. Kaur & D. Goyal. 2019. Algae-based metallic nanoparticles: Synthesis, characterization and applications. *Journal of Microbiological Methods* 163: 105656.

Kim, S.K., N.V. Thomas & X. Li. 2011. Anticancer compounds from marine macroalgae and their application as medicinal foods. *Advances in Food and Nutrition Research* 64: 213–224.

K.J.P. 2017. Multi-functional silver nanoparticles for drug delivery: A review. *International Journal of Current Medical and Pharmaceutical Research* 9: 1–5.

Kumar, D.A., V. Palanichamy & S.M. Roopan. 2014. Green synthesis of silver nanoparticles using *Alternanthera dentata* leaf extract at room T and their antimicrobial activity. *Spectrochimica Acta - Part A: Molecular and Biomolecular Spectroscopy* 127: 168–171.

Kumar-Krishnan, S., E. Prokhorov, M. Hernández-Iturriaga, J.D. Mota-Morales, M. Vázquez-Lepe, Y. Kovalenko, L.C. Sanchez & G. Luna-Bárcenas. 2015. Chitosan/silver nanocomposites: Synergistic antibacterial action of silver nanoparticles and silver ions. *European Polymer Journal* 67: 242–251.

Kumar, P., M. Govindaraju, S. Senthamilselvi & K. Premkumar. 2013. Photocatalytic degradation of methyl orange dye using silver (Ag) nanoparticles synthesized from Ulva Lactuca. *Colloids and Surfaces, Part B: Biointerfaces* 103: 658–661.

Kumar, P., S. Senthamilselvi, A. Lakshmipraba, K. Premkumar, R. Muthukumaran, P. Visvanathan, R. Ganeshkumar & M. Govindaraju. 2012. Efficacy of bio-synthesized silver nanoparticles using *Acanthophora spicifera* to encumber biofilm formation. *Digest Journal of Nanomaterials and Biostructures* 7: 511–522.

Kumar, P., S. Senthamil Selvi, A. Lakshmi Prabha, K. Prem Kumar, R. Ganeshkumar & M. Govindaraju. 2012. Synthesis of silver nanoparticles from *Sargassum tenerrimum* and screening phytochemicals for its antibacterial activity. *Nano Biomedicine and Engineering* 4(1): 12–16.

Kumar, S.S., Y. Kumar, M. Khan, J. Anbu & E. De Clercq. 2011. Antihistaminic and antiviral activities of steroids of *Turbinaria conoides. Natural Product Research* 25(7): 723–729.

Lengke, M.F., B. Ravel, M.E. Fleet, G. Wanger, R.A. Gordon & G. Southam. 2006. Mechanisms of gold bioaccumulation by filamentous cyanobacteria from gold (III) chloride complex. *Environmental Science and Technology* 40(20): 6304–6309.

Lengke, M.F., M.E. Fleet & G. Southam. 2006. Morphology of gold Nano particles synthesized by filamentous cyanobacteria from gold(I) –Thiosulfate and gold(III)–chloride complexes. *Langmuir: The ACS Journal of Surfaces and Colloids* 22(6): 2780–2787.

Lengke, M.F., M.E. Fleet & G. Southam. 2007. Biosynthesis of silver Nano particles by filamentous cyanobacteria from a silver (I) nitrate complex. *Langmuir: The ACS Journal of Surfaces and Colloids* 23(5): 2694–2699.

LewisOscar, F., S. Vismaya, M. Arunkumar, N. Thajuddin, D. Dhanasekaran & C. Nithya. 2016. Algal nanoparticles: Synthesis and biotechnological potentials. *Algae Organic Imminent Biotechnology* 7: 157–182.

Liu, B., J. Xie, J. Lee, Y. Ting & J.P. Chen. 2005. Optimization of high-yield biological synthesis of single-crystalline gold nanoplates. *Journal of Physical Chemistry. Part B, Condensed Matter, Materials, Surfaces, Interfaces and Biophysical* 109(32): 15256–15263.

Li, X., K. Schirmer, L. Bernard, L. Sigg, S. Pillai & R. Behra. 2015. Silver nanoparticle toxicity and association with the alga *Euglena gracilis. Environmental Science: Nano* 2(6): 594–602.

Madhiyazhagan, P., K. Murugan, A.N. Kumar, T. Nataraj, D. Dinesh, C. Panneerselvam, J. Subramaniam, P.M. Kumar, U. Suresh, M. Roni, M. Nicoletti, A.A. Alarfaj, A. Higuchi, M.A. Munusamy & G. Benelli. 2015. *Sargassum muticum* synthesized silver nanoparticles: An effective control tool against mosquito vectors and bacterial pathogens. *Parasitology Research* 114(11): 4305–4317.

Madhiyazhagan, P., K. Murugan, A.N. Kumar, T. Nataraj, J. Subramaniam, B. Chandramohan, C. Panneerselvam, D. Dinesh, U. Suresh, M. Nicoletti, M.S. Alsalhi, S. Devanesan & G. Benelli. 2017. One pot synthesis of silver nanocrystals using the seaweed *Gracilaria edulis*: Biophysical characterization and potential against the filariasis vector *Culex quinquefasciatus* and the midge *Chironomus circumdatus. Journal of Applied Phycology* 29(1): 649–659.

Mallmann, E.J.J., F.A. Cunha, B.N. Castro, A.M. Maciel, E.A. Menezes & P.B.A. Fechine. 2015. Antifungal activity of silver nanoparticles obtained by green synthesis. *Revista do Instituto de Medicina Tropical de Sao Paulo* 57(2): 165–167.

Maneerung, T., S. Tokura & R. Rujiravanit. 2008. Impregnation of silver nanoparticles into bacterial cellulose for antimicrobial wound dressing. *Carbohydrate Polymers* 72(1): 43–51.

Manimegalai, G., S.S. Kumar & C. Sharma. 2011. Pesticide mineralization in water using silver nanoparticles. *International Journal of Chemical Sciences* 9: 1463–1471.

Manivasagan, P., J. Venkatesan, K. Senthilkumar, K. Sivakumar & S.K. Kim. 2013. Biosynthesis, antimicrobial and cytotoxic effect of silver nanoparticles using a novel Nocardiopsis sp. MBRC-1. *BioMed Research International* 2013: 287638.

Marchiol, L., A. Mattiello, F. Poš´ci´c, C. Giordano & R. Musetti. 2014. In vivo synthesis of nanomaterials in plants: Location of silver nanoparticles and plant metabolism. *Nanoscale Research Letters* 9(1): 101.

Mata, Y., E. Torres, M. Blazquez, A. Ballester, F. González & J. Munoz. 2009. Gold (III) biosorption and bioreduction with the brown alga *Fucus vesiculosus. Journal of Hazardous Materials* 166(2–3): 612–618.

Meléndez-Martínez, A.J., C.M. Stinco, P.M. Brahm & I.M. Vicario. 2014. Analysis of carotenoids and tocopherols in plant matrices and assessment of their in vitro antioxidant capacity. *Methods Mol Biol* 1153(1): 77–97.

Meléndez-Martínez, A.J., P. Mapelli-Brahm & C.M. Stinco. 2018. The colourless carotenoids phytoene and phytofluene: From dietary sources to their usefulness for the functional foods and nutricosmetics industries. *Journal of Food Composition and Analysis* 67: 91–103.

Mishra, A., S.K. Tripathy & S.I. Yun. 2011. Bio-Synthesis of Gold and Silver Nanoparticles from Candida guillier mondii and *their antimicrobial* effect against pathogenic bacteria. *Journal of Nanoscience and Nanotechnology* 11(1): 243–248.

Mohamed, A., A. Fouda & M. Elgamal. 2017. Enhancing of cotton fabric antibacterial properties by silver nanoparticles synthesized by new Egyptian strain *Fusarium keratoplasticum* A1-3. *Egyptian Journal of Chemistry* 60 (Conference Issue The 8th International Conference of The Textile Research Division, *National Research Centre, Cairo 12622, Egypt*), 63–71.

Mohanpuria, P., N.K. Rana & S.K. Yadav. 2008. Biosynthesis of nanoparticles: Technological concepts and future applications. *Journal of Nanoparticle Research* 10(3): 507–517.

Mubarak Ali, D., J. Arunkumar, K.H. Nag, K.A. SheikSyedIshack, E. Baldev, D. Pandiaraj & N. Thajuddin. 2013. Gold nanoparticles from Pro and eukaryotic photosynthetic microorganisms—Comparative studies on synthesis and its application on biolabelling. *Colloids and Surfaces, Part B: Biointerfaces* 103: 166–173.

Murphy, C.J., A.M. Gole, J.W. Stone, P.N. Sisco, A.M. Alkilany, E.C. Goldsmith & S.C. Baxter. 2008. Gold nanoparticles in biology: Beyond toxicity to cellular imaging. *Accounts of Chemical Research* 41(12): 1721–1730.

Murugesan, S., S. Bhuvaneswari & V. Sivamurugan. 2017. Green synthesis, characterization of silver nanoparticles of a marine red alga *Spyridia fusiformis* and their antibacterial activity. *International Journal of Pharmaceutical Sciences and Research* 9(5): 192–197.

Muthukumar, H., S.K. Palanirajan, M.K. Shanmugam & S.N. Gummadi. 2020. Plant extract mediated synthesis enhanced the functional properties of silver ferrite nanoparticles over chemical mediated synthesis. *Biotechnological Reports* 26: e00469.

Muthuvel, A., K. Adavallan, K. Balamurugan & N. Krishnakumar. 2014. Biosynthesis of gold nanoparticles using *Solanum nigrum* leaf extract and screening their free radical scavenging and antibacterial properties. *Biomedical Previous Nutrition* 4(2): 325–332.

Nabikhan, A., K. Kandasamy, A. Raj & N.M. Alikunhi. 2010. Synthesis of antimicrobial silver nanoparticles by callus and leaf extracts from saltmarsh plant, *Sesuvium portulacastrum* L. *Colloids and Surfaces, Part B: Biointerfaces* 79(2): 488–493.

Nadeem, M., B.H. Abbasi, M. Younas, W. Ahmad & W.T. Khan. 2017. A review of the green syntheses and anti-microbial applications of gold nanoparticles. *Green Chemistry Letters and Reviews* 10(4): 216–227.

Nakkala, J.R., R. Mata, A.K. Gupta & S.R. Sadras. 2014. Biological activities of green silver nanoparticles synthesized with *Acorous calamus* rhizome extract. *European Journal of Medicinal Chemistry* 85: 784–794.

Nakkala, J.R., R. Mata, E. Bhagat & S.R. Sadras. 2015. Green synthesis of silver and gold nanoparticles from *Gymnema sylvestre* leaf extract: Study of antioxidant and anticancer activities. *Journal of Nanoparticle Research* 17(3): 151.

Namvar, F., S. Azizi, M.B. Ahmad, K. Shameli, R. Mohamad, M. Mahdavi & P.M. Tahir. 2015. Green synthesis and characterization of gold nanoparticles using the marine macroalgae *Sargassum muticum*. *Research on Chemical Intermediates* 41(8): 5723–5730.

Narayanan, K.B. & N. Sakthivel. 2011. Green synthesis of biogenic metal nanoparticles by terrestrial and aquatic phototrophic and heterotrophic eukaryotes and biocompatible agents. *Advances in Colloid and Interface Science* 169(2): 59–79.

Nirmala Grace, A. & K. Pandian. 2007. Antibacterial efficacy of aminoglycosidic antibiotics protected gold nanoparticles—A brief study. *Colloids Surface Physicochemical Engineering Aspects* 297(1–3): 63–70.

Nishanthi, R., S. Malathi & P. Palani. 2019. Green synthesis and characterization of bioinspired silver, gold and platinum nanoparticles and evaluation of their synergistic antibacterial activity after combining with different classes of antibiotics. *Materials Science and Engineering C: Biomimetic Materials Sensors and Systems* 96: 693–707.

Noah, N. 2019. Chapter 6: Green synthesis: Characterization and application of silver and gold nanoparticles. In Shukla, A.K. & Iravani, S., Eds. *Green Synthesis, Characterization and Applications of Nanoparticles*. Micro and Nano Technologies. Elsevier, Amsterdam, The Netherlands, pp. 111–135.

Nogueira, L.F.B., E.J. Guidelli, S.M. Jafari & A.P. Ramos. 2020. Green synthesis of metal nanoparticles by plant extracts and biopolymers. In Jafari, S.M., Ed. *Handbook of Food Nanotechnology*. Academic Press, Cambridge, MA, pp. 257–278.

Omara, A.E.D., T. Elsakhawy, T. Alshaal, H. El-Ramady, Z. Kovács & M. Fári. 2019. Nanoparticles: A novel approach for sustainable agro-productivity. *Environment, Biodiversity and Soil Security* 3(2019): 29–62.

Omomowo, I., V. Adenigba, S. Ogunsona, G. Adeyinka, O. Oluyide, A. Adedayo & B. Fatukasi. 2020. Antimicrobial and antioxidant activities of algal-mediated silver and gold nanoparticles. *IOP Conference Series: Materials Science and Engineering* 805(1): 012010.

Oza, G., S. Pandey, A. Mewada, G. Kalita, M. Sharon, J. Phata, W. Ambernath & M. Sharon. 2012. Facile biosynthesis of gold nanoparticles exploiting optimum pH and T of fresh water algae *Chlorella pyrenoidusa. Advances in Applied Science Research* 3: 1405–1412.

Padalia, H., P. Moteriya & S. Chanda. 2015. Green synthesis of silver nanoparticles from marigold flower and its synergistic antimicrobial potential. *Arabian Journal of Chemistry* 8(5): 732–741.

Palit, S. & C.M. Hussain. 2020. Chapter 1: Functionalization of nanomaterials for industrial applications: Recent and future perspectives. In Mustansar Hussain, C., Ed. *Handbook of Functionalized Nanomaterials for Industrial Applications.* Micro and Nano Technologies. Elsevier, Amsterdam, The Netherlands, pp. 3–14.

Pardhi, D.M., D. SenKaraman, J.W.W. Timonen, Q. Zhang, S. Satija, M. Mehta, M. Charbe, N. McCarron, P.A. Tambuwala, M.M. Tambuwala, H.A. Bakshi, P. Negi, A.A. Aljabali, K. Dua, D.K. Chellappan, A. Behera, K. Pathak, R.B. Watharkar, J. Rautio & J.M. Rosenholm. 2020. Anti-bacterial activity of inorganic nanomaterials and their antimicrobial peptide conjugates against resistant and non-resistant pathogens. *International Journal of Pharmaceutics* 586: 119531.

Parial, D., H.K. Patra, P. Roychoudhury, A.K. Dasgupta & R. Pal. 2012. Gold nanorod production by cyanobacteria—A green chemistry approach. *Journal of Applied Phycology* 24(1): 55–60.

Painuli, S., P. Semwal, A. Bacheti, R.K. Bachheti & A. Husen. 2020. Nanomaterials from nonwood forest products and their applications. In Husen, A. & Jawaid, M., Eds. *Nanomaterials for Agriculture and Forestry Applications.* Elsevier, pp. 15–40. https://doi.org/10.1016/B978-0-12-817852-2.00002-0.

Parial, D., H.K. Patra, A.K. Dasgupta & R. Pal. 2012. Screening of different algae for green synthesis of gold nanoparticles. *European Journal of Phycology* 47(1): 22–29.

Parial, D. & R. Pal. 2014. Green synthesis of gold nanoparticles using cyanobacteria and their characterization. *Indian Journal of Applied Research* 4(1): 69–72.

Parial, D. & R. Pal. 2015. Biosynthesis of monodisperse gold nanoparticles by green alga Rhizoclonium and associated biochemical changes. *Journal of Applied Phycology* 27(2): 975–984.

Park, J.H., G. von Maltzahn, L. Zhang, M.P. Schwartz, E. Ruoslahti, S.N. Bhatia & M.J. Sailor. 2008. Magnetic iron oxide nanoworms for tumor targeting and imaging. *Advanced Materials* 20(9): 1630–1635.

Patel, V., D. Berthold, P. Puranik & M. Gantar. 2015. Screening of cyanobacteria and microalgae for their ability to synthesize silver nanoparticles with antibacterial activity. *Biotechnological Reports* 5: 112–119.

Pereira, T.M., V.L.P. Polez, M.H. Sousa & L.P. Silva. 2020. Modulating physical, chemical, and biological properties of silver nanoparticles obtained by green synthesis using different parts of the tree *Handroanthus heptaphyllus* (Vell.) Mattos. *Colloids and Interface Science Communications* 34: 100224.

Poggetti, A., P. Battistini & P.D. Paolo. 2016. Nanosurfaces scaffold and magnetic nanoparticles to direct the neuronal growth process: Future strategies for peripheral nerve regeneration. *Journal of Orthopaedic Case Reports* 6(1): 3.

Prasad, R., A. Bhattacharyya & Q.D. Nguyen. 2017. Nanotechnology in sustainable agriculture: Recent developments, challenges, and perspectives. *Frontiers in Microbiolgy* 8: 1–13.

Prasad, R., R. Pandey & I. Barman. 2016. Engineering tailored nanoparticles with microbes: Quo vadis? *Wiley Interdisciplinary Reviews. Nanomedicine and Nanobiotechnology* 8(2): 316–330.

Prasad, T.N., V.S.R. Kambala & R. Naidu. 2013. Phyconanotechnology: Synthesis of silver nanoparticles using brown marine algae *Cystophora moniliformis* and their characterisation. *Journal of Applied Phycology* 25(1): 177–182.

Priyadharshini, R.I., G. Prasannaraj, N. Geetha & P. Venkatachalam. 2014. Microwave-mediated extracellular synthesis of metallic silver and zinc oxide nanoparticles using macro-algae (*Gracilaria edulis*) extracts and its anticancer activity against human PC3 cell lines. *Applied Biochemistry and Biotechnology* 174(8): 2777–2790.

Pugazhendhi, A., D. Prabakar, J.M. Jacob, I. Karuppusamy & R.G. Saratale. 2018. Synthesis and characterization of silver nanoparticles using *Gelidium amansii* and its antimicrobial property against various pathogenic bacteria. *Microbial Pathogens* 114: 41–45.

Pulz, O. & W. Gross. 2004. Valuable products from biotechnology of microalgae. *Applied Microbiology and Biotechnology* 65(6): 635–648.

Rafique, M., I. Sadaf, M.S. Rafique & M.B. Tahir. 2017. A review on green synthesis of silver nanoparticles and their applications. *Artificial Cells, Nanomedicine and Biotechnology* 45(7): 1272–1291.

Rajakumar, G. & A. Abdul Rahuman. 2011. Larvicidal activity of synthesized silver nanoparticles using *Eclipta prostrata* leaf extract against filariasis and malaria vectors. *Acta Tropica* 118(3): 196–203.

Rajasulochana, P., P. Krishnamoorthy & R. Dhamotharan. 2012. Potential application of Kappaphycus alvarezii in agricultural and pharmaceutical industry. *Journal of Chemical and Pharmaceutical Research* 4: 33–37.

Rajathi, F.A.A., C. Parthiban, V.G. Kumar & P. Anantharaman. 2012. Biosynthesis of antibacterial gold nanoparticles using brown alga, *Stoechospermum marginatum* (kützing). Spectrochim. *Spectrochimica Acta - Part A: Molecular and Biomolecular Spectroscopy* 99: 166–173.

Rajeshkumar, S., C. Kannan & G. Annadurai. 2012. Green synthesis of silver nanoparticles using marine brown algae *Turbinaria conoides* and its antibacterial activity. *International Journal of Pharmacy and Biological Sciences* 3: 502–510.

Rajeshkumar, S., C. Kannan & G. Annadurai. 2012. Synthesis and characterization of antimicrobial silver nanoparticles using marine brown seaweed *Padina tetrastromatica. Drug Invention Today* 4: 511–513.

Rajeshkumar, S., C. Malarkodi, G. Gnanajobitha, K. Paulkumar, M. Vanaja, C. Kannan & G. Annadurai. 2013. Seaweed-mediated synthesis of gold nanoparticles using *Turbinaria conoides* and its characterization. *Journal of Nanostructure in Chemistry* 3(1): 44.

Rajeshkumar, S., C. Malarkodi, K. Paulkumar, M. Vanaja, G. Gnanajobitha & G. Annadurai. 2014. Algae mediated green fabrication of silver nanoparticles and examination of its antifungal activity against clinical pathogens. *Metals and Materials International* 1(1): 692643.

Ramakrishna, M., D.R. Babu, R.M. Gengan, S. Chandra & G.N. Rao. 2016. Green synthesis of gold nanoparticles using marine algae and evaluation of their catalytic activity. *Journal of Nanostructure in Chemistry* 6(1): 1–13.

Ramakritinan, C., S. Shankar, M. Anand & A. Kumaraguru. 2020. Biosynthesis of silver, gold and bimetallic alloy (Ag: Au) Nanoparticles from green alga, Lyngpya sp. In *Proceedings of the 3rd National Conference on Nanaomaterials and Nanotechnology*, Lucknow. *Int J Mindshare* 1(1): 174–187.

Ramesh, A., P. Tamizhdurai, S. Gopinath, K. Sureshkumar, E. Murugan & K. Shanthi. 2019. Facile synthesis of core-shell nanocomposites au catalysts towards abatement of environmental pollutant rhodamine b. *Heliyon* 5(1): e01005.

Rao, P.H., R.R. Kumar, B.G. Raghavan, V.V. Subramanian & V. Sivasubramanian. 2011. Application of phycoremediation technology in the treatment of wastewater from a leather processing manufacturing facility. *Water S. Part A* 37(1): 7–14.

Rashmi, B.N., S.F. Harlapur, B. Avinash, C.R. Ravikumar, H.P. Nagaswarupa, M.R. Anil Kumar, K. Gurushantha & M.S. Santosh. 2020. Facile green synthesis of silver oxide nanoparticles and their electrochemical, photocatalytic and biological studies. *Inorganic Chemistry Communications* 111: 107580.

Ribeiro, C.A.S., L.J.C. Albuquerque, C.E. de Castro, B.L. Batista, A.L.M. de Souza, B.L. Albuquerque, M.S. Zilse, I.C. Bellettini & F.C. Giacomelli. 2019. One-pot synthesis of sugar-decorated gold nanoparticles with reduced cytotoxicity and enhanced cellular uptake. *Colloids and Surfaces. Part A: Physicochemical and Engineering Aspects* 580: 123690.

Sahayaraj, K., S. Rajesh & J. Rathi. 2012. Silver nanoparticles biosynthesis using marine alga *Padina pavonica* (linn.) and its microbicidal activity. *Digest Journal of Nanomaterials and Biostructures* 7: 1557–1567.

Sahoo, P.C., F. Kausar, J.H. Lee & J.I. Han. 2014. Facile fabrication of silver nanoparticle embedded CaCO3 microspheres via microalgae-templated CO2 biomineralization: Application in antimicrobial paint development. *RSC Advances* 4(61): 32562–32569.

Saifuddin, N., C.Y. Nian, L.W. Zhan & K.X. Ning. 2011. Chitosan-silver nanoparticles composite as point-of-use drinking water filtration system for household to remove pesticides in water. *Asian Journal of Biochemistry* 6(2): 142–159.

Salari, Z., F. Danafar, S. Dabaghi & S.A. Ataei. 2016. Sustainable synthesis of silver nanoparticles using macroalgae *Spirogyra varians* and analysis of their antibacterial activity. *Journal of Saudi Chemical Society* 20(4): 459–464.

Salem, S.S., M.M. Fouda & A. Fouda. 2020. Antibacterial, cytotoxicity and larvicidal activity of green synthesized selenium particles using *Penicillium corylophilum. Journal of Cluster Sciences* 32(2): 351–361.

Sani, M. & A. Tatiana. 2017. Síntesis y Caracterización de Nanopartículas de Plata a Partir de Varios Extractos Pigmentados de dos Plantas para Su Aplicación en Celdas Solares Híbridas. Bachelor's Thesis, Facultad de Ingeniería, Universidad de las Fuerzas Armadas: Latacunga, Ecuador.

Sasidharan, D., T.R. Namitha, S.P. Johnson, V. Jose & P. Mathew. 2020. Synthesis of silver and copper oxide nanoparticles Using Myristica fragrans fruit extract: Antimicrobial and catalytic applications. *Sustainable Chemistry and Pharmacy* 16: 100255.

Schmid, G. 2011. *Nanoparticles: From Theory to Application*. Hoboken, NJ: John Wiley & Sons, p. 8.

Selvam, G.G. & K. Sivakumar. 2015. Phycosynthesis of silver nanoparticles and photocatalytic degradation of methyl orange dye using silver (Ag) nanoparticles synthesized from *Hypnea musciformis* (Wulfen) JV Lamouroux. *Applied Nanoscience* 5(5): 617–622.

Senapati, S., A. Syed, S. Moeez, A. Kumar & A. Ahmad. 2012. Intracellular synthesis of gold nanoparticles using alga Tetraselmis kochinensis. *Materials Letters* 79: 116–118.

Shakibaie, M., H. Forootanfar, K. Mollazadeh-Moghaddam, Z. Bagherzadeh, N. Nafissi-Varcheh, A.R. Shahverdi & M.A. Faramarzi. 2010. Green synthesis of gold nanoparticles by the marine microalga *Tetraselmis suecica*. *Biotechnology and Applied Biochemistry* 57(2): 71–75.

Sharma, A., S. Sharma, K. Sharma, S.P.K. Chetri, A. Vashishtha, P. Singh, R. Kumar, B. Rathi & V. Agrawal. 2016. Algae as crucial organisms in advancing nanotechnology: A systematic review. *Journal of Applied Phycology* 28(3): 1759–1774.

Sheng, Z. & Y. Liu. 2011. Effects of silver nanoparticles on wastewater biofilms. *Water Research* 45(18): 6039–6050.

Sicard, C., R. Brayner, J. Margueritat, M. Hémadi, A. Couté, C. Yéprémian, C. Djediat, J. Aubard, F. Fiévet, J. Livage & T. Coradin. 2010. Nano-gold biosynthesis by silica-encapsulated micro-algae: A "living" bio-hybrid material. *Journal of Materials Chemistry* 20(42): 9342–9347.

Singaravelu, G., J. Arockiamary, V.G. Kumar & K. Govindaraju. 2007. Anovel extracellular synthesis of monodisperse gold nanoparticles using marine alga, *Sargassum wightii* Greville. *Colloids and Surfaces, Part B: Biointerfaces* 57(1): 97–101.

Singh, C.R., K. Kathiresan & S. Anandhan. 2015. A review on marine based nanoparticles and their potential applications. *African Journal of Biotechnology* 14(18): 1525–1532.

Singh, M., R. Kalaivani, S. Manikandan, N. Sangeetha & A. Kumaraguru. 2013. Facile green synthesis of variable metallic gold nanoparticle using *Padina gymnospora*, a brown marine macroalga. *Applied Nanoscience* 3(2): 145–151.

Sinha, S.N., D. Paul, N. Halder, D. Sengupta & S.K. Patra. 2015. Green synthesis of silver nanoparticles using fresh water green alga *Pithophora oedogonia* (Mont.) Wittrock and evaluation of their antibacterial activity. *Applied Nanoscience* 5(6): 703–709.

Sirelkhatim, A., S. Mahmud, A. Seeni, N.H.M. Kaus, L.C. Ann, S.K.M. Bakhori, H. Hasan & D. Mohamad. 2015. Review on zinc oxide nanoparticles: Antibacterial activity and toxicity mechanism. *Nano-Micro Letters* 7(3): 219–242.

Sk, I., M.A. Khan, A. Haque, S. Ghosh, D. Roy, S. Homechuadhuri & A. Alam. 2020. Synthesis of Gold and Silver nanoparticles using *Malva verticillata* leaves extract: study of Gold nanoparticles catalysed reduction of nitro-schiff bases and antibacterial activities of Silver nanoparticles. *Current Research in Green and Sustainable Chemistry* 3: 100006.

Sreeprasad, T.S. & T. Pradeep. 2013. Noble metal nanoparticles. In Vajtai, R., Ed. *Springer Handbook of Nanomaterials*. Springer, Berlin/Heidelberg, pp. 303–388.

Suganya, K.U., K. Govindaraju, V.G. Kumar, T.S. Dhas, V. Karthick, G. Singaravelu & M. Elanchezhiyan. 2015. Blue green alga mediated synthesis of gold nanoparticles and its antibacterial efficacy against gram positive organisms. *Materials Science and Engineering C: Biomimetic Materials Sensors and Systems* 47: 351–356.

Supraja, N., T. Prasad, M. Soundariya & R. Babujanarthanam. 2016. Synthesis, characterization and dose dependent antimicrobial and anti-cancerous activity of phycogenic silver nanoparticles against human hepatic carcinoma (HepG2) cell line. *AIMS Bioengineering* 3(4): 425–440.

Sushma, D. & S. Richa. 2015. Use of nanoparticles in water treatment: A review. *International Research Journal of Environmental Science* 4: 103–106.

Taghavizadeh Yazdi, M.E., A. Hamidi, M.S. Amiri, R. Kazemi Oskuee, H.A. Hosseini, A. Hashemzadeh & M. Darroudi. 2019. Eco-friendly and plant-based synthesis of silver nanoparticles using *Allium giganteum* and investigation of its bactericidal, cytotoxicity, and photocatalytic effects. *Materials Technology* 34(8): 490–497.

Takeuchi, M.T., M. Kojima & M. Luetzow. 2014. State of the art on the initiatives and activities relevant to risk assessment and risk management of nanotechnologies in the food and agriculture sectors. *International Food Research Journal* 64: 976–981.

Teodoro, K.B.R., F.M. Shimizu, V.P. Scagion & D.S. Correa. 2019. Ternary nanocomposites based on cellulose nanowhiskers, silver nanoparticles and electrospun nanofibers: Use in an electronic tongue for heavy metal detection. *Sensors and Actuators, Part B: Chemical* 290: 387–395.

Thakkar, K.N., S.S. Mhatre & R.Y. Parikh. 2010. Biological synthesis of metallic nanoparticles. *Nanomedicine: Nanotechnology, Biology, and Medicine* 6(2): 257–262.

Thangamani, N. & N. Bhuvaneshwari. 2019. Green synthesis of gold nanoparticles using *Simarouba glauca* leaf extract and their biological activity of micro-organism. *Chemistry Physical Letters* 732: 136587.

Vadlapudi, V. & R. Amanchy. 2017. Synthesis, characterization and antibacterial activity of Silver Nanoparticles from Red Algae, *Hypnea musciformis*. *International Journal of Advanced Biological and Biomedical Research* 11: 242–249.

Vaidyanathan, R., K. Kalishwaralal, S. Gopalram & S. Gurunathan. 2010. Nanosilver—The burgeoning therapeutic molecule and its green synthesis. *Advances in Biotechnology* 28: 940.

van Horssen, R., T.L. Ten Hagen & A.M. Eggermont. 2006. TNF-in cancer treatment: Molecular insights, antitumor effects, and clinical utility. *Oncologist* 11(4): 397–408.

Varma, R.S. 2012. Greener approach to nanomaterials and their sustainable applications. *Current Opinion in Chemical Engineering* 1(2): 123–128.

Venkatesan, J., P. Manivasagan, S.K. Kim, A.V. Kirthi, S. Marimuthu & A.A. Rahuman. 2014. Marine algae-mediated synthesis of gold nanoparticles using a novel *Ecklonia cava*. *Bioprocess and Biosystems Engineering* 37(8): 1591–1597.

Vigneshwaran, N., N. Ashtaputre, P. Varadarajan, R. Nachane, K. Paralikar & R. Balasubramanya. 2007. Biological synthesis of silver nanoparticles using the fungus *Aspergillus flavus*. *Materials Letters* 61(6): 1413–1418.

Vijayaraghavan, K. & T. Ashokkumar. 2017. Plant-mediated biosynthesis of metallic nanoparticles: A review of literature, factors affecting synthesis, characterization techniques and applications. *Journal of Environmental Chemical Engineering* 5(5): 4866–4883.

Vijayan, S.R., P. Santhiyagu, M. Singamuthu, N. Kumari Ahila, R. Jayaraman & K. Ethiraj. 2014. Synthesis and characterization of silver and gold nanoparticles using aqueous extract of seaweed, *Turbinaria conoides*, and their antimicrofouling activity. *Science World Journal* 1(1): 938272.

Vinci, G. & M. Rapa. 2019. Noble metal nanoparticles applications: Recent trends in food control. *Bioengineering* 6(1): 10.

Vivek, M., P.S. Kumar, S. Steffi & S. Sudha. 2011. Biogenic silver nanoparticles by Gelidiella acerosa extract and their antifungal effects. *Avicenna Journal of Medicinal Biotechnology* 3(3): 143.

Wang, Y.W., H. Tang, D. Wu, D. Liu, Y. Liu, A. Cao & H. Wang. 2016. Enhanced bactericidal toxicity of silver nanoparticles by the antibiotic gentamicin. *Environmental Science Nanotechnology* 3(4): 788–798.

Wei, D. & W. Qian. 2008. Facile synthesis of Ag and Au nanoparticles utilizing chitosan as a mediator agent. *Colloids and Surfaces, Part B: Biointerfaces* 62(1): 136–142.

Wang, L., C. Hu & L. Shao. 2017. The antimicrobial activity of nanoparticles: Present situation and prospects for the future. *International Journal of Nanomedicine* 12: 1227.

Xie, J., J.Y. Lee, D.I. Wang & Y.P. Ting. 2007. Identification of active biomolecules in the high-yield synthesis of single-crystalline gold nanoplates in algal solutions. *Small* 3(4): 672–682.

Yadi, M., E. Mostafavi, B. Saleh, S. Davaran, I. Aliyeva, R. Khalilov, M. Nikzamir, N. Nikzamir, A. Akbarzadeh, Y. Panahi & M. Milani. 2018. Current developments in green synthesis of metallic nanoparticles using plant extracts: A review. *Artificial Cells, Nanomedicine and Biotechnology* 46: S336–S343.

Yoon, H.S., K.M. Müller, R.G. Sheath, F.D. Ott & D. Bhattacharya. 2006. Defining the major lineages of red algae (Rhodophyta) 1. *Journal of Phycology* 42(2): 482–492.

Yousefzadi, M., Z. Rahimi & V. Ghafori. 2014. The green synthesis, characterization and antimicrobial activities of silver nanoparticles synthesized from green alga *Enteromorpha flexuosa* (wulfen). *Jacob Georg Agardh Materials Letters* 137: 1–4.

Yugay, Y.A., R.V. Usoltseva, V.E. Silant'ev, A.E. Egorova, A.A. Karabtsov, V.V. Kumeiko, S.P. Ermakova, V.P. Bulgakov & Y.N. Shkryl. 2020. Synthesis of bioactive silver nanoparticles using alginate, fucoidan and laminaran from brown algae as a reducing and stabilizing agent. *Carbohydrate Polymers* 245: 116547.

Zhang, L., Y. Mazouzi, M. Salmain, B. Liedberg & S. Boujday. 2020. Antibody-gold nanoparticle bioconjugates for biosensors: Synthesis, characterization and selected applications. *Biosensors and Bioelectronics* 165(1): 112370.

8 Essential Oils from Medicinal Plants and Their Role in Nanoparticles Synthesis, Characterization, and Applications

Rakesh Kumar Bachheti, Limenew Abate Worku, Yilma Hunde, Mesfin Getachew Tadesse, Archana Bachheti, D.P. Pandey, Ashutosh Sharma, Meseret Zebeaman, and Azamal Husen

CONTENTS

8.1 INTRODUCTION

Essential oils (EOs) are a mixture of various phytochemicals; they are highly volatile chemical constituents of plants with a strong and characteristic aroma (Bayala et al. 2014). The flowers, leaves, fruits, buds, seeds, rhizomes, barks, and roots of plants are the major sources of EOs (Shaaban et al. 2012). According to Tongnuanchanand and Benjakul (2014), EOs contain terpenes, mainly monoterpenes and sesquiterpenes, aromatic hydrocarbons such as aldehydes, alcohols, phenols, methoxy derivatives, and some other hydrocarbons. These multifunctional chemical constituents enable EOs to be applied as antiseptic, anthelmintic, antipruritic, analgesic, anti-inflammatory, local anaesthetic, antispasmodic, and therapeutic agents (Umaru et al. 2019). Some studies also demonstrated that EOs possessed antiviral activities. For example, Pedro et al. (2013) revealed that EOs of *Santolina insularis* have showed effective virucidal activity against HSV-1. Plant EOs are also employed in the drug, perfume, and food industries (Bhavaniramya et al. 2019). The application of EOs in beautifying agents, detergents, and the soap and scent industries is common. However,

DOI: 10.1201/9781003213727-8

the use of EOs in these industries is a great concern from their financial point of view (Naeem et al. 2019; Edris 2007; Tongnuanchan and Benjakul 2014).

The chemical constituents of EOs are highly volatile, unstable, and sensitive to environmental factors such as heat, light, and oxygen during their handling process. Similarly, their applications are significantly limited because of their high volatility, strong sensory, poor water solubility, and instability (Jin et al. 2019). Thus, to alleviate these problems, different scholars blend EOs with nanoparticles (NPs) to test synergetic activities and develop novel products (Theerthavathy et al. 2019). Applying nanotechnology generally improves the physicochemical properties of EOs and their bioactivity (Jin et al. 2019). Encapsulation of EOs components into nanopolymers was reported to enhance their stability against light oxidation (Bilia et al. 2014). Hence, the green synthesis method has received much attention over other NPs synthesis methods by different researchers (Prakasham et al. 2014). For instance, gold nanoparticles (AuNPs) synthesized using *Curcuma pseudomontana* EOs have shown cytotoxicity activity against breast carcinoma cells T47D (Muniyappan and Nagarajan 2014). Similarly, AuNPs synthesized using the assistance of EOs of *Anacardium occidentale* and *Mentha piperita* have demonstrated catalytic and antifungal activities, respectively (Sheny et al. 2012; Thanighaiarassu et al. 2014). Silver nanoparticles (AgNPs) synthesized using plant EOs were also reported for environmental, antimicrobial, and antioxidant applications.

Some recent studies displayed that nanoformulation of EOs isolated from neem, garlic, *Artemisia arborescens*, and *Lippia sidoides* were applied as insecticidal agents. The nanoformulation technique has been found to increase the shelf life of the EOs by preventing the premature degradation of their active chemical constituents (Bilia et al. 2014). Similarly, nanopesticide emulsion of EOs in water effectively delivers the active ingredients of EOs against pests in the pest control (Lai et al. 2006; Sahayaraj et al. 2017). Ultraviolet-visible spectroscopy (UV-vis), Fourier-transform infrared spectroscopy (FTIR), transmission electron microscopy (TEM), scanning electron microscope (SEM), energy dispersive X-ray analysis (EDS), X-ray diffraction (XRD), and dynamic light scattering (DLS) are the common instruments usually used for monitoring and characterizing NP formation (Nancy and Elumalai 2019). This chapter aims to summarize the current information on synthesized NPs using EOs obtained from medicinal plants, their characterization techniques, and their potential applications.

8.2 USES OF ESSENTIAL OILS OF PLANTS AND THEIR IMPORTANT PHYTOCHEMICAL CONSTITUENTS

EOs derived from plants have different uses in the preparation of shampoos, soaps, fragrances, beauty care agents, and cleansing gels. They are also used as potential medicines in aroma-based therapies or transporters for drugs to the target cell (Naeem et al. 2019). The Food and Drug Administration approved that EOs are safe to be used as flavours and preservatives in food. For instance, citrus EOs are very important food and beverage additives. EOs of citrus are also used as antimicrobial agents within the food industry (Palazzolo et al. 2013). Another significant application of EOs of plants is that they have some ecological roles such as healing wounds caused by plants, repelling herbivorous animals to defend plants, and inhibiting microbial growth (Tássio et al. 2020). Some literature suggests that EOs of medicinal plants are applied as the treatment for cancers of the skin, lungs, stomach, and liver (Palazzolo et al. 2013). EOs of flowers are usually used to attract pollinators such as bees, birds, and animals, which is important for a plant species' perpetuation and reproduction (Tássio et al. 2020). They have also demonstrated pesticides and insecticidal activities. For example, the EO obtained from *Nepeta cateria* is highly repellent to bees, mosquitoes, and other flying insects (Anand et al. 2011). It could be an alternative pesticide source (Lingan 2018). Monoterpenes, sesquiterpenes, and their oxygenated derivatives are the EO chemical components responsible for their characteristic aroma and biological activities (Brusotti et al. 2014). The medicinal, fragrant, and food additive uses of EOs of medicinal plants mainly originated from the presence of terpenoids (Mohammed et al. 2018). Terpenes were reported as the main chemical constituents of EOs used for their anticancer and insecticidal activities (Gershenzon and Dudareva 2007). EOs are commonly used for their pesticidal activities. Some of the main constituents of EOs that enable these activities are 1,8- limonene, anethole, carvone, thymol, menthol, α-pinene, limonene, and linalool (Figure 8.1) (Dambolena et al. 2016). EOs are also known for antibacterial activity due to phytochemicals such as terpene, limonene, terpenoids, carvacrol, thymol, menthol, citral, geraniol eugenol, and cinnamaldehyde (Figure 8.2) found in EOs (Pedro et al. 2013).

EOs can be a complex mixture of non-polar and polar compounds with potential characteristic odour, flavour, or scent (Masango 2005;Benchaar et al. 2008). Currently, about 3000 EOs are known and 300 are commercially available. Some 17,500 aromatic flora or plants have been reported to possess EOs. Some plant families that contain EOs are Lamiaceae, Rutaceae, Myrtaceae, Zingiberaceae, and Asteraceae (Regnault-Roger et al. 2012). The main constituents of EOs are monoterpenes, sesquiterpenes, aromatic, and aliphatic hydrocarbons (Porres-Martínez et al. 2015). The most abundant and structurally diverse phytochemicals of EOs are terpenes. They account for 90% of the EOs and form the most important family of natural products in plants (Tongnuanchan and Benjakul 2014). Monoterpenes and sesquiterpenes have two and three isoprene hydrocarbon units, respectively. They contain chemical constituents such as acid, phenol, aldehyde (e.g., sinensal and citronellal), ester (e.g., γ-tepinyl acetate and cedryl acetate), alcohol (e.g., α-bisabolol and geraniol), ketones (e.g., *p*-vetivone and menthone), or lactone which are the classes of oxygenated hydrocarbons (Tássio et al. 2020). EOs also contain non-terpenic compounds as minor components (Dhifi et al. 2016). The dominant composition of EOs is usually from terpenoids (Mohammed et al. 2018) (Table 8.1).

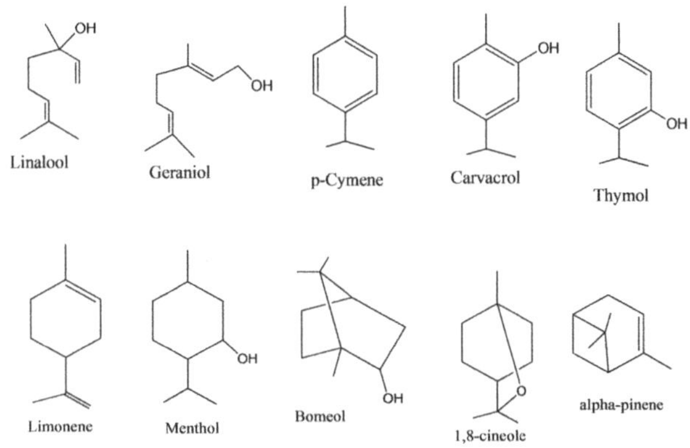

FIGURE 8.1 The structures of terpenes that have pesticidal properties.

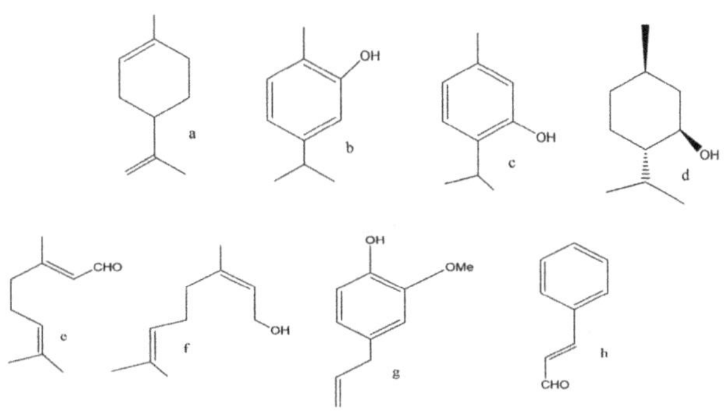

FIGURE 8.2 Structure of phytochemicals derived from essential oils with known antimicrobial activities: (a) limonene and terpenoids, (b) carvacrol, (c) thymol, (d) menthol, (e) citral, (f) geraniol, (g) eugenol, (h) cinnamaldehyde.

TABLE 8.1
Chemical Composition of Essential Oils

Type of constituent of EOs	Specific example	Reference
Non-terpenoid	Eugenol, cinnamaldehyde, and safrole	
Terpenoids	Ascaridol, menthol, sistosterol, and beta-carotene	Bayala et al. (2014)
Monoterpenoid	Beta-mycene, limonene, apha-phelladrene, alpha-terpinene, linalool, nerol, citronellol, carvacrol, tymol, citronella, carvone, alpha-thujone, ascaridol, menthol, sistosterol, geranial (E-citral), beta-caroline, myrcene, Z-*beta*-ocimen, E-*beta* ocimene, terpineol, geraniol, linalool, citronellol, nerol, ,neral (Z-citral), citronellal, α-pinene, camphor, menthona, piperitenone, pulegone, terpinen-4-ol, α-thujone, and Z-isocitral	Bayala et al. (2014) Dhifi et al. (2016) Tássio et al. (2020)
Sesquiterpenes	Zingiberene, humulene, beta-bisabolene, ar-curcumene, trans alpha-bergamotene, gama-carinene, caryophyllene, germacrene, elemane, germacrane, humulane, bergamotane, cadinane, edesmane, eremophilane, muurolane, amorphane, daucane, himalachane, guaiane, caryophyllane, cupranane, copaane, chamigrane, aromadendrane, bourbonane, silphiperfolane, cedreane, seychellane, patchoulane, (E,E)-farnesol, caryophyllene oxide, γ-eudesmol, and δ-cadineneandbisabolane	Bayala et al. (2014) Dhifi et al. (2016) Tássio et al. (2020)
Aromatic compound	Styrene, cinnamyl alcohol, charvicol, eugenol, anethol, safrolvaniline, cinnamaldehyde, thymol, and methyl cinnamate	Chen et al. (2013) Tássio et al. (2020)

Monoterpenoids and sesquiterpenoids are mainly responsible for metal NPs synthesis, especially AgNP production (Javed and Mashwani 2020). For example, Thanighaiarassu et al. (2018) displayed the synthesis of green AgNPs by reducing Ag^+ ions into Ag^0 using citral as a bioreductant and evaluated and determined its significant activity against human pathogenic fungal species.

8.3 ESSENTIAL OILS-BASED NANOPARTICLES SYNTHESIS AND THEIR CHARACTERIZATION

Nowadays, green NPs synthesis is an outstanding biocompatibility method, and hence, it has attracted the attention of many scholars from different perspectives (Husen et al. 2019; Painuli et al. 2020; Bachheti et al. 2021b). Green NPs can be produced through a simple, effective, economic, and eco-friendly biological synthesis technique (Maurya et al. 2016; Bachheti et al. 2020a). The principles of green chemistry such as selection of green solvent, eco-friendly reducing agent, and non-toxic metal should be considered before production of NPs (Baruwati and Varma 2009; Bachheti et al. 2020b). Plant extract-based NPs are monodispersed and more stable materials and natural plant products take less time to reduce metal ions. Plant EOs have unique chemical properties that can mediate the synthesis of NPs (Iravani et al. 2014). Mojtabaet al. (2019) synthesized Fe_3O_4-MgONPs based on both chemical and green methods. The synthesized NPs were characterized by using VSM, FESEM, EDAX, TEM, BET, MS, and HPLC. The synthesized NPs have a spherical shape with an average size of about 10–15 nm. BET analysis reveals the synthesized NPs have a mean pore diameter of 7.57 nm, a total pore volume of 0.3614 $cm^3\ g^{-1}$, and a specific surface area of 190.92 $m^2\ g^{-1}$.

Muniyappan and Nagarajan (2014) used a mixture of ethanol and water to dissolve *Curcuma pseudomontana* EO and the solution was slowly added to $HAuCl_4$ solution to synthesize AuNPs.

Then, synthesized green AuNPs were characterized by using UV-Vis, FTIR, SEM, and HRTEM methods. The obtained result exhibited excellent homogeneity with an average diameter of 20 nm with long-standing stability. Sheny et al. (2012) also reported the synthesis of AuNPs using EO isolated from the fresh leaves of *Anacardium occidental*. The synthesized AuNPs were characterized using UV-Vis, FTIR, and TEM methods and TEM measurements exhibited hexagonal shape with an average size of 36 nm at room temperature. Vilas and Mathew (2016) also synthesized Au and Au/Ag alloy NPs using EOs of *Coleus aromaticus* as bioreductant and UV-Vis, FTIR, TEM, and XRD techniques were employed for the characterization of NPs. The UV-Vis spectra of the synthesized NPs solution were measured in the wavelength range of 200–800 nm performed at a resolution of 1 nm. TEM analysis showed irregular shaped NPs with an average size of about 28 nm. Furthermore, FTIR spectra confirmed the presence of C–O, O–H and C=C functional groups at different vibration frequencies in the green NP sample.

Additionally, Dzimitrowicz et al. (2016) synthesized AuNPs using aqueous extracts of plants and EOs of *Eucalyptus globulus* and *Rosmarinus officinalis*. GC-MS, UV-Vis, FTIR, TEM, and EDS methods were used for the characterization of EOs and synthesized AuNPs. GC-MS analysis was performed for the determination of the major chemical constituents of the aqueous extracts and EOs of *Eucalyptus globulus* and *Rosmarinus officinalis*. The average sizes of AuNPs produced by aqueous extract of *Eucalyptus globulus* leaf and its EO were determined using TEM analysis as 12.8 nm and 42.2 nm. Similarly, TEM analysis showed the average size of AuNPs synthesized using aqueous leaf extracts and EO of *Rosmarinus officinalis* as 8.66 nm and 60.7 nm in diameter, respectively. EDS analysis detected oxygen, carbon, and gold in all tested NPs. FTIR identified the major functional groups of chemical constituents present in EOs which are responsible for the reduction of metal ions to metal free state in metal NPs.

Melo et al. (2020) revealed the synthesis of AgNPs using *Thymus vulgar* EO under different pH values. UV-Vis spectroscopy showed the absorbance of AgNPs at 415 nm and 440 nm. TEM analysis has revealed the synthesized AgNPs have a diameter of 40 nm and a spherical shape. DLS analysis confirmed that the average diameter synthesized AgNPs was about 90 nm for all pH values tested and XPS analysis demonstrated a crystalline synthesized AgNPs. Gonzalez-Rivera et al. (2017) reported that green synthesized AgNPs mediated by EO of rosemary which is used as renewable bioreductant. The total time consumed for the synthesis of AgNPs was about 2–30 min. The average particle size of synthesized AgNPs was found to be 7–18 nm in diameter with spherical and round morphologies. Thanighaiarassu et al.'s (2018) study indicated that the UV-Vis absorbance of synthesized green AgNPs using citral was 440 nm. SEM and TEM analysis also revealed uniformly dispersed AgNPs with spherical shapes. XRD analysis also confirmed the crystalline nature of AgNPs. Similarly, the FTIR peak indicated the presence of amide and methylene on synthesized AgNPs. Rahimi-Nasrabadi et al. (2014) displayed green AgNPs using EO obtained from *Eucalyptus leucoxylon* leaves. The optimum time and temperature required for the formation of these AgNPs were 120 min and 23°C. UV-Vis, SEM, TEM, and XRD analyses were used for the characterization of the synthesized AgNPs which showed a spherical shape with an average diameter of about 50 nm (Table 8.2).

8.4 APPLICATION OF ESSENTIAL OILS-BASED NANOPARTICLES

8.4.1 Antimicrobial Activities

The EOs of medicinal plants have demonstrated significant antimicrobial activities against pathogenic microorganisms. However, their high sensitivity to oxygen, moisture, heat, and light and low water solubility limited the utilization of EOs. To solve these existing problems, many modern technologies have emerged as safe delivery methods to protect EOs from degradation (Basavegowda et al. 2020). For instance, the advancement of nanoencapsulation as the promising candidate for the applicability of EOs as antimicrobial agents. However, nowadays most nanoencapsulation is

TABLE 8.2
Different Nanoparticle Properties and Characterization Techniques

NPs	Source of essential oils (EOs)	Characterization techniques	Shape	Size (nm)	Key references
Chitosan	*Thymus vulgaris*	FTIR, TGA, DTG, SEM, and zeta potential	Spherical	518.5–716.8	Ghahfarokhi et al. (2016)
Chitosan	*Cyperus articulatus*	UV-Vis, GC-MS, SEM, and FTIR	Rounded	Average 151	Kavaz et al. (2018)
Chitosan	*Torreya grandis*	XRD, DLS, FTIR, and zeta potential	Spherical	144.1–349.6	Wu et al. (2018)
Gold	*Nigella sativa*	UV-Vis, FTIR, TEM, and XRD	Spherical	15.6–28.4	Manju et al. (2016)
eHAp	Eucalyptus, frankincense, tea tree, wintergreen	UV-Vis, FTIR, PL, FESEM, and XRD	Rodlike	100–150	Gundewadi et al. (2021)
Gold	*Ferula persica*	UV-Vis, TEM, FTIR, and DLS	Spherical	Average 37.05	Hosseinzadeha et al. (2020)
Gold	*Anacardium occidentale*	UV-Vis, TEM, and FTIR	Hexagonal	Average 36	Sheny et al. (2012)
Gold	*Eucalyptus urograndis*	UV-Vis, GC-MS, and FTIR	Spherical	…….	Thesing et al. (2018)
Gold	*Menthapiperita*	UV-Vis, SEM-EDX, XRD, and FTIR	Irregular shape	~120	Thanighaiarassu et al. (2014)
Gold/Silver	*Coleus aromaticus*	UV-Vis, TEM, XRD, and FTIR	Irregular	Average 28	Vilas and Mathew (2016)
Zein	Thymol/carvacrol	DLS, SEM, and FTIR	–	100–500	Wu et al. (2012)
Silver	*Rosemarinus officinalis*	FTIR, SEM, and EDX	Small dot round	Average 52	Arassu et al. (2018a)
Silver	*Cuminum cyminum*	UV-Vis and TEM	Spherical	25–45	Keerthiga et al. (2019)
Silver	*Pelargonium graveolens*	SEM, TEM, FTIR, and XRD	–	~164	Arassu et al. (2018b)
Silver	*Aspergillus flavus*	UV-Vis, FT-IR, XRD, SEM with EDS, and HR-TEM	Small round shape	~29	Ramya et al. (2018)
Silica	Pepper fragrant	TEM, AFM, GC-MS, PSD, XRD, and zeta potential	Inerratic morphology	Average 717	Jin et al. (2019)

conducted by using liposomes, solid lipid NPs, nanoemulsions, and polymeric NPs (Pedro et al. 2013). Various reports showed that nanocomplexes were used to enhance the bactericidal and fungicidal activities of EOs (Chouhan et al. 2017; Bachheti et al. 2021b; Bachheti et al. 2019a). Manju et al.'s (2016) study revealed the synthesis of AuNPs by using EOs of *Nigella sativa* (NsEO-AuNPs). The study found out it inhibited the growth of *Staphylococcus aureus* and *Vibrio harveyii* bacterial strains by decreasing the hydrophobicity index of the bacteria by 78% and 46%, respectively. Au and Au/AgNPs synthesis with aid of EO of *Coleus aromaticus* was reported with its antibacterial activities. The zone of inhibition of growth of Gram-negative bacteria (*Escherichia coli*) and the Gram-positive bacteria (*Staphylococcus aureus)* using these NPs were measured using agar well diffusion method. A solution of 150 µL of Au/Ag NPs showed significant inhibition zone (28 mm) against *E. coli* as compared to positive control (Vilas and Mathew 2016). Similarly, synthesized AuNPs by using EO of *Menthapiperita* demonstrated high antifungal activity against *Candida albicans, Candida tropicalis*, and *Candida kefyr.* Some studies supported the eco-friendly and promising potential antifungal activity of NPs tailored by bioinspired or biological routes (Thanighaiarassu et al. 2014). Evaluation of a synergic effect of eucalyptus leaf EO combined with its AgNPs on antimicrobial activities was conducted and showed high inhibition of growth of *E. coli, S. enterica, B. subtilis* and MRSA. This is because lower doses and lower toxicity of medicines make the combined therapy effective (Heydari et al. 2017).

Cinnamaldehyde, a constituent of EO, demonstrated an intense synergistic activity with its AgNPs against *Bacillus cereus* and *Clostridium perfringens.* The rapid bactericidal action exerted by this combination of antimicrobial agents was observed from bacterial kill curve analysis (Ghosh et al. 2013). Nancy and Elumalai's (2019) study showed testing antifungal activity on pathogens using *Pelargonium graveolens* EO with its AgNPs: *Pelargonium graveolens* EO-mediated AgNPs also inhibited clinically isolated fungi growth such as *Candida kefyr, Candida tropicalis*, and Candida albicans. *Zanthoxylum ovalifolium* leaf EO-assisted AgNPs synthesis were applied for the screening of antimicrobial activities (Theerthavathy et al. 2019). Arassu et al. (2018a) evaluated the antifungal activities of synthesized AgNPs against human pathogens such as *Aspergillus flavus, Aspergillus niger, Candida albicans, Candida tropicalis*, and *Candida kefry.* In the research work conducted by Oroojalian et al. (2017), the phenolic constituent of EO from *Teucrium polium* appeared to possess a strong antibacterial activity. Barrera-Ruiz et al.'s (2020) studies showed the encapsulation of EOs of *Cinnamomum zeylanicum, Thymus vulgaris*, and *Schinusmolle* with NPs of chitosan by using the ionotropic gelation method. The antimicrobial activity of these encapsulated EOs with their NPs against *Enterococcus* sp., *Escherichia coli, Staphylococcus aureus, Pseudomonas aeruginosa*, and *Klebsiella pneumonia* was evaluated. The studied chitosan and EOs alone displayed a lower antimicrobial activity effect than chitosan NPs with EOs. From another research finding, EOs of *Cyperus articulates*-aided synthesized chitosan NPs showed significant antibacterial activity. These NPs inhibited the growth of *Staphylococcus aureus* and *Escherichia coli* even at low concentration (5 mg/mL MIC) compared to normal *Cyperus articulates* EOs (10 mg/mL MIC) (Kavaz et al. 2018).

Mojtaba et al. (2019) also studied antifungal performance of Fe3O4–MgONPs which were synthesized by using EO of nutmeg. The results showed that modification of Fe3O4-MgONPs with EO of nutmeg displayed superior antifungal activity with the following average diameter of inhibition: *Trichophyton verrucosum* (13.7 mm), *Candida albicans* (24 mm), *Cryptococcus neoformans* (18.6 mm), *Aspergillus flavous* (20.5 mm), and *Epidermophyton floccodum* (21.14 mm). Zein NPs were used to encapsulate thymol and caravel which are components of EOs that have similar chemical structure. The encapsulation was done following liquid–liquid dispersion method. These encapsulated zein NPs showed *Escherichia coli* reduction of 0.8–1.8 log CFU/mL treated at different pH (Wu et al. 2012). A study on the EO of pepper fragrant (PFEO) that was functionalized by mesoporous silica (MCM-41) NPs revealed that nanoencapsulation of PFEO could be a potential organic delivery agent in the food industry for antimicrobial activity enhancement. The antibacterial activity of PFEO and EO functionalized NPs (EONs) against *Salmonella enterica, Staphylococcus aureus*, *Escherichia coli*, and *Listeria monocytogenes* was tested. The results showed that minimum bactericidal concentration (MBC) of EONs used against some Gram-negative bacteria species decreased more than PFEO. The results of atomic force microscopy (AFM) also confirmed that EONs had a significant inhibitory effect on *Escherichia coli* by disrupting its cell membrane structure (Jin et al. 2019).

8.4.2 Antioxidant Activities

Nowadays, EOs are used as natural antioxidant agents because of their magnificent antioxidant properties. Hence, EOs as antioxidant agents are a main area of interest for researchers (Amorati et al. 2013). The chemicals most responsible for their antioxidant activity are phenolic and another secondary metabolite (Bhavaniramya et al. 2019). However, the antioxidant activity of EOs is limited because they have low water solubility, are easy to degrade, and have a short time of availability to exert the required bioactivity due to high volatility (Shoji and Nakashima 2004). EO nanoencapsulation can possibly solve these challenges by enhancing their bioavailability and solubility, protecting them from chemical and thermal degradation and controlling the delivery at the required site and time targets (Tao et al. 2014).

Ghahfarokhi et al. (2016) exhibited the enhancement of antioxidant activity of thyme EO by encapsulation in chitosan NPs (CSNP). The encapsulation of thyme EO in CS-NP using an emulsion-ionic gelation crosslinking method was reported. The antioxidant activity of encapsulated thyme EO in CSNPs has been proven to show superior effect than free thyme EO. This report also revealed potential applications of these NPs as delivery agents in medicine and food due to their considerable antioxidant activity and thermal stability. EO encapsulation with natural polymers enhances their biological activity and overcomes the gaps related to low stability and poor water solubility for efficient utilization in food systems. Vilas also synthesized silver nanocrystals by using the EOs of *Coleus aromaticus*. Their antioxidant activity was also tested at physiological pH and 373 K using free radical scavenging agents such as superoxide, hydroxyl groups, hydrogen peroxide, nitric oxide, and 2,2-diphenylpicrylhydrazyl (DPPH). These results showed that the reducing power of the free radical of EOs mediated silver nanocrystals.Muniyappan and Nagarajan (2014) reported the synthesized AuNPs that induced and functionalized with EO of *Curcuma pseudomontana*. Synthesized AuNPs have a strong antioxidant activity and are used as effective reducing agents in inducing the immediate passivation of AuNPs. AgNPs were also synthesized by mediating EO of *Cuminum cyminum* spice. The study concluded that synthesized AgNPs can be used as promising antioxidant agents (Keerthiga et al. 2019).

Kavaz et al.'s (2018) study displayed that EOs of *Cyperus articulates* (CPEO) loaded with CSNPs (CPEO-CSNPs) were successfully synthesized by using an oil-in-water mixture and ionic gelation methods. At the beginning, CSNPs and CPEO-CSNPs had less radical-scavenging activity than CPEO but CPEO-CSNPs became a stronger antioxidant over a prolonged period of time. Theerthavathy et al. (2019) synthesized AgNPs using the EO of the *Zanthoxylum ovalifolium* leaf and applied for the screening of their antioxidant potential. At a concentration of 100 μg/mL, AgNPs showed the maximum percentage of free radical scavenging tests for DPPH and hydrogen peroxide as 89.61% and 84.92%, respectively. The synthesized AuNPs using EO of *Eucalyptus sp.* strongly suggested the existence of a more potent reducing agent such as α-terpinyl acetate in the *Eucalyptus urograndis* with higher antioxidant capacity (Thesing et al. 2018). The antioxidant activity of EO of *Eucalyptus* sp. was also evaluated by using DPPH, β-carotene, and linoleic acid as reducing power. The results confirmed the higher antioxidant activity was due to the phenolic contents of the EO of the plant species (Rahimi-Nasrabadi et al. 2014). Zein NPs encapsulated with thymol and carvacrol EO were examined for their antioxidant activities at different pH. Encapsulation was done by the liquid–liquid dispersion method. DPPH and ferric ion spectrophotometric assay were used for the antioxidant activity of encapsulated zein NPs. As Wu et al. (2012) reported, the reduction of DPPH was observed in the range of 24.8–66.8% based on their formulation and more than 65% of hydroxyl free radicals were quenched by encapsulated NPs. The synthesis of EO-mediated NPs with their potential applications are shown in Table 8.3.

8.4.3 Anticancer Activities

Ali et al.'s 2020 study on the comparative anticancer activity tests showed that nonencapsulated EOs of *Origannum glandulosum* were more effective than free EOs. The nonencapsulated EOs also demonstrated more cytotoxic effects on liver cancer cell line Hep-G2 at lower concentration (54.93 μg/mL) than free EOs at 131.6 μg/mL concentration.

Various studies of synthesis of NPs using EO of *Ferula persica* were applied for their potential anticancer activities. For instance, the synthesized AuNPs demonstrated in vitro anticancer activity against murine colon carcinoma (CT26) cells such as cytotoxic, apoptotic, and antiproliferative effects (Hosseinzadeha et al. 2020). Some research reports on the green synthesis of AuNPs using EO of NsEO-AuNPs also revealed the synthesized NPs highly inhibited A549 lung cancer cells compared to bulk Au and EO of *Nigella sativa*-based MTT assay. The IC_{50} values of bulk Au, EO of *N. sativa*, and NsEO-AuNPs were determined as 87.2 μg mL^{-1}, 64.15 μg mL^{-1}, and 28.37 μg mL^{-1}, respectively (Manju et al. 2016). AuNPs and EOs of *Curcuma pseudomontana* also showed effective anticancer activity with 80% cell viability at a concentration of 25–159 μL. The anticancer activity evaluation was carried out using MTT assay against ductal breast carcinoma cells T47D and L929 cells line (Muniyappan and Nagarajan 2014).

Attallah et al. (2020) reported the new anticancer nanoparticulate agent composed of jasmine oil (JO) and pectin/chitosan composite nanoparticles (Pec/CSNPs) as encapsulating materials. The anticancer activity evaluation results showed that the encapsulated JO was almost 13 times more potent than free JO. In contrast, the treatment of normal cells with Pec/CSNPs had not affected the viability of cells except some of its enhancement. According to the research finding of Kavaz et al. (2018), CSNPs loaded with EO of rhizome of *Cyperus articulatus* also revealed stronger cytotoxic activity against MDAMB-231 breast cancer cells after 48 h. Additionally, CSNPs mediated by EO of *Zataria multiflora* (ZEO) showed significant anticancer effect against breast cancer cells (Salehi et al. 2019). EOs isolated from eucalyptus, frankincense, tea tree, and wintergreen were successfully incorporated into erbium-doped hydroxyapatite (eHAp) pellets by using the vacuum filtration technique with a Buckner funnel and used to treat cancer cells. Nanocomposites demonstrated the potent cytotoxic activity on breast cancer cells. Visualization of NP internalization in the cells was done by using the fluorescent property of eHAp (Gundewadi et al. 2021).

8.4.4 Catalytic Activities

Monodispersed and hexagonal AuNPs that prepared using EO of the fresh leaves of *Anacardium occidentale* were found to possess potential catalytic activity in reduction of nitro aromatic hydrocarbons to amino aromatic hydrocarbons (Sheny et al. 2012). Veisi et al. (2019) reported a green, efficient, cost effective, and simple technique for AgNPs synthesis using the EO of orange peel as a reducing, stabilizing, and capping agent. EOs of orange peel-mediated AgNPs were reported as air-stable and an effective catalyst in the A3-coupling reaction of aldehydes, amines, and alkynes to form propargylamines through a single-step reaction. High reaction yields were obtained and the AgNPs catalyst was recycled up to seven times with nearly no decrease in performance.

8.4.5 Miscellaneous Applications

Different characterization of the encapsulated EOs of thyme (TO) in CSNPs with an emulsion revealed that encapsulation in CSNPs improved TO's thermal stability. Ghahfarokhi et al.'s (2016) research report showed that the encapsulated TO resisted a higher temperature (318–325.4°C) than free TO (170°C). The insecticidal activity of polyethylene glycol (PEG)-coated NPs loaded with

TABLE 8.3
Metal/Metal Oxide Nanoparticles Synthesized from Essential Oils and Various Applications

Metal/metal oxide NPs	Plant name (family)	Synthesis condition	Shape	Size (nm)	Characterization methods	Responsible chemical constituents	Applications	Key reference
AgNPs	*Zataria multiflora* (Laminaceae)	Purchased	Not reported	40	Not reported	Carvacrol and thymol	Antibacterial activity	Sheikholeslami et al. (2016)
AgNPs	*Cuminum cyminum* (Apiaceae)	90 mL of 1 mM $AgNO_3$ solution was mixed with 10 mL of cumin oil extract and the solution and the purified pellet was collected and dried at 60°C for 2 hours	Spherical and pseudo spherical	25–45	UV-Vis, and TEM	Phenolic compounds	Antioxidant activity	Keerthiga et al. (2019)
AgNPs	*Zanthoxylum ovalifolium* (Rutaceae)	50 mL of 5×10^{-3} M aqueous solution of $AgNO_3$ was mixed with 1 mL of essential oil extract at room temperature and kept in the dark for 24 hours	Irregular	30–50	UV-Vis, and TEM	Phenolic and flavonoid compounds	Antioxidant and antimicrobial activities	Theerthavathy et al. (2019)
AgNPs	*Thymus vulgaris* (Lamiaceae) and *Szygiumaromaticum*	Using an electrochemical method with a sacrificial anode	Quasi-spherical	65	UV-Vis, FTIR, TEM, and DLS	Not stated	Antibacterial activity	Cinteza et al. (2018)
AgNPs	*Coleus aromaticus* (?)	3 mL of the diluted oil to 30 mL of boiling 2.14×10^{-4} M $AgNO_3$ solution. The pH of the solution is set at 7 using 0.1 M NaOH solution	Spherical	26–28	UV-Vis, FTIR, TEM, and XRD	Carvacrol, thymol eugenol	Antibacterial and catalytic activities	Vilas and Mathew (2016)
AgNPs	*Rosmarinus officinalis* (Lamiaceae)	Starting from silver nitrate and silver acetate under MW and conventional heatingatatmospheric pressure and reaction times ranging from 2 to 30 minutes	Spherical to rod-like	7–28	UV-Vis, SEM, and TEM	Oxygenated monoterpenes, hydrocarbon monoterpenes, and 1,8-cineole	Antibacterial and antioxidant activities	Gonzalez-Rivera et al. (2017)
AgNPs	*Rosmarinus officinalis* (Lamiaceae)	10 mL of essential oil was added into 90 mL of 1 mM $AgNO_3$ solution and the reaction mixture was kept at room temperature until colour change	Dot like round	52	UV-Vis, FTIR, and EDX	Not stated	Antifungal activity	ThanghaiArassu et al. (2018)

(Continued)

TABLE 8.3 (CONTINUED)
Metal/Metal Oxide Nanoparticles Synthesized from Essential Oils and Various Applications

Metal/metal oxide NPs	Plant name (family)	Synthesis condition	Shape	Size (nm)	Characterization methods	Responsible chemical constituents	Applications	Key reference
AgNPs	Geraniol oil	To 90 mL of 2 mM $AgNO_3$, 2 mL of plant leaf EO compound geraniol was added drop by drop at neutral pH	Not stated clearly (It was reported as round shape)	29	UV-Vis, FTIR, XRD, SEM with EDS, HR-TEM, and fluorescence spectroscopy	A complex network of antioxidant metabolites and enzymes, ferulic acid, and chlorogenic acid	Antifungal, antioxidant and anticancer activities	Nambikkairaj and Thanighaiarassu (2018)
AgNPs	Orange peel essential oil	2 mL of the EO of *orange peel* was added to 10 mL of 0.003 M aqueous solution of $AgNO_3$ with constant stirring at 70°C for 48 hours	Spherical	2.76	UV-Vis, TEM FE-SEM, EDS, XRD, and TGA	D-limonene	Catalytic agent	Veisi et al. (2019)
AgNPs	Clove oil	1 mM AgNO3 was dissolved in distilled water, and then 50 mL of concentration of the dissolved eugenol was adjusted to pH 7.5 and poured into 50 mL of $AgNO_3$ solution. The final solutions were placed on a shaker and heated to 60°C for 2 hours and then 40°C for 1 hour	Spherical	101	UV-Vis, SEM, EDS, and XRD	Eugenol	Antibacterial, anticancer, and antifungal activities	Abed and Othman (2019)
AgNPs	*Curcuma zedoaria* (Zingiberaceae)	*C. zedoaria* EO in polysorbate 20 was dissolved with deionized water. The pH value of the solution was adjusted to 7 using 0.1 M NaOH solution. The solution drops were dripped slowly into boiling 5 mM $AgNO_3$ solution under continuous stirring until colour change was observed	Globular	Not stated	UV/Vis, SEM, EDX, XRD, and FTIR	Not stated	Larvicidal activity	Sutthanont et al. (2019)
AgNPs	*Pelargonium graveolens* (Geraniaceae)	6 mL of EO was added to 80 mL of 2 mM $AgNO_3$ solution and the magnetic stirrer is kept inside the container and started to run vigorously on the hot plate for 15 minutes	Not stated	122 and 164	UV-Vis, SEM, TEM, FTIR, XRD, EDX, and fluorescence spectroscopy	Vinblastine and vincristine	Antifungal activity	Arassu et al. (2018a)
AgNPs	*Pelargonium graveolens* (Geraniaceae)	10 mL of plant leaf EO was added to 90 mL of 1 mM $AgNO_3$ solution under continuous stirring with a magnetic stirrer for 15 minutes	Predominately spherical	122 and 164	UV-Vis, XRD, FTIR, TEM, SEM-EDS, and fluorescence spectroscopy	Not stated	Antifungal activity	Nancy and Elumalai (2019)

(Continued)

TABLE 8.3 (CONTINUED)
Metal/Metal Oxide Nanoparticles Synthesized from Essential Oils and Various Applications

Metal/metal oxide NPs	Plant name (family)	Synthesis condition	Shape	Size (nm)	Characterization methods	Responsible chemical constituents	Applications	Key reference
AgNPs	*Origanum vulgare* (Lamiaceae)	Not stated	Spherical	77.68	SEM	Carvacrol and thymol	Antibacterial activity	Scandorieiro et al. (2016)
AgNPs	*Syzygiumaromaticum* (Myrtaceae)	An EO diluted in acetone at 1:170 and a stock solution of 0.31 mmol.L^{-1} $AgNO_3$ was prepared. Then, four aliquots of this solution were collected, and their pH was adjusted to 7, 8, 9, and 10 with a 0.1 M NaOH solution in order to evaluate the synthesis under different conditions. A 30-mL aliquot of each solution was heated under constant magnetic stirring, and 2 mL of the diluted EO was added in drops into the boiling solution and remained under heating and vigorous stirring for 30 minutes	Predominantly spherical	27–94	UV-Vis, SEM, TEM, and XRD	Eugenol	Antimicrobial activity	Diniz et al. (2020)
AgNPs	Clove, niaouli and mandarin essential oils	Two solutions were prepared: one obtained by dissolving 0.5 g of $AgNO_3$ in 100 mL of distilled water, and the other, the bioreductant, by dissolving 1 g of D-glucose and 4 g of NaOH in 100 mL of distilled water, under continuous stirring at 80°C. Then, they were mixed by dripping the bioreductant solution into the $AgNO_3$ solution. The final product was centrifuged, washed, and dried for 48 hours	Spherical	Av. 69	FTIR, SEM, and XRD	Not stated	Antimicrobial and antibiofilm activities (nanocoated wound dressings)	Vasile et al. (2020)
AuNPs	*Coleus aromaticus*	40 mL of 2.49 × 10^{-4} M $HAuCl_4$ solution, 11.5 mL of the diluted EO was added slowly with vigorous stirring at 100°C and at pH 7	Predominantly spherical	28	UV-Vis, TEM, and HR-TEM	Carvacrol	Antibacterial, antifungal, and antioxidant activities	Vilas and Mathew (2016a)
AuNPs	*Nigella sativa* (Ranunculaceae)	30 mL of 1 mM aqueous 2.5 × 10^{4} M $HAuCl_4$ solution at 100°C with continuous stirring and boiled for 1 minute	Spherical	15.6–28.4	UV-Vis, TEM, XRD, and FTIR	Not stated	Antibacterial, antibiofilm, and cytotoxic activities	Manju et al. (2016)

(Continued)

TABLE 8.3 (CONTINUED)
Metal/Metal Oxide Nanoparticles Synthesized from Essential Oils and Various Applications

Metal/metal oxide NPs	Plant name (family)	Synthesis condition	Shape	Size (nm)	Characterization methods	Responsible chemical constituents	Applications	Key reference
AuNPs	*Rosmarinus officinalis* (Lamiaceae)	A heated 1% water-ethanol solution of natural EO was mixed with $HAuCl_4.4H_2O$ at 95°C	Predominately spherical	60.7 ± 60.6	UV-Vis, TEM, EDX, and ATR-FTIR	Camphor, α-pinene, D-limonene, eucalyptol, borneol, boryl acetate, β-phellandrene, and caryophyllene	Biomedical application	Dzimitrowicz et al. (2016)
AuNPs	*Eucalyptus globulus* (Myrtaceae)	A heated 1% water-ethanol solution of natural EO was mixed with $HAuCl_4.4H_2O$ at 95°C	Predominately spherical	42.2 ± 42.0	UV-Vis, TEM, EDX, and ATR-FTIR	α-Pinene, γ-terpinene, eucalyptol, D-limonene, α-phellandrene, β-phellandrene, and β-ocimene	Biomedical application	Dzimitrowicz et al. (2016)
AuNPs	*Eucalyptus dunnii* (Myrtaceae)	By diluting 3.5 μL of $HAuCl_4$ and 80 μL of the methanolic (1:1, v/v) EO solution in 10 mL of ultrapure water. The resulted solution was kept under stirring and heated at 45°C until the colour change was observed	Spherical	Not stated	UV-Vis	1,8-Cineole, aromadendrene, α-Pinene, and α-terpineol	Antioxidant activity	Thesing et al. (2018)
AuNPs	*Eucalyptus urograndis* (Myrtaceae)	By diluting 3.5 μL of $HAuCl_4$ and 80 μL of the methanolic (1:1, v/v) EO solution in 10 mL of ultrapure water. The resulted solution was kept under stirring and heated at 45°C until the colour change was observed	Spherical	Not stated	UV-Vis	1,8-Cineole, α-Pinene, and α-terpinyl acetate	Antioxidant activity	Thesing et al. (2018)
AuNPs	*Ferula persica* (Apiaceae)	1.5 mL of the EO was mixed with 34 mL of $HAuCl_4$ solution (1 mmol/mL) and incubated at room temperature for 24 hours	Spherical	37.05	UV-Vis, FTIR, TEM, DLS, and XRD	α-Pinene, (Z)-1-propenyl sec-butyl disulphide, and β-pinene	Anticancer activity	Hosseinzadeh et al. (2020)

(Continued)

TABLE 8.3 (CONTINUED)
Metal/Metal Oxide Nanoparticles Synthesized from Essential Oils and Various Applications

Metal/metal oxide NPs	Plant name (family)	Synthesis condition	Shape	Size (nm)	Characterization methods	Responsible chemical constituents	Applications	Key reference
AuNPs	*Curcuma pseudomontana* (Zingiberaceae)	All reactions were carried out at room temperature. EO (0.1 mL) was dissolved in ethanol: water solution (2:8, (10 mL)). Subsequently, the mixture (3 mL) was added to $HAuCl_4$ solution (1 mM, 20 mL) and stirred for 30 minutes. However, high temperature of 95°C was essential to produce the AuNPs	Spherical	20	UV-Vis, FTIR, SEM, and HR-TEM	1,5-Dimethyl-4-hexenyl)-4-methyl β-elemenone, germacrone and pseudocumenol, 2-(4-methoxy phenyl) N,N trimethyl-1-pyrrolamine	Antioxidant activity	Muniyappan and Nagarajan, (2014)
AuNPs	*Anacardium occidentale*	About 2 mL of EO solution was added to 30 mL of aqueous 2.5×10^4 M $HAuCl_4$ solution at 100°C with continuous stirring and boiled for 1 minute	Hexagonal	36	UV-Vis, FTIR, and TEM	Terpenoids	Catalytic activity	Sheny et al. (2012)
CuNPs	*Thymus daenensis*	0.4 g of Cu(II) chloride was dissolved in 100 mL deionized water and then, a solution of 2 g of L-ascorbic acid in 100 mL of deionized water was added slowly at room temperature to the reaction mixture and stirred and heated to refluxing for 24 hours	Spherical	100–250	FTIR and SEM	*p*-Cymene, *γ*-terpinene, isoborneol, thymol, carvacrol, and trans-caryophyllene	Antifungal activity	Weisany et al. (2019)
CuNPs	*Anethum graveolens* (Apiaceae)	0.4 g of Cu(II) chloride was dissolved in 100 mL deionized water and then a solution of 2 g of L-ascorbic acid in 100 mL of deionized water was added slowly at room temperature to the reaction mixture and stirred and heated to refluxing for 24 hours	Spherical	100–250	FTIR and SEM	α-Phellandrene, *p*-cymene, cyclohexanone, and l-carvone, apiol	Antifungal activity	Weisany et al. (2019)

(Continued)

TABLE 8.3 (CONTINUED)
Metal/Metal Oxide Nanoparticles Synthesized from Essential Oils and Various Applications

Metal/metal oxide NPs	Plant name (family)	Synthesis condition	Shape	Size (nm)	Characterization methods	Responsible chemical constituents	Applications	Key reference
Fe_3O_4–MgO	*Myristica fragrans* (Myristicaceae)	Fe_3O_4–MgO solution and 200 mL of highly concentrated nutmeg EO were mixed and then, stirred for 30 minutes at 80°C ntil gaining a homogenous suspension. The resulting Fe_3O_4–MgO nanoparticles were separated using a strong magnet	Spherical	10–15	FTIR, FE-SEM, VSM,EDAX, and BET	Chloregenic acid, rutin, quercetin, carvacrol, hesperidin, hesperetin, and rosmarinic acid	Antibacterial, and antifungal activities	Mojtaba et al. (2019)
ZnO-Ag nanocomposite	*Zingiber zerumbet* (Zingiberaceae)	About 30 mL of the ginger EO solution was dissolved into 100 mL of ethanol under gentle stirring at room temperature. Then, 0.3 mM ($Zn(Ac)_2.2H_2O$) was added to react with EO solution in an water bath system under continuous stirring at 80°C. Stirring was continued for 3 hours to obtain a light cream colour solid. After the reaction was performed and the cream solid product appeared, 0.10 mM $AgNO_3$ solution was added into the suspension, and the reaction was further preceded for 30 minutes. The product was collected by centrifugation and washed with ethanol to remove extra EO solution and then dried at 10°C for 2 hours	Hexagonal	23	UV–Vis, FTIR, EDX, TEM, and WXRD	Zerumbone, humulene, and camphene	Antibacterial activity	Saad and Prud'homme, (2016)

garlic EO was examined against adult *Tribolium castaneum* using the melt-dispersion method in which over 80% pest control efficacy was observed for them after five months against the adult *Trbolium castaneum*. This effect is apparently due to the slow and persistent release of the active components of EOs from NPs (Yang et al. 2009). However, free EO of garlic revealed only 11% pest control efficacy at similar conditions.

Various types of EOs nanoformulations including EOs of neem, garlic, *Artemisia arborescens* L., and *Lippiasidoides* have shown prolonged shelf-life. This nanoformulation technique is used to prevent the premature degradation of active ingredients of EOs without affecting the non-target organisms under insecticidal application (Martín et al. 2010). Lai et al. (2006) introduced a new delivery system for ecological pesticides by incorporating EO of *Artemisia arborescens* L. into solid-lipid NPs (SLNPs). This study showed that SLNPs were suitable potential carriers for ecological pesticides in agriculture. It also has high physical stability and an excellent capability to reduce the volatility of EOs. Paula et al. (2011) synthesized and characterized microspheres composed of chitosan and cashew tree gum which was applied as EO of *Lippia sidoides* carriers and used as insecticidal agents. In another study, EO of *Nigella sativa*-mediated AuNPs effectively inhibited the biofilm formation of *Staphylococcus aureus* and *Vibrio harveyii* which decreased their hydrophobicity index by 78% and 46%, respectively (Manju et al. 2016).

8.5 CONCLUSIONS

EOs can be obtained from various plants as well as different parts of plant species. EOs of medicinal plants have a wide range of applications, such as maintaining and promoting health, and preventing and treating some pathogenic diseases. EOs of plants have been also used to produce NPs and applied as simple, cost-effective, and eco-friendly approaches to the production of innovative and novel nanomaterials. Low water solubility, stability, high volatility, and some side effect properties of EOs have limited their application. To overcome these challenges, EO-mediated syntheses of NPs are very important. For instance, NPs encapsulated by using EOs revealed the increment of water solubility and chemical stability and the reduction of rapid volatility and degradation of chemical constituents of EOs. Encapsulation of NPs with EOs also improves the activities of EOs by enhancing their bioavailability and efficacy nature. Therefore, nanoencapsulation of EOs is used as an efficient and effective approach to application of EOs. Generally, encapsulated NPs with EOs, EO-mediated NPs, nanoformulation of EOs, and nanoemulsion of EOs are used as stronger antimicrobial, antioxidant, anticancer, and catalytic agents than free EOs.

REFERENCES

Abed, M.S., A.S. Abed & F.M. Othman. 2019. Green synthesis of silver nanoparticles from natural compounds: Glucose, eugenol and thymol. *Journal of Advanced Research in Fluid Mechanics and Thermal Sciences 60*(1): 95–111.

Ali, H., A. Al-khalifa, A. Aouf, H. Boukhebti & A. Farouk. 2020. Effect of nanoencapsulation on volatile constituents, and antioxidant and anticancer activities of Algerian Origanumglandulosum Desf. essential oil. *Scientific Report* 10(1): 1–9.

Amorati, R., M. Foti & L. Valgimigli. 2013. Antioxidant activity of essential oils. *Journal of Agriculture and Food Chemistry* 61(46): 10835–10847.

Anand, A.K., M. Mohan, S.Z. Haider & A. Sharma. 2011. Essential oil composition and antimicrobial activity of three *Ocimum* species from Uttarakhand (India). *International Journal of Pharmacy and Pharmaceutical Sciences* 3(3): 1–3.

Arassu, R., B. Nambikkairaj & D. Ramya. 2018a. Green synthesis of silver nanoparticles and characterization using plant leaf essential oil *Rosemarius officinalis* and their antifungal activity against human pathogenic fungi. *Journal of Scientific Research in Pharmacy* 7(11): 138–144.

Arassu, R., B. Nambikkairaj & D. Ramya. 2018b. *Pelargonium graveolens* plant leaf essential oil mediated green synthesis of silver nanoparticles and its antifungal activity against human pathogenic fungi. *Journal of Pharmacognosy and Phytochemistry* 7: 1778–1784.

Attallah, O., A. Shetty, F. Elshishiny & W. Mamdouh. 2020. Essential oil loaded pectin/chitosan nanoparticles preparation and optimization via Box–Behnken design against MCF-7 breast cancer cell lines. *RSC Advances* 10(15): 8703–8708.

Bachheti, A., A. Sharma, R.K. Bachheti, A. Husen & V.K. Mishra. 2019. Plant-mediated synthesis of copper oxide nanoparticles and their biological applications. In *Nanomaterials and Plant Potential*. Cham: Springer, pp. 221–237. https://doi.org/10.1007/978-3-030-05569-1_8.

Bachheti, R.K., A. Sharma, A. Bachheti, A. Husen, G.M. Shanka & D.P. Pandey. 2020a. Nanomaterials from various forest tree species and their biomedical applications. In Husen A., Jawaid M. (eds.), *Nanomaterials for Agriculture and Forestry Applications*. Elsevier Inc., Cambridge, MA, USA, pp. 81–106. https://doi.org/10.1016/B978-0-12-817852-2.00004-4.

Bachheti, R.K., L. Abate, A. Bachheti, A. Madhusudhan & A. Husen. 2021. Algae-, fungi-, and yeast-mediated biological synthesis of nanoparticles and their various biomedical applications. In *Handbook of Greener Synthesis of Nanomaterials and Compounds*. Elsevier, pp. 701–734. https://doi.org/10.1016/B978-0-12-821938-6.00022-0.

Bachheti, R.K., Y. Godebo, A. Bachheti, M.O. Yassin & A. Husen. 2020b. Root-based fabrication of metal and or metal-oxide nanomaterials and their various applications. In Husen A., Jawaid M. (eds.), *Nanomaterials for Agriculture and Forestry Applications*. Elsevier Inc., Cambridge, MA, pp. 135–166. https://doi.org/10.1016/B978-0-12-817852-2.00006-8.

Baramati, B. & R. Varma. 2009. High value products from waste: Grape pomace extract- a three-in-one package for the synthesis of metal nanoparticles. *ChemSusChem: Chemistry & Sustenblity Energy & Material* 2(11): 1041–1044.

Barrera-Ruiz, D., G. Cuestas-rosas, R. Sánchez-mariñez, M. Álvarez-ainza, G. Moreno-ibarra, A. López-meneses, M. Plascencia-jatomea & M. Cortez-rocha. 2020. Antibacterial activity of essential oils encapsulated in chitosan nanoparticles. *Food Science and Technology* 40: 568–573. http://orcid.org/0000-0003-0231-0523.

Basavegowda, N., J. Patra & K.H. Baek. 2020. Essential oils and mono/bi/tri-metallic nanocomposites as alternative sources of antimicrobial agents to combat multidrug-resistant pathogenic microorganisms: An overview. *Molecules* 25: 1–24.

Bayala, B., I.H. Bassole, R. Scifo, C. Gnoula, L. Morel, J.A. Lobaccaro & J. Simpore. 2014. Anticancer activity of essential oils and their chemical components – A review. *American Journal of Cancer Research* 4(6): 591–607.

Benchaar, C., S. Calsamiglia, A.V. Chaves, G.R. Fraser, D. Colombatto, T.A. McAllister & K.A. Beauchemin. 2008. A review of plant-derived essential oils in ruminant nutrition and production. *Animal Feed Science and Technology* 145(1–4): 209–228. https://doi.org/10.1016/j.anifeedsci.2007.04.014.

Bhavaniramya, S., S. Vishnupriya, M. Al-aboody, R. Vijayakumar & D. Baskaran. 2019. Role of essential oils in food safety: Antimicrobial and antioxidant applications. *Grain & Oil Science and Technology* 2: 49–55.

Bilia, A.R., C. Guccione, B. Isacchi, C. Righeschi, F. Firenzuoli & M.C. Bergonzi. 2014. Essential oils loaded in nanosystems: A developing strategy for a successful therapeutic approach. *Evidence-based Complementary and Alternative Medicine* 2014: 1–15.

Brusotti, G., I. Cesari, A. Dentamaro, G. Caccialanza & G. Massolini. 2014. Isolation and characterization of bioactive compounds from plant resources: The role of analysis in the ethnopharmacological approach. *Journal of Pharmaceutical and Biomedical Analysis* 87: 218–228. https://doi.org/10.1016/j.jpba.2013.03.007.

Chen, Y., C. Zhou, Z. Ge,Y. Liu, W. Feng, S. Li, G. Chen & A. Wei. 2013. Composition and potential anticancer activities of essential oils obtained from myrrh and frankincense. *Oncology Letters* 6: 1140–1146.

Chouhan, S., K. Sharma & S. Guleria. 2017. Antimicrobial activity of some essential oils: Present status and future perspectives. *Medicines* 4: 1–21.

Cinteza, L.O., C. Scomoroscenco, S.N. Voicu, C.L. Nistor, S.G. Nitu, B. Trica, M.L. Jecu & C. Petcu. 2018. Chitosan-stabilized Ag nanoparticles with superior biocompatibility and their synergistic antibacterial effect in mixtures with essential oils. *Nanomaterials* 8(10): 826. https://doi.org/10.3390/nano8100826.

Dambolena, J.S., M.P. Zunino, J.M. Herrera, R.P. Pizzolitto, V.A. Areco and J.A. Zygadlo. 2016. Terpenes: Natural products for controlling insects of importance to human health – A structure-activity relationship study. *Psyche* 2016. https://doi.org/10.1155/2016/4595823.

Dhifi, W., S. Bellili, S. Jazi, N. Bahloul & W. Mnif. 2016. Essential oils' chemical characterization and investigation of some biological activities: A critical review. *Medicines* 3(4): 1–16.

Diniz, F.R., A.P. Maia, L.R. Andrade, L.N. Andrade, M. Vinicius, F. da Silva, C.B. Corrêa, C. de Albuquerque, P. da Costa, S.R. Shin, S. Hassan, E. Sanchez-Lopez, E.B. Souto & P. Severino. 2020. Silver nanoparticles-composing alginate/gelatine hydrogel improves wound healing in vivo. *Nanomaterials* 10: 390. https://doi.org/10.3390/nano10020390.

Dzimitrowicz, A., S. Berent, A. Motyka, P. Jamroz, K. Kurcbach, W. Sledz & P. Pohl. 2016. Comparison of the characteristics of gold nanoparticles synthesized using aqueous plant extracts and natural plant essential oils of *Eucalyptus globulus* and *Rosmarinus officinalis*. *Arabian Journal of Chemistry*: 1–11. http://dx.doi.org/10.1016/j.arabjc.2016.09.007.

Edris, A.E. 2007. Pharmaceutical and therapeutic potentials of essential oils and their individual volatile constituents: A review. *Phytotherapy Research* 21(4): 308–323. https://doi.org/10.1002/ ptr.2072.

Gershenzon, J. & N. Dudareva. 2007. The function of terpene natural products in the natural world. *Nature Chemical Biology* 3: 408–414.

Ghahfarokhi, M., M. Barzegar, M. Sahari & M. Azizi. 2016. Enhancement of thermal stability and antioxidant activity of thyme essential oil by encapsulation in chitosan nanoparticles. *Journal of Agricultural Science and Technology* 18: 1781–1792.

Ghosh, I., S. Patil, T. Sharma, S. Srivastava, R. Pathania & N. Navani. 2013. Synergistic action of cinnamaldehyde with silver nanoparticles against spore-forming bacteria: A case for judicious use of silver nanoparticles for antibacterial applications. *International Journal of Nanomedicine* 8: 4721–4731.

Gonzalez-Rivera, J., C. Duce, V. Ierardi, L. Longo, A. Spepi, M. Tine & C. Ferrari. 2017. Fast and eco–friendly microwave-assisted synthesis of silver nanoparticles using rosemary essential oil as renewable reducing agent. *ChemistrySelect* 2: 2131–2138.

Gundewadi, G., S.G. Rudra, R. Gogoi, T. Banerjee, S.K. Singh , S. Dhakate & A. Gupta. 2021. Electrospun essential oil encapsulated nanofibers for the management of anthracnose disease in Sapota. *Industrial Crops and Products* 170: 113727. https://doi.org/10.1016/j.indcrop.2021.113727.

Heydari, M., M. Mobini & M. Salehi. 2017. The synergic activity of eucalyptus leaf oil and silver nanoparticles against some pathogenic bacteria. *Archieve Pediatric Infectious Diseases* 5: 1–6.

Hosseinzadeha, N., T. Shomalia, S. Hosseinzadehb, F. Fardc, M. Pourmontaserib & M. Fazelia. 2020. Green synthesis of gold nanoparticles by using *Ferula persica* Willd. gum essential oil: Production, characterization and in vitro anticancer effects. *Journal of Pharmacy and Pharmacology*: 1–13. http://dx.doi.org/10.1111/jphp.13274.

Husen, A., Q.I. Rahman, M. Iqbal, M.O. Yassin & R.K. Bachheti. 2019. Plant-mediated fabrication of gold nanoparticles and their applications. In *Nanomaterials and Plant Potential*. Cham: Springer, pp. 71–110. https://doi.org/10.1007/978-3-030-05569-13.

Iravani, S., H. Korbekandi, S.V. Mirmohammadi & P.S.Zolfaghari. 2014. Synthesis of silver nanoparticles: Chemical, physical and biological methods. *Research Pharmaceutical Science* 9: 385–406.

Javed, B. & Z. Mashwani. 2020. Photosynthesis of colloidal nanosilver from *Mentha longifolia* and *Mentha arvensis*: Comparative morphological and optical characterization. *Microscopy Research & Techniques* 83(11): 1299–1307. https://doi.org/10.1002/jemt.23518.

Jin, L., J. Teng, L. Hu, X. Lan, Y. Xu, J. Sheng, Y. Songa & M. Wang. 2019. Pepper fragrant essential oil (PFEO) and functionalized MCM-41 nanoparticles: Formation, characterization, and bactericidal activity. *Journal of Science and Food Agriculture* 99: 5168–5175.

Kavaz, D., M. Idris & C. Onyebuchi. 2018. Physiochemical characterization, antioxidative, anticancer cells proliferation and food pathogens antibacterial activity of chitosan nanoparticles loaded with Cyperus articulatus rhizome essential oils. *International Journal of Biological Micromolecules* 123: 837–845. https://doi.org/10.1016/j.ijbiomac.2018.11.177.

Keerthiga, N., R. Anitha, S. Rajeshkumar & T. Lakshmi. 2019. Antioxidant activity of *Cumin* oil mediated silver nanoparticles. *Pharmacognesis Journal* 11: 787–789.

Lai, F., S. Wissing, R. Müller & A. Fadda. 2006. *Artemisia arborescens* L essential oil–loaded solid lipid nanoparticles for potential agricultural application: Preparation and characterization. *AAPS PharmSciTech* 7: 1–9.

Lingan, K. 2018. A review on major constituents of various essential oils and its application. *Science Translational Medicines* 8: 2161–1025.

Manju, S., B. Malaikozhundan, S. Vijayakumar, S. Shanthi, A. Jaishabanu, P. Ekambaram & B. Vaseeharan. 2016. Antibacterial, antibiofilm and cytotoxic effects of *Nigella sativa* essential oil coated gold nanoparticles. *Microbial Pathogenesis* 91: 129–135. https://doi:10.1016/j.micpath.2015.11.021.

Martín, Á., S. Varona, A. Navarrete & M.J. Cocero. 2010. Encapsulation and co-precipitation processes with supercritical fluids: Applications with essential oils. *The Open Chemical Engineering Journal* 4: 31–41.

Masango, P. 2005. Cleaner production of essential oils by steam distillation. *Journal of Cleaner Production* 13(8): 833–839.

Maurya, S., A. Bhardwaj, K. Gupta, S. Agarwal, A. Kushwaha, V. Chaturvedi, R. Pathak, R. Gopal, K. Uttam & A. Soingh. 2016. Green synthesis of silver nanoparticles using Pleurotus and bactericidal activity. *Cell Molecular Biology* 62: 131.

Melo, A., M. Maciel, W. Sganzerla, A. Almeida, R. Armas, M. Machado, C. Rosa, M. Nunes, F. Bertoldi & P. Barreto. 2020. Antibacterial activity, morphology, and physicochemical stability of biosynthesized silver nanoparticles using thyme (*Thymus vulgaris*) essential oil. *Materials Research Express* **7**: 15087. https://doi.org/10.1088/2053-1591/ab6c63.

Mohamed, S.H., M.S.M. Mohamed, M.S. Khalil, M. Azmy & M.I. Mabrouk. 2018. Combination of essential oil and ciprofloxacin to inhibit/eradicate biofilms in multidrug-resistant *Klebsiella pneumonia. Journal of Applied Microbiology* 125(1): 84–95. https://doi.org/10.1111/jam.13755.

Mojtaba, S., S. Alireza & S. Ramakrishna. 2019. Green synthesis of super magnetic Fe_3O_4-MgO nanoparticles via Nutmeg essential oil toward superior anti-bacterial and antifungal performance. *Journal of Drug Delivery Science and Technology* 54: 101352. https://doi.org/10.1016/j.jddst.2019.101352.

Muniyappan, N. & N. Nagarajan. 2014. Green synthesis of gold nanoparticles using *Curcumapseudomontana* essential oil, its biological activity and cytotoxicity against human ductal breast carcinoma cells T47D. *Journal of Environmental Chemical Engineering* 2: 2037–2044.

Naeem, A., M.A. Shabbir & M.R. Khan. 2019. Mango seed kernel fat as cocoa butter substitute suitable for the tropics. *Journal of Food Sciences* 84: 1315–1321. https://doi.org/10.1111/ 1750-3841.14614.

Nambikkairaj, B. & R.R. Thanighaiarassu. 2018. Green fabrication of silver nanoparticles and characterization with plant leaf essential oil compound Geraniol and their antifungal activity against human pathogenic fungi. *The Pharmaceutical innovation Journal* 7(11): 448–453. https://doi.org/10.1016/j.apsusc.2016.05 .052.

Nancy, B. & K. Elumalai. 2019. Synthesis of silver nanoparticles using *Pelargonium graveolens* essential oil and antifungal activity. *International Journal of Pharmacy and Biological Sciences* 9: 176–185.

Oroojalian, F., H. Orafaee & M. Azizi. 2017. Synergistic antibacterial activity of medicinal plants essential oils with biogenic silver nanoparticles. *Nanomedicine Journal* 4: 237–244.

Painuli, S., P. Semwal, A. Bacheti, R.K. Bachheti & A. Husen. 2020. Nanomaterials from nonwood forest products and their applications. In Husen A., Jawaid M. (eds.), *Nanomaterials for Agriculture and Forestry Applications*. Elsevier, pp. 15–40. https://doi.org/10.1016/B978-0-12-817852-2.00002-0.

Palazzolo, E., V.A. Laudicina & M.A. Germanà. 2013. Current and potential use of citrus essential oils. *Current Organic Chemistry* 17: 3042–3049.

Paula, H., F. Sombra, R. Cavalcante, F. Abreu & R. de Paula. 2011. Preparation and characterization of chitosan/cashew gum beads loaded with *Lippia sidoides* essential oil. *Material Science and Engineering C* 31(2): 173–178.

Pedro, A., E. Santo, C. Silva, C. Detoni & E. Albuquerque. 2013. The use of nanotechnology as an approach for essential oil-based formulations with antimicrobial activity. Microbial pathogens and strategies for combating them: Science, technology and education. In Méndez-Vilas A. (eds.), FORMATEX, pp. 1364–1374.

Porres-Martínez, M.E., E. González-Burgos, M.E. Carretero & M.P. GómezSerranillos. 2015. Major selected monoterpenes α-pinene and 1,8-cineole found in *Salvia lavandulifolia* (Spanish sage) essential oil as regulators of cellular redox balance. *Pharmaceutical Biology* 53(6): 921–929. https://doi.org/10.3109 /13880209.2014.950672.

Prakasham, R.S., B.S. Kumar, Y.S. Kumar & K.P. Kumar. 2014. Production and characterization of protein encapsulated silver nanoparticles by marine isolate *Streptomyces parvulus* SSNP11. *Indian Journal of Microbiology*: 1–11. https://doi.org/10.1007/s12088-014-0452-1.

Rahimi-Nasrabadi, M., S. Pourmortazavi, S. Shandiz, F. Ahmadi & H. Batooli. 2014. Green synthesis of silver nanoparticles using *Eucalyptus leucoxylon* leaves extract and evaluating the antioxidant activities of extract. *Natural Product Research* 28: 1964–1969.

Ramya, B.N. & R. Thanighaiarassu. 2018. Green fabrication of silver nanoparticles and characterization with plant leaf essential oil compound Geraniol and their antifungal activity against human pathogenic fungi. *The Pharmacology Innovation Journal* 7: 448–453.

Regnault-Roger, C., C. Vincent & J.T. Arnason. 2012. Essential oils in insect control: Low-risk products in a high-stakes world. *Annual Review Entomology* 57: 405–424.

Saad, W.S. & R.K. Prud'homme. 2016. Principles of nanoparticle formation by flash nanoprecipitation. *Nanotoday* 11(2): 212–227. https://doi.org/10.1016/j.nantod.2016.04.006.

Sahayaraj, K., M. Madasamy & S.A. Radhika. 2017. Insecticidal activity of bio-silver and gold nanoparticles against *Pericallia ricini* Fab. (Lepidoptera: Archidae). *Bionanoparticles for Pest Management* 9(1): 63–72.

Salehi, F., H. Behboudi, G. Kavoosi & S. Ardestani. 2019. Incorporation of *Zataria multiflora* essential oil into chitosan biopolymer nanoparticles: A nanoemulsion based delivery system to improve the in-vitro efficacy, stability and anticancer activity of ZEO against breast cancer cells. *International Journal of Biological Macromolecules* 143: 382–392. https://doi.org/10.1016/j.ijbiomac.2019.12.058.

Scandorieiro, S., L.C. de Camargo, C.A. Lancheros, S.F. Yamada-Ogatta, C.V. Nakamura, A.G. de Oliveira, C.G. Andrade, N. Duran, G. Nakazato & R.K. Kobayashi. 2016. Synergistic and additive effect of oregano essential oil and biological silver nanoparticles against multidrug-resistant bacterial strains. *Frontiers in Microbiology* 7: 760. https://doi.org/10.3389/fmicb.2016.00760.

Shaaban, H.A.E., A.H. El-Ghorab & T. Shibamoto. 2012. Bioactivity of essential oils and their volatile aroma components: Review. *Journal of Essential Oil Research* 24(2): 203–212. https://doi.org/10.1080/10412905.2012.659528.

Sheikholeslami, S., S.E. Mousavi, H.R.A. Ashtiani, S.R.H. Doust & S.M. Rezayat. 2016. Antibacterial activity of silver nanoparticles and their combination with *Zataria multiflora* essential oil and methanol extract. *Jundishapur Journal of Microbiology* 9(10): 36070. https://doi.org/10.5812/jjm.36070.

Shiny, D., J. Mathew & D. Philip. 2012. Synthesis characterization and catalytic action of hexagonal gold nanoparticles using essential oils extracted from *Anacardium occidentale. Spectrochimica Acta Part A: Molecular and Biomolecular Spectroscopy* 97: 306–310.

Shoji, Y. & H. Nakashima. 2004. Nutraceutics and delivery systems. *Journal of Drug Targeting* 12: 385–391.

Sutthanont, N., S. Attrapadung & S. Nuchprayoon. 2019. Larvicidal activity of synthesized silver nanoparticles from *Curcuma zedoaria* essential oil against *Culex quinquefasciatus. Insects* 10(1): 27. https://doi.org/10.3390/insects10010027.

Tao, F., L. Hill, Y. Peng & C. Gomes. 2014. Synthesis and characterization of β-cyclodextrin inclusion complexes of thymol and thyme oil for antimicrobial delivery applications. *LWT Food Science and Technology* 59: 247–255.

Tássio, R., J. Leite, S. Mesquita, S. Bezerra, B. Silveira, C. Mesquita, C. GomesRibeiro, M. Vilanova, M.C. de Sousa Ribeiro, M. Amaral & F. Coutinho. 2020. Seasonal variation in the chemical composition and biological activity of the essential oil of *Mesosphaerum suaveolens* (L.) Kuntze. *Industrial Crops and Products* 153: 112600. https://doi.org/10.1016/j.indcrop.2020.112600

Thanighaiarassu, R., B. Nambikkairaj & D. Ramya. 2018. Green synthesis of silver nanoparticles and characterization using plant leaf essential oil compound citral and their antifungal activity against human pathogenic fungi. *Journal of Pharmacognosy and Phytochemistry* 7: 902–907.

Thanighaiarassu, R., P. Sivamai, R. Devika & B. Nambikkairaj. 2014. Green synthesis of gold nanoparticles characterization by using plant essential oil *Menthapiperita* and their antifungal activity against human pathogenic fungi. *Journal of Nanomedicine and Nanotechnology* 5: 1–6.

Theerthavathy, B., S. Arakhanum, B. Kumar & S. Kiran. 2019. Antioxidant and antimicrobial activities of silver nanoparticle of essential oil extracts from leaves of *Zanthoxylum ovalifolium. European Journal of medicinal plants* 29: 1–11.

Thesing, A., J. Nascimento, R. Jacob & J. Santos. 2018. Eucalyptus oil-mediated synthesis of gold nanoparticles. *Journal of Chemistry and Chemical Engineering* 12: 52–59. https://doi.org/10.1016/j.indcrop.2019.02.031.

Tongnuanchan, P. & S. Benjakul. 2014. Essential oils: Extraction, bioactivities and their uses for food preservation. *Journal of Food Sciences* 79(7): 1231–1249. https://doi.org/ 10.1111/1750-3841.12492.

Umaru, I.J., A.F. Badruddin & H.A. Umaru. 2019. Phytochemical screening of essential oils and antibacterial activity and antioxidant properties of *Barringtonia asiatica* (L) leaf extract. *Biochemistry Research International* 2019: 1–7. https://doi.org/10.1155/2019/7143989.

Vasile, B.S., A.C. Birca, M.C. Musat & A.M. Holban. 2020. Wound dressings coated with silver nanoparticles and essential oils for the management of wound infections. *Materials* 13(7): 1682.

Veisi, H., N. Dadres, P. Mohammadi & S. Hemmati. 2019. Green synthesis of silver nanoparticles based on oil-water interface method with essential oil of orange peel and its application as nanocatalyst for A3 coupling. *Materials Science & Engineering* 105: 1–9.

Vilas, V., D. Philip & J. Mathew. 2016. Biosynthesis of Au and Au/Ag alloy nanoparticles using *Coleus aromaticus* essential oil and evaluation of their catalytic, antibacterial and antiradical activities. *Journal of Molecular Liquids* 221: 179–189. https://doi.org/10.1016/j.molliq.2016.05.066.

Weisany, W., S. Samadi, J. Amini, S. Hossaini, S. Yousefi & F. Maggi. 2019. Enhancement of the antifungal activity of thyme and dill essential oils against *Colletotrichum nymphaeae* by nano-encapsulation with copper NPs. *Industrial Crops and Products*132: 213–225.

Wu, Y., Y. Luo & Q. Wang. 2012. Antioxidant and antimicrobial properties of essential oils encapsulated in zein nanoparticles prepared by liquid-liquid dispersion method. *LWT Food Science and Technology* 48: 283–290.

Wu, J., Q. Shu, Y. Niu, Y. Jiao & Q. Chen. 2018. Preparation, characterization, and antibacterial effects of chitosan nanoparticles embedding essential oil synthesized in ionic liquid containing system. *Journal of Agriculture and Food Chemistry* 66(27): 7006–7014. https://doi:10.1021/acs.jafc.8b01428.

Yang, F.L., X.G. Li, F. Zhu & C.L. Lei. 2009. Structural characterization of nanoparticles loaded with garlic essential oil and their insecticidal activity against *Tribolium castaneum* (Herbst) (Coleoptera: Tenebrionidae). *Journal of Agriculture and Food Chemistry* 57: 10156–10162.

9 Medicinally Important Seed Extract and Seed Oil-Mediated Nanoparticles Synthesis and Their Role in Drug Delivery and Other Applications

Limenew Abate, Megersa Bedo Megra, Yenework Nigussie, Meseret Zebeaman, Mesfin Getachew Tadesse, Archana Bachheti, Rakesh Kumar Bachheti, and Azamal Husen

CONTENTS

9.1 INTRODUCTION

Nanoparticles (NPs) are essential for developing future sustainable technologies that benefit humanity and the environment (Abisharani et al., 2019a; Bachheti et al., 2020). Their physical and chemical properties differ due to their small size and huge surface area to volume ratio (Iravani, 2011; Bachheti et al., 2021; Abate et al., 2022). Due to this high surface area to volume ratio property, NPs have a high ability to react quickly with other particles (Bachheti et al., 2019). NPs can be formed using biological, physical, and chemical methods. However, the physical and chemical methods are potentially dangerous, complicated, and expensive (Prabhu and Poulose, 2012). The study of NPs has grown in importance over the past two decades as one of the most significant areas of modern

DOI: 10.1201/9781003213727-9

materials science. Due to the unique properties of NPs, including their catalytic behaviours, anticancer, antibacterial, and optical properties, they have received significant interest in medicine, biology, and electronics (He et al., 2017).

The biological method, often known as the 'green synthesis method', has recently drawn more attention due to the expanding demand for ecologically friendly NP synthesis technology (Geetha et al., 2013). Because green production of NPs is easy, affordable, and environmentally friendly; currently, the biomedical, environmental, and industrial sectors are interested in the green manufacture of NPs for use in various applications (Ahmed et al., 2016b).

Seed extracts and seed oil-mediated NPs are green and less costly methods that give more control over particle size and make it possible to make more diverse nanomaterials, such as bimetallic and core-shell NPs (Sharma et al., 2019). They have different applications. *Ocimum basilicum* seed extract-coated magnetic NPs (Fe3O4@BSM-CPX) are also used as a drug delivery system (Rajan et al., 2017b). Silver nanoparticles (AgNPs) mediated by seed extract of *Phoenix dactylifera* were used as chemical stabilizing and reducing agents because they are cost effective, nonhazardous, and eco-friendly, an advancement over physical and chemical methods. The AgNPs are also used in antibacterial activities for efficient infection control and disease prevention caused by MRSA and other microorganisms that resist drugs (Ansari and Alzohairy, 2018). *Butyrospermum paradoxum* and *Cyperus esculentus* seed extracts may be a useful starting point for manufacturing AgNPs with antifungal and strong antibacterial activities. Reducing sugar, terpenoids, tannin, sterols, saponins, phenols, flavonoids, cardiac glycosides, carbohydrates, and alkaloids, were all found in the two samples after the extracts underwent phytochemical analysis (Ajayi et al., 2015). Fourier transform infrared (FTIR), scanning electron microscopy (SEM), ultraviolet-visible (UV-Vis), and high-resolution transmission electron microscopy (HRTEM) were used to characterize the physicochemical features of this medicinal application (Abisharani et al., 2019b).

9.2 MEDICINALLY IMPORTANT SEED OIL PHYTOCHEMICALS

Phytochemicals are bioactive substances found in low, moderate, and high concentrations in plants, including grains and vegetables. They offer important protection against numerous degenerative diseases in humans and animals (Sam et al., 2017). Some of the phytochemicals in seed and seed oils responsible for therapeutic activities are cuminaldehyde, cymene, terpenoids, substituted pyrazine, pinene, p-cymene, safranal, terpinene, 2-methoxy-3-methyl pyrazine, 2-methoxy-3-secbutylpyrazine, 2-ethoxy-3 isopropylpyrazine, etc. (Keerthiga et al., 2019).

In most research work, using GC-MS analysis, the main oil ingredients were α-pinene (2.67%), γ-terpinene (2.82%), camphor (6.16%), and linalool (79.12%) (Kacmaz and Gul, 2021). Pumpkin seed oil contains fatty acids such as oleic acid (37%), linoleic acid (37.5%), and palmitic acid (11–15.5%). These compounds can reduce cholesterol levels in the body, improve abnormal prostate cancer, and reduce its impact (Hataminia and Farhadian, 2016). Plant seed oil showed anticancer, diabetes, back pain, and rheumatism properties. It also treats many diseases such as fever, migraine, cold, asthma, and headache. *Nigella sativa* seed oil is used as an immune stimulant, hepatoprotective and renal protector. There are many phytochemicals in *Nigella sativa* seed, mainly nigellone, thymoquinone, thymoquinone, thymol, and thymohydroquinone (Kumar et al., 2019). Other significant chemicals identified during the GC-MS analysis of the essential oils (EOs) isolated from *Nigella sativa* seeds included pcymene, carvacrolterpineol, tanethol, pinene, and longline. The seeds of *Nigella sativa* are enriched with unsaturated fatty acids including linolenic acid as the main fatty acid, and oleic acid, dihomolinoleic acid, and eicosadienoic acid (Kumar et al., 2019). In another study, *Wrightia arborea* and *Wrightia tinctoria* seed oils were analyzed using the GC-MS method. Studies on phytochemistry identified the presence of substances with potential medical value like phytol, lupeol, campesterol, γ-tocopherol, and squalene, in *Wrightia arborea* and lupenone, betulin, lupeol, campesterol, γ-tocopherol, squalene, β-amyrin, and α-amyrin, in *Wrightia tinctoria* (Nagalakshmi and Murthy, 2015).

9.3 SEED AND SEED OIL NANOPARTICLES SYNTHESIS AND CHARACTERIZATION

9.3.1 Silver Nanoparticles

Researchers have become interested in metal NPs because of their distinctive properties and wide range of applications (Rajan et al., 2017b). Aqueous AgNPs were produced and mediated with lecithin obtained from the seed oil of *Citrulus vulgaris*, which was extracted using a Soxhlet extractor. Physico-chemical characteristics such as the peroxide value, acid value, and percentage yield were carried out in the oil to check the suitability for NP preparation. The reaction was controlled using FTIR and UV-visible spectroscopy. Atomic force microscopy (AFM) was used to examine the dispersed AgNPs' size and surface (Barnabas et al., 2021b). According to a study by Chahardoli et al. (2017), powder extract of *Nigella arvensis* L. seed was used to create AgNPs from silver nitrate ($AgNO_3$) aqueous solution. They used a strong surface plasmon resonance (SPR) band at 435 nm, UV-vis spectroscopy, TEM, FTIR, to characterize the AgNPs. The X-ray diffraction (XRD) showed that AgNPs formed during the current synthesis were crystalline in nature. In another research, the EO from coriander seeds was extracted and used to form AgNPs.

The AgNPs were characterized by FTIR, SEM, EDX, XRD, and UV-Vis spectrophotometry. The SEM image showed that homogenous and spherical AgNPs were formed and EDX showed the elemental and confirmed the AgNPs' formation distribution. The UV-Vis indicated the SPR of AgNPs at 437 nm. A distinctively strong peak was visible using XRD at 3.0 keV (Kacmaz and Gul, 2021). The use of aqueous seed extract from the *Jatropha curcas* plant is an eco-friendly method for rapidly producing AgNPs by Bar et al. (2009). At various $AgNO_3$ concentrations, they could produce stable AgNPs that were generally spherical and ranged in size from 15 to 50 nm. XRD, UV-Vis, and HRTEM spectroscopy methods characterize the resultant silver particles. An XRD analysis reveals that the particles have face-centred cubic geometry and are naturally crystalline. In other research work, the AgNPs that had been created were examined, and the production of the AgNPs was determined using characterization methods like TEM, ED, FTIR, XRD, and UV-Vis. The findings demonstrated that the spherical AgNPs ranged in size from 13 to 69 nm, with a maximum absorbance peak at 371 nm (Nayaka et al., 2020). A clove oil microemulsion technique was proposed for producing AgNPs. HRTEM, UV-Vis, and XRD were used to analyze the AgNPs. The clove oil found the synthesized AgNOs to be stabilized and reduced good stability and dispersibility were obtained. In the microemulsion, Ag^+ was converted to nano-Ag by being adsorbed on the oil–water interface by the clove oil's active groups (Gao et al., 2017). In addition, cumin oil-mediated AgNPs were formed, and their properties were studied using UV-Vis and TEM by Keerthiga et al. (2019). From $AgNO_3$ and seed extracts of cauliflower, saag, and turnip, a quick and environmentally friendly method for forming AgNPs was developed. AgNPs' optical absorbance at 425 nm revealed that SPR had caused the production of metallic silver. FTIR spectroscopy was used to validate the effective capping of biological macromolecules. The XRD pattern showed the development of face-centred cubic silver nanocomposite with an average crystal size of between 14 and 20 nm (Aadil et al., 2016). For the synthesis of AgNPs, a straightforward, quick, and microwave-irradiated technique was used using *Cuminum cyminum* and *Malus domestica* seed extracts. The photosynthesized AgNPs were characterized by zeta sizer analysis, FTIR, XRD, TEM, and UV-Vis. The analysis proved the monodisperse character of the AgNPs with the size of NPs found between from 5.46–20 to 1.84–20.57 nm with a face-centred cubic crystalline structure (Jahan et al., 2021).

9.3.2 Gold Nanoparticles

Using the seeds and oil of *Punica granatum* L., gold nanoparticles (AuNPs) were observed to have a rectangular shape, with smooth and elongated properties for anticancer and antioxidant activity tests. Plant extract-mediated AuNPs are in high demand by researchers due to their high surface area to volume ratio, compatibility with the human body, low toxicity, and surfaces (Botteon et al., 2021). Synthesized AuNPs were characterized using a zeta potential (ZP) analyzer, dynamic light scattering (DLS), SEM, EDX, XRD, and FTIR. A UV-Vis spectrophotometer found that the photo-induced synthesis of AuNPs produced particles with a distinctive peak at 525 nm and a ZP of +34 mV and 70 nm, respectively. The functional groups of AuNPs shifted from 3632.27 to 541.899 cm^{-1} in the FTIR spectra, to indicate different functional groups exultancy in *Punica granatum* and *Punica granatum* seed oil-capped AuNPs (Lydia et al., 2020). In another study, aqueous *Elettaria cardamomum* seed extract was also used effectively to produce AuNPs with NP sizes of 15.2 nm (Rajan et al., 2017b). *Abelmoschus esculentus* seed aqueous extract was used in another study to create AuNPs to test if they formed adequately. The UV-Vis spectra of the NPs indicated a peak at 536 nm. XRD confirmed the 62 nm-sized NPs' crystalline nature of $AuNP_S$. The FTIR finding made it abundantly evident that the extracts with OH as a functional group contribute to capping the formation of NPs. The 3D topological property of AuNPs can be seen via AFM. According to the restricted size range of 45–75 nm seen in field-emission scanning electron microscopy (FESEM) images, all particles were spherical (Jayaseelan et al., 2013).

9.3.3 Iron Oxide Nanoparticles

Iron oxide NPs have received much attention lately because of their distinctive qualities, including a high surface-to-volume ratio, superparamagnetism, and simple separation techniques. Magnetic NPs with the right surface chemistry have been created using various biological techniques for many applications; plant-mediated iron oxide NPs are the most significant in this regard (Ali et al., 2016). Using specific surface analysis techniques such as vibrating sample magnetometer (VSM), TEM, XRD, FTIR, and FESEM, The drug cephalexin was loaded onto magnetic NPs that were covered in basil seed mucilage (Fe3O4@BSM-CPX), as a unique drug delivery system (Rayegan et al., 2018). Using *Nigella sativa* seed extract as a potent reducing agent, iron oxide NPs were produced to evaluate their antibacterial and cytotoxic potential. ZP, DLS, XRD, FESEM, and FTIR investigations were used to characterize the produced iron oxide NPs (Al-Karagoly et al., 2022).

9.3.4 Carbon Nanoparticles

Carbon is one of the most prevalent elements on the earth. Numerous carbon-based NPs, including carbon nanotubes, carbon dots, and quantum dots, have emerged in recent years as one of the most lucrative classes of nanomaterials with a wide range of uses. Some studies have used plants to help synthesize carbon NPs (Mondala et al., 2021). With the aid of the most basic and affordable technology, carbon NPs were produced utilizing a variety of seed oils, including cardamom oil, lemon oil, eucalyptus oil, almond oil, linseed oil, and olive oil. Differential scanning calorimetry (DSC), thermo gravimetric analysis (TGA), XRD, SEM, and FTIR were used to characterize the material (Muhsan et al., 2019).

Using powder XRD methods and SEM, the average size of carbon NPs was discovered to be 4.70, 4.69, 3.46, 57, and 24 nm, respectively.

9.3.5 Other Nanoparticles

Chacko and Newton (2019) developed successful stable and hybrid NPs with the medication using various quantities of grape seed oil and jojoba oil. Their findings showed that the particles are in the micro range, are stable, and have a significant ZP (–48.2 mV). The MPS analysis of the chosen formulations revealed values of 306.7183.4 and 416.5289.3. In another study, black cumin seed oil-containing solid lipid NPs were made using the emulsification-ultrasonication technique. Particle size effects of several emulsifiers were investigated. The particle diameter of the synthesized material was confirmed to range between roughly 220 nm and 1.2 mm by the DLS characterization of solid lipid NPs. According to the results of SEM, the particles had a sphere-like form and a diameter between 250 and 900 nm (Suksuwan et al., 2021).

9.4 APPLICATION OF SEED EXTRACT AND SEED OIL-MEDIATED NANOPARTICLES

9.4.1 Antimicrobial Activities

NPs have the potential to be antibacterial through a variety of methods. The most prevalent antimicrobial action mechanisms have been identified as reactive oxygen species, NP adherence to microbial cells, and their penetration into the cells (Staroń and Długosz, 2021). Plant-mediated NPs have been used for antibacterial activities due to their nontoxic and economic concerns (Ahmed et al., 2016a). Numerous bioactive substances, including tannins, steroids, flavonoids, and alkaloids, are present in *Nigella sativa* extract. Concentrated *Nigella sativa* seed extract has potent antibacterial properties that are effective against various bacterial types (Kumar et al., 2019). AuNPs derived from *Elettaria cardamomum* seed oil are utilized as antibacterial agents (Rayegan et al., 2018). Neem seed oil and the corresponding alkyd resin were subjected to an antimicrobial assessment. More inhibition zones for antibacterial activity were found when the antimicrobial assessment of the paints containing AgNPs was performed (Magaji et al., 2018). AgNPs produced from *Nigella sativa* seeds were powerful antibacterial substances. By tagging NPs with secondary metabolites, the results demonstrated that an excellent environmentally friendly and nontoxic source is available for synthesizing AgNPs compared to standard chemical/physical procedures (Chahardoli et al., 2017). Disk diffusion antibacterial testing performed by Rayegan et al. (2018), revealed that cephalexin's antibacterial capabilities were increased when it was loaded onto a Fe3O4@BSM nanocarrier with basil seed mucilage. Iron oxide NPs synthesized from

Nigella sativa seed extract were tested for antimicrobial activity against *Escherichia coli* and *Staphylococcus aureus*. The result showed excellent antibacterial efficacy toward *Staphylococcus aureus* and *Escherichia coli* with 11.52, 0.58, and 12.34 mm inhibition zones, respectively (Al-Karagoly et al., 2022). Salman Muhsan, (2015) achieved effective antibacterial activity against numerous species, including *Proteus refrigerator*, *Pseudomonas aeruginosa*, *Streptococcus haemolyticus*, and *Staphylococcus aureus*. Kacmaz and Gul (2021) used the broth microdilution test to examine the antibacterial properties of the EOs and synthesized AgNPs form of coriander seed against *Escherichia coli* and *Staphylococcus aureus*. The results demonstrated that AgNPs could have a synergistic antibacterial effect when used in conjunction with EO. *Abelmoschus esculentus* seed extract was used to create AuNPs, which were then tested using the conventional well diffusion method for their antifungal efficacy against *Aspergillus niger*, *Candida albicans*, *Aspergillus flavus*, and *Puccinia graminis*. The AuNPs showed the most significant zone of inhibition against *Candida albicans* (18 mm) and *Puccinia graminis* (17 mm). The outcomes point to the synthesized AuNPs' potential for antifungal activity. It is established that AuNPs can provide excellent antifungal efficacy, making them extremely useful in developing medications to treat fungi-related illnesses (Jayaseelan et al., 2013). In one study, *Piper nigrum* and *Nigella sativa* aqueous seed extracts were used as reducing agents for the green manufacture AgNPs from $AgNO_3$ to evaluate anticancer, antiviral, and antibacterial activities. The results showed that *Piper nigrum* and *Nigella sativa* seed extract AgNPs inhibit Gram-positive and negative bacteria species.

Additionally, the results showed that the synthesized NPs acted effectively against the herpes simplex virus-1 and decreased viral load and mortality by exhibiting 83.23 and 94.54% of antiviral activity, respectively. *Nigella sativa* and *Piper nigrum* AgNPs have demonstrated good cytotoxic effects in in vitro experiments against hepatocellular cancer, with IC_{50} values of 7.12 and 4.98 g/mL, respectively (Mahfouz et al., 2020). Apple pulp and cumin seed aqueous extracts demonstrated antibacterial activity when combined with AgNPs. The result demonstrated that *Escherichia coli* (Gram-negative bacteria) and *Staphylococcus aureus* (Gram-positive bacteria) were susceptible to the synthesized AgNPs (Jahan et al., 2021) (Figure 9.1).

9.4.2 Antioxidant Activities

Antioxidant activities of plants can be enhanced when they are mediated by NPs such as gold and silver (Stozhko et al., 2019). AuNPs synthesized using the seed oil of *Punica granatum* L. showed antioxidant activity in pharmaceutical, nutraceutical, and food applications (Lydia et al., 2020). Chitosan NPs in chia seeds (*Salvia hispanica* L.) showed antioxidant behaviour with significant advantages for the food sector and human health. The findings will enable the selection of chitosan particles rich in chia extracts for use in the pharmaceutical, food, and edible film industries and food packaging (Morales-Olán et al., 2021). AgNPs mediated by cumin oil were easily biosynthesized and showed good antioxidant efficacy compared to standard. The author of this study came to the conclusion that AgNPs mediated by cumin seed oil have the potential to be used as powerful antioxidants.

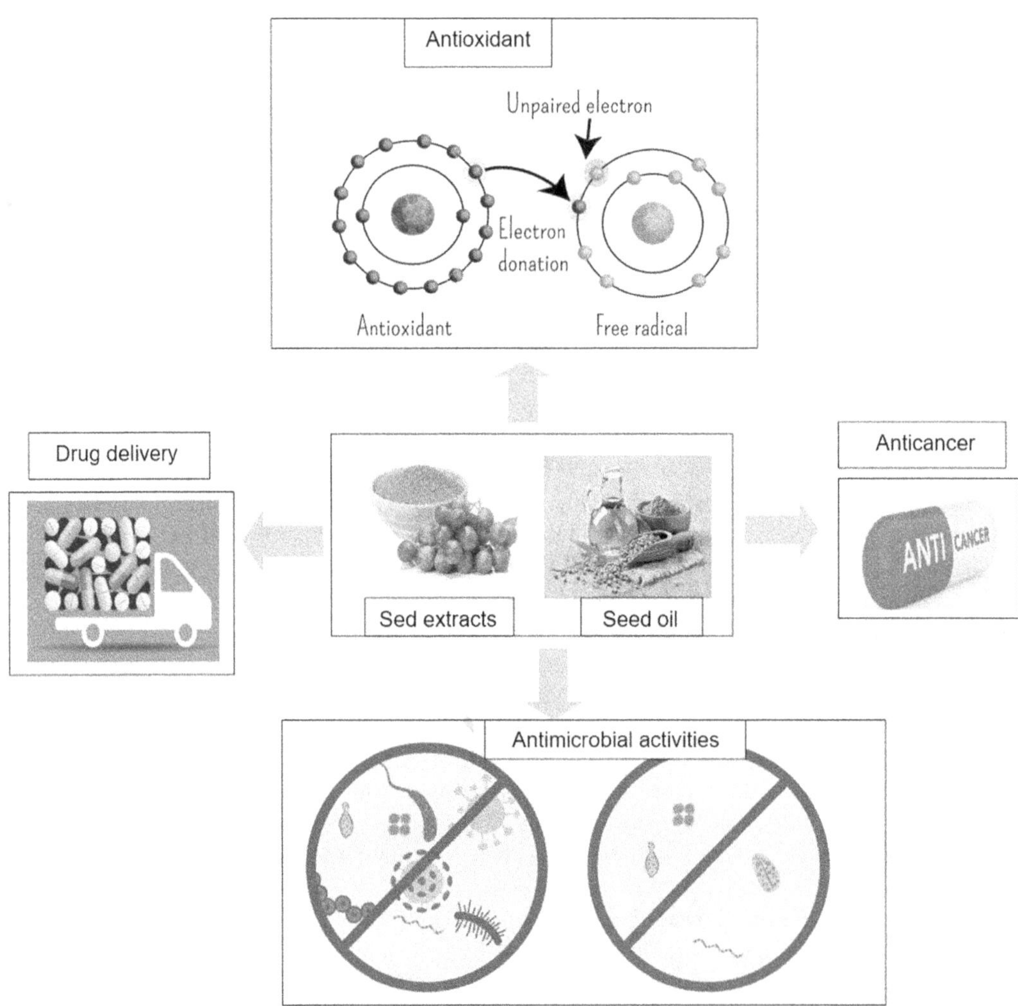

FIGURE 9.1 The application of seed extract and seed oil nanoparticles.

Additionally, it can be applied in numerous medicinal applications and large-scale manufacturing processes where an antioxidant is needed (Keerthiga et al., 2019). Additionally, AgNPs mediated by *Cuminum cyminum* seed oil can potentially be used as powerful antioxidant activities, according to Keerthiga et al. (2019). Additionally, antioxidant activity was demonstrated by aqueous extract of cumin and *Malus domestic* seed extracts in AgNPs. AgNPs synthesized through cumin extract showed larger free radical scavenging activities of 2,2'-azino-bis(3-ethylbenzothiazoline-6-sulfonic acid (ABTS) (96.43 ± 0.78% at 160 μg/mL) compared to *Malus domestic*. However, both AgNPs showed excellent free radical scavenging activities of DPPH with a value of 27.84 ± 0.56% and 13.12 ± 0.32% from cumin and *Malus domestic* extracts, respectively (Jahan et al., 2021).

9.4.3 Anticancer Activities

Plant-mediated NP preparation for anticancer activities is a well-known research area. AgNPs and AuNPs mediated by plant extracts are mainly researched for their anticancer activities (Sun et al., 2019). Toxic substances adsorbed on the surfaces of pre-processed $AgNP_S$ might cause problems in biological applications. Phytochemicals such as alkaloids, flavonoids, and other phenolic components found in *Trigonella foenum-graceum* seeds are associated with AgNPs' anticancer effects (Varghese et al., 2019). Lecithin is effectively used for AgNPs. Seed oil derived from *Citrullus vulgaris* can be used as a lecithin substitute to alleviate the high demand for lecithin in corn and soybeans, which is utilized as an anticancer activity (Barnabas et al., 2021a). Following the environmentally friendly manufacture of AgNPs from an aqueous solution of $AgNO_3$ using *Nigella arvensis* L. seed extract, two human cancer cell lines (MCF-7) and a human colorectal adenocarcinoma cell line (HT-29) demonstrated excellent cytotoxicity (Chahardoli et al., 2017). Using an aqueous seed extract from *Putranjiva roxburghii*, an investigation focused on the environmentally friendly manufacture of AgNPs and an analysis of the anticancer properties of the NPs.

The generated AgNPs were examined for possible tumour-suppressing qualities and anticancer activity on a human breast cancer cell line. The anticancer activity of the generated AgNPs was assessed using the 96 well plates mean transit time cell proliferation test methods. With an IC_{50} of 72.32 g/mL, the results showed that the human breast cancer cell line's viability was decreased at doses between 12.5 and 200 g/L (Nayaka et al., 2020). *Nigella Sativa* seed-derived AgNPs successfully showed exceptional cytotoxic activities on a breast cancer cell line (MCF-7) and a human colorectal adenocarcinoma cell line (HT-29) by labelling NPs with secondary metabolites (Chahardoli et al., 2017). Anticancer medications are made from black cumin seed extract containing active ingredients such as thymoquinone, which has shown potential anticancer activity in cytotoxicity studies at low concentrations (Gnanasekaran et al., 2021). A study by Lydia et al. (2020) found that the characterization and formation of AuNPs from the seed oil of *Punica granatum* L. showed potent anticancer properties towards colon and lung cancer, with cell viability found between 83.3 to 28.42% and from 80.3 to 25%, respectively. A good antitumour activity was confirmed against a variety of tumour cells using *Annona squamosa* seed oil, which is obtained as a by-product during extraction of *Annonaceous acetogenin*. The NPs made from *Annona squamosa* seed oil showed strong antitumour activities both in vivo and in vitro. The NPs have shown more enhanced antitumour efficacy than plant seed extract without NPs. In 4T1 tumour-bearing mice, the NPs generated from *Annona squamosa* seed oil (15 mg/kg) exhibited the most significant tumour suppression rate at 69.8%, higher than the non-NP plant seed extract solution (52.7%, 135 mg/kg, $p < 0.05$) (Ao et al., 2022).

9.4.4 Drug Delivery Activities

Drug delivery activities through NPs mediated by plants are the most well-known in different research areas (Yang and Merlin, 2019). Black cumin seeds are used to make black seed oil (BSO), which possesses hepatoprotective, antioxidant, and anti-inflammatory properties examined to improve hepatoprotective and hypotriglyceridemic functions by developing a self-emulsifying drug

TABLE 9.1
Metal and Metal Oxide Nanoparticles Synthesis, Particles Size, Shapes, and Their Application

Plant name	Plant material used	NP	Size (nm)	Shape	Application	Key reference
Annona squamosa	Seed extract	Ag	22	Spherical	Excellent catalytic activity against Coomassie brilliant blue dye degradation	Jose et al. (2021)
Azadirachta indica	Seed oil	Ag	98	—	Antifungal activity	Muddukrishnaiah (2019)
Avicennia marina	Seed extract	Ag	5–10	—	Antibacterial activity	Naidu et al. (2019)
Azadirachta indica	Seed extract	CuO	41 ± 21	—	Nutrition, growth, and seed germination are all improved as a result of this treatment	Jahagirdar et al. (2020)
Bixa orellana	Seed extract	TiO_2	13 ± 2	Spherical	—	Isacfranklin et al. (2021)
Bunium persicum	Seed extract	Ag	35–70	—	Inhibited urease and tyrosinase	Khan et al. (2022)
Cuminum cyminum L.	Seed extract		& 143232#x2053;100	Spherical	Effective against human breast cancer cells (IC50 = 1.25 μg/mL)	Dinparvar et al. (2020)
Cuminum cyminum	Seed extract	TiO_2	15.17	—	—	Mathew et al. (2021)
Elettaria cardamomum	Seed extract	Au	16.6	Spherical	Anticancer and antioxidant activities	Rajan et al. (2017a)
Elettaria cardamom	Seed extract	CuO	1–100	—	—	Venkatramanan et al. (2020)
Lactuca sativa	Seed extract	ZnO	50	—	Effect on the process of seed germination	Rawashdeh et al. (2020)
Dimocarpus longan	Seed extract	ZnO	10–100	Hexagonal	Orange II (70%) decolorization, methylene blue (MB) decolorization, and methyl orange decolorization (80%)	Chankaew et al. (2019)
Morinda citrifolia	Seed extract	Ag	3	Spherical	Antibacterial activity	Morales-Lozoya et al. (2021)
Moringa oleifera	Seed extract	Ag	—	Spherical	Wound contraction and tissue growth wall are both improved	Mehmood et al. (2016)
pomegranate seed	Seed oil	Au	12	Spherical	Antimicrobial activity	Botteon et al. (2021); Kumari and Philip (2013)
Punica granatum	Seed extract	Iron oxide	25–55	Semi spherical	Degradation was efficient toward reactive blue (95.08% after 56 min)	Karpagavinayagam and Vedhi (2019)
Rosa canina	Seed extract	Ag	150	Rod and spherical	—	Saygi and Usta (2021)
Salvia hispanica L.	Seed extract	Ag	7	Spherical	Antibacterial	Hernández-Morales et al. (2019)
Trachyspermum ammi	Seed extract	TiO_2	16.63	Spherical and spheroidal	—	Perveen et al. (2021)
Triticum aestivum	Seed extract	CuO	22 ± 1.5	Spherical	Catalytic activity for the elimination of 4-nitrophenol has been described (97.6% after 5 days)	Buazar et al. (2019)

delivery system (SEDDS) for BSO. By generating micelles with a mean droplet size of 183 nm and at least 2.6 times greater dispersibility than BSO, SEDDS-BSO improved the dispersion behaviour of BSO in water (Halder et al., 2021). The BSO (*Nigella sativa* L.) self-nanoemulsifying drug delivery system (SNEDDS) was developed by Priani et al. (2020) to improve oral bioavailability. The optimization of SNEDDS was carried out using various comparisons between cosurfactant, surfactant, and oil. When evaluated for global size determination, thermodynamic stability, robustness, dispersibility, emulsification duration, and transmittance percentage, the formulated SNEDDS displayed a good drug delivery performance. Cephalexin (CPX) was loaded onto basil seed mucilage-coated magnetic NPs (Fe3O4@BSM-CPX), a novel drug delivery system that was created, described, and tested for its antibacterial properties. The disk diffusion test demonstrated that loading CPX on the Fe_3O_4@BSM nanocarrier not only had a detrimental impact on the composition and functionality of the medicine in drug delivery application but also significantly showed in enhancement antibacterial characteristics (Rayegan et al., 2018) (Table 9.1).

9.5 CONCLUSION AND FUTURE PROSPECTS

The biosynthesis of metal NPs and metal oxide NPs such as Ag, Au, Ce, Fe, Se, Si, Ti, Zn, ZnO, TiO_2, CuO, CeO_2, SnO_2, Fe_3O, and MgO prepared by mediated seed and seed oil NPs has shown different pharmaceutical uses in the field of nanotechnology as an anticancer, antibacterial, antioxidant, antifungal, and environmental remediation agent. The bioactive compounds such as linoleic acid, palmitic acid, nigellone, thymoquinone, thymoquinone, thymol, thymohydroquinone, dihomolinoleic acid, and eicosadienoic acid are present in plant seed extract and its oil. These phytochemicals are responsible for forming stable NPs with smaller sizes (1–100 nm) with NPs shaped mostly as hexagonal spheres. FTIR, SEM, UV-Vis, and HRTEM were used to evaluate the physicochemical properties of the phytochemicals. However, synthesis pathways of the phytochemicals used to mediate NP synthesis are still unknown. Not enough investigations have been made into seed and seed oil-mediated NP synthesis, characterization, and application. Thus, in depth, critical studies should be carried out on the mechanism and pharmaceutical activities in this area.

REFERENCES

Aadil, K. R., Barapatre, A., Meena, A. S. & Jha, H. 2016. Hydrogen peroxide sensing and cytotoxicity activity of Acacia lignin stabilized silver nanoparticles. *International Journal of Biological Macromolecules*, 82, 39–47.

Abate, L., Tadesse, M. G., Bachheti, A. & Bachheti, R. K. 2022. Traditional and phytochemical bases of herbs, shrubs, climbers, and trees from Ethiopia for their anticancer response. *BioMed Research International*, 2022.

Abisharani, J., Devikala, S., Kumar, R. D., Arthanareeswari, M. & Kamaraj, P. 2019a. Green synthesis of TiO2 nanoparticles using Cucurbita pepo seeds extract. *Materials Today: Proceedings*, 14, 302–307.

Abisharani, J., Devikala, S., Kumar, R. D., Arthanareeswari, M. & Kamaraj, P. 2019b. Green synthesis of TiO2 nanoparticles using Cucurbita pepo seeds extract. *Materials Today: Proceedings*, 14, 302–307.

Ahmed, S., Ahmad, M., Swami, B. L. & Ikram, S. 2016a. A review on plants extract mediated synthesis of silver nanoparticles for antimicrobial applications: A green expertise. *Journal of Advanced Research*, 7(1), 17–28.

Ahmed, S., Ahmad, M., Swami, B. L. & Ikram, S. 2016b. A review on plants extract mediated synthesis of silver nanoparticles for antimicrobial applications: A green expertise. *Journal of Advanced Research*, 7(1), 17–28.

Ajayi, I. A., Raji, A. A. & Ogunkunle, E. O. 2015. Green synthesis of silver nanoparticles from seed extracts of Cyperus esculentus and Butyrospermum paradoxum. *IOSR Journal of Pharmacy and Biological Sciences*, 10(4), 76–90.

Al-Karagoly, H., Rhyaf, A., Naji, H., Albukhaty, S., Almalki, F. A., Alyamani, A. A., Albaqami, J. & Aloufi, S. 2022. Green synthesis, characterization, cytotoxicity, and antimicrobial activity of iron oxide nanoparticles using Nigella sativa seed extract. *Green Processing and Synthesis*, 11(1), 254–265.

Ali, A., Zafar, H., Zia, M., Ul Haq, I., Phull, AR, Ali, JS & Hussain, A. 2016. Synthesis, characterization, applications, and challenges of iron oxide nanoparticles. *Nanotechnology, Science and Applications*, 9, 49–67.

Ansari, M. A. & Alzohairy, M. A. 2018. One-pot facile green synthesis of silver nanoparticles using seed extract of Phoenix dactylifera and their bactericidal potential against MRSA. *Evidence-Based Complementary and Alternative Medicine*, 26, 1860280.

Ao, H., Lu, L., Li, M., Han, M., Guo, Y. & Wang, X. 2022. Enhanced solubility and antitumor activity of Annona squamosa Seed oil via nanoparticles stabilized with TPGS: Preparation and in vitro and in vivo evaluation. *Pharmaceutics*, 14(6), 1232.

Bachheti, R. K., Abate, L., Bachheti, A., Madhusudhan, A. & Husen, A. 2021. Algae-, fungi-, and yeast-mediated biological synthesis of nanoparticles and their various biomedical applications. In B. Kharisov & Oxana Kharissova (Eds), *Handbook of Greener Synthesis of Nanomaterials and Compounds* (pp. 701–734). Elsevier.

Bachheti, R. K., Fikadu, A., Bachheti, A. & Husen, A. 2020. Biogenic fabrication of nanomaterials from flower-based chemical compounds, characterization and their various applications: A review. *Saudi Journal of Biological Sciences*, 27(10), 2551–2562.

Bachheti, R. K., Konwarh, R., Gupta, V., Husen, A. & Joshi, A. 2019. Green synthesis of iron oxide nanoparticles: Cutting edge technology and multifaceted applications. In A. Husen & M. Iqbal (Eds), *Nanomaterials and Plant Potential*. Springer.

Bar, H., Bhui, D. K., Sahoo, G. P., Sarkar, P., Pyne, S. & Misra, A. 2009. Green synthesis of silver nanoparticles using seed extract of Jatropha Curcas. *Physicochemical and Engineering Aspects*, 348(1–3), 212–216.

Barnabas, H. L., Ngoshe, A. & Gidigbi, J. A. 2021a. Synthesis and characterization of stable aqueous dispersion of silver nanoparticle from *Citrullus vulgaris* seed oil via lecithin. *Spectroscopy*, 5, 8.

Barnabas, H. L., Ngoshe, A. & Gidigbi, J. A. 2021b. Synthesis and characterization of stable aqueous dispersion of silver nanoparticle from Citrullus vulgaris seed oil via lecithin. *Spectroscopy*, 2(2), 16–25.

Botteon, C., Silva, L., Ccana-Ccapatinta, G., Silva, T., Ambrosio, S., Veneziani, R., Bastos, J. & Marcato, P. 2021. Biosynthesis and characterization of gold nanoparticles using Brazilian red propolis and evaluation of its antimicrobial and anticancer activities. *Scientific Reports*, 11(1), 1–16.

Buazar, F., Sweidi, S., Badri, M. & Kroushawi, F. 2019. Biofabrication of highly pure copper oxide nanoparticles using wheat seed extract and their catalytic activity: A mechanistic approach. *Green Processing and Synthesis*, 8(1), 691–702.

Chacko, A. C. & Newton, A. M. 2019. Synthesis and characterization of valacyclovir HCl hybrid solid lipid nanoparticles by using natural oils. *Recent Patents on Drug Delivery and Formulation*, 13(1), 46–61.

Chahardoli, A., Karimi, N. & Fattahi, A. 2017. Biosynthesis, characterization, antimicrobial and cytotoxic effects of silver nanoparticles using Nigella arvensis seed extract. *Iranian Journal of Pharmaceutical Research: IJPR*, 16(3), 1167.

Chankaew, C., Tapala, W., Grudpan, K. & Rujiwatra, A. 2019. Microwave synthesis of ZnO nanoparticles using longan seeds biowaste and their efficiencies in photocatalytic decolorization of organic dyes. *Environmental Science and Pollution Research International*, 26(17), 17548–17554.

Dinparvar, S., Bagirova, M., Allahverdiyev, A. M., Abamor, E. S., Safarov, T., Aydogdu, M. & Aktas, D. 2020. A nanotechnology-based new approach in the treatment of breast cancer: Biosynthesized silver nanoparticles using Cuminum cyminum L. seed extract. *Journal of Photochemistry and Photobiology B: Biology*, 208, 111902.

Gao, H., Yang, H. & Wang, C. 2017. Controllable preparation and mechanism of nano-silver mediated by the microemulsion system of the clove oil. *Results in Physics*, 7, 3130–3136.

Geetha, R., Ashokkumar, T., Tamilselvan, S., Govindaraju, K., Sadiq, M. & Singaravelu, G. J. C. N. 2013. Green synthesis of gold nanoparticles and their anticancer activity. *Cancer Nanotechnology*, 4(4), 91–98.

Gnanasekaran, P., Roy, A., Natesh, N. S., Raman, V., Ganapathy, P. & Arumugam, M. K. 2021. Removal of microbial pathogens and anticancer activity of synthesized Nano-thymoquinone from Nigella sativa seeds. *Environmental Technology and Innovation*, 24, 102068.

Halder, S., Islam, A., Muhit, M. A., Shill, M. C. & Haider, S. S. 2021. Self-emulsifying drug delivery system of black seed oil with improved hypotriglyceridemic effect and enhanced hepatoprotective function. *Journal of Functional Foods*, 78, 104391.

Hataminia, F. & Farhadian, N. 2016. The synthesis of iron oxide nanoparticles coated with pumpkin seed oil fatty acids. *International Congress on Nanoscience and Nanotechnology (ICNN 2016).*

He, Y., Wei, F., Ma, Z., Zhang, H., Yang, Q., Yao, B., Huang, Z., Li, J., Zeng, C. & Zhang, Q. J. R. A. 2017. Green synthesis of silver nanoparticles using seed extract of Alpinia katsumadai, and their antioxidant, cytotoxicity, and antibacterial activities. *RSC Advances*, 7(63), 39842–39851.

Hernández-Morales, L., Espinoza-Gómez, H., Flores-López, L. Z., Sotelo-Barrera, E. L., Núñez-Rivera, A., Cadena-Nava, R. D., Alonso-Núñez, G. & Espinoza, K. A. 2019. Study of the green synthesis of silver nanoparticles using a natural extract of dark or white Salvia hispanica L. seeds and their antibacterial application. *Applied Surface Science*, 489, 952–961.

Iravani, S. 2011. Green synthesis of metal nanoparticles using plants. *Green Chemistry*, 13(10), 2638–2650.

Isacfranklin, M., Yuvakkumar, R., Ravi, G., Kumar, P., Saravanakumar, B., Velauthapillai, D., Alahmadi, T. A. & Alharbi, S. A. 2021. Biomedical application of single anatase phase TiO2 nanoparticles with addition of Rambutan (Nephelium lappaceum L.) fruit peel extract. *Applied Nanoscience*, 11(2), 699–708.

Jahagirdar, A. S., Shende, S., Gade, A. & Rai, M. 2020. Bioinspired synthesis of copper nanoparticles and its efficacy on seed viability and seedling growth in mungbean (Vigna radiata L.). *Current Nanoscience*, 16(2), 246–252.

Jahan, I., Erci, F. & Isildak, I. 2021. Rapid green synthesis of non-cytotoxic silver nanoparticles using aqueous extracts of'Golden Delicious' apple pulp and cumin seeds with antibacterial and antioxidant activity. *SN Applied Sciences*, 3(1), 1–14.

Jayaseelan, C., Ramkumar, R., Rahuman, A. A. & Perumal, P. 2013. Green synthesis of gold nanoparticles using seed aqueous extract of Abelmoschus esculentus and its antifungal activity. *Industrial Crops and Products*, 45, 423–429.

Jose, V., Raphel, L., Aiswariya, K. & Mathew, P. 2021. Green synthesis of silver nanoparticles using Annona squamosa L. seed extract: Characterization, photocatalytic and biological activity assay. *Bioprocess and Biosystems Engineering*, 44(9), 1819–1829.

Kacmaz, B. & Gul, S. 2021. *Antimicrobial Activities of Coriander Seed Essential Oil and Silver Nanoparticles*. https://doi.org/10.21203/rs.3.rs-526332/v1.

Karpagavinayagam, P. & Vedhi, C. 2019. Green synthesis of iron oxide nanoparticles using Avicennia marina flower extract. *Vacuum*, 160, 286–292.

Keerthiga, N., Anitha, R., Rajeshkumar, S. & Lakshmi, T. 2019. Antioxidant activity of cumin oil mediated silver nanoparticles. *Pharmacognosy Journal*, 11(4), 787–789.

Khan, I., Bawazeer, S., Rauf, A., Qureshi, M. N., Muhammad, N., Al-Awthan, Y. S., Bahattab, O., Maalik, A. & Rengasamy, K. R. 2022. Synthesis, biological investigation and catalytic application using the alcoholic extract of Black Cumin (Bunium Persicum) seeds-based silver nanoparticles. *Journal of Nanostructure in Chemistry*, 12(1), 59–77.

Kumar, A., Singh, D., Rehman, H., Sharma, N. R. & Mohan, A. 2019. Antibacterial, antioxidant, cytotoxicity and qualitative phytochemical evaluation of seed extracts of Nigella sativa and its silver nanoparticles. *IJPSR 10*, 11, 4922–4931.

Kumari, M. M. & Philip, D. 2013. Facile one-pot synthesis of gold and silver nanocatalysts using edible coconut oil. *Spectrochimica Acta Part A: Molecular and Biomolecular Spectroscopy*, 111, 154–160.

Lydia, D. E., Khusro, A., Immanuel, P., Esmail, G. A., Al-Dhabi, N. A. & Arasu, M. V. 2020. Photo-activated synthesis and characterization of gold nanoparticles from Punica granatum L. seed oil: An assessment on antioxidant and anticancer properties for functional yoghurt nutraceuticals. *Journal of Photochemistry and Photobiology B: Biology*, 206, 111868.

Magaji, A., Musa, H., Abubakar, A. & Umar, H. 2018. Synthesis, characterization and antimicrobial evaluation of silver nanoparticles embedded alkyd resin derived from neem seed Oil. *IOSR Journal of Applied Chemistry (IOSR-JAC)*, 11(2), 13–20.

Mahfouz, A. Y., Daigham, G. E., Radwan, A. M. & Mohamed, A. A. 2020. Eco-friendly and superficial approach for synthesis of silver nanoparticles using aqueous extract of Nigella sativa and Piper nigrum L seeds for evaluation of their antibacterial, antiviral, and anticancer activities a focus study on its impact on seed germination and seedling growth of Vicia faba and Zea mays. *Egyptian Pharmaceutical Journal*, 19(4), 401.

Mathew, S. S., Sunny, N. E. & Shanmugam, V. 2021. Green synthesis of anatase titanium dioxide nanoparticles using Cuminum cyminum seed extract; effect on Mung bean (Vigna radiata) seed germination. *Inorganic Chemistry Communications*, 126, 108485.

Mehmood, A., Murtaza, G., Bhatti, T. M., Kausar, R. & Ahmed, M. J. 2016. Biosynthesis, characterization and antimicrobial action of silver nanoparticles from root bark extract of Berberis lycium Royle. *Pakistan Journal of Pharmaceutical Sciences*, 29(1), 131–137.

Mondala, R., Yilmaz, D. & Mandalad, A. 2021. Green synthesis of carbon nanoparticles: Characterization and their biocidal properties. *Handbook of Greener Synthesis of Nanomaterials and Compounds*, 2, 277–306.

Morales-Lozoya, V., Espinoza-Gómez, H., Flores-López, L. Z., Sotelo-Barrera, E. L., Núñez-Rivera, A., Cadena-Nava, R. D., Alonso-Nuñez, G. & Rivero, I. A. 2021. Study of the effect of the different parts of Morinda citrifolia L.(noni) on the green synthesis of silver nanoparticles and their antibacterial activity. *Applied Surface Science*, 537, 147855.

Morales-Olán, G., Luna-Suárez, S., Figueroa-Cárdenas, J. D. D., Corea, M. & Rojas-López, M. 2021. Synthesis and characterization of chitosan particles loaded with antioxidants extracted from chia (Salvia hispanica L.) seeds. *International Journal of Analytical Chemistry*, 2021. https://doi.org/10.1155/2021/5540543.

Muddukrishnaiah, K. & Shilpa, V. P. 2019. Synergistic effect of silver nanoparticles Produced by green synthesis and neem oil (Azadirachta Indica) against human pathogenic Candida. *Albicansacta Scientific Agriculture*, 3(6), 143–145.

Muhsan, M. S., Nadeem, S., Hassan, A. U., Din, A. M. U., Shahida, S. & Ali, S. 2019. Synthesis of carbon nanoparticles by using seed oils. *Pak. J. Sci. Ind. Res. Ser. A: Phys. Sci.* 62(1), 1–7.

Nagalakshmi, M. & Murthy, K. S. R. 2015. Phytochemical profile of crude seed oil of Wrightiatinctoria R. BR. and Wrightiaarborea (Dennst.) Mabb. by GC-MS. *International Journal of Pharmaceutical Sciences Review and Research*, 31(2), 46–51.

Naidu, K. S. B., Murugan, N., Adam, J. & Sershen. 2019. Biogenic synthesis of silver nanoparticles from avicennia marina seed extract and its antibacterial potential. *BioNanoScience*, 9(2), 266–273.

Nayaka, S., Bhat, M., Chakraborty, B., Pallavi, S., Airodagi, D., Muthuraj, R., Halaswamy, H., Dhanyakumara, S., Shashiraj, K. & Kupaneshi, C. 2020. Seed extract-mediated synthesis of silver nanoparticles from Putranjiva roxburghii Wall., phytochemical characterization, antibacterial activity and anticancer activity against MCF-7 Cell Line. *Indian Journal of Pharmaceutical Sciences*, 82, 2, 260–269.

Perveen, K., Husain, F. M., Qais, F. A., Khan, A., Razak, S., Afsar, T., Alam, P., Almajwal, A. M. & Abulmeaty, M. M. 2021. Microwave-assisted rapid green Synthesis of gold nanoparticles using seed extract of Trachyspermum ammi: ROS mediated biofilm inhibition and anticancer activity. *Biomolecules*, 11(2), 197.

Prabhu, S. & Poulose, E. K. 2012. Silver nanoparticles: Mechanism of antimicrobial action, synthesis, medical applications, and toxicity effects. *International Nano Letters*, 2(1), 1–10.

Priani, S., Maulidina, S., Darusman, F., Purwanti, L. & Mulyanti, D. 2020. Development of self nano emulsifying drug delivery system for black seed oil (Nigella sativa L.). *Journal of Physics: Conference Series*. IOP Publishing, 012022.

Rajan, A., Rajan, A. R. & Philip, D. 2017a. Elettaria cardamomum seed-mediated rapid synthesis of gold nanoparticles and its biological activities. *OpenNano*, 2, 1–8.

Rajan, A., Rajan, A. R. & Philip, D. 2017b. Elettaria cardamomum seed-mediated rapid synthesis of gold nanoparticles and its biological activities. *OpenNano*, 2, 1–8.

Rawashdeh, R. Y., Harb, A. M. & Alhasan, A. M. 2020. Biological interaction levels of zinc oxide nanoparticles; lettuce seeds as case study. *Heliyon*, 6(5), e03983.

Rayegan, A., Allafchian, A., Sarsari, I. A. & Kameli, P. 2018. Synthesis and characterization of basil seed mucilage coated Fe3O4 magnetic nanoparticles as a drug carrier for the controlled delivery of cephalexin. *International Journal of Biological Macromolecules*, 113, 317–328.

Salman Muhsan, M. 2015. *Synthesis of Carbon Nanoparticles by Using Seed Oils and Evaluation of Their Biological Applications*. Lahore: University of Management and Technology.

Sam, S., Udosen, I. & Esenowo, G. 2017. The phytochemical and physicochemical properties of the seed and seed oil of Perseagratissima miller and Chrysophyllumalbidumg. Don. *Global Journal of Pure and Applied Sciences*, 23(2), 245–248.

Saygi, K. O. & Usta, C. J. M. C. 2021. Rosa canina waste seed extract-mediated synthesis of silver nanoparticles and the evaluation of its antimutagenic action in Salmonella typhimurium. *Materials Chemistry and Physics*, 266, 124537.

Sharma, G., Kumar, A., Sharma, S., Naushad, M., Dwivedi, R. P., Alothman, Z. A. & Mola, G. T. 2019. Novel development of nanoparticles to bimetallic nanoparticles and their composites: A review. *Journal of King Saud University-Science*, 31(2), 257–269.

Staroń, A. & Długosz, O. 2021. Antimicrobial properties of nanoparticles in the context of advantages and potential risks of their use. *Journal of Environmental Science and Health – Part A*, 56(6), 680–693.

Stozhko, M., Bukharinova, M., Khamzina, E. I., Tarasov, A. V., Vidrevich, M. B. & Brainina, K. Z. 2019. The effect of the antioxidant activity of plant extracts on the properties of gold nanoparticles. *Nanomaterials*, 9(12), 1655.

Suksuwan, A., Arour, Z., Santiworakun, N. & Dahlan, W. 2021. Preparation and characterization of black seed oil loaded solid lipid nanoparticles for topical formulations: A preliminary study. *AIP Conference Proceedings*. AIP Publishing LLC, 020003.

Sun, B., Hu, N., Han, L., Pi, Y., Gao, Y. & Chen, K. 2019. Anticancer activity of green synthesised gold nanoparticles from Marsdenia tenacissima inhibits A549 cell proliferation through the apoptotic pathway. *Artificial Cells, Nanomedicine and Biotechnology*, 47(1), 4012–4019.

Varghese, R., Almalki, M. A., Ilavenil, S., Rebecca, J. & Choi, K. C. 2019. Silver nanoparticles synthesized using the seed extract of Trigonella foenum-graecum L. and their antimicrobial mechanism and anticancer properties. *Saudi Journal of Biological Sciences*, 26(1), 148–154.

Venkatramanan, A., Ilangovan, A., Thangarajan, P., Saravanan, A. & Mani, B. 2020. Green synthesis of copper oxide nanoparticles (CuO NPs) from aqueous extract of seeds of Eletteria cardamomum and its antimicrobial activity against pathogens. *Current Biotechnology*, 9(4), 304–311.

Yang, C. & Merlin, D. 2019. Nanoparticle-mediated drug delivery systems for the treatment of IBD: Current perspectives. *International Journal of Nanomedicine*, 14, 8875.

10 The Function of Medicinally Significant Tree Bark in Nanoparticle Production and Applications

Sankar Hari Prakash and Selvaraj Mohana Roopan

CONTENTS

ABBREVIATIONS

AE *Albizia lebbeck* bark extract
ATC ammonium thiocyanate
BHT butylated hydroxytoluene
BR Briggs Rauscher
CNTs carbon nanotubes
CV cyclic voltammetry
CVD chemical vapour deposition
DPPH 1,1-diphenly-2-picrylhydrazine
FRAP ferric reducing cell reinforcement power
FTC Folin–Ciocalteu strategy
GCD galvanostatic charge discharge
MDR multi drug resistance
MIC maximum inhibitory concentration
ORAC oxygen extremist absorbance limit
TBA thiobarbituric corrosive technique

DOI: 10.1201/9781003213727-10

10.1 INTRODUCTION

Nanotechnology plays a significant function and can be employed at the nanoscale; it has a wide range of applications as well. It is portrayed as controlling or reconstructing of matter at the sub-atomic and atomic scale in the 1–100 nm size range (Robinson and Hsu 2017). Norio Taniguchi first coined the phrase 'nanotechnology' in 1974 (Naik and Selukar 2015), defining it as 'the processing of separation, consolidation, and deformation of materials by a single atom or molecule'. However, in the early 2000s, nanotechnology was explored extensively and finally used in regular practice, compared with 1997–1998. The demand for highly functionalized biomaterials is increasing as the area of biomedical engineering grows. In spite of the remarkable diversity and complexity of living systems, they all have the ability to react to different climatic conditions, which is vital for sustaining normal operations in the long term (Aflori 2021).

Researchers in a variety of sectors have been driven to nanomaterials because they have fundamentally altered the microstructure. These unusual electrical and optical features are attributed to nanomaterials' surface area, quantum-size, and macro-quantum tunnelling effects. Moreover, when paired with large-scale assembly processes, these nanoparticles should be able to meet the demand for widely used, high-performance electronics. These are used extensively in many industries, like sensing, photovoltaics, field-effect transistors, and photodetectors (Yin et al. 2020).

Nanomaterials are typically classified by the following factors: morphology, size, shape, composition, uniform, or agglomeration, as illustrated in Figure 10.1. Nanoparticles have three different surface morphologies: crystalline, spherical, and flat. Nanoparticles can similarly be separated into four classifications dependent on the direction in which electrons travel (Jeyaraj 2019).

10.2 DIMENSIONS OF NANOPARTICLES

10.2.1 0-Dimension

All measurements (x, y, z) are at the nanoscale, i.e., no measurement is bigger than 100 nm. It includes nanospheres and nanoclusters. Specialists have been reading 0-D nanomaterials for many years. Numerous physical and substance strategies have been created to make 0-D nanomaterials with exact measurements. A few exploration groups effectively made 0-D nanomaterial including uniform molecule exhibits (quantum dabs), heterogeneous molecule clusters, shell quantum dots, onions, empty circles, and nanolenses (Tiwari et al. 2012).

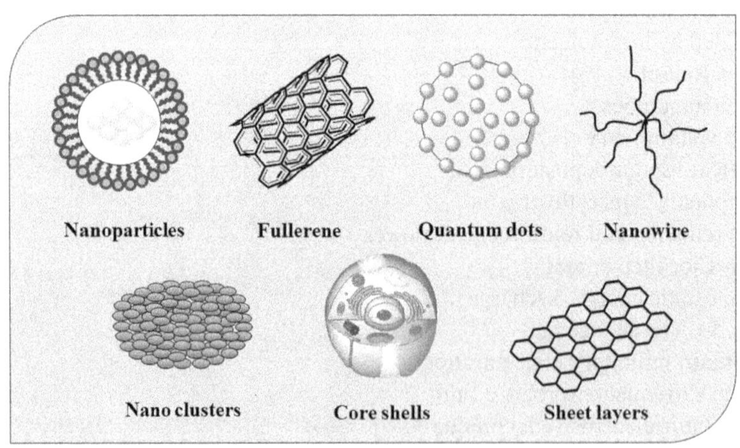

FIGURE 10.1 Dimensions of nanomaterials.

10.2.2 1-Dimension

Here, the materials with two measurements (x, y) are at nanoscale and one is outside the nanoscale. This prompts thin moulded nanomaterials. 1-D nanomaterials have excellent properties, like mechanical strength, light absorption, harvesting, enhancing the flexible nanogenerator, low thermal conductivity, and high electronic conductivity (Han and Ho 2013). Some 1-D nanostructured resources are nanowires, nanotubes, and nanofibres (multi-wall carbon nanotubes, single-wall carbon nanotubes, etc.) because of their fascinating features and wide range of practical uses. The improvement of innovative approaches for the precise synthesis of 1-D nanomaterials has been extensively studied. Even though 1-D nanomaterials have unique features, they can be used in various fields, including microelectronics, magnetism, optics, and catalysis. Researchers are continually aiming to produce new scientific research and future applications in 1-D nanomaterial development (Zhou et al. 2013).

10.2.3 2-Dimension

These materials have one dimension in the nanoscale and the other two are outside the nanoscale. The 2-D nanomaterials show plate-like shapes. They incorporate nanofilms, nanolayers, and nanosheets. 2-D nanoparticles are generally utilized in drug/quality conveyance, biosensors, multimodal imaging, antibacterial specialists, engineered science, and disease treatment in view of their mechanical conduct, biosimilarity, and debasement rate. 2-D nanomaterials are the thinnest materials and have the largest surface area, indicating that they have huge reservoirs and anchoring sites for efficiently loading and delivering therapeutic agents. Light, sonication, and magnetic properties, as well as biological behaviours such as endocytosis, biodistribution, biodegradation, and excretion, are all distinct physio-chemical features of planar nanostructured materials (Hu et al. 2019).

10.2.4 3-Dimension

These are nanomaterials that are not restricted to the nanoscale in any measurement. These materials have three optional measurements over 100 nm. They include nanoparticle dispersions, nanowire and nanotube bundles, and multinanolayers (polycrystals), in which the 0-D, 1-D, and 2-D structural components are in close contact and form interfaces (Anton Paar, website). 3-D carbon nanotubes (CNTs) are carbon allotropes with a cylinder-shaped nanostructure. They are made up of one (single-walled CNT) or multiple (multi-walled CNT) carbon layers wrapped around a hollow core with a graphitic structure. Both the core and the wall have diameters in the nanometre range, although the overall length of the tubes is often significantly larger. CNT composites distributed in matrix materials (e.g., polymers) exhibit a variety of intriguing and unique features. As a result, they might be valuable in a variety of domains, including materials science, electronics, optics, and others (Jang et al. 2019).

There are two different techniques utilized for the creation of nanomaterials, top-down and bottom-up, to make nanoparticles with positive properties, like shape, size, and movement. The top-down approach involves lithographic processes, ball milling, sputtering, etching, laser ablation, and so on. Bottom-up synthesis involves laser pyrolysis, chemical vapour deposition (CVD), reduction method, sol-gel method, and so on. These methods are successful in synthesizing metallic nanoparticles, but their cost of production remains the main obstacle. As a result, a low cost and eco-friendly approach is needed, which allows for the elimination of a potentially harmful substance while simultaneously increasing the amount of a stabilizing one used to create nanoparticles. To synthesize green nanoparticles, all of these criteria (Kumar et al. 2020), shown in Figure 10.2, must be met.

The green method is the best option. It is essential to follow the 12 principles illustrated in Figure 10.3 in sequence to reduce the negative effects of nanomaterial creation while boosting the safety of nanotechnology (Shafey 2020).

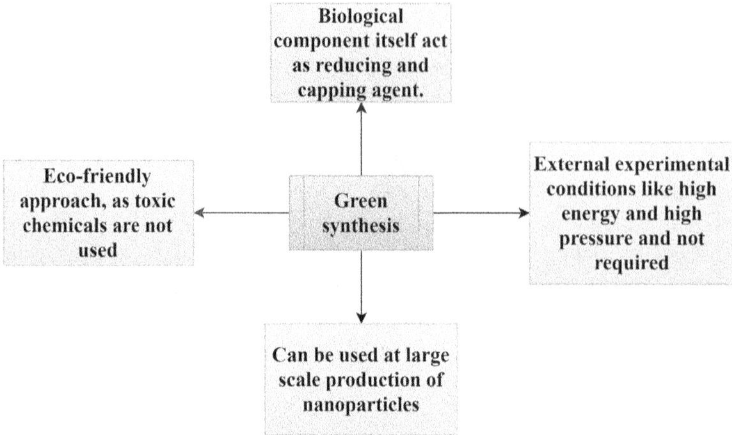

FIGURE 10.2 Key merits of green synthesis.

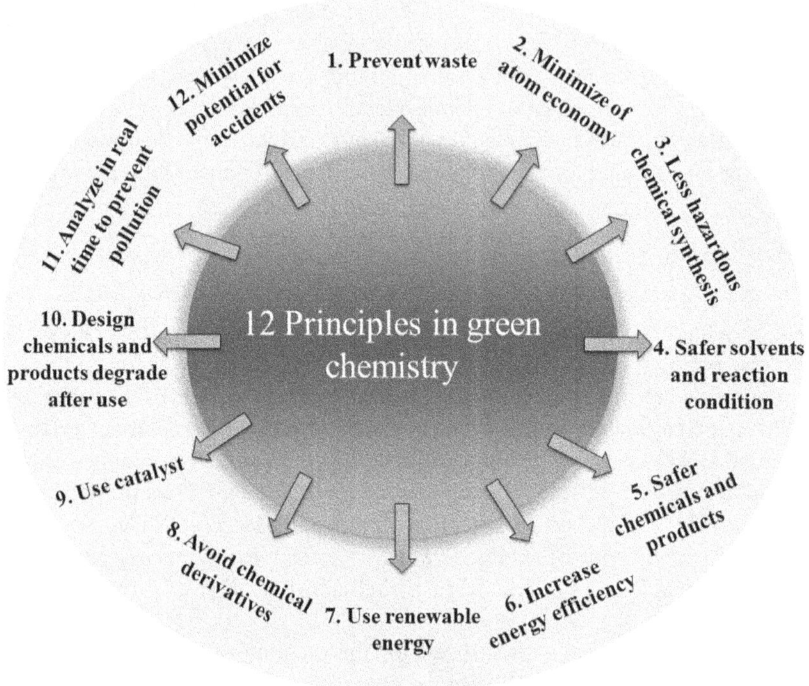

FIGURE 10.3 Twelve principles of green chemistry.

In nanoparticle synthesis, green reducing agents such as plant extracts, sea algae, or microorganisms are used (Husen et al. 2019; Bachheti et al. 2021). Additionally, the process calls for aseptic conditions and is harmful to the environment in some waste products. A simple, successful process involving the combination of plant extracts with metallic salt solutions involves the synthesis of nanoparticles. The reaction occurs in a short time, from minutes to hours, at room temperature. This technique is indeed simple, and the environmental consequences are decreased. It can be a revaluation of a waste product in some circumstances (Martínez-Cabanas 2021). Plants for the production of metallic nanoparticles are described as bioreactors. Metal nanoparticle production in conifers has some unusual features. The green synthesis of metal nanoparticles is an effective and unproblematic method of biological synthesis which is ecologically and economically benign. Metallic nanoparticles contain both pure metal and metal oxides that have numerous applications (Painuli et al. 2020; Bachheti 2020, 2020a; Bhardwaj et al. 2020).

Green production of metal nanoparticles using a green source has been an exciting feature as the practice is eco-friendly and safe. Plants and plant products have recently emerged as viable sources for the fabrication of nanostructured materials. For the manufacture of various metal nanoparticles, extracts originating from leaves, bark, stems, roots, and flowers have been widely used. The biogenic manufacture of monodispersed nanoparticles with uniform shape, on the other hand, remains a barrier (Sarwar et al. 2021). Employing various plant resources, much research has recently documented the production of nanoparticles such as TiO_2, Ag, CuO, Au. In this chapter, the medicinal importance of tree bark extract used for nanomaterials synthesis and its potential assay applications is described.

10.3 SYNTHESIS OF BARK EXTRACT-MEDIATED NANOPARTICLES

The combination of nanomaterials utilizes three sorts of techniques called actual strategies, compound techniques, and biological techniques. Physical methods are generally electrochemical, ultrasonic, laser ablation, and evaporation-condensation methods. Chemical methods are etching, mechanical milling, sputtering, and thermal decomposition. After this, the biological method is often called green synthesis and uses plant materials, algae, yeasts, and actinomycetes (Khandel et al. 2018). The bark contains the cortex, which refers to all tissues outside of the vascular cortex, as well as the parallel meristem that produces optional xylem within and the auxiliary phloem, the deepest layer of the cortex, located on the outside of the stem (Rosell 2018).

Currently, the paper industry generates a large amount of residue from the exploitation of woody vascular plants. Usually, a large amount of woody bark is found among the wood scraps in the forest. This waste is frequently used for heating or as a low-cost energy source in pulp mills, although these types of operations are inefficient and can lead to environmental problems (Tanase et al. 2019). However, some research work on nanomaterial synthesis is shown in Figure 10.4, using tree bark extract with various kinds of synthetic precursors.

Iravani and Zolfaghari (2013) have investigated the green method of silver nanoparticles (AgNPs) using bark extract from *Pinus eldarica* as a capping agent. In this synthesis, optimization by different concentration parameters of silver nitrate ($AgNO_3$), different amounts of the substrate bark extract with phosphate buffer, and the various pH range were carried out. UV absorption of colloidal suspensions absorption spectra were measured at a wavelength of approximately 430 nm. In addition, TEM analysis was implemented on designated specimens to study the formation of the AgNPs with dimensions of 10–40 nm and mainly spherical in shape. Furthermore, In 2017 Olajire et al. (2017) developed green synthesis of platinum nanoparticles from *Alchornea laxiflora* bark extract for oxidative desulphurization of oil samples. Desulphurization of chloroplatinic acid with an aqueous solution of bark extract as a bioreduction process at a reaction temperature of 100°C and they confirmed that the characterized nanoparticles, including ultraviolet (UV), Fourier transform infrared spectroscopy (FTIR), high-resolution transmission electron microscopy (HR-TEM), energy dispersive X-ray (EDX), and X-ray diffraction (XRD) spectroscopic analyses, were considered for the synthesis of platinum nanoparticles. Platinum-based nanoparticles measured by HR-TEM varied in size from 3.68 to 8.77 nm, with only a mean particle size of about 5.93 ± 1.43 nm. The resulting platinum nanoparticles were indeed stabilized by compounds and proteins via hydroxyl and carbonyl groups, according to the XRD pattern, which exhibits a face centred cubic (fcc) structure.

Supraja et al. (2016) developed biosynthesis of *Boswellia ovalifoliolata* stem bark extract was utilized to make zinc oxide nanoparticles (ZnONPs). At the point when the making of ZnO was signified from the UV spectrometer range in 230 nm, the bark separate was treated with 1 mM of zinc nitrate arrangement in an aqueous medium. The FTIR analysis reported that primary and

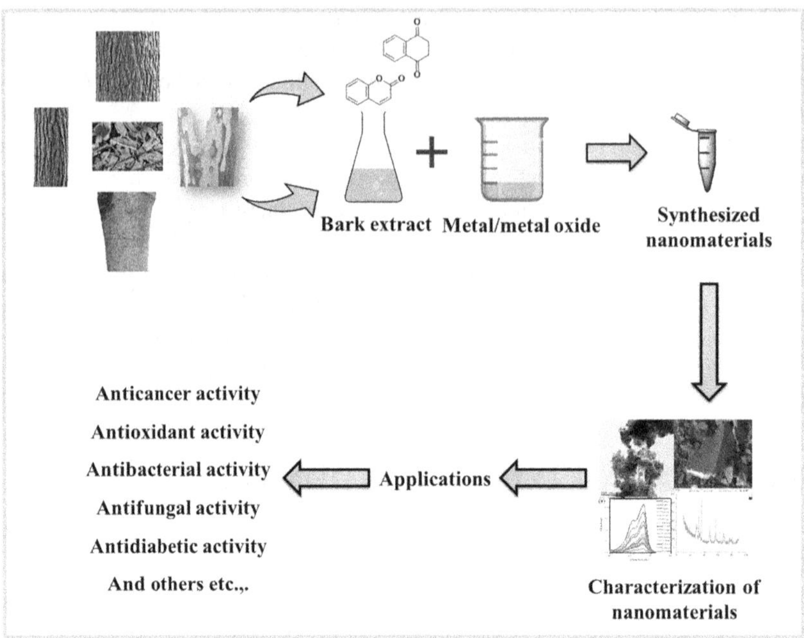

FIGURE 10.4 Bark-mediated nanoparticles synthesis pathway.

secondary amine groups are confirmed ZnONPs and the morphology and crystalline phase was proved in TEM analysis with 20.3 nm size. Furthermore, they used the dynamic light scattering (DLS) technique of zeta potential (ZP) at the values of 4.8 mV. Saha et al. (2018) investigated clarified ZnONPs produced using *Terminalia arjuna* bark concentrate and zinc acetic acid derivation ($ZnC_4H_6O_4$) as a precursor. The creators led a relative report utilizing different methods like sonication, the wet chemical method, and hydrothermal method. The zinc oxide sample made from hydrothermal treatment has an elevated absorbance in the UV-Vis region, indicating that it has superstructure development. We may arrive at that conclusion since the SEM investigation of the zinc oxide specimen reveals a spherical framework. Moreover, the excellent absorption region observed in the hydrothermal synthesis method of ZnO NPs was confirmed from the UV-spectroscopy, and the particle size in the spherical structure reported in the SEM analysis. Further characterization of the XRD, particle size, TEM, and surface area analysis techniques were used to characterize the structural morphology and conformation of ZnONPs. Pawar et al. (2012) investigated gold nanoparticles (AuNPs) using *Ficus nitida* bark extracts. The bark was discovered to become a capping and reducing agent that may reduce auric ions (Au^+ to Au^0) quickly. The AuNPs were characterized and confirmed using UV-Vis spectroscopy and STM.

10.4 ANTIMICROBIAL ASSAY OF TREE BARK-MEDIATED SYNTHESIS OF NANOMATERIALS

The biosynthetic pathway of nanomaterials is expounded in plant disease control; pesticides have been used in an initial portion of pests in agricultural crops without affecting the plants themselves, as well as the symbiotic flora, fauna, and humans. Even so, the hazardous use of these pesticide residues is a major source of public concern and an environmental crisis (Garg et al. 2014). Besides that, the advent and formation of alternative pathogens is an ongoing issue, and chemical treatment is both costly and ineffective. The latest advances in the field of green methods have propelled research to tackle microbial microorganisms (Naikoo et al. 2021).

Metallic nanoparticles from biosynthetic processes, like silver, copper, gold, and zinc, have been demonstrated to be successful against Gram-positive and Gram-negative organisms including *Bacillus subtilis* and *Escherichia coli*. These nanoparticles hindered the development of pathogenic fungi such *as Aspergillus niger, Fusarium oxysporum*, and *Aspergillus fumigatus*, as well as many other pathogenic microbes. Nanoparticles have also emerged as the new option to address antimicrobial properties. This effect is attributed to the size and high surface/area proportion of nanoparticles, which favour their communication with plant pathogens membranes (Ali et al. 2020). *Cochlospermum religiosum* aqueous stem bark extract of AgNPs showed the highest toxicity to *Staphylococcus* followed by *Pseudomonas*, *Escherichia coli*, and *Bacillus* and the lowest toxicity towards *Proteus*. Whereas in fungal species the highest inhibition zone against *Aspergillus flavus* followed by *Rhizopus*, *Fusarium*, and *Curvularia*, and the minimum inhibition zone was observed against *Aspergillus niger* species. The outcome of this study could be useful for the development of value-added products from the indigenous medicinal plants of India for nanotechnology-based biomedical applications (Sasikala et al. 2015).

Mandal et al. (2013) investigated the antimicrobial properties of *Terminalia arjuna* bark concentrate, and starter phytochemical uncovered the event of lively mixtures, for example, lactones, phytosterol, phenolic compounds, flavonoids, glycosides, and tannins in high fixations.

The antimicrobial action has been viewed as more prominent against two Gram-negative microscopic organisms, *Escherichia coli* and *Klebsiella pneumoniae*, than two Gram-positive microbes, *Staphylococcus aureus* and *Streptococcus mutans*. Human pathogenic bacteria were tested in two concentrations (50% and 100%). For antibacterial activity, the zone of inhibition of *Terminalia arjuna* bark extract was compared to standards such as chloramphenicol (Mandal et al. 2013). Supraja (2018) synthesized ZnONPs from the *Alstonia scholaris* stem bark extract for antimicrobial action and was tested against microbes, Gram-negative, and Gram-positive bacteria separated from potable water PVC pipeline microbes using the disc diffusion method. The findings revealed that ZnONPs possessed higher antifungal and antimicrobial properties against every microbe evaluated in the in-disc diffusion experiment (170, 100, and 50 ppm). Furthermore, the cytotoxicity of biologically synthesized ZnONPs has been evaluated against *Alstonia scholaris* to determine the LC_{50} with 90% confidence intervals (Supraja et al. 2018). Furthermore, Ansari et al. (2020) developed ZnONPs from the source of *Cinnamomum verum* bark extract and investigated the potential application for antimicrobial activity. Green synthesized ZnONPs reduced the improvement of *Escherichia coli* and *Staphylococcus aureus* with the least conceivable inhibitory fixation minimum inhibitory concentration (MIC) of 125 $\mu g\ mL^{-1}$ and 62.5 $\mu g\ mL^{-1}$, in both. The outcomes demonstrated that the pre-arranged ZnONPs can possibly be utilized as an antibacterial property against bacterial substances.

Ramanathanet et al. (2018) also investigated the *Solanum trilobatum* extract-embedded with AgNPs for antimicrobial activity. Disk diffusion and well diffusion assays were used to evaluate the effectiveness of *Solanum trilobatum* bark extract-capped AgNPs in terms of bacterial growth suppression. *Solanum trilobatum* extract-roofed AgNPs were discovered to have improved antibacterial capabilities, involving inhibition of growth of Gram-positive and Gram-negative bacteria, as well as fungal diseases. Those findings indicated the possibility for indigenous medicinal herbs to be used in the domain of nanotechnology.

Nayak et al. (2016) have reported that the biosynthetic potentials of *Azadirachta indica* and *Ficus benghalensis* bark extracts aimed at the manufacture of AgNPs. Ag produced demonstrated dose-dependent antiproliferative efficacy against the MG-63 osteosarcoma cell line. As a result, these AgNPs can be consumed in a wide range of healing agents against osteosarcoma and bacteria. Grains with ZP values above +30 mV and below −30 mV are deemed stable for colloidal distribution due to the lack of steric stabilization. In 2001, MEM enriched with 1% penicillin-streptomycin solution and 10% FBS, 3000 cells/well were planted in 96 well plates at a thickness of 3000 cells relying upon the greatest development rate and developed for 24 h at 37°C in a hatchery with 5% CO_2.

Of the various types of nanomaterials synthesized, the bark extract received excellent reports in past years in the field of antimicrobial assay. Prasad and Swamy (2013) examined the *Syzygium cumini* bark utilized for synthesis of AgNPs for antibacterial growth activity. Ontong et al. (2019) synthesized AgNPs from *Senna alata* bark extract. *Garcinia kola* bark extract embedded with synthesis of AgNPs for antibacterial activity was explored by Akintelu et al. (2020). Tanase et al. (2020), also used *Picea abies* L. bark extract-mediated synthesis of AgNPs for antimicrobial activity. Sudarsan et al. (2021) produced AgNPs using *Terminalia arjuna* bark extract and Sarwar et al. (2021) described citric acid-mediated green synthesis of copper nanoparticles using cinnamon bark extract.

10.5 ANTIOXIDANT ACTIVITY USING TREE BARK-BASED NANOMATERIALS

In nanomaterials, one of the most important activities is as an antioxidant which includes metal/metal oxide-based nanomaterials. Indicate intrinsic redox action, is frequently linked with radical trapping, superoxide dehydrogenase, and catalase capabilities by grafting low molecular weight antioxidants on redox inert nanomaterials (Valgimigli et al. 2018). According to some studies, multiple global plant species have therapeutic benefits; most of the medicinal herbs have excellent antioxidant activity. Antioxidants lower oxidative stress in tissues, making them effective as therapy for many human disorders such as cancer, cardiovascular disease, and inflammatory disease (Abate and Yayinie 2018). Antioxidants have phenolic groups like flavonoids, tannins, and phenolic acids, they have a variety of biological benefits as a result of their antioxidant activity, including anti-inflammatory, anticarcinogenic, and antiatherosclerotic properties (Abate et al. 2017).

Evaluation of the antioxidant properties was separated by several methods, such as 1,1-diphenyl-2-picrylhydrazine (DPPH), reticence of linoleic acid peroxidation, ferric reducing cell reinforcement power (FRAP), oxygen extremist absorbance limit (ORAC) test, Briggs Rauscher (BR) technique, thiobarbituric corrosive (TBA) strategy, ammonium thiocyanate (ATC) technique, etc. (Abate et al. 2017). Appropriately, the phenolic content was surveyed utilizing the Folin–Ciocalteu strategy (FTC), and the absolute flavonoid content, the all-out flavanol content, the complete phenol content, and the tannin content were totally assessed by these techniques (Abate et al. 2018). Although there are numerous ways to determine antioxidant action, it is critical to have a reliable and quick technique. Whereas each approach has positive impacts and limitations, it has been discovered that the DPPH and 2,2-azinobis(3-ethylbenzthiazoline-6-sulphonic acid) (ABTS) procedures are the most popular and efficient; these were developed and refined recently (Krishnaiah et al. 2011). Mahapatra et al. (2020) examined the antioxidant activity of AgNPs from the *Albizia lebbeck* bark extract (AE). AgNP-AE is energetic against many multi-drug resistance (MDR) medical strains. The antibacterial impact of AgNP-AE was performed utilizing the disk dissemination strategy, broth weakening test, and agar technique and acted in inconsistency of 14 MDR disengages, just as their base inhibitory action (MIC) and least bactericidal focus (MBC). The AgNP-AE showed critical bioactivity even against tried examples with a measurement scope of 10.4–19.5 mm and MIC of 0.5–0.84 M. It has free-extremist rummaging activity at 15–20 M, making it harmless up to 50 M (0.085–0.143 g). In spite of the fact that the zone width for AE is 10.2–15 mm and the MIC is 128–1024 g/mL, 100% passing was seen at 0.50 M after 2–4 h of treatment because of cell architectural damage.

Mohamed et al. (2012) reported five various extracts of *Terminal arjuna* bark tested for potential antioxidant properties. Using appropriate assay systems, the ability of five bark extracts to scavenge DPPH radical, NO radical, hydrogen peroxide radical, and reduce the power evaluation, cupric limiting antioxidant capacity, was assessed as well as associated to usual and synthetic antioxidants. Spectroscopy analysis was used to measure total antioxidant activity of phenolic and flavonoid levels. In terms of DPPH free radical scavenging activity, methanol extract had the highest IC_{50} value of 6.34 g/mL, followed by petroleum ether value of 7.76 and ethanol value of 25.63, individually, as opposed to butylated hydroxytoluene (BHT) of 8.816 and ascorbic acid with a value of 5.698 respectively. Despite an increase of concentration, the five fractional extracts demonstrated excellent degradation efficiency and cupric reducing capacity, with methanol extract gaining the highest position. The results show that *Terminal arjuna* is a possible source of natural antioxidants and so has the potential to prevent numerous reactive species ailments.

Moreover, Suganthy et al. (2018) synthesized AuNPs from the same *Terminal arjuna* ethanol bark extract for neuroprotective capabilities. UV-visible spectroscopy revealed AuNPs-specific surface plasmon resonance (SPR) absorption on 536 nm. The development of fcc spherical crystalline, and triangular shaped AuNPs with a size of about 20–50 nm was demonstrated by XRD, TEM, and FESEM investigations. AgNPs have a mean size of 30 nm and thus are viable at 45 mV, according to ZP and DLS experiments. In vitro antimicrobial investigations, AuNPs had the highest reducing power and DPPH radical scavenging activity. The non-lethality of these nanomaterials permits us to propose them as a decent contender for the development of medications helpful in the treatment of neurodegenerative sicknesses such as Alzheimer's disease (Suganthy et al. 2018).

Das et al. (2019) developed AgNPs from *Xylocarpus granatum* (XG-AgNPs) bark extract and *Avicennia officinalis* leaf extract for in vitro antioxidant, antidiabetic, and anti-inflammatory assays. The maximum absorption range in the UV range at 420 nm, with correlates to their own SPR. FTIR study demonstrates phenolic group capping, which contributes to the stability of manufactured AgNPs. AgNPs had higher IC_{50} values for DPPH scavenging, protein denaturation, and superoxide, with IC_{50} ranging 0.21, 0.14, and 0.32 mg/mL, respectively. Furthermore, XG-AgNPs inhibited a-amylase and a-glucosidase more effectively than AO-AgNPs. It is logical to argue that *Xylocarpus granatum* bark extract and *Asparagus officinalis* leaf extract were employed effectively for the manufacture of naturally important AgNPs that perhaps be used in pharmaceutical fields. Amrulloh et al. (2021) investigated the biosynthesis of magnesium oxide nanoparticles (MgONPs) employing *Moringa oleifera* bark extract as a green reducing agent. A UV-Vis spectrum was used to confirm the MgONPs formation. XRD measurement revealed the spherical crystal structure of MgONPs. Utilizing SEM and TEM pictures, the normal molecule size of MgONPs was estimated to be somewhere in the range of 60–100 nm. DPPH testing was utilized to assess the cell reinforcement action of magnesium oxide, with ascorbic corrosive filling in as a positive control. The concentration of DPPH, BEM, and magnesium oxide has a direct relationship with their free radical capture activity. DPPH is a steady atom that takes electrons or hydrogen from BEM or MgONPs. This measure is often used to decide the cancer prevention agent action of synthetic substances found in restorative plant extricates. MgONPs regularly have more grounded cell reinforcement action than BEM.

Gauthami et al. (2015) synthesized AgNPs via cinnamon bark extracts. The physical characteristics of the AgNPs was validated using UV-Vis spectra, SEM, and XRD. The nanoparticle size dispersion ranged from 30 to 150 nm. Cinnamon nanoparticles outperformed crude cinnamon extract in the total antioxidant assay. To decide the level of superoxide anion scavenging capacity of the nanoparticles, a superoxide anion revolutionary searching measure was performed. The ascorbic acid with cinnamon nanoparticles had an 82% radical scavenging, while the ascorbic acid and crude cinnamon extract had a 23% radical scavenging.

10.6 ANTICANCER ACTIVITY USING TREE BARK-BASED NANOMATERIALS

Nanoparticles of metals and metal oxides are a well-developed source for utilization in biomedicinal fields. Metallic nanoparticles are being used as medicine for humans to diagnose cancer. Functionalization of nanomaterials is important and this functionalization aids in the attachment of biomolecules such as antibody, polynucleotides, and peptides, as well as ligands and anticancer medicines (Aswathanarayan et al. 2018). Forest debris, particularly bark, is an important source of bioactive chemicals. Phenolic substances such as phenolics, flavonoids, lignans, and stilbenes can trap and scavenge free radicals, hence preventing cell ageing and protecting cells against carcinogens. Plants contain around 8000 phenolic chemicals, and their antioxidant action has been demonstrated in numerous studies. The findings revealed that the presence of more phenolic chemicals was connected to its anticancer efficacy. It has been demonstrated in the literature that combining two or three polyphenols has a combinatorial effect in cancer therapy (Burlacu and Tanase 2021).

Sowmyya and Vijaya (2018) reported the production of AgNPs using *Soymida febrifuga* stem bark for anticancer activity. The antibacterial effectiveness of the AgNPs against four bacterial strains was investigated. Two Gram-positive bacteria, *Staphylococcus aureus* and *Bacillus subtilis*, and two Gram-negative bacteria, *Campiglossa putrida* and *Escherichia coli* were tested using the disc diffusion method. The disinfected petri plates were filled with the prepared nutritional agar media, which was then allowed to harden. After that the agar medium was dried, and each bacterial strain was injected onto separate agar plates and dispersed evenly. A sterile filter paper disc containing various quantities of colloidal AgNPs, namely 5, 10, 15, and 20 mL, was inserted on infected Petri plates, including a control sample. As a control sample, ampicillin 10 mg/10 mL was put to use at 37°C and all of the plates were incubated. The inhibition zone was examined and measured, which seemed as a clear zone surrounding the discs.

Jobie et al. (2021) have investigated ZnONPs by utilizing the bark extracts of *Amygdalus scoparia* in an environmentally friendly and cost-effective manner, and subsequently employed as antifungal, antibacterial, anticancer, and antidiabetic medicines. Zn^{2+} ion capping, reduction, and stability of ZnONPs are explained by the elemental analysis of ZnONPs and bioactive chemicals identified in *Artemisia scoparia* extract. In comparison to antibiotic standards, the ZnONPs showed an exceptional inhibitory effect against *Enterobacter aerogenes*, *Staphylococcus aureus*, *Fusarium thapsinum*, *Escherichia coli*, *Fusarium semitectum*, and *Pleomorphomonas oryzae* when compared to the Streptozotocin (STZ)-induced diabetic group and other treatment groups. ZnONPs (30 mg/kg) significantly increased insulin levels and decreased aspartate transaminase (AST), alanine transaminase (ALT), and blood glucose levels in diabetic rats ($P = 0.05$). When compared to STZ-induced diabetic mice, ZnONPs and extract-treated mice showed significantly higher rates of IR, GluT2, and GCK transcription and significantly lower levels of TNFa transcription

Oves et al. (2021) have investigated that nano formulation-based antibacterial medicines are an alternate treatment for drug-resistant illnesses. AgNPs were synthesized using a *Conocarpus lancifolius* fruit extract and have a spherical dimension of 26.28 nm on average. Antibacterial activity against both bacteria has been assessed to be zone of inhibition of 18 and 24 mm. In order to determine the efficiency of AgNPs for bacterial isolates, they were incubated for 24 h at 60 μg/mL. AgNPs have high action against the fungal pathogen *Aspergillus flavus* and *Rhizopusus stolonifera* which cause major illnesses. The nanoparticles were biodegradable and had latent anticancer effects towards MDA MB-231 cells after treatment for 24 h.

10.7 OTHER APPLICATIONS FOR TREE BARK-MEDIATED NANOPARTICLES

Venkatesan et al. (2019) investigated *Boswellia ovalifoliolata* bark extract which was employed to make carbon dots (CDs) using a hydrothermal carbonization process. For instance, fluorescent CDs were made using *Boswellia ovalifoliolata* bark extract, and they can be used for metal ion sensing and bioimaging. With 10.2% quantum efficiency, CDs exhibit excellent stability and a strong green emission. The average diameter of the CDs was 4.16 nm. The synthesized CDs showed remarkable sensitivity and selectivity towards Fe^{3+}, with a linear relationship up to a concentration of 500 M and a detection limit of 0.41 M (S/N = 3). The CDs also have a high level of radical scavenging activity with a 65 g/mL for IC_{50}. Overall, the findings show that Fe^{3+} sensing might be accomplished using CDs generated from natural carbon sources.

The GO chemical reduction technique is commonly used, according to Manchala et al. (2019), nevertheless, this pathway frequently contains very toxic reducing chemicals that are dangerous towards both people and the ecosystem. Electrochemical performance is evaluated using cyclic voltammetry (CV) and galvanostatic charge discharge (GCD). At a current density of 2 A g^{-1}, The e-graphene capacitors have a high specific capacitance of 239 F g^{-1} and a high energy density of 71 W h kg^{-1}, according to GCD studies. The creation of 14 layers of graphene has been confirmed by TEM and AFM photographs. The powerful couplings between eucalyptus polyphenols and graphene are confirmed by the e-stable graphene's homogeneous dispersion in aqueous and non-aqueous solvents

Lang et al. (2019) analyzed the hotmelt adhesive formulation, torrefied birch bark materials, which was used as a tackifier. They tested three birch bark-originated tackifiers in tandem with a regularly employed ethylene vinyl acetate (EVA) copolymer for adhesive and thermomechanical properties. The high glass transition temperature (T_g) tackifier resin (66°C) was produced in a rotary tubular furnace in a continuous process of production at 240°C. High product yields of roughly 25% (w/w) were achieved after solvent extraction of the torrefied bark using acetone. Tackifying chemicals in birch bark transitions from a stiff, brittle substance below Tg to a low viscosity fluid at higher temperatures over several decades, followed by a shift in storage and loss modulus

10.8 CONCLUSION

Nanoparticles are essential because of their biocompatibility and their ability to act as antibacterial, antioxidant, and anticancer agents. In this study, we investigated the production of nanoparticles using various bark extracts: phytochemical components such as polyphenols, flavonoids, alkaloids, and tannins. Extracts from the bark can be used to produce AgNPs, ZnONPs, and AuNPs. Nanomaterials can be 0-D, 1-D, 2-D, or 3-D, depending on the overall shape. Nanoparticles through the bark can be used for antibacterial activity. The antioxidant activity of nanoparticles produced by mediator phytochemicals from the bark of plants can be used to treat a variety of diseases. The results of this review may help develop value-added products from traditional medicinal plants for nanotechnology-based biomedical applications.

REFERENCES

Abate, L., A. Abebe and A. Mekonnen. 2017. Studies on antioxidant and antibacterial activities of crude extracts of *Plantago lanceolata* leaves. *Chemistry International* 3: 277–287.

Abate, L. M. and M. Yayinie. 2018. Effect of solvent on antioxidant activity of crude extracts of *Otostegia integrifolia* leave. *Chemistry International* 4(3): 183–188.

Aflori, M. 2021. Smart nanomaterials for biomedical applications – A review. *Nanomaterials* 11(2): 1–33.

Akintelu, S. A., F. A. Olugbeko Folorunso, A. K. Oyebamiji and A. S. Folorunso. 2020. Characterization and pharmacological efficacy of silver nanoparticles biosynthesized using the bark extract of *Garcinia kola*. *Journal of Chemistry* 2020: 1–7.

Ali, M. A., T. Ahmed, W. Wu, A. Hossain, R. Hafeez, M. M. I. Masum, Y. Wang, Q. An, G. Sun and B. Li. 2020. Advancements in plant and microbe-based synthesis of metallic nanoparticles and their antimicrobial activity against plant pathogens. *Nanomaterials* 10(6): 1–24.

Amrulloh, H., A. Fatiqin, W. Simanjuntak, H. Afriyaniand and A. Annissa. 2021. Bioactivities of nano-scale magnesium oxide prepared using aqueous extract of *Moringa oleifera* Leaves as green agent. *Advances in Natural Sciences: Nanoscience and Nanotechnology* 12(1): 1–8.

Ansari, M. A., M. Murali, D. Prasad, M. A. Alzohairy, A. Almatroudi, M. N. Alomary, A. C. Udayashankar, S. B. Singh, S. M. M. Asiri and B. S. Ashwini. 2020. *Cinnamomum verum* bark extract mediated green synthesis of zno nanoparticles and their antibacterial potentiality. *Biomolecules* 10(2): 1–14.

Bachheti, R. K., A. Fikadu, B. Archana and H. Azamal. 2020. Biogenic fabrication of nanomaterials from flower-based chemical compounds, characterization and their various applications: A review. *Saudi Journal of Biological Sciences* 27(10): 2551–2562.

Bachheti, R. K., L. Abate, A. Bachheti, A. Madhusudhan and A. Husen. 2021. Algae-, fungi-, and yeast-mediated biological synthesis of nanoparticles and their various biomedical applications. In *Handbook of Greener Synthesis of Nanomaterials and Compounds*. Elsevier, pp. 701–734. https://doi.org/10.1016/B978-0-12-821938-6.00022-0.

Bachheti, R. K., Y. Godebo, A. Bachheti, M. O. Yassin and H. Azamal. 2020a. Root-based fabrication of metal/metal-oxide nanomaterials and their various applications. In Husen A., Jawaid M. (eds) *Nanomaterials for Agriculture and Forestry Applications*. Elsevier, pp. 135–166. https://doi.org/10.1016/B978-0-12-817852-2.00006-8.

Bai Aswathanarayan, J., R. Rai Vittal and U. Muddegowda. 2018. Anticancer activity of metal nanoparticles and their peptide conjugates against human colon adenorectal carcinoma cells. *Artificial Cells, Nanomedicine, and Biotechnology* 46(7): 1444–1451.

Bhardwaj, K., D. S. Dhanjal, A. Sharma, E. Nepovimova, A. Kalia, S. Thakur, S. Bhardwaj, C. Chopra, R. Singh and R. Verma. 2020. Conifer-derived metallic nanoparticles: Green synthesis and biological applications. *International Journal of Molecular Sciences* 21(23): 1–22.

Burlacu, E. and C. Tanase. 2021. Anticancer potential of natural bark products – A review. *Plants* 10(9): 1895.

Das Mahapatra, A., C. Patra, J. Mondal, C. Sinha, P. Chandra Sadhukhan and D. Chattopadhyay. 2020. Silver nanoparticles derived from albizialebbeck bark extract demonstrate killing of multidrug-resistant bacteria by damaging cellular architecture with antioxidant activity. *Chemistry Select* 5(15): 4770–4777.

Das, S. K., S. Behera, J. K. Patra and H. Thatoi. 2019. Green synthesis of sliver nanoparticles using *Avicennia officinalis* and *Xylocarpus granatum* extracts and in vitro evaluation of antioxidant, antidiabetic and anti-inflammatory activities. *Journal of Cluster Science* 30(4): 1103–1113.

Gauthami, M., N. Srinivasan, N. Goud, K. Boopalan and K. Thirumurugan. 2015. Synthesis of silver nanoparticles using *Cinnamomum zeylanicum* bark extract and its antioxidant activity. *Nanoscience and Nanotechnology – Asia* 5(1): 2–7.

Han, N. and J. C. Ho. 2013. *One-Dimensional Nanomaterials for Energy Applications*, Second Edition. Hong Kong: Elsevier Ltd.

Hu, T., X. Mei, Y. Wang, X. Weng, R. Liangand and M. Wei. 2019. Two-dimensional nanomaterials: Fascinating materials in biomedical field. *Science Bulletin* 64(22): 1707–1727.

Husen, A., Q. I. Rahman, M. Iqbal, M. O. Yassin and R. K. Bachheti. 2019. Plant-mediated fabrication of gold nanoparticles and their applications. In *Nanomaterials and Plant Potential*. Cham: Springer, pp. 71–110 https://doi.org/10.1007/978-3-030-05569-1_3.

Iravani, S. and B. Zolfaghari. 2013. Green synthesis of silver nanoparticles using *Pinus eldarica* bark extract. *BioMed Research International* 2013: 1–5.

Jang, L. W., J. Shim, D. I. Son, H. Cho, L. Zhang, J. Zhang, M. Menghini, J. P. Locquetand and J. W. Seo. 2019. Simultaneous growth of three-dimensional carbon nanotubes and ultrathin graphite networks on copper. *Scientific Reports* 9(1): 1–9.

Jeyaraj, M., S. Gurunathan, M. Qasim, M. H. Kang and J. H. Kim. 2019.A comprehensive review on the synthesis, characterization, and biomedical application of platinum Nanoparticles. *Nanomaterials* 9(12): 2–41.

Kaur, H. and H. Garg. 2014. Pesticides: Environmental impacts and management strategies. *Pesticide - Toxic Aspects* 8: 187.

Khandel, P., R. K. Yadaw, D. K. Soni, L. Kanwar and S. K. Shahi. 2018. Biogenesis of metal nanoparticles and their pharmacological applications: Present status and application prospects. *Springer Berlin Heidelberg* 8(3): 217–254.

Krishnaiah, D., R. Sarbatly and R. Nithyanandam. 2011. A review of the antioxidant potential of medicinal plant species. *Food and Bio-Product Process* 89(3): 217–233.

Kumar, H., K. Bhardwaj, K. Kuča, A. Kalia, E. Nepovimova, R. Vermaand and D. Kumar. 2020. Flower-based green synthesis of metallic nanoparticles: Applications beyond Fragrance. *Nanomaterials* 10(4): 766.

Lang, J., B. Winkeljann, O. Lielegand and C. Zollfrank. Continuous synthesis and application of novel, archaeoinspired tackifiers from birch bark waste. *ACS Sustainable Chemistry and Engineering* 7(15): 13157–13166.

Manchala, S., V. S. R. K. Tandava, D. Jampaiah, S. K. Bhargava and V. Shanker. 2019. Novel and highly efficient strategy for the green synthesis of soluble graphene by aqueous polyphenol extracts of Eucalyptus bark and its applications in high-performance supercapacitors. *ACS Sustainable Chemistry and Engineering* 7(13): 11612–11620.

Mandal, S., A. Patra, A. Samanta, S. Roy, A. Mandal, T. Das Mahapatra, S. Pradhan, K. Das and D. K. Nandi. 2013. Analysis of phytochemical profile of *Terminalia arjuna* bark extract with antioxidative and antimicrobial properties. *Asian Pacific Journal of Tropical Biomedicoine* 3(12): 960–966.

Martínez-cabanas, M., M. López-garcía, P. Rodríguez-barro, T. Vilariño, P. Lodeiro, R. Herrero, J. L. Barriadaand and M. E. S. de Vicente. 2021. Antioxidant capacity assessment of plant extracts for green synthesis of nanoparticles. *Nanomaterials* 11(7): 1–14.

Mohammad, S., A. Sadika, I. H. Md., A. H. Md. and A. B. Mohiuddin. 2012. Evaluation of in vitro antioxidant activity of bark extracts of *Terminalia arjuna*. *Journal of Medicinal Plants Research* 6(39): 5286–5298.

Naik, A. B. and N. B. Selukar. 2015. Role of nanotechnology in medicine. *Everyman's Science* 44(3): 151–153.

Naikoo, G. A., M. Mustaqeem, I. U. Hassan, T. Awan, F. Arshad, H. Salim and A. Qurashi. 2021. Bioinspired *and green synthesis of nanoparticles from plant extracts with antiviral and antimicrobial properties*: A critical review. *Journal of Saudi Chemical Society* 25(9): 101304.

Nayak, D., S. Ashe, P. R. Rauta, M. Kumari and B. Nayak. 2016. Bark extract mediated green synthesis of silver nanoparticles: Evaluation of antimicrobial activity and antiproliferative response against *Osteosarcoma. Material Science and Engineering. Part C* 58: 44–52.

Norouzi Jobie, F., M. Ranjbar, A. Hajizadeh Moghaddam and M. Kiani. 2021. Green synthesis of zinc oxide nanoparticles using *Amygdalus scoparia*spach stem bark extract and their applications as an alternative antimicrobial, anticancer, and anti-diabetic agent. *Advanced Powder Technology* 32(6): 2043–2052.

Olajire, A. A., G. O. Adeyeyand and R. A. Yusuf. 2017. *Alchornealaxiflora* bark extract assisted green synthesis of platinum nanoparticles for oxidative desulphurization of model oil. *Journal of Cluster Science* 28(3): 1565–1578.

Ontong, J. C., S. Paosen, S. Shankarand and S. P. Voravuthikunchai. 2019. Eco-Friendly synthesis of silver nanoparticles using *Senna alata* Bark extract and its antimicrobial mechanism through enhancement of bacterial membrane degradation. *Journal of Microbiological Methods* 165: 105692.

Oves, M., M. Ahmar Rauf, M. Aslam, H. A. Qari, H. Sonbol, I. Ahmad, G. S. Zaman and M. Saeed. 2021. Green synthesis of silver nanoparticles by *Conocarpuslancifolius* plant extract and their antimicrobial and anticancer activities. *Saudi Journal of Biological Sciences* 29(1): 460–471.

Painuli, S., P. Semwal, A. Bacheti, R. K. Bachheti and A. Husen. 2020. Nanomaterials from nonwood forest products and their applications. In Husen A., Jawaid M. (eds) *Nanomaterials for Agriculture and Forestry Applications*. Elsevier, pp. 15–40. https://doi.org/10.1016/B978-0-12-817852-2.00002-0.

Pawar, J. and P. Vinchurkar. 2012. Synthesis and characterization of gold and silver nanoparticles by using leaves and bark extract of *Ficus nitida. International Journal of Advances in Management, Technology and Engineering Sciences* 1(5): 120–124.

Prasad, R. and V. S. Swamy. 2013. Antibacterial activity of silver nanoparticles synthesized by bark extract of *Syzygiumcumini. Journal of Nanoparticles* 2013: 1–6.

Ramanathan, S., S. C. B. Gopinath, P. Anbu, T. Lakshmipriya, F. H. Kasim and C. G. Lee. 2018. Eco-friendly synthesis of *Solanum trilobatum* extract-capped silver nanoparticles is compatible with good antimicrobial activities. *Journal of Molecular Structure* 1160: 80–91.

Robinson, P. R. and C. S. Hsu. 2017. Introduction to petroleum technology. In *Springer Handbook of Petroleum Technology*. Cham: Springer, pp. 1–83.

Rosell, A. J. 2018. Bark in woody plants: Understanding the diversity of a multifunctional structure. *Materials Chemistry and Physics* 59(3): 535–547.

Saha, R., S. Karthik, K. S. Balu, R. Suriyaprabha, P. Siva and V. Rajendran. 2018. Influence *of the various synthesis methods on the ZnO nanoparticles property made using the bark extract of* Terminalia Arjuna. *Materials Chemistry and Physics* 209: 208–216.

Sarwar, N., U. Bin Humayoun, M. Kumar, S. F. A. Zaidi, J. H. Yoo, N. Ali, D. I. Jeong, J. H. Lee and D. H. Yoon. 2021. Citric acid mediated green synthesis of copper nanoparticles using cinnamon bark extract and its multifaceted applications. *Journal of Cleaner Production* 292: 125974.

Sasikala, A., M. L. Rao, N. Savithramma and T. N. V. K. V. Prasad. 2015. Synthesis of silver nanoparticles from stem bark of *Cochlospermum religiosum* (L.) Alston: An important medicinal plant and evaluation of their antimicrobial efficacy. *Applied Nanoscience* 5(7): 827–835.

Shafey, A. M. El. 2020. Green synthesis of metal and metal oxide nanoparticles from plant leaf extracts and their applications: A review. *Green Processing and Synthesis* 9(1): 304–339.

Sowmyya, T. and G. V. Lakshmi. 2018. Spectroscopic investigation on catalytic and bactericidal properties of biogenic silver nanoparticles synthesized using *Soymidafebrifuga*aqueous stem bark extract. *Journal of Environmental Chemical Engineering* 6(3): 3590–3601.

Sudarsan, S., M. K. Shankar, A. K. B. Motatis, S. Shankar, D. Krishnappa, C. D. Mohan, K. S. Rangappa, V. K. Gupta and C. N. Siddaiah. 2021. Green synthesis of silver nanoparticles by *Cytobacillus firmus* isolated from the stem bark of *Terminalia arjuna* and their antimicrobial activity. *Biomolecules* 11(2): 1–16.

Suganthy, N., V. Sri Ramkumar, A. Pugazhendhi, G. Benelli and G. Archunan. 2018. Biogenic synthesis of gold nanoparticles from *Terminalia arjuna* bark extract: Assessment of safety aspects and neuroprotective potential via antioxidant, anticholinesterase, and antiamyloidogenic effects. *Environmental Science and Pollution Research International* 25(11): 10418–10433.

Supraja, N., T. N. V. K. V. Prasad, A. D. Gandhi, D. Anbumani, P. Kavitha and R. Babujanarthanam. 2018. Synthesis, Characterization and evaluation of antimicrobial efficacy and brine shrimp lethality assay of *Alstoniascholaris* stem bark extract mediated ZnONPs. *Biochemistry and Biophysics Reports* 14: 69–77.

Supraja, N., T. N. V. K. V. Prasad, T. G. Krishna and E. David. 2016. Synthesis, characterization, and evaluation of the antimicrobial efficacy of *Boswellia ovalifoliolata* stem bark-extract-Mediated zinc oxide nanoparticles. *Applied Nanoscience* 6(4): 581–590.

Tanase, C., L. Berta, A. Mare, A. Man, A. I. Talmaciu, I. Roşca, E. Mircia, I. Volfand and V. I. Popa. 2020. Biosynthesis of silver nanoparticles using aqueous bark extract of *Piceaabies* L. and their antibacterial activity. *European Journal of Wood and Wood Products* 78(2): 281–291.

Tanase, C., S. Cosarcăand and D. L. Muntean. 2019. A critical review of phenolic compounds extracted from the bark of woody vascular plants and their potential biological activity. *Molecules* 24(6): 1182.

Tiwari, J. N., R. N. Tiwari and K. S. Kim. 2012. Zero-Dimensional, one-dimensional, two-dimensional and three-dimensional nanostructured materials for advanced electrochemical energy devices. *Progress in Materials Science* 57(4): 724–803.

Valgimigli, L., A. Baschieriand and R. Amorati. 2018. Antioxidant activity of nanomaterials. *Journal of Materials Chemistry* B6(14): 2036–2051.

Venkatesan, G., V. Rajagopalan and S. N. Chakravarthula. 2019. *Boswellia ovalifoliolata* bark extract derived carbon dots for selective fluorescent sensing of Fe^{3+}. *Journal of Environmental Chemical Engineering* 7(2): 103013.

Yin, J., Y. Huang, S. Hameed, R. Zhou, L. Xie and Y. Ying. 2020. Large scale assembly of nanomaterials: Mechanisms and applications. *Nanoscale* 12(34): 17571–17589.

Zhao, Y., H. Hong, Q. Gong and L. Ji. 2013. 1D nanomaterials: Synthesis, properties, and applications. *Journal of Nanomaterials* 2013 .

11 Medicinally Important Plant Roots and Their Role in Nanoparticles Synthesis and Applications

Yilma Hunde Gonfa, Fekade Beshah Tessema, Mesfin Getachew Tadesse, Archana Bachheti, and Rakesh Kumar Bachheti

CONTENTS

ABBREVIATIONS

AFM Atomic force microscopy
DLS Dynamic light scattering
EDS Energy-dispersive spectroscopy
EDX Energy-dispersive X-rays
FT-IR Fourier transform infrared
MNPs Metal nanoparticles
NPs Nanoparticles
SEM Scanning electron microscopy
TEM Transmission electron microscopy
XRD X-ray diffraction
UV-Vis Ultraviolet-visible
ZP Zeta potential

DOI: 10.1201/9781003213727-11

11.1 INTRODUCTION

Plant roots are one of the main sources of phytochemicals employed in the green synthesis of nanoparticles (NPs). The roots of most plants are commonly utilized to treat various health ailments. They have been reported to be effectively used to treat diseases such as pathogenic bacteria, fungus, inflammation, ulcers, tumours, and allergies (Khairullah et al. 2020). Plant-mediated NPs are widely employed in the health, environmental, and industrial sectors (Zare et al. 2020; Bachheti et al. 2020; 2021). This technique is safer, energy efficient, less toxic, eco-friendly, and more economical than its chemically and physically synthesized counterparts (Singh et al. 2018; Bachheti et al. 2020a; Nisar et al. 2022).

Metal nanoparticles (MNPs) were reported to be a promising alternative to modern drugs. They interact with a microorganism's cell organelles such as DNA, enzymes, ribosomes, and lysosomes, affecting the cells' permeability, oxidative stress, protein activation, gene expression, and enzyme activation (Singh et al. 2020). The unique properties of NPs are because of their high surface-to-volume ratio, biocompatible surface properties, low toxicity, and physicochemical stability (Husen et al. 2019; Bachheti et al. 2020, 2020b; Sidorowicz et al. 2021; Poudel et al. 2022). Plant extracts, including roots, are highly utilized in synthesizing green NPs with diverse applications in agriculture, biomedicine, catalysis, biosensing, wastewater treatment, and cosmetic areas (Mahakham et al. 2016; Bachheti et al. 2019, 2019a; Bhardwaj et al. 2020; Painuli et al. 2020).

11.2 OVERVIEW OF PHYTOCHEMICAL CONTENT AND THERAPEUTIC USES OF ROOT EXTRACTS

The root extract of *Withania somnifera* is rich in alkaloids and essential oils. It is reported for its sedative, treatment of nervous disorders, epilepsy curing, antidiabetic, antistress, anti-inflammatory, and antiageing properties in animal models and clinical tests (Alam et al. 2012; Umadevi et al. 2012; Jonathan et al. 2015; Ali et al. 2017). The root extracts of *W. somnifera* were traditionally used in Ayurveda, Siddha, and Unani medicines for many formulation practices (Palliyaguru et al. 2016). In particular, the extract of the root of *W. somnifera* was reported to improve memory and cognition functions (Choudhary et al. 2017). Withanolides were the major phytochemical constituents mainly reported from the root of plant species and potentially responsible for these biological activities (Bhushan et al. 2021).

The Indians, Egyptians, Chinese, Greeks, and Romans widely practised utilizing roots and rhizomes of *Glycyrrhiza glabra* to treat disorders of the respiratory tract and digestive system (Dhanani et al. 2017). The root of liquorice (*G. glabra*) is commonly used for the treatment of diabetic, ulcer, allergic, and inflammatory effects due to the presence of main bioactive compounds such as glycyrrhizic acid, liquiritin, and liquiritigenin (Hosein Farzaei et al. 2015; Yu et al. 2015; Srivastava et al. 2019). *G. glabra*'s root extract also revealed significant antibacterial activities against both Gram-positive and Gram-negative bacterial strains (Reda et al. 2021). Some flavonoids were reported as the potential chemical components isolated from the root part of *G. glabra* and used as antibacterial, antitumour, and antioxidant agents (Sedighi et al. 2017).

Various literature has reported *Raphanus sativus* extracts for their potential anticancer and neuroprotective effects due to their being rich in various phenolic compounds (Do et al. 2021). The root extract of *Piper methysticum* contains kava lactones responsible for treating anxiety disorders, stress disorders, nervous tension, restlessness, gonorrhoea, menstrual pain, tuberculosis, and respiratory tract infections (Martin et al. 2014; Xuan et al. 2008). The phytochemicals from hexane extract from the root of *R. sativus* exhibited substantial inhibition of the growth of Henrietta Lacks (HeLa) cells with the percent of inhibition range of 40–95% (Beevi et al. 2010). Using an in vivo model, *Beta vulgaris* root juice significantly reduced blood glucose levels and increased its impact on the glycaemic and insulin responses (Mirmiran et al. 2020).

Many research reports demonstrated that the roots of *Zingiber officinale* are used to prevent vomiting, diarrhoea, headache, abdominal discomfort, common cold, and Alzheimer's disease. The roots of this plant contain bioactive compounds and are used as antioxidant, anti-inflammatory, antimicrobial, antidiabetic, antihypertensive, cardio protective, neuroprotective, antiobesity, anti-migraine, and anticancer agents (Ahmed et al. 2021; Talebi et al. 2021).

The root extract of *Wasabia japonica* using 80% aqueous methanol revealed potential activities against cancer cells and neuro-inflammatory effects (Park et al. 2022). Similarly, its root paste showed numerous activities such as antigen toxic, anti-inflammatory, antioxidant, and atopic dermatitis, antiplatelet, anticancer, and antiobesity (Szewczyk et al. 2021). Methanol extract of rhizomes of *Boesenbergia pandurata* after being partitioned between $CHCl_3$ and H_2O, the chloroform fraction was subjected to column chromatographic and preparative thin-layer chromatography (TLC) separation techniques. Then, the isolated flavanone derivatives were assayed for their preferential cytotoxicity against human pancreatic cancer cell line (Nguyen et al. 2020).

The root extract of *Boesenbergia rotunda*, which had been reported to be rich in flavonoids and chalcone derivatives, displayed wound-healing acceleration effects using an animal model (Rosdianto et al. 2020). The roots of *Valeriana officinalis* contain medicinally important secondary metabolites for treating blood, circulatory, mental, digestive, and urinary tract disorders (Nandhini et al. 2018). From the root extract of *Curcuma longa,* the most active polyphenolic constituent known as curcumin possesses a wide range of biological activities such as antifungal, antidiabetic, antioxidant, anti-inflammatory, anticancer, antiallergic, antiprotozoal, and antibacterial activities (Kumari et al. 2022). Essential oils isolated from the roots and rhizomes of *Ligusticum porter* showed significant antinociceptive activity (Juárez-Reyes et al. 2014).

Ginger (*Zingiber officinale*) based NPs were used as an additive to fish feed to improve cognitive behaviours (Talebi et al. 2021). *Rumex* species are herbs that have their root extracts applied in folk medicine to treat jaundice, fever, bleeding, constipation, dislocated bones, headache, and inflammation (Farooq et al. 2013; Bekele 2016; Mishra et al. 2018; Shaikh et al. 2018; Tan et al. 2019; Gonfa et al. 2021). The roots of *Hydrastis canadensis* have traditionally been profiled to treat gastrointestinal irritation, pathogenic bacteria, and skin and eye infections (Leyte-Lugo et al. 2017).

11.3 NANOPARTICLE FABRICATION FROM MEDICINAL PLANT ROOTS AND THEIR CHARACTERIZATION

NPs have generated enormous research interest in recent times due to their nanoscale size (less than 100 nm in three extension diameters), various shapes (spherical, triangular, pyramidal, cubic, hexagonal, irregular, star, flower, oval), and potential applications in the wide field of science and technology (Joseph and Mathew 2014; Zare et al. 2020). NPs can be obtained either from inorganic (metal, metal oxide, and quantum dots NPs) or organic (fullerenes, nanotubes, etc.) based materials (Lawrence 2019). By 2021, the market demand for NPs was reported to have reached USD16.8 billion, from which silver nanoparticles (AgNPs) accounted for 17.86% of the total (Poudel et al. 2022). The ability of plant root extracts to reduce and stabilize NPs is due to the presence of a wide spectrum phytochemicals such as steroids, saponins, tannins, alkaloids, flavonoids, terpenoids, phenols, amides, glycosides, thiamine, carboxylic acids, carbohydrates, and proteins (Cavalcanti et al. 2013; El-Seedi et al. 2019; Karthik et al. 2022; Moradi et al. 2021). For example, green NP synthesis results from the reduction of metal ions to metal by plant extract (Chugh et al. 2021). However, optimization of reaction parameters such as pH, the concentration of the salt solution, the volume of salt solution, volume of plant extract solution, ratio of the reactants, the temperature of the reaction, and time of reaction are required to generate homogenous NPs of identical size and shape (Bachheti et al. 2019). Plant-based NPs have the advantages of availability, biocompatibility, and biodegradability (Li et al. 2019) (Figure 11.1).

The mechanism of phytochemicals from plant roots displays the conversion of metallic ions to MNPs as a result of the redox process. Phytochemicals act as biorecing, capping, and stabilizing agents (El-Seedi et al. 2019). Various characterization techniques are employed to determine

FIGURE 11.1 Important medicinal roots of plants used for nanoparticles synthesis.

and confirm the fabrication of MNPs. The chemical and physical properties of MNPs determine the effectiveness and efficiency of NPs in different areas of application (Siddiqi and Husen 2017) (Figure 11.2).

Characterization is an important step in the synthesis of NPs to determine their morphology, size, shape, purity, charge, density, dispersity, surface area, and surface chemistry using various instruments (Bamal et al. 2021; Huq et al. 2022). Ultraviolet-visible (UV-Vis) spectra determination is employed to confirm the formation of NPs under a specific set of conditions (Joseph and Mathew 2014). Surface plasmon resonance (SPR) absorptions are the characteristic optical properties observed for the particular MNPs using UV-Vis spectroscopy instruments (Hijau et al. 2015). Fourier transform infrared (FT-IR) spectroscopic techniques are essential for determining functional groups of phytochemicals responsible for NP bioreducing, capping, and stabilization (Alagesan and Venugopal 2019). The scanning electron microscopy (SEM) method is used to determine NPs' size, shape, and morphology (Shanmugapriya et al. 2017). Transmission electron microscopy (TEM) images provide the synthesized NPs' size, shape, and size distribution (Mahakham et al. 2016; Venugopal et al. 2017). X-ray diffraction (XRD) characterizes the crystal structure, phase purity, lattice constants, and geometry of NPs (Eisa et al. 2019; Raveesha et al. 2021). Elemental analysis of synthesized NPs is performed using energy-dispersive spectroscopy (EDS) (Haider et al. 2020). Similarly, energy-dispersive X-rays (EDX) are also used to confirm the purity, dispersity, and elemental composition of NPs (Haq et al. 2021). Dynamic light scattering (DLS) analysis examines the NPs' size distribution and zeta potential (ZP) determination. The ZP of NPs values i.e., ZP

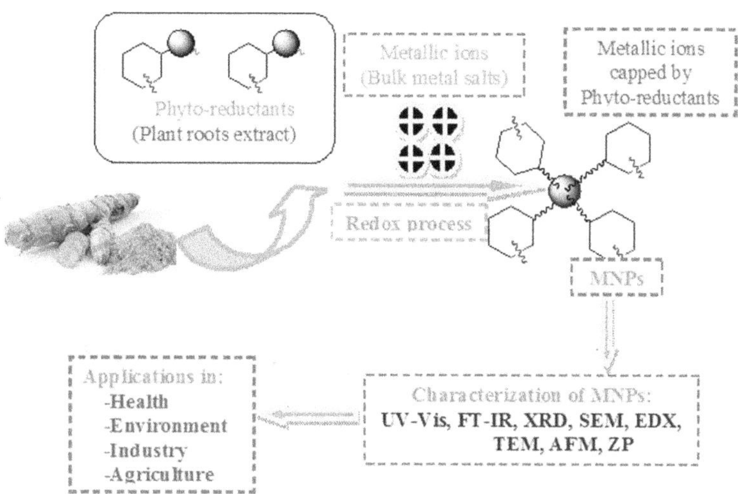

FIGURE 11.2 Scheme describing the mechanism of plant extract-mediated metal nanoparticles synthesis.

values > +25 mV or < −25 mV, reveal potential surface charges and high degrees of long-term stability of the dispersion of MNPs (Menon et al. 2019; Raafat et al. 2021; Sorbiun et al. 2018). Reports on medicinal plant root extracts mediated NP synthesis, characterization, and their applications are summarized in Table 11.1.

11.4 APPLICATIONS OF NANOPARTICLES FROM ROOT EXTRACTS

Biological NPs offer many potential applications due to their large surface area to volume ratio, size, and shape; they are less toxic, cost effective, and environmentally friendly. Green NPs possess multifunctional applications in different areas such as antibacterial, antifungal, antiviral, antioxidant, anticancer, catalytic, and some other activities (Velmurugan et al. 2014; Jain et al. 2019; Kuppusamy et al. 2016; Latif et al. 2019; Lawrence 2019; Raveesha et al. 2021). Some examples of applications of various green NPs are provided as follows.

11.4.1 Antimicrobial Activities

MNPs or metal oxide nanoparticles (MONPs) were reported to possess antimicrobial activity against various bacteria and fungi, for example, pathogenic microorganisms affecting human and crop organs (Netala et al. 2018). Green MNPs synthesized from plant roots were reported to reveal promising antimicrobial activity (Saeed Al-Zahrani 2019). Many reports also showed that C, Cu, Ag, Au, Fe, Zn, Mn, Pd, Ti, Si, ZnO, TiO_2, SiO_2, and SeO_2 NPs synthesized using plant root extracts and used for their potential antimicrobial activities (Göl et al. 2020; Khan et al. 2022). Microwave-assisted biosynthesized AgNPs from rhizome extract of *Alpinia galanga* were reported to have antimicrobial activity (Joseph and Mathew 2014). The NPs of some metal oxides and bimetallics such as CuO, NiO, and Cu-Ni hybrids were also reported for their practical antibacterial activities against Gram-positive and Gram-negative bacterial strains, including *Pseudomonas aurigenosa, Ptoteus vulgaris,* and *Escherichia coli* (Faisal et al. 2021). Rhizome extract of plant-based α-Fe_2O_3 NPs also demonstrated potent antibacterial activity against bacterial strains such as *Streptococcus mutans, Bacillus cereus, Proteus vulgaris,* and *Serratia marcences* (Raveesha et al. 2021). The binding of MNPs with sulphur residues from the virus's glycoproteins was reported to inhibit the virus interaction with the receptors and the entry into the host cell (Bamal et al. 2021). AgNPs, copper nanoparticles (CuNPs), and zinc oxide nanoparticles (ZnONPs) synthesized from ginger

TABLE 11.1
Summary of Some Roots of Plant-Mediated Nanoparticles Synthesis, Characterization, and Their Applications

Plant botanical name	Nanoparticles	Size (nm)	Shape	Characterization techniques	Applications	Key reference
Acorus calamus	AgNPs	15–25	Spherical	UV-Vis, SEM, and TEM	Antioxidant activity	Pipriya et al. (2018)
	ZnONPs	50–100	Hexagonal	UV-Vis, SEM-EDX, and TEM	Antimicrobial and anticancer activities	Vakayil et al. (2021)
Allium sativum	AgNPs	1.7–18.2	Spherical	UV-Vis, Raman spectroscopy, XPS, TEM, and XRD	Colorimetric detection of mercury (II) ions and tin (II) ions	Paw et al. (2021)
Alpinia galangal	AgNPs	20.82 ± 1.8	Spherical	UV-Vis., FT-IR, XRD, and HR-TEM	Catalytic and antimicrobial activities	Joseph and Mathew (2014)
Angelica spp.	CuO QDs NPs	5–10	Spherical	UV-Vis, FT-IR, SEM, and TEM	Antiproliferation and treats osteoporosis	Hu Z. et al. (2019)
Berberis vulgaris	AgNPs	58.6–92.8	Octahedral and spherical	UV-Vis, FT-IR, XRD, TEM, PCCS, EDS, and ZP	Antibacterial and antioxidant activities	Salayová et al. (2021)
Beta vulgaris	ZnNPs	52–76	Hexagonal	XRD and TEM	UV-filtering, antifungal, antibacterial, and catalytic activities	Khan et al. (2022)
	ZnONPs	52–76	Hexagonal wurtzite	UV-Vis, XRD, and SEM	Catalytic degradation of organic dyes and antioxidant activities	(Pavan Kumar et al. 2015)
Chlorophytum borivilianum	AgNPs	30–50	Spherical	UV-Vis, FT-IR, XRD, DLS, SEM, and TEM	Antibacterial activity	Singh et al. (2018)
Cichorium intybus	AgNPs	17.17	Spherical	UV-Vis, FT-IR, TEM, SEM-EDS, and ZP	Breast cancer treatment	Behboodi et al. (2019)
Curcuma longa	CuONPs, NiONPs, CuNiNPs	27.72 23.13 17.38	Spherical	UV-Vis, FT-IR, XRD, TEM, SEM-EDX, and TGA	Antimicrobial, antiparasitic and cytotoxic activities	Faisal et al. (2021)
Curcuma pseudomontana	AuNPs	20	Spherical	UV-Vis, FT-IR, SEM, and HR-TEM	Cytotoxic, antibacterial, anti-inflammatory, and antioxidant activities	Muniyappan and Nagarajan (2014)
Glycyrrhiza glabra	AgNPs	20	Spherical	UV-Vis, SEM, TEM, and AFM	Antimicrobial, cytotoxic activity, and anti-inflammatory activities	Suwannakul et al. (2018)
	AuNPs	2.647–16.25	Spherical	UV-Vis, FT-IR, XRD, DLS, and SEM-EDX,	Antimicrobial, antioxidant, and anticancer activities	Al-Radadi (2021)

(Continued)

TABLE 11.1 (CONTINUED)
Summary of Some Roots of Plant-Mediated Nanoparticles Synthesis, Characterization, and Their Applications

Plant botanical name	Nanoparticles	Size (nm)	Shape	Characterization techniques	Applications	Key reference
Ipomoea pescaprae	AgNPs	10–50	-	UV-Vis, FT-IR, and SEM	Antimicrobial activity	El-Seedi et al. (2019)
Panax vietnamensis	AgNPs	3–20 nm	-	UV-Vis, FT-IR, SEM-EDX, SAED, TEM, TGA-DSC, and XRD	Antibacterial and catalytic reduction activities	Tran et al. (2021)
Parthenium hysterophorus	AgNPs	-	Spherical	UV-Vis, FT-IR, and SEM	Larvicidal agent as mosquito control	Mondal et al. (2014)
Raphanus sativus	AgNPs	10–30	Spherical	UV-Vis, FT-IR, XRD, SEM-EDX, TEM, and AFM	Antibacterial activity	Singh T. et al. (2017)
	NiONPs	12.7–51.42	Irregular	UV-Vis, FT-IR, XRD, TEM, DRS, EDX, and FT-IR	Antibacterial and antioxidant activities	Haq et al. (2021)
Rhodiola imbricata	AgNPs	37–42	Spherical	UV-Vis, FT-IR, FESEM, EDX, TEM, DLS, and ZP	Seed germination, catalytic reduction, antioxidant, and cytotoxic activities	Kapoor et al. (2022)
Rhodiola rosea	AgNPs	10	Spherical	UV-Vis and TEM	Antioxidant and catalytic reduction activities	Hu et al. (2021)
Salvadora persica	AgNPs	35–40	Spherical	UV-Vis, FT-IR, TEM, FE-SEM, DLS, and PXRD	Antimicrobial activity	Arshad et al. (2021)
Valeriana officinalis	AgNPs	20	-	SEM and TEM	Catalytic activity in organic dyes degradation	Hariram and Vivekanandhan (2018)
	ZnONPs	40–50	-	SEM and TEM	Catalytic activity in organic dyes degradation	
Withania somnifera	TiO_2NPs	50–90	Aggregate of spherical and square	UV-Vis, FT-IR, XRD, SEM-EDS, TEM, DLS, TGA, and BET	Inhibition of microbial biofilms formation and cytotoxic activity against cancer cells	Al-Shabib et al. (2020)
Zingiber officinale	Ag/Ni bi-metallic NPs	70–88	Spherical	UV-Vis, FT-IR, SEM-EDX, and XRD	Textile dye removal	Nomura (2019)
	AgNPs	10–25	Spherical	UV-Vis, FT-IR, TEM, and XRD	Degradation of textile dyes	Mehata (2021)
	SeNPs	100–150	Spherical	UV-Vis, FT-IR, SEM, TEM, DLS AFM, and ZP	Antimicrobial and antioxidant activities	Menon et al. (2019)

extracts showed practical antifungal activities against mycotoxigenic fungi (Raafat et al. 2021). There are different arguments about the mechanism of action of NPs against microbial agents. For example, some literature claims that NPs are attached to the microbial cell membrane, interact with sulphur-containing proteins and phosphorus-containing groups in DNA, and finally cause the death of microbial cells (Sorbiun et al. 2018). Some literature also revealed that NPs generate oxidative stress, resulting in reactive oxygen species (ROS) responsible for the oxidation of protein and DNA of pathogenic microbes and causing the breakdown of their cell walls (Raveesha et al. 2021).

11.4.2 Antioxidant Activities

Antioxidants are free-radical scavengers protecting oxidizable molecules against autoxidation (Zare et al. 2020). *Garcinia*-mediated NPs showed the ability to scavenge free radicals of 2,2-diphenyl-1-picrylhydrazyl (DPPH) free radicals, which showed their potential antioxidant activities (Sarip et al. 2022). Green synthesized AgNPs from *Allium ampeloprasum* showed significant antioxidant activity (Jalilian et al. 2020). Gold nanoparticles (AuNPs) and AgNPs synthesized from the roots of *Scutellaria baicalensis* were evaluated for their antioxidant activity using DPPH assay. The results revealed that NPs show stronger antioxidant activity than the extract of the corresponding root (Chen et al. 2020). Green nickel oxide nanoparticles (NiONPs) synthesized using the root extract of *Raphanus sativus* showed strong antioxidant activity using 2,2′-azino-bis (3-ethylbenzothiazoline-6-sulphonic acid) (ABTS) radical scavenging assay (Haq et al. 2021). The root extract of *Beta vulgaris* was reported with potential phytochemicals, including phenolic acids, flavonoids, and betalains. The root extract of *B. vulgaris* assisted the synthesis of potential antioxidant ZnONPs (Pavan Kumar et al. 2015).

11.4.3 Anticancer Activities

Titanium dioxide nanoparticles (TiO_2NPs) synthesized based on the root extract of *Glycyrrhiza glabra* were evaluated for their promising selective inhibition of African green monkey kidney epithelial (Vero) and human laryngeal carcinoma cell line (HEp-2) (Bavanilatha et al. 2019). *Curcuma* spp. rhizome extract-mediated AgNPs assays revealed the decreasing percentage viability of the human colon cancer cell line, i.e., HT-29 cell (Jain et al. 2022). *G. glabra* root/rhizome-assisted AgNPs demonstrated significant anticancer activity against HeLa cells (Jain et al. 2019). The cytotoxicity of AuNPs synthesized using *Glycyrrhiza glabra* roots extract were also examined against the HepG-2 and MCF-7 cell lines by 3-4,5 dimethylthiazol-(2yl) 2-5 diphenyltetrazolium bromide (MTT) assay method. Results demonstrated the half-maximal inhibitory concentration (IC_{50}) at a 23 μg/ml concentration towards the HepG-2 cell line while 50 μg/ml towards the MCF-7 cell line. This evaluation showed that the synthesized AuNPs marked significant anticancer activity (Al-Radadi 2021).

11.4.4 Catalytic Activities

Various MNPs or MONPs were reported for their outstanding catalytic activities in different areas by transferring electrons from electron donor species sodium borohydride ($NaBH_4$) to electron acceptor species methylene blue (MB) and stabilizing the system by reducing the activation energy. Microwave-assisted biosynthesized AgNPs using the rhizome extract of *Alipinia galangal* demonstrated an efficient reduction of organic pollutants such as organic dyes (Joseph and Mathew 2014). AgNPs synthesized using the root extract of *Zingiber officinale* displayed significant catalytic activity in the degradation of p-nitrophenol to less toxic organic product p-aminophenol and MB, methyl red, and safranin O (Barman et al. 2020; Eisa et al. 2019). Similarly, the AuNPs were synthesized using root extract of *Z. officinale*. They demonstrated promising catalytic reduction of pesticides, p-nitrophenol, and industrial organic dyes such as eosin blue (EB), brilliant cresyl blue (BCB), and erichrome black T (EBT) (Gargi et al. 2018). Biosynthesized ZnONPs using *Z. officinale* and *A. sativum* root extracts were reported for their potential catalytic reduction of MB (Haider et al. 2020). Green TiO_2NPs are an efficient photocatalytic agent used in water treatment to filter out the organic pollutants from water reservoirs (Khan et al. 2022).

11.4.5 Miscellaneous Applications

Biosynthesized ZnONPs using *Raphanus sativus* root extract revealed good activity for wound-healing treatment (Kiran Kumar et al. 2019). AgNPs synthesized using the root extract of *G. glabra* were reported to treat gastric ulcers with significant activity (Tarannum et al. 2019). AuNPs synthesized using *Alpinia galanga* root extract were evaluated and showed promising growth enhancement activity in maize seed germination (Mahakham et al. 2016). *Curcuma longa* root extract-mediated ZnONPs exhibited their potential antileishmanial activity (Gul et al. 2019). The apoptotic pathways evaluation in the animal model showed an effective down-regulation result for AgNPs synthesized from *Alpinia officinarum* root extract (Zhang et al. 2019). The root extract of *Parthenium*-mediated AgNPs were reported for their potential mosquitocidal activity (Mondal et al. 2014).

11.5 CONCLUSION AND FUTURE PROSPECTS

The roots of medicinal plants such as *Withania somnifera*, *Glycyrrhiza glabra*, *Raphanus sativus*, and *Zingiber officinale* are widely known in traditional medicine. Much recent literature has confirmed that the extract of the roots of plants and their different pure compounds are used as bioreducing, stabilizing, and capping agents during the synthesis of NPs. Even though plant-mediated MNP synthesis using Ag and Au is widely common, other MNPs such as Pd, Pt, Cu, and Fe are less explored. The advancement of MNP production through the biosynthesis processes encourages their potential applications as antimicrobial, antioxidant, anti-inflammatory, antidiabetic, and antiulcer agents due to their greener, economical, and eco-friendly approach. It is well known that the reaction mechanisms for phytochemicals (bioreductants) in the green synthesis of MNPs are not yet clear. The complexity and diverse nature of the phytochemicals of plants are the sources of this problem. Commonly, secondary metabolites like alkaloids, flavonoids, terpenoids, tannins, phenols, and other organic compounds are expected, and some of these are even reported. The cheap, eco-friendly, safe, and green nature of plant-mediated biogenic synthesis attracts tremendous research interest in fabricating a wide range of nanomaterials. The vast and potential applications of green NPs in various areas are still developing and offer promising opportunities to explore.

REFERENCES

Ahmed, S.H.H., T. Gonda, and A. Hunyadi. 2021. Medicinal chemistry inspired by ginger: Exploring the chemical space around 6-gingerol. *RSC Advances* 11(43): 26687–26699. https://doi.org/10.1039/d1ra04227k.

Alagesan, V., and S. Venugopal. 2019. Green synthesis of selenium nanoparticle using leaves extract of *Withania somnifera* and its biological applications and photocatalytic activities. *BioNanoScience* 9(1): 105–116. https://doi.org/10.1007/s12668-018-0566-8.

Alam, N., M. Hossain, M.I. Khalil, M. Moniruzzaman, S.A. Sulaiman, and S.H. Gan. 2012. Recent advances in elucidating the biological properties of *Withania somnifera* and its potential role in health benefits. *Phytochemistry Reviews* 11(1): 97–112. https://doi.org/10.1007/s11101-011-9221-5.

Ali, K., M. Shuaib, M. lyas, and F. Hussain. 2017. Medicinal uses of chemical extracts from *Withania somnifera* and its antimicrobial activity: A mini-review. *PSM Microbiol* 2(1): 20–23.

Al-Radadi, N.S. 2021. Facile one-step green synthesis of gold nanoparticles (AuNp) using licorice root extract: Antimicrobial and anticancer study against HepG2 cell line. *Arabian Journal of Chemistry* 14(2): 1–25. https://doi.org/10.1016/j.arabjc.2020.

Al-Shabib, N.A., F.M. Husain, F.A. Qais, N. Ahmad, A. Khan, A.A. Alyousef, M. Arshad, S. Noor, J.M. Khan, P. Alam, T.H. Albalawi, and S.A. Shahzad. 2020. Phyto-mediated synthesis of porous titanium dioxide nanoparticles from *Withania somnifera* root extract: Broad-spectrum attenuation of biofilm and cytotoxic properties against HepG2 Cell Lines. *Frontiers in Microbiology* 11: 1–13. https://doi.org/10.3389/ fmicb.2020.01680.

Arshad, H., M.A. Sami, S. Sadaf, and U. Hassan. 2021. *Salvadora persica* mediated synthesis of silver nanoparticles and their antimicrobial efficacy. *Scientific Reports* 11(1): 1–11. https://doi.org/10.1038/s41598-021-85584-w.

Bachheti, A., A. Sharma, R.K. Bachheti, A. Husen, and V.K. Mishra 2019a. Plant-mediated synthesis of copper oxide nanoparticles and their biological applications. In *Nanomaterials and Plant Potential*. Springer, Cham, pp. 221–237. https://doi.org/10.1007/978-3-030-05569-1_8.

Bachheti, A., R.K. Bachheti, L. Abate, and Azamal Husen. 2021a. Current status of Aloe-based nanoparticle fabrication, characterization and their application in some cutting-edge areas. *South African Journal of Botany*. https://doi.org/10.1016/j.sajb.2021.08.021.

Bachheti, R.K., A. Fikadu, Archana Bachheti, and Azamal Husen. 2020. Biogenic fabrication of nanomaterials from flower-based chemical compounds, characterization and their various applications: A review. *Saudi Journal of Biological Sciences* 27(10): 2551–2562.

Bachheti, R.K., A. Sharma, A. Bachheti, A. Husen, G.M. Shanka, and D.P. Pandey. 2020b. Nanomaterials from various forest tree species and their biomedical applications. In Husen A., Jawaid M. (eds) *Nanomaterials for Agriculture and Forestry Applications*. Elsevier, pp. 81–106. https://doi.org/10.1016/B978-0-12-817852-2.00004-4.

Bachheti, R.K., L. Abate, A. Bachheti, A. Madhusudhan, and A. Husen. 2021. Algae-, fungi-, and yeast-mediated biological synthesis of nanoparticles and their various biomedical applications. In *Handbook of Greener Synthesis of Nanomaterials and Compounds*. Elsevier, pp. 701–734. https://doi.org/10.1016/B978-0-12-821938-6.00022-0.

Bachheti, R.K., R. Konwarh, V. Gupta, and A. Husen. 2019. Nanomaterials and plant potential. *Nanomaterials and Plant Potential*: 1–21. https://doi.org/10.1007/978-3-030-05569-1.

Bachheti, R.K., Y. Godebo, A. Bachheti, M.O. Yassin, and Azamal Husen. 2020a. Root-based fabrication of metal/metal-oxide nanomaterials and their various applications. In Husen A., Jawaid M. (eds) *Nanomaterials for Agriculture and Forestry Applications*. Elsevier, pp. 135–166. https://doi.org/10.1016/B978-0-12-817852-2.00006-8.

Bamal, D., A. Singh, G. Chaudhary, M. Kumar, M. Singh, N. Rani, P. Mundlia, and A.R. Sehrawat. 2021. Silver nanoparticles biosynthesis, characterization, antimicrobial activities, applications, cytotoxicity and safety issues: An updated review. *Nanomaterials* 11(8): 1–40. https://doi.org/10.3390/nano11082086.

Barman, K., D. Chowdhury, and P.K. Baruah. 2020. Bio-synthesized silver nanoparticles using *Zingiber officinale* rhizome extract as efficient catalyst for the degradation of environmental pollutants. *Inorganic and Nano-Metal Chemistry* 50(2): 57–65. https://doi.org/10.1080/24701556.2019.1661468.

Bavanilatha, M., L. Yoshitha, S. Nivedhitha, and S. Sahithya. 2019. Bioactive studies of TiO_2 nanoparticles synthesized using *Glycyrrhiza glabra*. *Biocatalysis and Agricultural Biotechnology* 19: 1–17. https://doi.org/10.1016/j.bcab.2019.101131.

Beevi, S.S., L.N. Mangamoori, M. Subathra, and J.R. Edula. 2010. Hexane extract of *Raphanus sativus* L. Roots inhibits cell proliferation and induces apoptosis in human cancer cells by modulating genes related to apoptotic pathway. *Plant Foods for Human Nutrition* 65(3): 200–209. https://doi.org/10.1007/s11130-010-0178-0.

Behboodi, S., F. Baghbani-Arani, S. Abdalan, and S.A. Sadat Shandiz. 2019. Green engineered biomolecule-capped silver nanoparticles fabricated from *Cichorium intybus* extract: In vitro assessment on apoptosis properties toward human breast cancer (MCF-7) cells. *Biological Trace Element Research* 187(2): 392–402. https://doi.org/10.1007/s12011-018-1392-0.

Bekele, T. 2016. *Bioprospecting Potential of Rumex Abyssinicus for Access and Benefit-Sharing Rumex abyssinicus*. Access and Benefit Sharing Directorate, pp. 1–6.

Bhardwaj, K., D.S. Dhanjal, A. Sharma, E. Nepovimova, A. Kalia, S. Thakur, S. Bhardwaj, C. Chopra, R. Singh, R. Verma, D. Kumar, P. Bhardwaj, and K. Kuča. 2020. Conifer-derived metallic nanoparticles: Green synthesis and biological applications. *International Journal of Molecular Sciences* 21(23): 1–22. https://doi.org/10. 3390/ijms21239028.

Cavalcanti, R.N., T. Forster-Carneiro, M.T.M.S. Gomes, M.A. Rostagno, J.M. Prado, and M.A.A. Meireles. 2013. Uses and applications of extracts from natural sources. *RSC Green Chemistry* 21: 1–57. https://doi.org/10.1039/9781849737579-00001.

Chen, L., Y. Huo, Y.X. Han, J.F. Li, H. Ali, I. Batjikh, J. Hurh, J.Y. Pu, and D.C. Yang. 2020. Biosynthesis of gold and silver nanoparticles from *Scutellaria baicalensis* roots and in vitro applications. *Applied*

Physics A: Materials Science and Processing 126(6): 1–12. https://doi.org/10.1007/s00339-020-03603-5.

Choudhary, D., S. Bhattacharyya, and S. Bose. 2017. Efficacy and Safety of Ashwagandha (*Withania somnifera* (L.) Dunal) Root Extract in Improving Memory and Cognitive Functions. *Journal of Dietary Supplements* 14(6): 599–612. https://doi.org/10.1080/ 19390211.2017.1284970.

Chugh, D., V.S. Viswamalya, and B. Das. 2021. Green synthesis of silver nanoparticles with algae and the importance of capping agents in the process. *Journal of Genetic Engineering and Biotechnology* 19(1): 1–21. https://doi.org/10.1186/s43141-021-00228-w.

Dhanani, T., S. Shah, N.A. Gajbhiye, and S. Kumar. 2017. Effect of extraction methods on yield, phytochemical constituents and antioxidant activity of *Withania somnifera*. *Arabian Journal of Chemistry* 10: S1193–S1199. https://doi.org/10.1016/j.arabjc.2013.02.015.

Do, M.H., M. Kim, S.Y. Choi, P. Lee, Y. Kim, and J. Hur. 2021. Wild radish (*Raphanus sativus* var. hortensis f. raphanistroides) root extract protects neuronal cells by inhibiting microglial activation. *Applied Biological Chemistry* 64(1). https://doi.org/10.1186/ s13765-021-00604-7.

Eisa, W.H., M.F. Zayed, B. Anis, L.M. Abbas, S.S.M. Ali, and A.M. Mostafa. 2019. Clean production of powdery silver nanoparticles using *Zingiber officinale*: The structural and catalytic properties. *Journal of Cleaner Production* 241: 118398. https://doi.org/10.1016/j.jclepro.2019.118398.

El-seedi, H. R., R.M. El-shabasy, S.A.M. Khalifa, A. Saeed, A. Shah, R. Shah, F.J. Iftikhar, M.M. Abdel-daim, A. Omri, N.H. Hajrahand, J.S.M. Sabir, X. Zou, M.F. Halabi, W. Sarhan, and W. Guo. 2019. Metal nanoparticles fabricated by green chemistry using natural extracts: Biosynthesis, mechanisms, and applications. *RSC Advances* 9(42): 24539–24559. https://doi.org/10.1039/c9ra02225b.

Faisal, S., N.S. Al-Radadi, H. Jan, S.A. Abdullah, S. Shah, M. Rizwan, Z. Afsheen, Z. Hussain, M.N. Uddin, M. Idrees, and N. Bibi. 2021. *Curcuma longa* mediated synthesis of copper oxide, nickel oxide and Cu-Ni bimetallic hybrid nanoparticles: Characterization and evaluation for antimicrobial, anti-parasitic and cytotoxic potentials. *Coatings* 11(7): 1–22. https://doi.org/10.3390/coatings11070849.

Farooq, U., S.A. Pandith, M.I. Singh Saggoo, and S.K. Lattoo. 2013. Altitudinal variability in anthraquinone constituents from novel cytotypes of *Rumex nepalensis* Spreng-a high value medicinal herb of North-Western Himalayas. *Industrial Crops and Products* 50: 112–117. https://doi.org/10.1016/j.indcrop.2013.06.044.

Gargi, D., H. Dipankar, and M. Atanu. 2018. Synthesis of gold colloid using *Zingiber officinale*: Catalytic study. *Nano Mat Chem-Bio Dev* 1(1): 1–14.

Göl, F., A. Aygün, A. Seyrankaya, T. Gür, C. Yenikaya, and F. Şen. 2020. Green synthesis and characterization of *Camellia sinensis* mediated silver nanoparticles for antibacterial ceramic applications. *Materials Chemistry and Physics* 250: 123037. https://doi.org/10.1016/j.matchemphys.2020.123037.

Gonfa, Y.H., F. Beshah, M.G. Tadesse, and A. Bachheti. 2021. Phytochemical investigation and potential pharmacologically active compounds of *Rumex nepalensis*: An appraisal. *Beni-Suef University Journal of Basic and Applied Sciences* 10(18): 1–11. https://doi.org/10.1186/s43088-021-00110-1.

Gudalwar, B. R., W. A. Panchale, J. V. Manwar, M. G. Nimbalwar, N. A. Badukale, and R. L. Bakal. 2021. Pharmacognosy, phytochemistry and clinical applications of traditional medicinal plants as memory-booster. *GSC Advanced Research and Reviews* 8(2): 19–29. https://doi.org/10.30574/gscarr.2021.8.2.0155.

Gul, S., Y.Z.H.Y. Hashim, N.I.M. Puad, and N. Samsudin. 2019. Fabrication and characterization of plant mediated green zinc nanoparticles for antileishmanial properties. *International Journal of Recent Technology and Engineering* 8(2): 5743–5749. https://doi.org/10.35940/ijrte.B3503.078219.

Haider, A., M. Ijaz, M. Imran, M. Naz, H. Majeed, J.A. Khan, M.M. Ali, and M. Ikram. 2020. Enhanced bactericidal action and dye degradation of spicy roots' extract-incorporated fine-tuned metal oxide nanoparticles. *Applied Nanoscience (Switzerland)* 10(4): 1095–1104. https://doi.org/10.1007/s13204-019-01188-x.

Haq, S., S. Dildar, M. Ben Ali, A. Mezni, A. Hedfi, M.I. Shahzad, N. Shahzad, and A. Shah. 2021. Antimicrobial and antioxidant properties of biosynthesized of NiO nanoparticles using *Raphanus sativus* extract. *Materials Research Express* 8(5): 1–12. https://doi.org/10.1088/2053-1591/abfc7c.

Hariram, M., and S. Vivekanandhan. 2018. Phytochemical process for the functionalization of materials with metal nanoparticles: Current trends and future perspectives. *Chemistry Select* 3(48): 13561–13585. https://doi.org/10.1002/slct.201802748.

Hijau, S., N. Perak, M. Ekstrak, R. Lengkuas, A. Galanga, A.A. Azmi, and N.M. Ahyat. 2015. Green synthesis of silver nanoparticles using rhizome extract of galangal, *Alpinia galanga*. *Malaysian Journal of Analytical Sciences* 19(6): 1187–1193.

Hosein Farzaei, M., R. Rahimi, F. Farzaei, and M. Abdollahi. 2015. Traditional medicinal herbs for the management of diabetes and its complications: An evidence-based review. *International Journal of Pharmacology* 11(7): 874–887. https://doi.org/10.3923/ijp.2015.874.887.

Hu, D., X. Yang, W. Chen, Z. Feng, C. Hu, F. Yan, X. Chen, D. Qu, and Z. Chen. 2021. *Rhodiola rosea* rhizome extract-mediated green synthesis of silver nanoparticles and evaluation of their potential antioxidant and catalytic reduction. activities. *ACS Omega* 6(38): 24450–24461. https://doi.org/10.1021/acsomega.1c02843.

Hu, Z., Y. Tang, Z. Yue, W. Zheng, and Z. Xiong. 2019. The facile synthesis of copper oxide quantum dots on chitosan with assistance of phyto-angelica for enhancing the human osteoblast activity to the application of osteoporosis. *Journal of Photochemistry and Photobiology B: Biology* 191: 6–12. https://doi.org/10.1016/j.jphotobiol.2018.11.009.

Husen, A., Q.I. Rahman, M. Iqbal, M.O. Yassin, and R.K. Bachheti. 2019. Plant-mediated fabrication of gold nanoparticles and their applications. In *Nanomaterials and Plant Potential*. Springer, Cham, pp. 71–110. https://doi.org/10.1007/978-3-030-05569-1_3.

Huq, M.A., M. Ashrafudoulla, M.M. Rahman, S.R. Balusamy, and S. Akter. 2022. Green Synthesis and Potential antibacterial applications of bioactive silver nanoparticles: A review. *Polymers* 14(4): 1–22. https://doi.org/10.3390/polym14040742.

Jain, A., P. Jain, P. Soni, A. Tiwari, and S.P. Tiwari. 2022. Design and characterization of silver nanoparticles of different species of curcuma in the treatment of cancer using human colon cancer cell line (HT-29). *Journal of Gastrointestinal Cancer* 0123456789: 1–6. https://doi.org/10.1007/s12029-021-00788-7.

Jain, S., N. Saxena, M.K. Sharma, and S. Chatterjee. 2019. Metal nanoparticles and medicinal plants: Present status and future prospects in cancer therapy. *Materials Today: Proceedings* 31: 662–673. https://doi.org/10.1016/j.matpr.2020.06.602.

Jalilian, F., A. Chahardoli, K. Sadrjavadi, A. Fattahi, and Y. Shokoohinia. 2020. Green synthesized silver nanoparticle from *Allium ampeloprasum* aqueous extract: Characterization, antioxidant activities, antibacterial and cytotoxicity effects. *Advanced Powder Technology* 31(3): 1323–1332. https://doi.org/10.1016/j.apt.2020.01.011.

Jonathan, G., R. Rivka, S. Avinoam, H. Lumír, and B. Nirit. 2015. Hypoglycemic activity of withanolides and elicitated *Withania somnifera*. *Phytochemistry* 116(1): 283–289. https://doi.org/10.1016/j.phytochem.2015.02.029.

Joseph, S., and B. Mathew. 2014. Microwave assisted biosynthesis of silver nanoparticles using the rhizome extract of *Alpinia galanga* and evaluation of their catalytic and antimicrobial. activities. *Journal of Nanoparticles* 2014: 1–9. https://doi.org/10.1155/ 2014/967802.

Juárez-Reyes, K., G.E. Ángeles-López, I. Rivero-Cruz, R. Bye, and R. Mata. 2014. Antinociceptive activity of *Ligusticum porteri* preparations and compounds. *Pharmaceutical Biology* 52(1): 14–20. https://doi.org/10.3109/13880209.2013.805235.

Kapoor, S., H. Sood, S. Saxena, and O.P. Chaurasia. 2022. Green synthesis of silver nanoparticles using *Rhodiola imbricata* and *Withania somnifera* root extract and their potential catalytic, antioxidant, cytotoxic and growth-promoting activities. *Bioprocess and Biosystems Engineering* 45(2): 365–380. https://doi.org/10.1007/s00449-021-02666-9.

Karthik, C., K.A. Punnaivalavan, S.P. Prabha, and D.G. Caroline. 2022. Multifarious global flora fabricated photosynthesis of silver nanoparticles: A green nanoweapon for antiviral approach including SARS-CoV-2. In *International Nano Letters* (Issue 0123456789). Springer, Berlin Heidelberg. https://doi.org/10.1007/s40089-022-00367-z.

Khairullah, A.R., T.I. Solikhah, A.N.M. Ansori, A. Fadholly, S.C. Ramandinianto, R. Ansharieta, A. Widodo, K.H.P. Riwu, N. Putri, A. Proboningrat, M.K.J. Kusala, B.W. Rendragraha, A.R.S. Putra, and A. Anshori. 2020. A review of an important medicinal plant: *Alpinia galanga* (l.) willd. *Systematic Reviews in Pharmacy* 11(10): 387–395. https://doi.org/10.31838/srp.2020.10.62.

Khan, F., M. Shariq, M. Asif, M.A. Siddiqui, P. Malan, and F. Ahmad. 2022. Green nanotechnology: Plant-mediated nanoparticle synthesis and application. *Nanomaterials* 12(4): 13–22. https://doi.org/10.3390/nano12040673.

Kiran Kumar, A.B.V., E.S. Saila, P. Narang, M. Aishwarya, R. Raina, M. Gautam, and E.G. Shankar. 2019. Biofunctionalization and biological synthesis of the ZnO nanoparticles: The effect of Raphanus sativus (white radish) root extract on antimicrobial activity against MDR strain for wound healing applications. *Inorganic Chemistry Communications* 100: 101–106. https://doi.org/10.1016/j.inoche.2018.12.014.

Kumari, A., C. Prasad, and R. Kumar. 2022. Essential oil and curcumin content in different varieties of Turmeric (*Curcuma longa* L.). *The Pharma Innovation Journal* 11(1): 841–844.

Kuppusamy, P., M.M. Yusoff, G.P. Maniam, and N. Govindan. 2016. Biosynthesis of metallic nanoparticles using plant derivatives and their new avenues in pharmacological applications -An updated report. *Saudi Pharmaceutical Journal* 24(4): 473–484. https://doi.org/10.1016/j.jsps.2014.11.013.

Latif, M.S., Abbas, S., F. Kormin, and MK. Mustafa. 2019. Green synthesis of plant-mediated metal nanoparticles: The role of polyphenols. *Asian Journal of Pharmaceutical and Clinical Research* 12: 75–84. https://doi.org/10. 22159/ajpcr.2019.v12i7.33211.

Lawrence, A. 2019. A review on nanotechnology and plant mediated metal nanoparticles and their applications. *International Journal of Scientific Research and Review* 8(6): 269–286.

Leyte-Lugo, M., E.R. Britton, D.H. Foil, A.R. Brown, D.A. Todd, J. Rivera-Chávez, N.H. Oberlies, and N.B. Cech. 2017. Secondary metabolites from the leaves of the medicinal plant goldenseal (*Hydrastis canadensis*). *Phytochemistry Letters* 20: 54–60. https://doi.org/10.1016/j.phytol.2017.03.012.

Li, R., J. He, H. Xie, W. Wang, S.K. Bose, Y. Sun, J. Hu, and H. Yin. 2019. Effects of chitosan nanoparticles on seed germination and seedling growth of wheat (*Triticum aestivum* L.). *International Journal of Biological Macromolecules* 126: 91–100. https://doi.org/10.1016/j.ijbiomac.2018.12.118.

Mahakham, W., P. Theerakulpisut, S. Maensiri, S. Phumying, and A.K. Sarmah. 2016. Environmentally benign synthesis of phytochemicals-capped gold nanoparticles as nanopriming agent for promoting maize seed germination. *Science of the Total Environment* 573: 1089–1102. https://doi.org/10.1016/j .scitotenv.2016.08.120.

Martin, A.C., E. Johnston, C. Xing, and A.D. Hegeman. 2014. Measuring the chemical and cytotoxic variability of commercially available kava (*Piper methysticum* G. Forster). *PLOS ONE* 9(11). https://doi.org /10.1371/journal.pone.0111572.

Mehata, M.S. 2021. Green route synthesis of silver nanoparticles using plants/ginger extracts with enhanced surface plasmon resonance and degradation of textile dye. *Materials Science and Engineering B: Solid-State Materials for Advanced Technology* 273: 115418. https://doi.org/10.1016/j.mseb.2021.115418.

Menon, S., S.D. Shrudhi, H. Agarwal, and V.K. Shanmugam. 2019. Efficacy of biogenic selenium nanoparticles from an extract of ginger towards evaluation on antimicrobial and antioxidant activities. *Colloids and Interface Science Communications* 29(November 2018): 1–8. https://doi.org/10.1016/j.colcom.2018 .12.004.

Mirmiran, P., Z. Houshialsadat, Z. Gaeini, Z. Bahadoran, and F. Azizi. 2020. Functional properties of beetroot (*Beta vulgaris*) in management of cardio-metabolic diseases. *Nutrition and Metabolism* 17(1): 1–15. https://doi.org/10.1186/s12986-019-0421-0.

Mishra, A.P., M. Sharifi-Rad, M.A. Shariati, Y.N. Mabkhot, S.S. Al-Showman, A. Rauf, B. Salehi, M. Župunski, M. Sharifi-Rad, P. Gusain, J. Sharifi-Rad, H.A.R. Suleria, and M. Iriti. 2018. Bioactive compounds and health benefits of edible *Rumex* species-A review. *Cellular and Molecular Biology* 64(8): 27–34. https://doi.org/10. 14715/cmb/2018.64.8.5.

Mondal, N.K., A. Chowdhury, U. Dey, P. Mukhopadhya, S. Chatterjee, K. Das, and J.K. Datta. 2014. Green synthesis of silver nanoparticles and its application for mosquito control. *Asian Pacific Journal of Tropical Disease* 4: 1–7. https://doi.org/10.1016 /S2222-1808(14)60440-0.

Moradi, F., S. Sedaghat, O. Moradi, and S. Arab Salmanabadi. 2021. Review on green nano-biosynthesis of silver nanoparticles and their biological activities: with an emphasis on medicinal plants. *Inorganic and Nano-Metal Chemistry* 51(1): 133–142. https://doi.org/10. 1080/24701556.2020.1769662.

Muniyappan, N., and N.S. Nagarajan. 2014. Green synthesis of gold nanoparticles using *Curcuma pseudomontana* essential oil, its biological activity and cytotoxicity against human ductal breast carcinoma cells T47D. *Journal of Environmental Chemical Engineering* 2(4): 2037–2044. https://doi.org/10.1016 /j.jece.2014.03.004.

Nandhini, S., K.B. Narayanan, and K. Ilango. 2018. Valeriana officinalis: A review of its traditional uses, phytochemistry and pharmacology. *Asian Journal of Pharmaceutical and Clinical Research* 11(1): 36. https://doi.org/10.22159/ajpcr.2017.v11i1.22588.

Netala, V.R., S. Bukke, L. Domdi, F. Soneya, G.S. Reddy, M.S. Bethu, V.S. Kotakdi, K.V. Saritha, and V. Tartte. 2018. Biogenesis of silver nanoparticles using leaf extract of *Indigofera hirsuta* L. and their potential biomedical applications (3-in-1 system). *Artificial Cells, Nanomedicine and Biotechnology* 46(supl): 1138–1148. https://doi.org/10.1080/21691401.2018.1446967.

Nguyen, M.T.T., H.X. Nguyen, T.H. Le, T.N. Van Do, P.H. Dang, T. Van Pham, T.T.M. Giang, S. Sun, M.J. Kim, A.M. Tawila, A.M. Omar, S. Awale, and N.T. Nguyen. 2020. A new flavanone derivative from the rhizomes of *Boesenbergia pandurata*. *Natural Product Research*: 1–7. https://doi.org/10.1080/14786419 .2020.1837822.

Nisar, M., U. Haq, G.M. Shah, A. Gul, A.I. Foudah, M.H. Alqarni, H.S. Yusufoglu, M. Hussain, H.M. Alkreathy, I. Ullah, A.M. Khan, S. Jamil, M. Ahmed, and R.A. Khan. 2022. Biogenic synthesis of silver nanoparticles using *Phagnalon niveum* and its in vivo antidiabetic effect against alloxan-induced diabetic Wistar rats. *Nanomaterials* 12: 1–18.

Nomura, K. 2019. Self-dual Leonard pairs photocatalytic activity of Ag / Ni bimetallic nanoparticles on textile dye removal. *Green Processing and Synthesis* 8: 895–900.

Palliyaguru, D.L., S.V. Singh, and T.W. Kensler. 2016. Withania somnifera: From prevention to treatment of cancer. *Molecular Nutrition and Food Research* 60(6): 1342–1353. https://doi.org/10.1002/mnfr.201500756.

Park, J.E., T.H. Lee, S.L. Ham, L. Subedi, S.M. Hong, S.Y. Kim, S.U. Choi, C.S. Kim, and K.R. Lee. 2022. Anticancer and anti-neuroinflammatory constituents isolated from the roots of Wasabia japonica. *Antioxidants* 11(3): 482.

Pavan Kumar, M.A., D. Suresh, H. Nagabhushana, and S.C. Sharma. 2015. Beta vulgaris aided green synthesis of ZnO nanoparticles and their luminescence, photocatalytic and antioxidant properties. *European Physical Journal Plus* 130(6): 1–7. https://doi.org/10.1140/epjp/i2015-15109-2.

Paw, R., M. Hazarika, P.K. Boruah, A.J. Kalita, A.K. Guha, M.R. Das, and C. Tamuly. 2021. Highly sensitive and selective colorimetric detection of dual metal ions (Hg2+and Sn2+) in water: An eco-friendly approach. *RSC Advances* 11(24): 14700–14709. https://doi.org/10.1039/d0ra09926k.

Painuli, S., P. Semwal, A. Bacheti, R.K. Bachheti, and A. Husen 2020. Nanomaterials from nonwood forest products and their applications. In Husen A., Jawaid M. (eds) *Nanomaterials for Agriculture and Forestry Applications*. Elsevier, pp. 15–40. https://doi.org/10.1016/B978-0-12-817852-2.00002-0.

Pipriya, S., N. Kundu, and U. Tiwari. 2018. Green synthesis, characterization and antioxidant activity of silver nanoparticles in extracts of *Acorus calamus* and *Agaricus bisporus. International Journal of Biochemistry Research and Review* 21(4): 1–15. https://doi.org/10.9734/ijbcrr/2018/41615.

Poudel, D.K., P. Niraula, H. Aryal, B. Budhathoki, S. Phuyal, R. Marahatha, and K. Subedi. 2022. Plant-mediated green synthesis of AgNPs and their possible applications: A critical review. *Journal of Nanotechnology* 2022: 1–24.

Raafat, M., A.S.A. El-Sayed, and M.T. El-Sayed. 2021. Biosynthesis and anti-mycotoxigenic activity of *Zingiber officinale* roscoe-derived metal nanoparticles. *Molecules* 26(8): 1–13. https://doi.org/10.3390/molecules26082290.

Raveesha, H.R., H.L. Bharath, D.R. Vasudha, B.K. Sushma, S. Pratibha, and N. Dhananjaya. 2021. Antibacterial and antiproliferation activity of green synthesized nanoparticles from rhizome extract of *Alpinia galangal* (L.) wild. *Inorganic Chemistry Communications* 132: 108854. https://doi.org/10.1016/j.inoche.2021.108854.

Reda, F.M., M.T. El-Saadony, T.K. El-Rayes, M. Farahat, G. Attia, and M. Alagawany. 2021. Dietary effect of licorice (*Glycyrrhiza glabra*) on quail performance, carcass, blood metabolites and intestinal microbiota. *Poultry Science* 100(8): 101266. https://doi.org/10.1016/j.psj.2021.101266.

Rosdianto, A.M., I.M. Puspitasari, R. Lesmana, and J. Levita. 2020. Bioactive compounds of *Boesenbergia* sp. and their antiinflammatory mechanism: A review. *Journal of Applied Pharmaceutical Science* 10(7): 116–126. https://doi.org/10.7324/JAPS.2020.10715.

Saeed Al-Zahrani, S. 2019. Silver nanoparticles (AgNPs) from plant extracts. *Life and Applied Sciences-AJSRP-Issue* 70: 70–95.

Salayová, A., Z. Bedlovičová, N. Daneu, M. Baláž, Z. Lukáčová Bujňáková, L. Balážová, and L. Tkáčiková. 2021. Green synthesis of silver nanoparticles with antibacterial activity using various medicinal plant extracts: Morphology and antibacterial efficacy. *Nanomaterials* 11(4): 1–20. https://doi.org/10.3390/nano11041005.

Sarip, N.A., N.I. Aminudin, and W.H. Danial. 2022. Green synthesis of metal nanoparticles using *Garcinia* extracts: A review. *Environmental Chemistry Letters* 20(1): 469–493. https://doi.org/10.1007/s10311-021-01319-3.

Sedighi, M., M. Bahmani, S. Asgary, F. Beyranvand, and M.R. Kopaei. 2017. A review of plant-based compounds and medicinal plants effective on atherosclerosis. *Journal of Research in Medical Sciences* 22: 1–16. https://doi.org/10.4103/1735-1995.202151.

Shaikh, S., V. Shriram, A. Srivastav, P. Barve, and V. Kumar. 2018. A critical review on Nepal Dock (*Rumex nepalensis*): A tropical herb with immense medicinal importance. *Asian Pacific Journal of Tropical Medicine* 11(7): 405–414. https://doi.org/10.4103/1995-7645.237184.

Shanmugapriya, R., A. Nareshkumar, K. Meenambigai, R. Kokila, A. Shebriya, K. Chandhirasekar, A.T. Manikandan, and C. Munusamy. 2017. Antifungal and insecticidal activities of *Raphanus sativus* mediated AgNPs against mango leafhopper, *Amritodus brevistylus* and its associated fungus, *Aspergillus niger. Journal of Entomological and Acarological Research* 49(1): 13–21. https://doi.org/10.4081/jear.2017.5953.

Siddiqi, K.S., and A. Husen. 2017. Recent advances in plant-mediated engineered gold nanoparticles and their application in biological system. *Journal of Trace Elements in Medicine and Biology* 40: 10–23. https://doi.org/10.1016/j.jtemb.2016.11.012.

Sidorowicz, A., T. Szymański, and J.D. Rybka. 2021. Photodegradation of biohazardous dye brilliant blue R using organometallic silver nanoparticles synthesized through a green chemistry method. *Biology* 10(8): 1–16. https://doi.org/10.3390/biology10080784.

Singh, A., P.K. Gautam, A. Verma, V. Singh, P.M. Shivapriya, S. Shivalkar, A.K. Sahoo, and S.K. Samanta. 2020. Green synthesis of metallic nanoparticles as effective alternatives to treat antibiotics resistant bacterial infections: A review. *Biotechnology Reports* 25: e00427. https://doi.org/10.1016/j.btre.2020.e00427.

Singh, S., R. Vyas, A. Chaturvedi, and R. Sisodia. 2018. Rapid photosynthesis of silver nanoparticles using *Chlorophytum borivilianum* root extract and its antimicrobial activity. *Journal of Pharmacognosy and Phytochemistry* 7(5): 1738–1744.

Singh, T., K. Jyoti, A. Patnaik, A. Singh, R. Chauhan, and S.S. Chandel. 2017. Biosynthesis, characterization and antibacterial activity of silver nanoparticles using an endophytic fungal supernatant of *Raphanus sativus*. *Journal of Genetic Engineering and Biotechnology* 15(1): 31–39. https://doi.org/10.1016/j.jgeb.2017.04.005.

Sorbiun, M., E.S. Mehr, Ramazani, and A.M. Malekzadeh. 2018. Biosynthesis of metallic nanoparticles using plant extracts and evaluation of their antibacterial properties. *Nanochemistry Research* 3(1): 1–16. https://doi.org/10.22036/ncr.2018.01.001.

Srivastava, M., G. Singh, S. Sharma, S. Shukla, and P. Misra. 2019. Elicitation enhanced the yield of glycyrrhizin and antioxidant activities in hairy root cultures of *Glycyrrhiza glabra* L. *Journal of Plant Growth Regulation* 38(2): 373–384. https://doi.org/ 10.1007/s00344-018-9847-2.

Suwannakul, S., S. Wacharanad, and P. Chaibenjawong. 2018. Rapid green synthesis of silver nanoparticles and evaluation of their properties for oral disease therapy. *Songklanakarin Journal of Science and Technology* 40(4): 831–839. https://doi.org/10.14456/sjst-psu.2018.112.

Szewczyk, K., W. Pietrzak, K. Klimek, M. Miazga-Karska, A. Firlej, M. Flisiński, and A. Grzywa-Celińska. 2021. Flavonoid and phenolic acids content and in vitro study of the potential anti-aging properties of *Eutrema japonicum* (Miq.) koidz cultivated in wasabi farm poland. *International Journal of Molecular Sciences* 22(12): 1–18. https://doi.org/10.3390/ijms22126219.

Talebi, M., S. İlgün, V. Ebrahimi, M. Talebi, T. Farkhondeh, H. Ebrahimi, and S. Samarghandian. 2021. *Zingiber officinale* ameliorates Alzheimer's disease and cognitive impairments: Lessons from pre-clinical studies. *Biomedicine and Pharmacotherapy* 133: 1–13. https://doi.org/10.1016/j.biopha.2020.111088.

Tan, W., H. Gao, T. Yang, W. Jiang, H. Zhang, X. Yu, and X. Tian. 2019. The complete chloroplast genome of a medicinal resource plant (*Rumex crispus*). *Mitochondrial DNA Part B: Resources* 4(2): 2800–2801. https://doi.org/10.1080/23802359.2019.1660254.

Tarannum, N., Divya, and Y.K. Gautam. 2019. Facile green synthesis and applications of silver nanoparticles: A state-of-the-art review. *RSC Advances* 9(60): 34926–34948. https://doi.org/10.1039/c9ra04164h.

Tran, M.T., L.P. Nguyen, D.T. Nguyen, T. Le Cam-Huong, C.H. Dang, T.T.K. Chi, and T.D. Nguyen. 2021. A novel approach using plant embryos for green synthesis of silver nanoparticles as antibacterial and catalytic agent. *Research on Chemical Intermediates* 47(11): 4613–4633. https://doi.org/10.1007/s11164-021-04548-x.

Umadevi, M., R. Rajeswari, C.S. Rahale, S. Selvavenkadesh, R. Pushpa, and K.P.S. Kumar. 2012. Traditional and medicinal uses of *Withania somnifera*. *The Pharmaceutical Innovation* 1(9): 102–110.

Vakayil, R., S. Muruganantham, N. Kabeerdass, M. Rajendran, A. Mahadeo palve, S. Ramasamy, T.S. Awad Alahmadi, H. Almoallim, V. Manikandan, and M. Mathanmohun. 2021. *Acorus calamus*-zinc oxide nanoparticle coated cotton fabrics show antimicrobial and cytotoxic activities against skin cancer cells. *Process Biochemistry* 111(P1): 1–8. https://doi.org/10.1016/j.procbio.2021.08.024.

Velmurugan, P., K. Anbalagan, M. Manosathyadevan, K.J. Lee, M. Cho, S.M. Lee, J.H. Park, S.G. Oh, K.S. Bang, and B.T. Oh. 2014. Green synthesis of silver and gold nanoparticles using *Zingiber officinale* root extract and antibacterial activity of silver nanoparticles against food pathogens. *Bioprocess and Biosystems Engineering* 37(10): 1935–1943. https://doi.org/10.1007/s00449-014-1169-6.

Venugopal, K., H. Ahmad, E. Manikandan, K. Thanigai Arul, K. Kavitha, M.K. Moodley, K. Rajagopal, R. Balabhaskar, and M. Bhaskar. 2017. The impact of anticancer activity upon *Beta vulgaris* extract mediated biosynthesized silver nanoparticles (Ag-NPs) against human breast (MCF-7), lung (A549) and pharynx (Hep-2) cancer cell lines. *Journal of Photochemistry and Photobiology, Part B: Biology* 173: 99–107. https://doi.org/10.1016/j.jphotobiol.2017.05.031.

Xuan, T.D., M. Fukuta, A.C. Wei, A.A. Elzaawely, T.D. Khanh, and S. Tawata. 2008. Efficacy of extracting solvents to chemical components of kava (*Piper methysticum*) roots. *Journal of Natural Medicines* 62(2): 188–194. https://doi.org/10.1007/s11418-007-0203-2.

Yu, J.Y., J.Y. Ha, K.M. Kim, Y.S. Jung, J.C. Jung, and S. Oh. 2015. Antiinflammatory activities of *licorice* extract and its active compounds, glycyrrhizic acid, liquiritin and liquiritigenin, in BV2 cells and mice liver. *Molecules* 20(7): 13041–13054. https://doi.org/10.3390/molecules200713041.

Zare, E.N., V.V.T. Padil, B. Mokhtari, A. Venkateshaiah, S. Wacławek, M. Černík, F.R. Tay, R.S. Varma, and P. Makvandi. 2020. Advances in biogenically synthesized shaped metal- and carbon-based nanoarchitectures and their medicinal applications. *Advances in Colloid and Interface Science* 283: 102236. https://doi.org/10.1016/j.cis. 2020.102236.

Zhang, Z., G. Xin, G. Zhou, Q. Li, V.P. Veeraraghavan, S. Krishna Mohan, D. Wang, and F. Liu. 2019. Green synthesis of silver nanoparticles from *Alpinia officinarum* mitigates cisplatin-induced nephrotoxicity via down-regulating apoptotic pathway in rats. *Artificial Cells, Nanomedicine and Biotechnology* 47(1): 3212–3221. https://doi.org/10.1080/21691401.2019.1645158.

12 Medicinally Important Flowers and Their Role in Nanoparticle Synthesis and Applications

Amit Kumar Mittal and Uttam Chand Banerjee

CONTENTS

12.1 INTRODUCTION

Nanotechnology has gained much more attention in the last decade due to its wide application in diverse areas like medicine, catalysis, energy, and materials (Nasrollahzadeh et al. 2019). An availability of nanoparticles has unlocked new technical applications in the field of medicine. The distinctive properties of these nanomaterials benefit humans in many areas including catalysis, medicines, antimicrobials, biosensors, drug delivery, and applications in electronics (Lekha et al. 2021). Nanomaterials are currently being utilized in pharmaceutical sciences as a tool for novel drug delivery systems. These nanomaterials can be used in drug delivery which may have multiple functions for diagnosis and therapeutics (Bhaumik et al. 2015). At present, nanomedicine has also opened new doors to treat some life-threatening diseases (Das et al. 2009). To achieve the targeted drug delivery system, the conjugation of nanoparticles with either an antibody or ligand is necessary (Kwon et al. 2012). Furthermore, there are some limitations such as conjugated drugs to be carried by the individual nanoparticle which is rather imperfect and their quantification is tedious (Rabanel et al. 2019). However, the toxic effect of these nanoparticles must be taken into consideration before using them in therapeutic applications (Arvizo et al. 2012).

The synthesizing of novel materials by a green route is an emergent field in nanotechnology (Mittal et al. 2013; Painuli et al. 2020; Husen et al. 2019; Bachheti et al. 2021a, b). In particular, nanoparticles with a small size and large surface area have great prospects in medicine as well as having numerous applications in chemical, biological, biomedical, pharmaceuticals, physics, and in electronics (Suttee et al. 2019). Recently iron, silver, zinc, cerium, nickel, copper, and gold nanoparticles have become common because of their applications and advantages (Radad et al. 2012). Currently, nanomaterials are being synthesized by various methods and with a diversity of precursors (Mittal et al. 2014).

Researchers have made an important advance towards the synthesis of nanomaterials by various sources, including physical, chemical, and biological methods (Mittal et al. 2013; Bachheti et al. 2020a). The conventional synthesis methods could not be energy efficient; they are environmentally unfriendly due to the usage of highly toxic chemicals and solvents (Duan et al. 2015). Chemical synthesis involves the use of chemical reducing agents that may be toxic and such nanoparticles become inappropriate for biomedical applications (Kalimuthu et al. 2020). This green bio-perspective synthesis is a replacement for chemical and physical methods. Biosynthesis and green chemistry are

DOI: 10.1201/9781003213727-12

interconnected with nanoscience and technology (El-Seedi et al. 2019; Bachheti et al. 2019a). This green method is eco-friendly, very easy to perform, economic, non-hazardous, and preferable over other methods. It is simple, convenient, environmentally friendly, and requires less reaction time (Nasrollahzadeh et al. 2020). In the last decade, the biosynthesis of nanomaterials using microbes or plant extracts has been explored and their synthesis via various routes is shown in Figure 12.1.

Flowers have various secondary metabolites including dyes, volatile substances, and other phenolics which may have insightful pharmacological actions (Alamgir 2018). These compounds can be exploited as reducing agents for the biosynthesis of a diversity of materials, such as gold, silver, copper, iron, selenium, zinc, etc. (Kumar et al. 2020). This chapter stresses the use of medicinally important flowers for the biosynthesis of nanoparticles and their plausible applications.

12.2 MEDICINALLY IMPORTANT FLOWERS TO SYNTHESIZE NANOPARTICLES

Flowers have distinctive chemical moieties that can be useful for the synthesis of nanoparticles (Kumar et al. 2020). Flower-mediated nanoparticle synthesis is beneficial compared to other biological methods (Kumar et al. 2020). A schematic diagram for the synthesis of various nanoparticles using flower extracts is shown in Figure 12.2. Reported literature for the green synthesis of nanoparticles has been summarized in Table 12.1. The plants and the plant parts like leaves, latex, stem, root, bark, and flowers have been shown to possess various biological activities including antioxidant, antibacterial, antiarthritic, antitumour and antidiabetic properties (Maheshwari 2019). The flowers

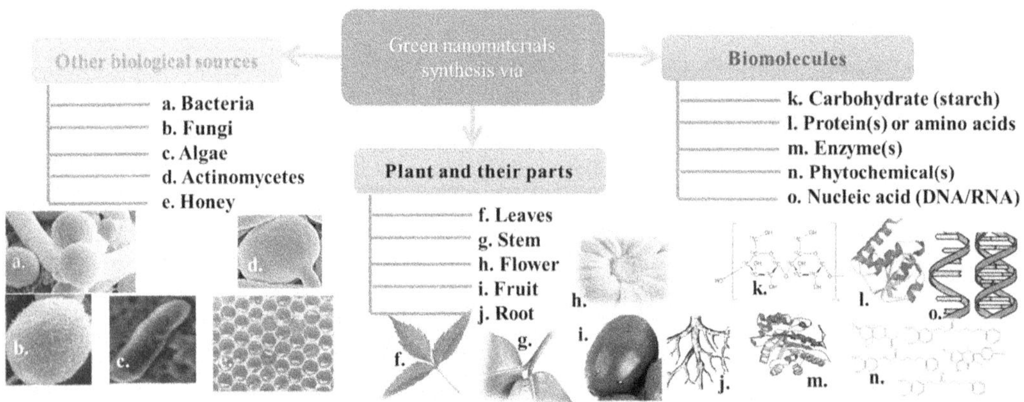

FIGURE 12.1 Various types of catalysts for the green synthesis of nanomaterials.

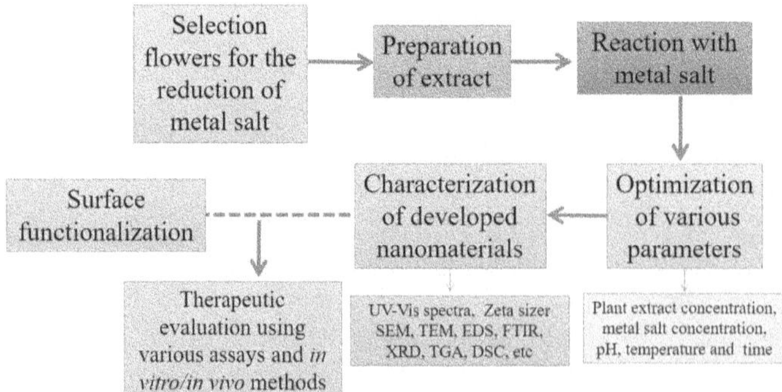

FIGURE 12.2 A general procedure for the synthesis of nanoparticles from flower extract and their characterization.

TABLE 12.1
Various Types of Flowers Used as Reducing and Stabilizing Agents for the Synthesis of Nanoparticles and Their Therapeutic Applications

Plant name	Part	Type of nanomaterials	Application	References
Fritillaria imperialis	Flower	Fe_3O_4	Catalytic activity	Veisi et al. (2021)
Abelmoschus esculentus	Flower	Silver	Anticancer and antibacterial activities	Devanesan and AlSalhi (2021)
Catharanthus roseus	Flower	Silver	Antioxidant, antimicrobial, and photocatalytic activity	Kandiah and Chandrasekaran (2021)
Allium Sativum	Flower	Silver	Antimicrobial, antioxidant, anti-inflammatory, and anticancer activity	Velsankar et al. (2020)
Canna indica, *Cosmos bipinnata*, and *Lantana camara*	Flowers	Silver	Antimicrobial activities	Cheng et al. (2020)
Aerva lanata	Flower	Silver	Cytotoxic, bactericidal, and photocatalytic activities	Kanniah et al. (2020)
Zephyranthe srosea	Flower	Silver oxide	Antibacterial, antioxidant, anti-inflammatory, and antidiabetic activities	Maheshwaran et al. (2020)
Hibiscus rosa-sinensis	Flower	Silver	–	Rao et al. (2020)
Syzygium aromaticum	Flower	Zinc oxide	Antifungal activities	Thimappa et al. (2019)
Clitoria ternatea	Flower	Gold	–	Chan et al. (2020)
Avicennia marina	Flower	Iron oxide	Methanol oxidation for fuel	Karpagavinayagam and Vedhi (2019)
Mangifera indica	Flower	Silver	Antibacterial activity	Ameen et al. (2019)
Fritillaria imperialis	Flower	Silver	Antibacterial activity	Hemmati et al. (2019)
Cassia auriculata	Flower	Silver	Catalytic activity	Muthu et al. (2017)
Mimosa pudica	Flower	Gold	Catalytic activity	Mapala and Pattabi (2017)
Madhuca longifolia	Flower	Silver	Antibacterial activity	Patil et al. (2018)
Moringa oleifera	Flower	Palladium	Catalytic and antimicrobial activities	Anand et al. (2016)
Nyctanthes arbortristis	Flower	Silver	Antibacterial and cytotoxic activity	Gogoi et al. (2015)
Plumeria alba (frangipani)	Flower	Silver	Catalytic, antibacterial, and cytotoxic activity	Mata et al. (2015)
Marigold	Flower	Silver	Antimicrobial activity	Padalia et al. (2015)
Gnidiaglauca	Flower	Gold	Catalytic	Ghosh et al. (2012)
Nyctanthes arbortristis	Flower	Gold	–	Das et al. (2013)
Lonicera japonica	Flower	Silver and gold	–	Nagajyothi et al. (2012)
Achillea biebersteinii	Flower	Silver	Antiangiogenic activity	Bahara et al. (2014)
Aloe vera	Flower	Copper	–	Karimi and Mohsenzadeh (2015)

comprise numerous bioactive compounds, including β-sitosterol, scopoletin, iridoidsisoplumericin, plumieride, plumieridecoumarate, and plumieridecoumarate glucoside. These compounds could be particularized for the higher medicinal value with the materials they synthesize (Mata et al. 2015).

Nanomaterials synthesized from the extracts of various flowers are of diverse size, shape, and surface areas and are characterized by ultraviolet-visible, X-ray diffraction, Fourier transform infrared, energy dispersive, differential light scattering, and Raman spectroscopy (Mittal and Banerjee2016). The range of the ultraviolet-visible spectroscopy wavelength, from 300 to 700 nm, illustrates the presence of nanomaterials of a size ranging from 2 to 100 nm (Velmurugan et al. 2015). The occurrence of ultraviolet-visible spectroscopy wavelength absorption of silver and gold nanoparticles occurs at 425 and 530 nm (Tetgure et al. 2015). An approximation of the size of synthesized nanoparticles, along with charges on the exterior of the materials is accompanied using zeta sizer analysis (Dubey et al. 2010). The microscopic structure of materials was evaluated using electron microscopy and composition of the element through energy dispersive X-ray analysis attached to the instrument. X–ray diffraction is executed to distinguish the shape and crystalline nature of the materials (Singh et al. 2016). Fourier transform infrared spectroscopy is used to identify the functional group present on the surface of materials such as amino, carboxyl, hydroxyls of flavonoids, and phenols of phytochemicals (Mittal et al. 2015).

12.3 APPLICATION OF MEDICINALLY IMPORTANT FLOWER-MEDIATED SYNTHESIZED NANOPARTICLES

Among the noble metal nanoparticles used in green synthesis, silver is widely preferred owing to its antibacterial, antifungal, larvicidal, antiparasitic, and catalytic functions (Dauthal et al. 2016). Hence, there is an upward prerequisite to advance a simple, inexpensive, eco-friendly, stable, size, and shape control method for the synthesis of nanoparticles (Fathima et al. 2018). Use of biological resources such as microorganisms and plant extracts could be an appropriate alternative (Remya et al. 2017). Furthermore, metal nanoparticles have shown positive results in wound healing, retinal therapies, diagnostic agents, and biosensor development and as pharmaceuticals along with other conventional uses in electronics, optics, catalysis, and Raman scattering (Azharuddinet al. 2019). Figure 12.3 shows a wide range of applications of nanomaterials synthesized by medicinal flowers. The fundamental uses of silver nanoparticles were strongly associated with the water treatment process as well as silver nanoparticle-containing filters which are highly effective and utilized against contamination in water treatment and filtration processes (Albukhari et al. 2019).

In 2021, Kandiah and Chandrasekaran reported the synthesis of green silver nanoparticles using *Catharanthus roseus* flower extracts and determined their antioxidant, antimicrobial, and photocatalytic activities. The nanoparticles showed higher antioxidation potential than their respective flower extracts and showed substantial antibacterial activity against Gram-negative *Escherichia coli* (Kandiah and Chandrasekaran 2021). In 2021 Devanesan and AlSalhi reported the biosynthesis of silver nanoparticles using the flower extract of *Abelmoschus esculentus* and the cytotoxicity and antimicrobial studies using those nanoparticles. *A. esculentus* flower extract-mediated silver nanoparticles exhibited excellent activities *in vitro* cytotoxic activity against cancer cell lines and antimicrobial effect on Gram-positive and Gram-negative pathogens (Devanesan and AlSalhi 2021). In 2021 Veisi et al. reported the bio-inspired synthesis of palladium nanoparticle-fabricated magnetic Fe_3O_4 nanocomposite by *Fritillaria imperialis* flower extract as a competent recyclable catalyst for the reduction of nitroarenes. They also reported that the nanocatalyst was retrieved using a bar magnet and recycled numerous times without considerable leaching or loss of activity (Veisi et al. 2021).

In 2020, Rao et al. reported the biosynthesis of silver nanoparticles using *Hibiscus rosa-sinensis* flower extract (Rao et al. 2020). In 2020, Kanniah et al. reported the green synthesis of multifaceted silver nanoparticles using the *Aerva lanata* flower extract and evaluated its cytotoxic, bactericidal, and photocatalytic activities (Kanniah et al. 2020). In 2020, Maheshwaran et al. reported the *in vitro* synthesis of silver oxide nanoparticles using *Zephyranthes rosea* flower extract by a greener

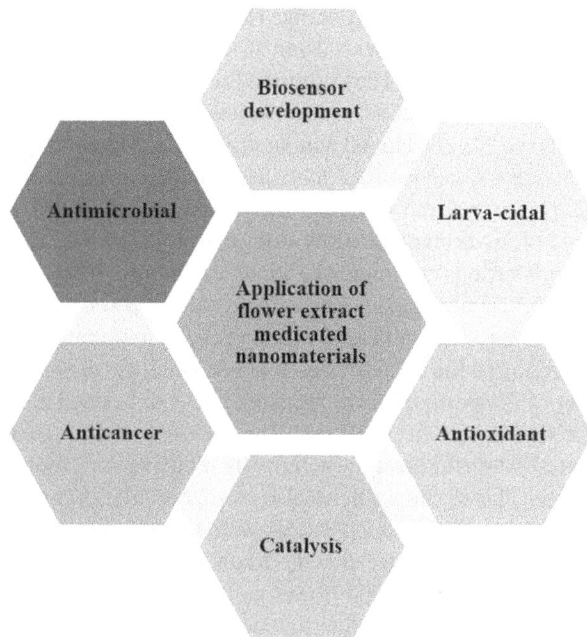

FIGURE 12.3 Various applications of nanomaterials synthesized by medicinal flowers.

approach and it was found to be a potential candidate for antibacterial, antioxidant, anti-inflammatory, and antidiabetic activities (Maheshwaran et al. 2020). In 2020, Velsankar et al. reported the eco-friendly phyto-assisted synthesis of silver nanoparticles using *Allium sativum* flower extract and found antimicrobial, antioxidant, anti-inflammatory, and anticancer activities (Velsankar et al. 2020). In 2020, Cheng et al. reported silver nanoparticle synthesis via three different plant flower extracts including *Canna indica* L., *Cosmos bipinnata* Cav., and *Lantana camara* L. and evaluated their antibacterial properties (Cheng et al. 2020).

In 2020, Chan et al. reported gold nanoparticle synthesis using aqueous extract of the *Clitoria ternatea* flower (Chan et al. 2020). In 2019, Karpagavinayagam and Vedhi reported a nontoxic method for iron oxide nanoparticle synthesis using a flower extract of *Avicennia marina* with a size range of 30–100 nm. Electrochemical studies were carried out to assess the redox behaviour of iron oxide nanoparticles which can be used in industrial applications including dye degradation and controlling environmental pollution (Karpagavinayagam and Vedhi 2019). In 2019 Ameen et al. reported the phyto-synthesis of silver nanoparticles using *Mangifera indica* flower extract and assessed their antibacterial activity. The morphology of the synthesized nanomaterials was sphere-shaped with a size of 10–20 nm. The nanomaterials exhibited extensive toxic effect on clinically important pathogens including Gram-positive (*Staphylococcus* sp.) and Gram-negative (*Klebsiella* sp., *Pantoea agglomerans*, and *Rahnella* sp.) bacteria (Ameen et al. 2019). When nanomaterials interact with the bacterial cells, they show antibacterial action due to the nanoparticles passing over the membrane of the microorganisms and inhibiting the normal metabolism of the cells. Furthermore, nanomaterials bind to the cellular organelles and genetic materials of the microbial cell, like DNA, RNA, ribosomes, lysosomes, and various enzymes. In 2019, Hemmati et al. reported the biosynthesis of silver nanoparticles using *Fritillaria* flower extract as the reducing, stabilizing, and capping agent in the aqueous phase and saw antibacterial activity against some human pathogens (Hemmati et al. 2019). in 2019, Thimappa et al. reported the biosynthesis of zinc oxide nanoparticles with *Syzygium aromaticum* flower bud extract and found a new application in controlling the growth of *Fusarium graminearum* (Thimappa et al. 2019).

In 2017, Muthu and Priya conducted green synthesis of silver nanoparticles using *Cassia auriculata* flower extract and detected the catalytic effect through reduction of highly polluted organic compounds such as 4-nitrophenol and methyl orange (Muthu and Priya 2017). In 2016, Anand et al. reported palladium nanoparticles biosynthesis using *Moringa oleifera* flower extract and evaluated their biological and catalytic properties. They reported the catalytic degradation of industrial effluent containing the organic toxic compounds, p-nitrophenol and methylene blue by nanoparticles and the reusability of the catalysts. The developed nanomaterials were found to be nontoxic and showed antimicrobial activity against *Enterococcus faecalis*, *Bacillus cereus*, *Staphylococcus aureus*, *Esherichia coli*, *Candida albicans*, and *Candida utilis* (Anand et al. 2016). In 2015, Padalia et al. reported the biosynthesis of silver nanoparticles from marigold flowers and found its synergistic antimicrobial potential with various commercial antibiotics against Gram-positive (*Staphylococcus aureus* and *Bacillus cereus*), Gram-negative (*Escherichia coli* and *Pseudomonas aeruginosa*) bacteria, and fungi (*Candida glabrata*, *Candida albicans*, and *Cryptococcae neoformans*) (Padalia et al. 2015). Mata et al. (2015) reported the catalytic and biological actions of silver nanoparticles synthesized from flower extract of *Plumeria alba* (frangipani). The synthesized nanoparticles were found to be spherical in shape with a size of 36.19 nm. The presence of face-centred cubic (fcc) phase structure of silver exhibited powerful catalytic activity by reducing 4-nitrophenol to 4-aminophenol within a short span of time. These nanoparticles also oxidized other organic dyes, methylene blue, and ethidium bromide. The developed nanoparticles were also found to be antioxidant, cytotoxic, and antibacterial against cancer and bacterial cells (Mata et al. 2015). Gogoi et al. (2015) reported silver nanoparticle synthesis using alcoholic flower extract of *Nyctanthes arbortristis* and did *in vitro* investigation of their antimicrobial and cytotoxic activities. The synthesized silver nanoparticles exhibited antimicrobial activity against the Gram-negative pathogenic strain of *Escherichia coli* and non-cytotoxic activity on mouse fibroblastic cell line (Gogoi et al. 2015).

Bahara et al. (2014) reported silver nanoparticle biosynthesis using *Achillea biebersteinii* flower extract and evaluated its antiangiogenic activity in the animal model led to a 50% reduction in the length and number of vessel-like structures (Bahara et al. 2014). In 2013 Karimi and Mohsenzadeh reported the rapid, green, and eco-friendly green synthesis of copper nanoparticles using A*loe vera* flower extract (Karimi and Mohsenzadeh 2015). Similarly, in 2012, Mittal et al. reported the free radical scavenging and antioxidant potentials of silver nanoparticles synthesized from *Rhododendron dauricum* flower extract (Mittal et al. 2012). In 2012, Ghosh et al. reported the synthesis of gold nanoparticles using *Gnidiaglauca* flower extract and chemo-catalytic potential in a reduction reaction of 4-nitrophenol to 4-aminophenol by $NaBH_4$ in aqueous phase (Ghosh et al. 2012). Similarly, in 2013, Das et al. reported gold nanoparticle biosynthesis using *Nyctanthes arbortristis* flower extract. In 2012, Nagajyothi et al. reported the rapid synthesis of silver and gold nanoparticles using *Lonicera japonica* flower extract as a reducing as well as capping agent (Das et al. 2013).

12.4 CONCLUSIONS AND PROSPECTS

Medicinally important flowers are able to develop a nontoxic, biocompatible, simple, rapid, eco-friendly, inexpensive, and nontoxic process for the synthesis of nanomaterials. The technique for the green synthesis of nanomaterials, especially via flowers, is an efficient process of nanoparticle synthesis. Phytochemicals such as flavonoids and phenolic compounds, having antioxidant properties present in flowers, are involved in the reduction and stabilization of metal ions as well as synthesis of nanoparticles. The synthesized nanomaterials have extensive properties such as antioxidant, antibacterial, antidiabetic, anticancer, and catalytic activities and can be further used in drug delivery applications.

REFERENCES

Alamgir, A. N. M. "Secondary metabolites: Secondary metabolic products consisting of C and H; C, H, and O; N, S, and P elements; and O/N heterocycles." In *Therapeutic Use of Medicinal Plants and their Extracts: Volume 2*, pp. 165–309. Springer, Cham, 2018.

Albukhari, Soha M., Muhammad Ismail, Kalsoom Akhtar, and Ekram Y. Danish. "Catalytic reduction of nitrophenols and dyes using silver nanoparticles@ cellulose polymer paper for the resolution of waste water treatment challenges." *Colloids and Surfaces A: Physicochemical and Engineering Aspects* 577 (2019): 548–561.

Ameen, Fuad, P. Srinivasan, T. Selvankumar, S. Kamala-Kannan, S. Al Nadhari, A. Almansob, T. Dawoud, and M. Govarthanan. "Phytosynthesis of silver nanoparticles using Mangiferaindica flower extract as bioreductant and their broad-spectrum antibacterial activity." *Bioorganic Chemistry* 88 (2019): 102970.

Anand, Krishnan, Charlette Tiloke, Alisa Phulukdaree, B. Ranjan, A. Chuturgoon, S. Singh, and Robert Moonsamy Gengan. "Biosynthesis of palladium nanoparticles by using Moringaoleifera flower extract and their catalytic and biological properties." *Journal of Photochemistry and Photobiology B: Biology* 165 (2016): 87–95.

Arvizo, Rochelle R., Sanjib Bhattacharyya, Rachel A. Kudgus, Karuna Giri, Resham Bhattacharya, and Priyabrata Mukherjee. "Intrinsic therapeutic applications of noble metal nanoparticles: Past, present and future." *Chemical Society Reviews* 41, no. 7 (2012): 2943–2970.

Azharuddin, Mohammad, Geyunjian H. Zhu, Debapratim Das, Erdogan Ozgur, Lokman Uzun, Anthony P.F. Turner, and Hirak K. Patra. "A repertoire of biomedical applications of noble metal nanoparticles." *Chemical Communications* 55, no. 49 (2019): 6964–6996.

Bachheti, A., A. Sharma, R.K. Bachheti, A. Husen & V. K. Mishra. Plant-mediated synthesis of copper oxide nanoparticles and their biological applications. In *Nanomaterials and Plant Potential*, pp. 221–237. Springer, Cham, 2019a. https://doi.org/10.1007/978-3-030-05569-1_8.

Bachheti, A., R. K. Bachheti, L. Abate & Azamal Husen. Current status of Aloe-based nanoparticle fabrication, characterization and their application in some cutting-edge areas. *South African Journal of Botany* (2021a). https://doi.org/10.1016/j.sajb.2021.08.021.

Bachheti, R.K., L. Abate, A. Bachheti, A. Madhusudhan & A. Husen. Algae-, fungi-, and yeast-mediated biological synthesis of nanoparticles and their various biomedical applications. In *Handbook of Greener Synthesis of Nanomaterials and Compounds*, pp. 701–734. Elsevier, 2021b. https://doi.org/10.1016/B978-0-12-821938-6.00022-0.

Bachheti, R.K., R. Konwarh, V. Gupta, A. Husen & Archana Joshi. Green synthesis of iron oxide nanoparticles: Cutting edge technology and multifaceted applications. In *Nanomaterials and Plant Potential*, pp. 239–259. Springer, Cham, 2019b. https://doi.org/10.1007/978-3-030-05569-1_9.

Bachheti, R.K., Y. Godebo, A. Bachheti, M. O. Yassin & Azamal Husen. Root-based fabrication of metal/metal-oxide nanomaterials and their various applications. In Husen A., Jawaid M. (eds) *Nanomaterials for Agriculture and Forestry Applications*, pp. 135–166. Elsevier, 2020a. https://doi.org/10.1016/B978-0-12-817852-2.00006-8.

Baharara, Javad, Farideh Namvar, Tayebe Ramezani, Nasrin Hosseini, and Rosfarizan Mohamad. "Green synthesis of silver nanoparticles using Achilleabiebersteinii flower extract and its anti-angiogenic properties in the rat aortic ring model." *Molecules* 19, no. 4 (2014): 4624–4634.

Bhaumik, Jayeeta, Amit Kumar Mittal, Avik Banerjee, Yusuf Chisti, and Uttam Chand Banerjee. "Applications of phototheranosticnanoagents in photodynamic therapy." *Nano Research* 8, no. 5 (2015): 1373–1394.

Chan, J. Z., R. Rasit Ali, K. Shameli, M. S. N. Salleh, K. X. Lee, and ED Mohamed Isa. "Green synthesis of gold nanoparticles using aqueous extract of clitoriaternatea flower." In *IOP Conference Series: Materials Science and Engineering* 808, no. 1: 012036. IOP Publishing, 2020.

Cheng, Hai-Jun, Hui Wang, and Jing-Ze Zhang. "Phytofabrication of silver nanoparticles using three flower extracts and their antibacterial activities against pathogen Ralstoniasolanacearum strain YY06 of bacterial wilt." *Frontiers in Microbiology* 11 (2020): 2110.

Das, Manasi, Chandana Mohanty, and Sanjeeb K. Sahoo. "Ligand-based targeted therapy for cancer tissue." *Expert Opinion on Drug Delivery* 6, no. 3 (2009): 285–304.

Das, Ratul Kumar, Nayanmoni Gogoi, and Utpal Bora. "Green synthesis of gold nanoparticles using Nyctanthesarbortristis flower extract." *Bioprocess and Biosystems Engineering* 34, no. 5 (2011): 615–619.

Dauthal, Preeti, and Mausumi Mukhopadhyay. "Noble metal nanoparticles: Plant-mediated synthesis, mechanistic aspects of synthesis, and applications." *Industrial & Engineering Chemistry Research* 55, no. 36 (2016): 9557–9577.

Devanesan, Sandhanasamy, and Mohamad S. AlSalhi. "Green synthesis of silver nanoparticles using the flower extract of abelmoschusesculentus for cytotoxicity and antimicrobial studies." *International Journal of Nanomedicine* 16 (2021): 3343.

Duan, Haohong, Dingsheng Wang, and Yadong Li. "Green chemistry for nanoparticle synthesis." *Chemical Society Reviews* 44, no. 16 (2015): 5778–5792.

Dubey, Shashi Prabha, Manu Lahtinen, and Mika Sillanpää. "Green synthesis and characterizations of silver and gold nanoparticles using leaf extract of Rosa rugosa." *Colloids and Surfaces A: Physicochemical and Engineering Aspects* 364, no. 1–3 (2010): 34–41.

El-Seedi, Hesham R., Rehan M. El-Shabasy, Shaden AM Khalifa, Aamer Saeed, Afzal Shah, Raza Shah, Faiza Jan Iftikhar et al. "Metal nanoparticles fabricated by green chemistry using natural extracts: Biosynthesis, mechanisms, and applications." *RSC Advances* 9, no. 42 (2019): 24539–24559.

Fathima, John Bani, ArivalaganPugazhendhi, Mohammad Oves, and Rose Venis. "Synthesis of eco-friendly copper nanoparticles for augmentation of catalytic degradation of organic dyes." *Journal of Molecular Liquids* 260 (2018): 1–8.

Ghosh, Sougata, Sumersing Patil, Mehul Ahire, Rohini Kitture, Deepanjali D. Gurav, Amit M. Jabgunde, Sangeeta Kale, et al. "Gnidiaglauca flower extract mediated synthesis of gold nanoparticles and evaluation of its chemocatalytic potential." *Journal of Nanobiotechnology* 10, no. 1 (2012): 1–9.

Gogoi, Nayanmoni, Punuri Jayasekhar Babu, Chandan Mahanta, and Utpal Bora. "Green synthesis and characterization of silver nanoparticles using alcoholic flower extract of Nyctanthesarbortristis and in vitro investigation of their antibacterial and cytotoxic activities." *Materials Science and Engineering: C* 46 (2015): 463–469.

Hemmati, Saba, Asra Rashtiani, Mohammad Mahdi Zangeneh, Pourya Mohammadi, Akram Zangeneh, and Hojat Veisi. "Green synthesis and characterization of silver nanoparticles using Fritillaria flower extract and their antibacterial activity against some human pathogens." *Polyhedron* 158 (2019): 8–14.

Husen, A., Q. I. Rahman, M. Iqbal, M. O. Yassin & R. K. Bachheti. Plant-mediated fabrication of gold nanoparticles and their applications. In *Nanomaterials and Plant Potential*, pp. 71–110. Springer, Cham, 2019. https://doi.org/10.1007/978-3-030-05569-1_3.

Kalimuthu, Kalishwaralal, ByungSeok Cha, Seokjoon Kim, and Ki Soo Park. "Eco-friendly synthesis and biomedical applications of gold nanoparticles: A review." *Microchemical Journal* 152 (2020): 104296.

Kandiah, Mathivathani, and Kavishadhi N. Chandrasekaran. "Green synthesis of silver nanoparticles using catharanthusroseus flower extracts and the determination of their antioxidant, antimicrobial, and photocatalytic activity." *Journal of Nanotechnology* 2021 (2021): 1–18.

Kanniah, Paulkumar, Jila Radhamani, Parvathiraja Chelliah, Natarajan Muthusamy, Emmanuel Joshua Jebasingh Sathiya Balasingh Thangapandi, JesiReeta Thangapandi, Subburathinam Balakrishnan, and Rajeshkumar Shanmugam. "Green synthesis of multifaceted silver nanoparticles using the flower extract of Aervalanata and evaluation of its biological and environmental applications." *ChemistrySelect* 5, no. 7 (2020): 2322–2331.

Karimi, Javad, and Sasan Mohsenzadeh. "Rapid, green, and eco-friendly biosynthesis of copper nanoparticles using flower extract of Aloe vera." *Synthesis and Reactivity in Inorganic, Metal-Organic, and Nano-Metal Chemistry* 45, no. 6 (2015): 895–898.

Karpagavinayagam, P., and C. Vedhi. "Green synthesis of iron oxide nanoparticles using Avicennia marina flower extract." *Vacuum* 160 (2019): 286–292.

Kumar, Harsh, Kanchan Bhardwaj, Kamil Kuča, Anu Kalia, Eugenie Nepovimova, Rachna Verma, and Dinesh Kumar. "Flower-based green synthesis of metallic nanoparticles: Applications beyond fragrance." *Nanomaterials* 10, no. 4 (2020): 766.

Kwon, IlKeun, Sang Cheon Lee, Bumsoo Han, and Kinam Park. "Analysis on the current status of targeted drug delivery to tumors." *Journal of Controlled Release* 164, no. 2 (2012): 108–114.

Lakshmeesha, Thimappa Ramachandrappa, Naveen Kumar Kalagatur, Venkataramana Mudili, Chakrabhavi Dhananjaya Mohan, Shobith Rangappa, Bangari Daruka Prasad, Bagepalli Shivaram Ashwini et al. "Biofabrication of zinc oxide nanoparticles with Syzygiumaromaticum flower buds extract and finding its novel application in controlling the growth and mycotoxins of Fusarium graminearum." *Frontiers in Microbiology* 10 (2019): 1244.

Lekha, D. Chandra, R. Shanmugam, K. Madhuri, L. Priyanka Dwarampudi, Mahendran Bhaskaran, Deepak Kongara, Jule Leta Tesfaye, N. Nagaprasad, V. L. Bhargavi, and Ramaswamy Krishnaraj. "Review on silver nanoparticle synthesis method, antibacterial activity, drug delivery vehicles, and toxicity pathways: Recent advances and future aspects." *Journal of Nanomaterials* 2021 (2021): 1–11.

Maheshwaran, G., A. Nivedhitha Bharathi, M. Malai Selvi, M. Krishna Kumar, R. Mohan Kumar, and S. Sudhahar. "Green synthesis of Silver oxide nanoparticles using ZephyranthesRosea flower extract and evaluation of biological activities." *Journal of Environmental Chemical Engineering* 8, no. 5 (2020): 104137.

Maheshwari, J. K. *Ethnobotany and Medicinal Plants of Indian Subcontinent*. Jodhpur, India: Scientific Publishers, 2019.

Mapala, K., and M. Pattabi. "Mimosa pudica flower extract mediated green synthesis of gold nanoparticles." *NanoWorld Journal* 3, no. 2 (2017): 44–50.

Mata, Rani, Jayachandra Reddy Nakkala, and Sudha Rani Sadras. "Catalytic and biological activities of green silver nanoparticles synthesized from Plumeriaalba (frangipani) flower extract." *Materials Science and Engineering: C* 51 (2015): 216–225.

Mittal, Amit Kumar, Abhishek Kaler, and Uttam Chand Banerjee. "Free radical scavenging and antioxidant activity of silver nanoparticles synthesized from flower extract of Rhododendron dauricum." *Nano Biomedicine & Engineering* 4, no. 3 (2012): 118–124.

Mittal, Amit Kumar, and Uttam Chand Banerjee. "Current status and future prospects of nanobiomaterials in drug delivery." In *Nanobiomaterials in Drug Delivery*, pp. 147–170. William Andrew Publishing, 2016.

Mittal, Amit Kumar, Debabrata Tripathy, Alka Choudhary, Pavan Kumar Aili, Anupam Chatterjee, Inder Pal Singh, and Uttam Chand Banerjee. "Bio-synthesis of silver nanoparticles using Potentillafulgens Wall. ex Hook. and its therapeutic evaluation as anticancer and antimicrobial agent." *Materials Science and Engineering: C* 53 (2015): 120–127.

Mittal, Amit Kumar, Jayeeta Bhaumik, Sanjay Kumar, and Uttam Chand Banerjee. "Biosynthesis of silver nanoparticles: Elucidation of prospective mechanism and therapeutic potential." *Journal of Colloid and Interface Science* 415 (2014): 39–47.

Mittal, Amit Kumar, Yusuf Chisti, and Uttam Chand Banerjee. "Synthesis of metallic nanoparticles using plant extracts." *Biotechnology Advances* 31, no. 2 (2013): 346–356.

Muthu, Karuppiah, and Sethuraman Priya. "Green synthesis, characterization and catalytic activity of silver nanoparticles using Cassia auriculata flower extract separated fraction." *Spectrochimica Acta Part A: Molecular and Biomolecular Spectroscopy* 179 (2017): 66–72.

Nagajyothi, P. C., Seong-Eon Lee, Minh An, and Kap-Duk Lee. "Green synthesis of silver and gold nanoparticles using Lonicera japonica flower extract." *Bulletin of the Korean Chemical Society* 33, no. 8 (2012): 2609–2612.

Nasrollahzadeh, Mahmoud, Mohaddeseh Sajjadi, Jaber Dadashi, and Hossein Ghafuri. "Pd-based nanoparticles: Plant-assisted biosynthesis, characterization, mechanism, stability, catalytic and antimicrobial activities." *Advances in Colloid and Interface Science* 276 (2020): 102103.

Nasrollahzadeh, Mahmoud, S. Mohammad Sajadi, Mohaddeseh Sajjadi, and Zahra Issaabadi. "An introduction to nanotechnology." In *Interface Science and Technology*, vol. 28, pp. 1–27. Elsevier, 2019.

Padalia, Hemali, Pooja Moteriya, and Sumitra Chanda. "Green synthesis of silver nanoparticles from marigold flower and its synergistic antimicrobial potential." *Arabian Journal of Chemistry* 8, no. 5 (2015): 732–741.

Painuli, S., P. Semwal, A. Bacheti, R.K. Bachheti & A. Husen. Nanomaterials from nonwood forest products and their applications. In Husen A., Jawaid M. (eds) *Nanomaterials for Agriculture and Forestry Applications*, pp. 15–40. Elsevier, 2020. https://doi.org/10.1016/B978-0-12-817852-2.00002-0.

Patil, Maheshkumar Prakash, Rahul Dheerendra Singh, Prashant Bhimrao Koli, Kalpesh Tumadu Patil, Bapu Sonu Jagdale, Anuja Rajesh Tipare, and Gun-Do Kim. "Antibacterial potential of silver nanoparticles synthesized using Madhucalongifolia flower extract as a green resource." *Microbial Pathogenesis* 121 (2018): 184–189.

Rabanel, Jean-Michel, Vahid Adibnia, Soudeh F. Tehrani, Steven Sanche, Patrice Hildgen, Xavier Banquy, and Charles Ramassamy. "Nanoparticle heterogeneity: An emerging structural parameter influencing particle fate in biological media?." *Nanoscale* 11, no. 2 (2019): 383–406.

Radad, Khaled, Mubarak Al-Shraim, Rudolf Moldzio, and Wolf-Dieter Rausch. "Recent advances in benefits and hazards of engineered nanoparticles." *Environmental Toxicology and Pharmacology* 34, no. 3 (2012): 661–672.

Rao, B. Lakshmeesha, G. Parameshwara Gouda, and C. S. Shivananda. "Green synthesis of silver nanoparticles using Hibiscus Rosa Sinensis flower extract." In *AIP Conference Proceedings* 2220, no. 1 (2020): 020103. AIP Publishing LLC.

Remya, V. R., V. K. Abitha, P. S. Rajput, A. V. Rane, and A. Dutta. "Silver nanoparticles green synthesis: A mini review." *Chemistry International* 3, no. 2 (2017): 165–171.

Singh, Priyanka, YeonJu Kim, Chao Wang, Ramya Mathiyalagan, and Deok Chun Yang. "The development of a green approach for the biosynthesis of silver and gold nanoparticles by using Panax ginseng root extract, and their biological applications." *Artificial Cells, Nanomedicine, and Biotechnology* 44, no. 4 (2016): 1150–1157.

Suttee, Ashish, Gurpal Singh, Nishika Yadav, Ravi PratapBarnwal, Neha Singla, Kirti S. Prabhu, and Vijay Mishra. "Drug eluting stent: A review on status of nanotechnology in pharmaceutical sciences." *International Journal of Drug Delivery Technology* 9, no. 1 (2019): 98–103.

Tetgure, Sandesh R., Amulrao U. Borse, Babasaheb R. Sankapal, Vaman J. Garole, and Dipak J. Garole. "Green biochemistry approach for synthesis of silver and gold nanoparticles using Ficusracemosa latex and their pH-dependent binding study with different amino acids using UV/Vis absorption spectroscopy." *Amino Acids* 47, no. 4 (2015): 757–765.

Veisi, Hojat, Bikash Karmakar, Taiebeh Tamoradi, Reza Tayebee, Sami Sajjadifar, Shahram Lotfi, Behrooz Maleki, and Saba Hemmati. "Bio-inspired synthesis of palladium nanoparticles fabricated magnetic Fe 3 O 4 nanocomposite over Fritillariaimperialis flower extract as an efficient recyclable catalyst for the reduction of nitroarenes." *Scientific Reports* 11, no. 1 (2021): 1–15.

Velmurugan, Palanivel, Min Cho, Sung-Sik Lim, Sang-Ki Seo, Hyun Myung, Keuk-Soo Bang, Subpiramaniyam Sivakumar, Kwang-Min Cho, and Byung-Taek Oh. "Phytosynthesis of silver nanoparticles by Prunusyedoensis leaf extract and their antimicrobial activity." *Materials Letters* 138 (2015): 272–275.

Velsankar, K., R. Preethi, P.S. Jeevan Ram, M. Ramesh, and S. Sudhahar. "Evaluations of biosynthesized Ag nanoparticles via Allium Sativum flower extract in biological applications." *Applied Nanoscience* 10 (2020): 3675–3691.

13 Green and Cost-Effective Nanoparticles Synthesis from Medicinally Important Aquatic Plants and Their Applications

R. Priyadharshini, A. Abirami, and S. Rajeshkumar

CONTENTS

13.1 INTRODUCTION

The human brain and imagination, born out of huge dreams, has often given rise to new science and technology in the 21st century. Even though human exposure to nanoparticles (NPs) has occurred throughout human history, once the golden era of nanotechnology began in the 1980s (Nouailhat 2010), the science of nanotechnology was advanced further when Iijima and another Japanese scientist developed carbon nanotubes. In the 21st century, there is increased interest in the different types of the newer branches and fields in nanoscience and nanotechnology (Bayda et al. 2019). Nowadays, the national technology initiative group manages the overall framework, the US President's cabinet-level National Science and Technology Council (NSTC) and related technology (National Research Council et al. 2010).

DOI: 10.1201/9781003213727-13

Nanotechnology has shown extreme results on medical equipment like drug delivery systems (DDS), diagnostic biosensors, and biomarkers along with imaging probes. Nanomaterials are now widely used in the food and cosmetics industries to improve production, packaging, shelf life, and bioavailability (Kamath, Nasim, and Rajeshkumar 2020). As a result of these issues, new scientific disciplines such as nanotoxicology and nanomedicine have emerged. Nanotoxicology studies the potential adverse health effects of NPs (Kumar et al. 2021).

Nanotechnology (in Greek nano is 'dwarf') is a multidisciplinary field focused on creating machines and devices in the nanoscale which measure around half the width of human DNA or one billionth (10^{-9}) (Mansoori and Soelaiman 2005). Many researchers believe that nanotechnology has a major impact on medicinal applications for the human race. Its revolutionary role at the molecular level brought benefits in nano diagnostics, local drug delivery, proteomics, genomics, and molecular biology in nanomedicine. Nanotechnology helps identify host cell pathogens and in vivo repair-and-replace host cells (Patil et al. 2008).

An overview of nanotechnology concepts entitled 'There is plenty of room at the bottom' was put forth by Richard P. Feynman, Nobel Laureate of the American Physical Society, in December 1959 (Feynman 1960). Taniguchi was the first to use the term 'nanotechnology' in 1974 (Taniguchi 1974)

13.2 NANOPARTICLES

NPs are a broad category of materials that include particulate compounds with at least one dimension of less than 100 nm. It depends on the overall shape of these materials in one dimension, two dimensions, and three dimensions. Researchers discovered that size could alter a substance's physiochemical qualities, such as its optical properties, emphasizing the significance of these materials (Nagumo and Yao 2021). The colours of 20-nm gold (Au), platinum (Pt), silver (Ag), and palladium (Pd) NPs are, respectively, wine-red, yellowish-grey, black, and dark-black (Bryan et al. 2016).

NPs are not simple molecules; they are made up of: (i) the surface layer, which can be functionalized with a variety of small molecules, metal ions, surfactants, and polymers; (ii) the interior layer, which can be functionalized with a variety of small molecules, metal ions, surfactants, and polymers; (iii) the shell layer, which is chemically and physically distinct from the core; and (iv) the core, which is the NP's central component and commonly refers to the NP itself. Depending on their morphology, size, and chemical characteristics, NPs are classified into several groups. Some of the most well-known classes of NPs are listed below, based on physical and chemical features (Chung et al. 2019).

Silver Nanoparticles: Silver nanoparticles (AgNPs) are a type of material with diameters ranging from 1 to 100 nm. The popularity of the AgNPs is studied in relation to their diverse properties (Cepoi et al. 2015). Because of their distinct characteristics, their behaviour has recently increased. Their physic, chemical, and biological properties, these have been shown to be the most effective due to their potent antibacterial properties (Bhuyar et al. 2020). Bacteria, viruses, and other eukaryotic microorganisms, of all the nanomaterials available, antimicrobial agents are the most common application for the sunscreen creams, water purification, and textiles industries. Toxicity, surface plasmon resonance, and electrical resistance are all recognized to be distinctive features of AgNPs (Korbekandi et al. 2014; Bhuyar et al. 2020). On the basis of these findings, extensive research has been carried out to determine their qualities and potential for a variety of applications, including antimicrobial agents in wound dressings, anticancer agents, electronic devices, and water treatment (Korbekandi et al. 2014).

Gold Nanoparticles: Gold (Au) has been used for therapeutic purposes from ancient Chinese medicine to modern medicine. Gold nanoparticles (AuNPs) are frequently used as desirable materials in many sectors because of their unique optical and physical features, such as surface plasmon oscillations. Many advancements in biological applications have recently been made in disease diagnostics and biocompatibility therapeutics. Many functionalizing agents, such as polymers, surfactants, ligands, dendrimers, medicines, DNA, and others, could be used to make AuNPs. RNA, proteins, peptides, and oligonucleotides are all examples of macromolecules. AuNPs could be made and used (Bongiovanni et al. 2021). The utilization of AuNPs and surface functionalization with a wide range of materials was discussed in this paper. They are targeting pharmaceuticals for photothermal treatment with molecules, increasing and enhancing AuNPs, lower cytotoxic effects in a variety of malignancies, gene therapy, and a variety of other illnesses. AuNPs would be a viable drug delivery mechanism and therapy (Scroccarello et al. 2021). Intensive research on AuNPs was carried out, taking advantage of their unique features, although a few negative toxic consequences of AuNPs were discovered, necessitating further investigation (Patra et al. 2020).

The NPs are generally classified into organic, inorganic, and carbon-based.

13.3 ORGANIC NANOPARTICLES

Organic NPs or polymers are generally known as dendrimers, micelles, liposomes, and ferritin. Some NPs, such as micelles and liposomes, have a hollow core, sometimes known as nanocapsules, and are sensitive to thermal and electromagnetic radiation such as heat and light (Voliani 2020). Organic NPs are frequently employed in biomedical fields, such as medication delivery systems, since they are effective and may be injected into specific body regions, a process known as targeted drug delivery (Geng et al. 2021). Apart from their usual properties like size, composition, surface shape, etc., their drug-carrying capacity, stability, and delivery systems, whether the entrapped drug or adsorbed drug system, define their area of application and efficiency (Prabaharan and Mano 2004; Geng et al. 2021).

13.4 INORGANIC NANOPARTICLES

Non-carbon NPs are known as inorganic NPs. Inorganic NPs are made up of metal and metal oxide-based NPs. (i) Metal-based NPs are synthesized from metals to nanometric sizes using destructive or constructive processes (Gittins and Caruso 2002). Almost all metals have NPs that can be synthesized. Aluminium (Al), cadmium (Cd), cobalt (Co), copper (Cu), gold (Au), iron (Fe), lead (Pb), silver (Ag), and zinc (Zn) are the most widely employed metals for NP synthesis. NPs have unique characteristics such as sizes ranging from 10 to 100 nm, surface characteristics such as high surface area to volume ratio, pore size, surface charge, and surface charge density, crystalline and amorphous structures, spherical and cylindrical shapes, and colour, as well as reactivity and sensitivity to environmental factors such as air, moisture, heat, and sunlight (Thota and Crans 2018). (ii) Metal oxide-based NPs are synthesized to change the properties of their respective metal-based NPs. For example, iron nanoparticles (FeNPs) quickly oxidize to iron oxide (Fe_2O_3) in the presence of oxygen at ambient temperature, increasing their reactivity. The improved reactivity and efficiency of metal oxide NPs are the fundamental reason for their creation (Badar and Ullah Khan 2020).

13.5 CARBON-BASED NANOPARTICLES

Carbon-based NPs are those that are entirely formed of carbon. Occasionally activated carbon at nano-size can all be classified as fullerenes, graphene, carbon nanotubes (CNT), carbon nanofibres, and carbon black (Bai et al. 2021).

13.5.1 Application in Different Fields

NPs are used or are being evaluated for use in many fields. The list below introduces a few of the uses under development (Figure 13.1).

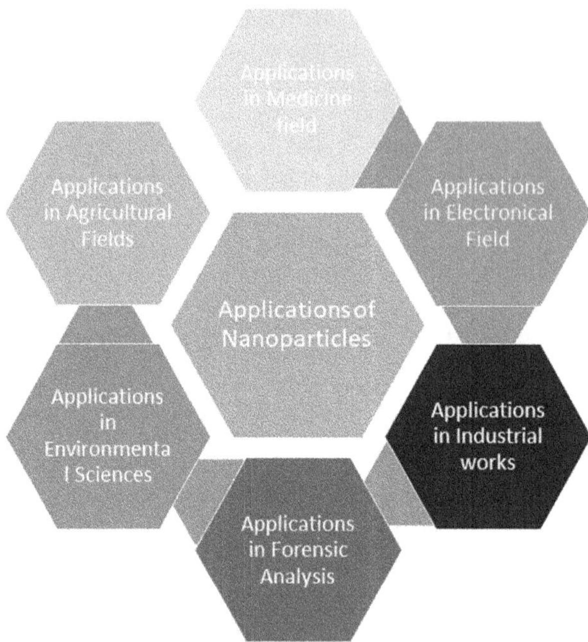

FIGURE 13.1 Applications of nanoparticles in various divisions.

13.5.1.1 Applications in Medicine

Thanks to nanotechnology, the use of NPs in drug delivery has benefited the medical field. NPs can be used to deliver medicine to specific cells. By placing the drug in the desired area and the required dosage, drug consumption, and adverse effects are considerably reduced (Mody et al. 2013). This technique saves time and money, cost, and side effects. Tissue engineering is reproducing and repairing damaged tissue (Malam, Lim, and Seifalian 2011). Tissue engineering can be used to replace traditional treatments like artificial implants and organ transplants (Zhang et al. 2010). The development of carbon nanotube scaffolds in bones is one such example.

13.5.1.2 Applications in Electronics

NPs with increased electrical conductivity are utilized to detect gases like NO_2 and NH_3. This is due to an increase in NP pores due to charge transfer from NPs to NO_2 as gas molecules bind them together, making them better gas sensors (Nunes et al. 2018). Separator plates in batteries should be made of NPs. Because of their foam-like (aerogel) shape, they can store significantly more energy than regular batteries. Because of their enormous surface area, batteries composed of nanocrystalline nickel and metal hydrides require less recharging and last longer (Nunes et al. 2018; Ahmad and Ezema 2019).

13.5.1.3 Applications in the Environment

For more than a decade, environmental remediation utilizing NPs, or nanoremediation, has effectively treated or decontaminated the air, water, and soil (Ciampi et al. 2021). Nanoremediation is one of the most successful options because it provides in situ treatment, eliminating the need to pump groundwater for treatment and excavate to reach the desired location. NPs are injected into the appropriate site and carried along with the groundwater flow, decontaminating the water by immobilizing pollutants (Seng et al. 2020; Ciampi et al. 2021).

13.5.2 Green Synthesis of Nanoparticles

Green synthesis has advantages over physical and chemical synthesis. It is less expensive, more environmentally friendly, and easier to scale up for large-scale synthesis because it does not require high pressure, temperature, energy, or harmful chemicals (Husen et al. 2019; Bachheti et al. 2019a; Painuli et al. 2020; Bachheti et al. 2020, 2020a, 2020b, 2021, 2022)

13.5.3 Biosynthesis of Nanoparticles: A Green Map

Various procedures for the synthesis of nano and microsized inorganic materials have contributed to establishing a relatively new and mostly untapped area of research centred on NP biosynthesis. Environment-friendly, non-toxic, and safe reagents are used in the green synthesis of NPs (Barzinjy et al. 2020). Phytoremediation in plants may be linked to the mechanism of NP manufacture in plants (Sanjay 2019) (Figure 13.2).

1. **Plants:** Due to the terrible implications that the world is facing and the limited time available to find effective answers, green chemistry has been developed as an alternative to the use of environmentally hazardous processes and products. For a greener synthesis of NPs, extracts from the neem (*A. indica*) plant were employed. The main benefit of utilizing neem leaves is that they are a widely available medicinal plant with antibacterial properties (Ijaz, Zafar, and Iqbal 2021). The synthesis of AuNPs was validated by colour changes and described by ultraviolet UV-visible (UV-Vis) spectroscopy using neem leaf extract. Surgical masks coated with AgNPs were found to have antibacterial effects in studies. After incubation, NP-coated masks reduced the number of viable *E. coli* and *S. aureus* cells by 100% (Ranu and Chattopadhyay 2009). Furthermore, none of the people who used the masks showed any signs of skin irritation, according to the study. The bactericidal action of aqueous extract of neem leaves (20% w/v) and 0.01 M silver nitrate ($AgNO_3$) solution in a 1:4 mixing ratio on cotton cloth against *E. coli* was investigated using the disc diffusion method (Nangare and Patil 2020).
2. **Bacteria:** Bacteria are one of the most important and necessary components of nature's ecosystem, performing important macro-level work at the micron scale. They are possibly the most versatile living species on Earth, having adapted to various environments ranging from high mountains to deep oceans, frigid ice areas to high-temperature volcanic zones (Soleimani and Habibi-Pirkoohi 2017). Bacteria can be found in the form of single

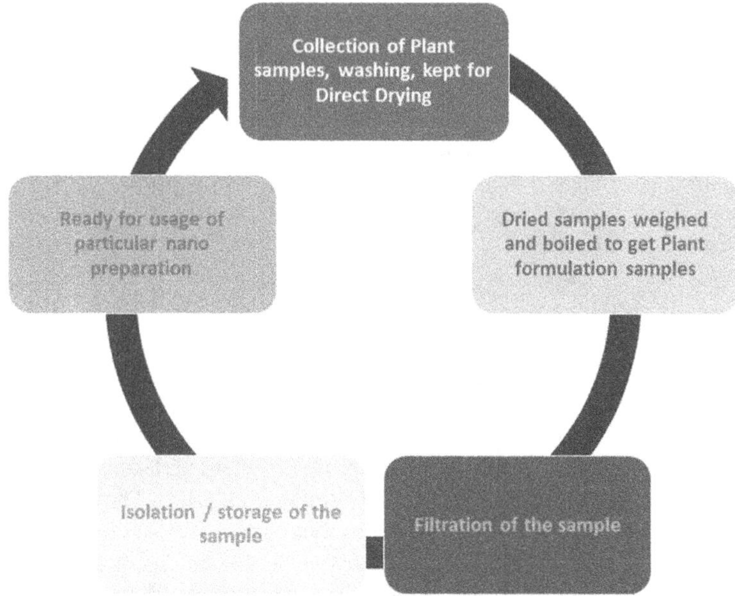

FIGURE 13.2 Green synthesis of nanoparticles.

cells (planktonic cells) or biofilms. Several NPs have been demonstrated to be more effective than traditional antibiotics at killing bacteria and removing biofilms (Singh, Kaushal, and Sodhi 2020). Antimicrobial NPs are classified as (i) metallic NPs (such as AuNPs and AgNPs), (ii) polymeric NPs (such as chitosan NPs), (iii) carbon-based NPs (such as graphene nanosheets), (iv) lipidic NPs (such as liposomes), (v) non-metallic inorganic NPs (such as silica (e.g., albumin NPs) (Pascu et al. 2021). AuNPs and AgNPs work against bacteria in a variety of ways. Their bacteriostatic/bactericidal effect may or may not be specific against a molecular target, but they largely demonstrate a summation of multiple molecular events that tend to block, alter, or disrupt diverse targets in planktonic bacteria and biofilms (Graves 2021; Pascu et al. 2021).

3. **Fungi:** Fungi are eukaryotic organisms that live in various ordinary environments and usually form decomposer organisms. Only roughly 70,000 species of fungi have been identified out of an estimated 1.5 million species on the planet. According to more recent research, around 5.1 million fungal species are identified using high-throughput sequencing technologies. The importance of investigating the role of fungi in nanobiotechnology is emphasized. Because of their tolerance and metal bioaccumulation ability, fungi have received much attention in researching the biological creation of metallic NPs (Jogee and Rai 2020; Naik and Shankar Naik 2020). The ease with which fungi can be scaled up is a distinct advantage of using them in NP manufacturing (for example, using a thin solid substrate fermentation approach). Because fungi are excellent secretors of extracellular enzymes, large-scale synthesis of enzymes is possible (Jogee and Rai 2020).
4. **Actinomycetes:** For their saprophytic behaviour and the generation of numerous bioactive secondary metabolites and extracellular enzymes, actinomycetes are recognized as superior groupings among commercially important microbial species. Actinomycetes can make NPs both inside and outside the cell (Sapkal and Sapkal 2017). The electrostatic attachment of Ag^+ ions to the negatively charged carboxylate groups in the enzyme located on the cell wall of mycelia causes intracellular production on the mycelia's surface. Silver nuclei are formed when the Ag^+ ions are reduced by enzymes in the cell wall (Sapkal and Sapkal 2017). Actinomycetes are well known for their ability to create antibiotic-like secondary metabolites/new chemical entities. Since actinomycetes are the principal source of antibacterial chemicals, there have been multiple reports of metal and metal oxide NP production and antimicrobial activity mediated by actinomycetes (Sapkal and Sapkal 2017; Kalaba et al. 2021).
5. **Bioactive compounds:** Nanotechnology is a branch of study that has been hailed as a promising method for preventing and treating a variety of human health problems. This has been emphasized as a therapeutic delivery mechanism for various pathophysiological circumstances. The industry has used this technology to look for new oral administration options based on bioactive component solubility properties that have been altered (Bolade, Williams, and Benson 2021).

13.6 MEDICINALLY IMPORTANT AQUATIC PLANTS

Aquatic plants comprise a diverse ecosystem with 168 genera under 87 families and 1022 species. Aquatic plants play an initial role in landscaping and the purification of water. Also have a role in bioenergy, biomass, and genomics (Hu et al. 2017). Aquatic plants are natural absorbers of contaminants over larger areas, and it is a resourceful and profitable method. Extensive absorption of contaminants is stored in their shoots and root portion (Pratas et al. 2014). They are grouped based on their pharmacological and ethnobotanical properties. Sculthorpe's types of aquatic plants are free-floating, water lettuce, water hyacinth, salcinia, duckweed, and emergent and submerged aquatic plants (Simonsen 1968; Ali et al. 2020). Major medicinal aquatic plants that prefer to survive in water include *Acorus calamus* L. (Araceae; sweet flag), *Bacopa monnieri* L. *Pennell* (Plantaginaceae),

Alternanthera sessilis L. (Amaranthaceae; sessile joyweed), *Centipeda minima* A. Braun, and *Ascheron* (Asteraceae; spreading sneezeweed *Enhydra fluctans* Lour. (Asteraceae; water cress), *Hedychium coronarium* J. Koenig (Zingiberaceae; white ginger lily), *Nelumbo nucifera* Gaertn. (Nelumbonaceae; sacred lotus), *Rotula aquatica* Lour. (Boraginaceae; aquatic rotala), *Nymphaea nouchali* Burn. F. (Nymphaeaceae; blue water lily), *Pistia stratiotes* L. (Araceae; water lettuce), *Centella asiatica* L. Urban (Apiaceae; centella), *Marsilea minuta* L. (Marsileaceae; dwarf water clover), *Ipomea aquatica* Forssk. (Convolvulaceae; water spinach). (Aasim et al. 2019; Arya et al. 2022). Flavones, flavanols, flavonoids, phenolic polyphenols, coumarins, tannin, terpenoids, lectins, alkaloids, and polypeptides compounds are the principal substances that avail its antimicrobial (Abreu et al. 2012), anticancer (Chaudhary et al. 2011), antibiotic, antifouling, antioxidant (Rai et al. 2006), anti-inflammatory, antimitotic, and cytotoxic activity (Goud et al. 2009)

Aquatic plants live both in marine environments and freshwater. These are phytoplankton or otherwise called macrophytes/macroalgae. They are categorized under unicellular algae, such as bryophytes, macroalgae, angiosperms, and pteridophytes which accumulate under aquatic conditions.

13.6.1 Medical Applications of Aquatic Plants

Bacopa monnieri L. Pennell is an aquatic herb called 'Brahmi' used in folk medicine as nootropics for cognitive processing, attention, memory, monoaminergic, and cholinergic functions of the brain. It interacts with the nervous system as an adaptogen in promoting neuronal communication. It acts as an antioxidant, anti-inflammatory, antimicrobial, and enhances the pharmacological effects on endocrine and gastrointestinal activities (Saha et al. 2020). The leaves and roots of the plant involved in the degradation of dyes reveal the phytoremediation ability of the (Shanmugam, Ahire, and Nikam 2020) *Alternanthera sessilis* plant, otherwise called the noxious weed of the Amaranthaceae family found in subtropical, subtropical areas, North and South America, is a whitish linear, bisexual herbaceous plant growing in sunlight and warm conditions. Noxious weed is rich in proteins, vitamins, carbohydrates, fibres, starch, calcium, amino acids, and phosphorus. It shows medicinal activity against hepatitis, tight chest, bronchitis, asthma, and lung disease. It also enhances antiulcer, antipyretic, antioxidant, antimalarial, prophylactic, antidiarrheal, and anti-inflammatory activity (Walter, Merish, and Tamizhamuthu 2014). *Centipeda minima* A. is a medicinal herb found in the Asian tropical area. Flavonoids, terpenoids, amides, fatty acids, and phenols are the secondary metabolites that enhance the medicinal effect against malaria, nasal allergy, cough, and asthma. It has biological outcomes like antiviral, anticancer, anti-inflammatory, antibacterial, and hepatoprotective effects (Linh et al. 2021; Huang et al. 2013). *Enhydra fluctuans* (helencha) is a herbaceous tropical herb belonging to the Asteraceae family. The herbal plant is sessile with oblong leaves, about 2–3 inches. The leaves are bitter in taste, which cures skin lesions, bronchitis, and are used as laxatives, leukoderma, nervous affection, and for treating smallpox (Ali et al. 2013). The phytochemical constituents of *Enhydra fluctuans* ethanolic crude extract are evidenced against antioxidant, antimicrobial, antidiarrhoeal, analgesic, cytotoxic, and hepatoprotective activity (Kamal et al. 2019). The perennial herb *Hedychium coronarium* J. Koenig distributed worldwide was used as a traditional perennial herbal drug. It is a tall, vigorous stem that is erect, hard, branched weed up to 5–6 feet tall. Its bioactive secondary metabolites include saponins, flavonoids, phenols, glycosides, volatile oils, etc., and it has medicinal properties that can be used against headache, bodyache, conjunctivitis, antidiabetic, rheumatism, swelling, vomiting, and snake bites (Pachurekar and Dixit 2017). *Nelumbo nucifera* Gaertn. is an aquatic flowering plant belonging to the Nelumbonaceae family. Water lily and sacred or Indian lotus are the common names used for this perennial plant. Well known in traditional medicine from ancient times for its pharmacological effects and antidiabetic, anti-inflammatory activity from triterpenoids of the rhizome extract. It also enhances antiobesity, anticancer, and antiangiogenic activity. Lotus embryos (seeds) help overcome insomnia, nervous disorder, arrhythmia, and hypertension (Paudel and Panth 2015). The leaf extract is effective against cholera, hyperdipsia, and diarrhoea. The medicinal properties are emphasized by its nuciferine component and other phytochemicals like glycosides and flavonoids. (Tungmunnithum, Pinthong, and Hano 2018). Rhodophyta or red seaweed are macroalgae in a group of 7000 species. It is a rich source of polysaccharides, vitamins, pigments, minerals, polyunsaturated fatty acids, and other phenolic compounds like bromophenols, phenolic acids, and flavonoids (Ismail, Alotaibi, and El-Sheekh 2020). The majority of these contents showed a promising role as antiviral, antitumour, antioxidant, antimicrobial, antidiabetic, anti-inflammatory, anticoagulant, analgesics and antiallergic agents (Abu-Khudir, Ismail, and Diab 2021) (Gómez-Guzmán et al. 2018). *Rotula aquatica* Lour., a small shrub branched among pebbles and rocks from the Boraginaceae family, otherwise called asanabheda. Tannins and polyphenolic compounds have hypoglycaemic, antibacterial, anti-inflammatory, anthelmintic, and antiurolithic properties (Bency et al. 2017). Pashanbhed is an essential component in traditional medicinal treatment against bladder and kidney stones, ulcers, skin and venereal diseases like syphilis, uterine disorders, and piles (Vysakh et al. 2016). *Nymphaea nouchali* Burn. belongs to the Nymphaeaceae family, is an aquatic perennial herb, and its crude extract has an antioxidant property, is antidiabetic, reduces inflammation, menorrhagia, urinary and liver disorders, is an aphrodisiac, and treats menstruation issues (Parimala and Shoba 2013). In folk medicine, it has its phytochemical constituents of flowers like flavones, gallic acid, quercetin, astragalin, phenols, tannins, saponins, glycosides, steroids, and alkaloids kaempferol (Raja, Sethiya and Mishra 2010). It has been reported that nymphaea flowers act against CCl_4 instigated liver damage (Das et al. 2012).

Pistia stratiotes L. belonging to the Araceae family, is floating aquatic lettuce with fan-shaped green leaves which is bitter in taste and odourless. It has antidermatophytic, antitubercular, antiseptic, antinociceptive, and antidysentery properties (Rahman et al. 2011). It also treats piles, ulcers, eczema, leprosy, and stomach disorder with its laxative capacity (Khan et al. 2014). It acts as a calcium channel blocker to decrease blood pressure (Achola, Indalo, and Munenge 1997). *Centella asiatica* L. Urban (gotu kola or Indian pennywort) is a native aquatic plant of South East Asia, Pakistan, Madagascar, Sri Lanka, South Africa, and Europe. It is used as a neuroprotective agent against neurotoxic dopamine in Parkinson's disease, warding off the formation of amyloid plaques in Alzheimer's disease and reducing oxidative stress (Orhan 2012; Hamidpour et al. 2015). This plant is mainly used for wound healing. It also has a role in treating skin lesions like lupus, leprosy, eczema, diarrhoea, amenorrhoea, and varicose ulcers. It treats genitourinary disorders, helps in improving cognition, and decreases stress. It was also called the 'miracle elixir of life' in China 2000 years ago (Gohil et al. 2010). Raw Indian pennywort and neurotrophic factors induce mesenchymal cell differentiation into neural cell lineage and Schwann cell differentiation (Omar et al. 2019) (Prakash, Jaiswal, and Srivastava 2017). *Marsilea minuta* L. is a herbal aquatic plant which possesses analgesic, antitussive, antiamnaesic, antimicrobial, antifertility, hepatoprotective, and antioxidant activity (Sajini et al. 2019). The aquatic and semiaquatic *Ipomea aquatica* Forssk, a plant native to China and South East Asian countries, is a nutritious vegetable rich in vitamins like A, B, C , K, E, and U. Vitamin U is S methyl methionine and treats intestinal and gastric disorders. It is also rich in phytochemical components like glycosides, steroids, tannins, phenols, saponins, lipids, amino acids, and minerals etc., and in treating piles, headache, ringworm infestation, sleeplessness, intestinal and liver disorders, diabetes (Malalavidhane, Wickramasinghe, and Jansz 2000), abscesses, constipation (Samuelsson et al. 1992), mental illness, high blood pressure (Vent 1987), and nose bleeds. It is an effective antimicrobial agent against *Pseudomonas aeruginosa*, *E. coli*, and *Bacillus subtilis* (Figure 13.3).

13.6.2 Aquatic Plant-Mediated Nanoparticles

13.6.2.1 Macroalgae-Mediated Nanoparticle Synthesis

Marine algae or macroalgae are the substrata in coastal shoreline areas attached to sand, rock, undersea, and dead plants ranging from green, red, or brown algae (Bhuyar et al. 2021). Kim and

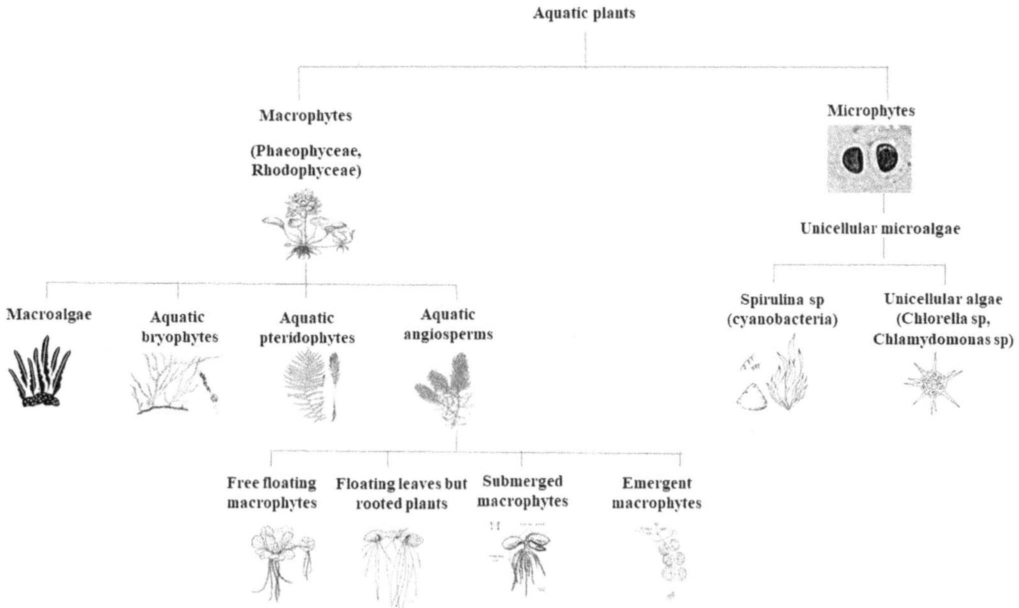

FIGURE 13.3 Classification of aquatic plants.

Wijesekara (2017) discussed that marine algae-based phlorotannins, sterols, carotenoids, and sulphated polysaccharides play a major role in anticancer activity. Bhuyar et al. (2020) synthesized aqueous-alcoholic extract-based AgNPs from *Padina* sp. Marine algae served as an effective antimicrobial agent against Gram-positive and -negative bacteria. The characterization of the synthesized AgNPs from marine algae is increased with a peak wavelength of 420–445 nm by UV-Vis spectrophotometer. X-ray spectroscopy confirmed AgNP purity with a percentage of 48.34% Ag. Field emission scanning electron microscope (FESEM) and scanning electron microscope (SEM) images were done to examine the morphology and size of NPs. The size of the AgNPs was uniform ranging between 25–60 nm. *Pseudomonas aeruginosa* and *Staphylococcus aureus* are susceptible with the zone of inhibition about 13.33 ± 0.76 mm and 15.17 ± 0.58 mm in diameter, which exhibits a potent antimicrobial activity (Bhuyar et al. 2020). *Sargassum muticum* aqueous microalgae-based green synthesis of AuNPs monitors as a capping or reductant agent. Growth of AuNP characterization was analysed by UV-Vis spectrophotometer during Au ion exposure to aqueous extract of *S. muticum* with no significant change in absorbance at 550 nm; transmission electron microscopy (TEM) with a mean particle size of 5.42 ± 1.18 nm. The capping surface of an anionic biocompound NPs with a measurement of −35.8 mV. Aqueous extract of *S. muticum* brown marine algae as a bioactive component reduces gold to AuNPs and is highly stable at ambient temperature and wide range pH (Namvar et al. 2015). Safaat et al. (2021) reported that the utilization of AgNP-sargassum is an excellent source of antibacterial and antilarvicidal properties. AgNPs eliminate aerosols, biological fluids of microorganisms, and their disease transmission. AgNPs about 30–80 nm in size penetrate the bacterial cell wall, followed by the release of silver ions, enacted with the thiol group attachment of proteins, accompanied by the elimination of DNA replication with the resultant inactivation of bacteria (Safaat, Tursiloadi, and Perisha 2021).

13.6.2.2 Aquatic Pteridophytes-Mediated Nanoparticle Synthesis

Antibacterial activity of aquatic pteridophytes is based on AgNP concoction by agar diffusion method against *K. pneumoniae*, *E. coli*, *Proteus vulgaris*, *Proteus vulgaris*, *Salmonella typhi*, *P. aeruginosa*, and *S. aureus*. The maximum ventures of *Acrostichum aureum* and *Cyclosorus interruptus* against pathogens are active and effective with AgNP-based synthesis (Pal 2012).

13.6.2.3 Aquatic Bryophyte-Mediated Nanoparticle Synthesis

Vimala et al. (2017) synthesized and characterized *Campylopus flexuosus*-based AgNPs. The characterization of the mixture showed a brownish-red colour change indicating AgNP formation. The resultant NP formation is characterized by Fourier-transform infrared (FTIR), UV-Vis, FESEM, zeta potential, X-ray diffraction (XRD) analysis, and particle size. UV-spectroscopy manifested the maximum level of absorption obtained at 436 nm. The average particle measurement of the NPs is 58 nm identified by XRD, FESEM, and FTIR analysis, which indicates the presence of carbonyl compound which acts as a capping reagent with the end result of the formation of NPs (Vimala et al. 2017).

Green synthesis of *Azolla pinnata*-based NPs and the efficacy of hydroethanolic *A. pinnata* extract concentration by Korbekandi et al. (2014) were done by drop coating onto the carbon- copper-coated grids by drop coating the NP silver suspensions for TEM preparation. The *A. pinnata* extract prepared was small with the size of 6.5 nm as the mean size and is seen as aggregates with biotransformation in the early time period. These NPs serve as antimicrobial and antiviral agents, especially in health, and have various medicinal usage. The potential NP is effective against *Pseudomonas aeruginosa, Pseudomonas aeruginosa, Vibrio cholerae, Bacillus subtilis, Escherichia coli, and Syphilis typhus* (Korbekandi et al. 2014).

Janthima et al. (2021) performed biogenesis of NPs by the root of *Azolla pinnata* plant cells. The plant is exposed to nickel (Ni) and iron (Fe) ions transformed into metal-based NPs after combined exposure of these metals for 12 h, where X-ray fluorescence revealed uptake of metals by the plant's roots. TEM revealed Ni and Fe NPs near the cell membrane, cell vacuoles, vascular cells, and cell walls of the roots. Electron diffraction analysis showed the existence of FeNPs as Fe_3O_4 and alpha Fe_3O_4 which perform the reducing activity through the enzymatic antioxidants (Janthima and Siri 2021).

Hajra and Mondal (2017) enacted a facile and easy method of AgNP biosynthesis where aquatic *Azolla pinnata* extract reduced $AgNO_3$ to AgNPs. A UV spectrophotometer detected the absorbance characteristic at 445 nm of the aquatic plant solution with salt solution. The capping NPs were unwrapped in SEM, TEM, and FTIR images which showed the spherical images of NPs (Hajra and Mondal 2017).

Rajagopal et al. (2021) evaluated *A. pinnata*'s catalytic attribute in antimicrobial, oxidation, and antioxidant through selenium NPs by the green synthesis method through X-ray photoelectron spectroscopy, scanning electron microscopy, XRD analysis, and FTIR analysis. The size of the crystalline structure of *A. pinnata*-based NPs was 36.45 nm, and they were spherical in shape. On oxidation to p-toluic acid yielded a percentage about 78.3%. At 500 µg/mL concentration maximum free scavenging activity was developed. Antimicrobial activity of *E. faecium, E. coli, C. albicans, and S. aureus* showed maximum inhibition zones of 15 ± 0.13, 17 ± 0.67, 15 ± 0.32, and 14 ± 0.93 mm with the formed Ap-SeNPs (Rajagopal et al. 2021).

Ravi et al. (2020) determined nanosynthesized *A. pinnata* AgNPs at different concentration efficacies against larvicide of the malarial parasite *Aedes aegypti* at six various concentrations were low mortality was recorded at 10 ppm with 7.5%, while 250 ppm was observed with 95% mortality at elevated concentration. After 24 h, a confidence interval of 95% was observed with LC_{95} and LC_{50} with 369.438 ppm and 121.570 ppm aiding its larvicidal property (Ravi et al. 2019).

Li et al. (2016) determined the extract of *Equisetum ramosissimum* biofunctional and protective activities against melanogenesis, melanoma, and oxidation. Inhibiting activity analysed by MTT assay (3-(4,5-dimethylthiazol-2-yl)-2,5-diphenyl tetrazolium bromide) exhibited a reduced melanoma growth with an inhibiting activity of three main melanoma A375.S2, A2058, and A375 with an inhibited production of melanin by the cells (Li et al. 2016).

Tayebee et al. (2018) evaluated *Equisetum arvense* as a catalyst to synthesize chromones. *E. arvense*, with its abundant silica resources as NPs, acts as an efficient nanohybrid catalyst (Tayebee et al. 2018).

Arokiyarahj et al. (2018) examined the efficacy of *Marsilea minuta* leaf methanol hexane extract and its food-borne bacterial pathogens, and its antioxidant activity. Effective against *Pseudomonas aeruginosa*, *Streptococcus mutans*, *and Streptococcus aureus* exhibited 17 mm of maximum inhibition. SEM images revealed distorted and swollen cells. *M. minuta* played an effective role in relieving oxygen stress by increasing cellular alkaline phosphatase, superoxide dismutase, lactate dehydrogenase, and protein leakage vandalizing the pathogens. Major chemical constituents, namely farnesol, esters, and benzoic acid, played an influential role against topoisomerase IV and TEM-72 protein (Arokiyaraj et al. 2018).

Bala and Sukhen et al. (2015) demonstrated *M. minuta*-mediated synthesis of AuNPs. Electron microscopy revealed the stable synthesized AuNPs of around ~25 nm. It enacted the catalytic property followed by the reduction to para-aminophenol from para-nitrophenol. AuNPs revealed better potency of antibacterial features against *S. aureus* and *E. coli* (Sukhen et al. 2015).

Balashanmugam et al. (2018) synthesized AuNPs via aqueous *Marsilea quadrifolia* extract. XRD, TEM, and energy dispersive x-ray (EDX) analysis revealed the spherical shape of AuNPs with red ruby colour, which signified AuNP formation. AuNPs are spherical and 20–40 nm in size. In UV-Vis spectroscopy, they attain their peak in surface plasmon resonance at 544 nm wavelength. Scavenging activity enabled excellent stability at 50 μg/mL. A549 and PA1 cell lines have been recorded using the in vitro cytotoxic activity with IC_{50} (inhibitory concentration) of about 52.015 μg/mL and 45.88 μg/mL. Hence, the biogenic extract is effective against ovarian and lung cancer (Balashanmugam, Mosachristas, and Kowsalya 2018).

Mani et al. (2019) delineated AgNP synthesis from *Marsilea quadrifolia.* The characterization of NPs was done by dynamic light scattering (DLS), TEM, UV-Vis spectroscopy, SEM, FTIR, and XRD. The XRD study revealed that the average size of the formed NP was 27 nm. The UV-Vis spectrum divulged the maximum absorption peak at 436 nm with AgNP formation. SEM and TEM revealed the spherical shape, an average size of about 19–30 nm, and agglomeration of NPs. FTIR revealed that phytochemical contents like carbohydrates, nitrogenous admixture, phenols, and proteins enact the stabilization and capping of NPs. Potent DLS study revealed an average formed AgNP size of about 62.1 nm and zeta potential value of −18.3 mV of the AgNPs. *M. quadrifolia* aqueous extract-based AgNPs exhibit vigorous inhibitory action against Gram-positive (*E. coli* and *S. aureus*) and Gram-negative bacterial pathogens. The maximum inhibition zone noted was about 16 mm at 60 μL concentration. It also evaluated the minimum inhibitory concentration against pathogens with the AgNP formation by *M. quadrifolia* with 46.90 ± 4.23 and 17.27 ± 2.66 against *S. aureus* and *E. coli* (Shukla and Iravani 2018).

Ramya Juliet et al. (2020) synthesized copper nanoparticles (CuNPs) from pteridophyte *M. quadrifolia* rhizome and antibacterial activity. The characterization of NPs is done by FTIR, XRD, UV-Vis spectroscopy, atomic force microscopy (AFM), and SEM. The XRD pattern revealed the crystalline structure of CuNPs with an average size of about 25.2 nm. UV-Vis spectra exhibited a band of absorption at 324 nm. SEM depicted the leaf-like pattern which confirms the biogenic formation of CuNP crystals. The AFM image displayed the gravel-like CuNP structure formation. Also, CuNPs were extremely toxic to *Streptococcus faecalis* and *Bacillus thuringiensis.* The FTIR measurements confirmed the presence of CuNPs with the functional group presence over stabilization and reduction of NPs (Ramya Juliet et al. 2020).

13.6.2.4 Aquatic Angiosperms-Based Nanoparticle Synthesis

An in vivo exposure of aquatic angiosperm interaction with AuNPs was hypothesized by Glenn et al. (2012). Submerged angiosperms like *Egeria densa*, *Myriophyllum simulans*, whereas free-floating angiosperms like *Azolla caroliniana* absorbed the water nutrients, which facilitated their absorption of NP suspension. Where each plant has been exposed to AuNPs at 250 μg/L of NPs for one day (24 h). Various investigations demonstrated the gold concentration of the different angiosperms. AuNPs were analysed at six points including 1, 3, 6, 12, 18, and 24 h through plasma-mass spectrometry. TEM root images of *M. simulans* absorbed 4 nm AuNPs and *A. caroliniana* roots absorbed 4–18 nm AuNPs. Henceforth, the study revealed that AuNP absorption is dependent on the species and size (Glenn, White, and Klaine 2012).

Anuradha et al. (2015) synthesized AuNPs from *Pistia stratiotes* L., an aquatic invasive weed effective in treating leprosy, ulcers, inflammation, eczema, throat, and stomach disorders. The presence of AuNPs was confirmed by XRD and EDAX study. UV-Vis spectra demonstrated the delineation after mixing the aquatic weed extract with gold solutions at room temperature with the rapid commencement of NP synthesis. The pistia weed's submerged and aerial parts resulted in the electron microscopic expression of spherical monodisperse AuNPs based on the extract and metal concentrations. Also, with a mixture of triangular, hexagonal, truncated, and pentagonal shaped NPs. The FTIR study enacted the reduction and stabilization of gold to AuNPs with the association of various polypeptide molecules (Anuradha, et al. 2015).

A study of aquatic free-floating angiosperms like *Ceratophyllum demersum* L.-based synthesis of zinc nanoparticles (ZnNPs) by Aravind et al. (2005) demonstrated the detoxification effects enacted by zinc, organic, and amino acid supplements. This experimental study performed with the metal supplements showed the oxidized and reduced estimated level of glutathione detoxification system along 200 μM Zn supplementation (Aravind and Prasad 2005).

Yuan et al. (2018) demonstrated the tolerance to the toxicity of AgNPs by enhancing *J. effusus* and *E. densa* by peroxidase and superoxide dismutase activities in invading free oxygen radicals. The results disclosed that membrane damage is higher in *E. densa* submerged plant than *J. effusus* with the rise in malondialdehyde concentrations, indicating that AgNPs exerted ample stress in macrophytic submerged angiosperms (Yuan et al. 2018).

13.6.2.5 Cyanobacteria (*Spirulina* sp.)-Based Nanoparticle Synthesis

Elkomy et al. (2020) synthesized AgNPs from *Phormidium formosum* cyanobacteria and it gained attention as a nontoxic, eco-friendly, and cost-effective biosynthetic agent. The chemical properties were analyzed by TEM, FTIR, and UV-Vis spectra. UV-Vis spectroscopy, where peak absorption is reached at 437 nm corresponding to AgNP resonance. Spherical shapes were obtained by TEM for AgNPs, with an average size of 1.83–26.15 nm. FTIR confirms the biocomponents of *P. formosum* which is responsible for NP synthesis. The outcome of the experiment confirmed that AgNP formation acts as a powerful microbial property (Elkomy 2020).

Algae and cyanobacterial species have the ability to absorb metallic ions. Cicci et al. (2017) synthesized AgNPs from *A. platensis* cyanobacteria cultures. The NP characterization is done based on the structure by SEM, stability, and size by nanosizer and composition by XRD. The study confirmed the spirulina had beneficial effects against microbes.

Cepoi et al. (2015) demonstrated the antiradical activity with the silver ion NP synthesis from *Nostoc linckia*, *Spirulina platensis*, and assessment, which revealed its protein, carbohydrate, phycobilin, and lipid components. The resulting analysis revealed the reduction of AgNPs after 24 h and 48 h with resultant phycobiliprotein, carbohydrate, and protein components during NP generation with decreased antiradical activity (Cepoi et al. 2015).

Hamouda et al. (2019) synthesized AgNPs from *Oscillatoria limnetic*. Characterization was done using FTIR, TEM, SEM, and UV-Vis spectroscopy. Quasi spherical NP formation was detected in TEM with a size range of 3.3–17.9 nm. Biosynthesized NP analysis by FTIR confirmed the presence of sulphur, phosphorous, and amino stabilizing agents and their attachment to silver. Henceforth, this study exhibited antibacterial effect against *B. cereus* and *E. coli* (multidrug-resistant bacteria) and the cytotoxic activity HCT-116 in colon cancer cell line and MCF-7 in breast carcinoma cell line with breast cancer cell line the inhibitory concentration of $IC5_0$–6.14 μg/mL and IC_{50}–5.36 μg/mL respectively (Hamouda et al. 2019).

Duong et al. (2016) analyzed the antipathogenic inhibitory effect against *M. aeruginosa* by synthesizing AgNPs. The characterization was observed with SEM and TEM with the resultant AgNP formed by the reduction method at ambient temperature with an average size of 10–18 nm with a peak in absorbance at 410 nm. The morphological and density change was observed in *M. aeruginosa* cells exposed to AgNPs for a period of 10 days, with a resultant decrease in inhibition of growth and cellular density around 98.7% (Duong et al. 2016).

AgNPs were synthesized using cyanobacteria extract and synthesized NPs were characterized by Fourier transform infrared spectroscopy, scanning electron microscope, transmission electron microscopy, and X-ray powder diffraction. Synthesized NPs were spherical with diameters ranging from 4.5 to 26 nm. AgNPs show effective anticancer against MCF-7, HepG2, and Caco-2 cancer cell lines and also show antimicrobial activities (Hamida et al. 2020).

Also, AgNPs were prepared from cyanobacteria, characterized by change in colour and their peak absorbance at 430–440 nm in UV-Vis spectra. TEM and SEM analysis revealed an oval shape with a particle size range of 23–34 nm. $AgNO_3$ reduction and formation are encouraged by FTIR with a peak range at 3400–3450 cm^{-1}. The zone of inhibition was measured as 4, 6.9, 3, 2 mm against *E. coli*, *S. paratyphi*, *S. aureus*, *K. pneumoniae* respectively. Cytotoxic activity of *C. turgidus*, *C. typicum* in MCF-7 with the IC_{50} range of 40.9 and 43.3 μg/mL and HepG2 cell line of breast carcinoma cell line with 55.7 and 20.8 μg/mL (Swetha et al. 2020).

Singh et al. (2020) synthesized AgNPs from *Leptolyngbya* sp. of cyanobacterium. This biosynthesis of AgNPs from cyanobacterium is a biocompatible, clean, and innocuous method. XRD, UV-Vis spectroscopy, EDS, and TEM characterized the NPs. AgNP formation was confirmed with the emergence of absorbance at 430 nm and colour change is observed in the UV-Vis spectrum. AgNPs appeared as crystalline structures with an average size range of 21–35 nm. An antibacterial effect was demonstrated with an inhibited growth at 10 mg/L^{-1} concentration against *E. coli* and *B. subtilis*. This study executed an alternative to chemical bactericides in pharmaceuticals (Singh, Kaushal, and Sodhi 2020).

Evaluation of the antibacterial and photocatalytic effects was demonstrated by Keskin et al. (2016) with the synthesis of AgNPs from cyanobacterium and characterized the NP formation with AT-FTIR which confirms the presence of proteins and its possibility as reducing agents, UV-Vis spectrophotometer where UV spectrum revealed peak resonance at 430–450 nm, SEM, TEM revealed the spherical shape of NP formation. Its photocatalytic property is enhanced with the reduction of methylene blue dye within 4 h confirming the ability of biosynthesized AgNPs (Keskin et al. 2016).

13.6.2.6 Unicellular Algae (*Chlorella* sp.)-Based Nanoparticle Synthesis

Annamalai et al. (2015) studied the synthesis of AuNPs from *Chlorella vulgaris*. The physical, optical, bactericidal, and chemical properties of AuNPs investigated its crystal structure, size, toxicity, and surface structure via TEM, SEM, FTIR, UV-Vis spectra, and XRD. AuNPs were spherical with a size range of 3–10 nm with that of phytochemical components like flavonoids, phenols, proteins, and peptides with a dual action of reduction of AuIII and its ability in inhibiting *S. aureus* and *C. albicans* (Annamalai and Nallamuthu 2015).

Soleimani et al. (2017) biosynthesized AgNPs from *Chlorella vulgaris* as powerful antimicrobial agents at the molecular level. The algal experiment was conducted with the suspension of algae incubated with $AgNO_3$ and its NP formation is then characterized by TEM, SEM, XRD, and EDS. Spherical rod-like NPs with a neural range of pH at alkaline level executed an inhibited *S. aureus* growth at 50 μL concentration and haemolysin alpha expresses an antagonist effect against pathogens (Soleimani and Habibi-Pirkoohi 2017).

Torabfam et al. (2020) synthesized AgNPs with the microwave mediated dried *C. vulgaris* extract and explained its antibacterial efficacy. The absorbance reached its peak at 431 nm in UV-Vis spectroscopy. Ten minutes of processing revealed highly stable spherical-shaped AgNPs and sizes of around 50 nm were obtained. FTIR analysis revealed carbonyl groups along with *C. vulgaris* and its significant role in NPs function. *C. vulgaris* along with formed AgNPs produced an effective antibacterial role at various concentration against *S. aureus*, *S. Typhi*, and *S. enterica* (Torabfam and Yüce 2020)

13.7 CONCLUSION

This chapter reviews nanotechnology, NP types, the application and biosynthesis of NPs, and medicinally important aquatic plants. Aquatic plants are abundantly available in our ecosystem with various biomedical applications. Aquatic plants in and around us have their own applications and potential role in exploration in medicine. The status of quality of aquatic plants can be improved by extensive systematic interpretation. The phytochemical metabolites of these aquatic plants based on NP synthesis, their availability, characteristics, and cost-effectiveness gain better therapeutic medical applications. The usage of cost-effective aquatic plants for the NP synthesis process is eco-friendly and more advantageous when compared to other green processes.

REFERENCES

Aasim, M. et al. (2019) 'Multiple uses of some important aquatic and semiaquatic medicinal plants', in: Ozturk M., Hakeem K. R. (eds) *Plant and Human Health, Volume 2: Phytochemistry and Molecular Aspects.* Springer International Publishing, pp. 541–577.

Abreu, A. C., McBain, A. J. and Simões, M. (2012) 'Plants as sources of new antimicrobials and resistance-modifying agents', *Natural Product Reports*, 29(9), pp. 1007–1021.

Abu-Khudir, R., Ismail, G. A. and Diab, T. (2021) 'Antimicrobial, antioxidant, and anti-tumor activities of Sargassum linearifolium and Cystoseira crinita from Egyptian Mediterranean coast', *Nutrition and Cancer*, 73(5), pp. 829–844.

Achola, K. J., Indalo, A. A. and Munenge, R. W. (1997) 'Pharmacologic activities of Pistia stratiotes', *International Journal of Pharmacognosy*, 35(5), pp. 329–333.

Ahmad, I. and Ezema, F. (2019) Graphene and its derivatives: Synthesis and applications. BoD – Books on Demand.

Ali, R. et al. (2013) 'Enhydra fluctuans Lour: A review', *Research Journal of Pharmacy and Technology*, 6(9), pp. 927–929.

Ali, S. et al. (2020) 'Application of floating aquatic plants in phytoremediation of heavy metals polluted water: A review', *Sustainability: Science, Practice, and Policy*, 12(5), p. 1927.

Annamalai, J. and Nallamuthu, T. (2015) 'Characterization of biosynthesized gold nanoparticles from aqueous extract of Chlorella vulgaris and their anti-pathogenic properties', *Applied Nanoscience*, 5(5), pp. 603–607.

Anuradha, J., Abbasi, T. and Abbasi, S. A. (2015) 'An eco-friendly method of synthesizing gold nanoparticles using an otherwise worthless weed pistia (Pistia stratiotes L.)', *Journal of Advertising Research*, 6(5), pp. 711–720.

Aravind, P. and Prasad, M. N. V. (2005) 'Cadmium-induced toxicity reversal by zinc in Ceratophyllum demersum L. (a free floating aquatic macrophyte) together with exogenous supplements of amino- and organic acids', *Chemosphere*, 61(11), pp. 1720–1733.

Arokiyaraj, S. et al. (2018) 'Chemical composition, antioxidant activity and antibacterial mechanism of action from Marsilea minuta leaf hexane: Methanol extract', *Chemistry Central Journal*, 12(1), p. 105.

Arya, A. K. et al. (2022) 'Ethnomedicinal use, phytochemistry, and other potential application of aquatic and semiaquatic medicinal plants', *Evidence-Based Complementary and Alternative Medicine*, 2022. DOI: 10.1155/2022/4931556.

Bachheti, R. K. et al. (2020) 'Biogenic fabrication of nanomaterials from flower-based chemical compounds, characterization and their various applications: A review', *Saudi Journal of Biological Sciences*, 27(10), pp. 2551–2562.

Bachheti, R. K. et al. (2019) 'Green synthesis of iron oxide nanoparticles: Cutting edge technology and multi-faceted applications', in: *Nanomaterials and Plant Potential.* Springer, pp. 239–259.

Bachheti, A. et al. (2019a) 'Plant-mediated synthesis of copper oxide nanoparticles and their biological applications', in *Nanomaterials and Plant Potential.* Springer, pp. 221–237.

Bachheti, R. K. et al. (2020a) 'Root-based fabrication of metal/metal-oxide nanomaterials and their various applications', in: Husen A., Jawaid M. (eds) *Nanomaterials for Agriculture and Forestry Applications.* Elsevier, pp. 135–166.

Bachheti, R. K. et al. (2020b) 'Nanomaterials from various forest tree species and their biomedical applications', in: Husen A., Jawaid M. (eds) *Nanomaterials for Agriculture and Forestry Applications.* Elsevier, pp. 81–106.

Bachheti, R. K. et al. (2021) 'Algae-, fungi-, and yeast-mediated biological synthesis of nanoparticles and their various biomedical applications', in: *Handbook of Greener Synthesis of Nanomaterials and Compounds.* Elsevier, pp. 701–734.

Bachheti, A. et al. (2022) 'Current status of aloe-based nanoparticle fabrication, characterization and their application in some cutting-edge areas', *South African Journal of Botany*, 147, pp. 1058–1069.

Badar, W. and Ullah Khan, M. A. (2020) 'Analytical study of biosynthesised silver nanoparticles against multi-drug resistant biofilm-forming pathogens', *IET Nanobiotechnology/ IET*, 14(4), pp. 331–340.

Bai, H. et al. (2021) 'A simple and sensitive nanogold RRS/abs Dimode sensor for trace as based on aptamer controlled nitrogen doped carbon dot catalytic amplification', *Molecules*, 26(19). DOI: 10.3390/molecules26195930.

Balashanmugam, P., Mosachristas, K. and Kowsalya, E. (2018) 'In vitro cytotoxicity and antioxidant evaluation of biogenic synthesized gold nanoparticles from Marselia quadrifolia on lung and ovarian cancer cells', *International Journal of Applied Pharmaceutics*, 10(5), pp. 153–158.

Barzinjy, A. A. et al. (2020) 'Green synthesis of the magnetite (Fe3O4) nanoparticle using Rhus coriaria extract: A reusable catalyst for efficient synthesis of some new 2-naphthol bis-Betti bases', *Inorganic and Nano-Metal Chemistry*, pp. 620–629. DOI: 10.1080/24701556.2020.1723027.

Bayda, S. et al. (2019) 'The history of nanoscience and nanotechnology: From chemical–physical applications to nanomedicine', *Molecules*, 25(1), p. 112.

Bency, B. T. et al. (2017) 'Rotula aquatica. Lour-A review on medicinal uses phytochemistry and pharmacological actions', *Indian Journal of Pharmaceutical and Biological Research*, 5(4), pp. 4–9.

Bhuyar, P. et al. (2020) 'Synthesis of silver nanoparticles using marine macroalgae Padina sp. and its antibacterial activity towards pathogenic bacteria', *Beni-Suef University Journal of Basic and Applied Sciences*, 9(1), p. 3.

Bhuyar, P. et al. (2021) 'Microalgae cultivation using palm oil mill effluent as growth medium for lipid production with the effect of CO2 supply and light intensity', *Biomass Conversion and Biorefinery*, pp. 1555–1563. DOI: 10.1007/s13399-019-00548-5.

Bolade, O. P., Williams, A. B. and Benson, N. U. (2021) 'Dataset on analytical characterization of bioactive components from and leaf extracts and their applications in nanoparticles biosynthesis', *Data in Brief*, 38, p. 107407.

Bongiovanni, G. et al. (2021) 'The fragmentation mechanism of gold nanoparticles in water under femtosecond laser irradiation', *Nanoscale Advances*, 3(18), pp. 5277–5283.

Bryan, W. W. et al. (2016) 'Preparation of THPC-generated silver, platinum, and palladium nanoparticles and their use in the synthesis of Ag, Pt, Pd, and Pt/Ag nanoshells', *RSC Advances*, pp. 68150–68159. DOI: 10.1039/c6ra10717f.

Cepoi, L. et al. (2015) 'Biochemical changes in cyanobacteria during the synthesis of silver nanoparticles', *Canadian Journal of Microbiology*, 61(1), pp. 13–21.

Chaudhary, H. et al. (2011) 'Evaluation of hydro-alcoholic extract of Eclipta alba for its anticancer potential: An in vitro study, *Journal of Ethnopharmacology*, pp. 363–367. DOI: 10.1016/j.jep.2011.04.066.

Chung, E. J., Leon, L. and Rinaldi, C. (2019) *Nanoparticles for Biomedical Applications: Fundamental Concepts, Biological Interactions and Clinical Applications*. Elsevier.

Ciampi, P. et al. (2021) 'A field-scale remediation of residual light non-aqueous phase liquid (LNAPL): Chemical enhancers for pump and treat', *Environmental Science and Pollution Research International*, 28(26), pp. 35286–35296.

Cicci, A. et al. (2017) 'Production and characterization of silver nanoparticles in cultures of the cyanobacterium A. platensis (Spirulina)', *Chemical Engineering Transactions*, 57, pp. 1405–1410.

Das, D. R. et al. (2012) 'Nymphaea stellata: A potential herb and its medicinal importance', *Journal of Drug Delivery and Therapeutics*, 2(3). DOI: 10.22270/jddt.v2i3.173.

Duong, T. T. et al. (2016) 'Inhibition effect of engineered silver nanoparticles to bloom forming cyanobacteria', *Advances in Natural Sciences: Nanoscience and Nanotechnology*, 7(3), p. 035018.

Elkomy, R. G. (2020) 'Antimicrobial screening of silver nanoparticles synthesized by marine cyanobacterium Phormidium formosum', *Iranian Journal of Microbiology*, 12(3), pp. 242–248.

Feynman, R. P. (1960) 'There's Plenty of Room at the Bottom', *Engineering and Science*, 23(5), 22–36.

Geng, P. et al. (2021) 'GSH-sensitive nanoscale Mn-sealed coordination particles as activatable drug delivery systems for synergistic photodynamic-chemo therapy', *ACS Applied Materials and Interfaces*, 13(27), pp. 31440–31451.

Gittins, D. I. and Caruso, F. (2002) 'Biological and physical applications of water-based metal nanoparticles synthesised in organic solution', *ChemPhysChem*, pp. 110–113. DOI: 10.1002/1439-7641(20020118)3:1<110::aid-cphc110>3.0.co;2-q.

Glenn, J. B., White, S. A. and Klaine, S. J. (2012) 'Interactions of gold nanoparticles with freshwater aquatic macrophytes are size and species dependent', *Environmental Toxicology and Chemistry / SETAC*, 31(1), pp. 194–201.

Gohil, K. J., Patel, J. A. and Gajjar, A. K. (2010) 'Pharmacological review on Centella asiatica: A potential herbal cure-all', *Indian Journal of Pharmaceutical Sciences*, 72(5), pp. 546–556.

Gómez-Guzmán, M. et al. (2018) 'Potential role of seaweed polyphenols in cardiovascular-associated disorders', *Marine Drugs*, 16(8). DOI: 10.3390/md16080250.

Goud, J. V., Suryam, A. and Charya, M. A. S. (2009) 'Biomolecular and phytochemical analyses of three aquatic angiosperms', *African Journal of Microbiology Research*. https://academicjournals.org/journal/AJMR/article-abstract/D47541B13644.

Graves, J. L., Jr (2021) *Principles and Applications of Antimicrobial Nanomaterials*. Elsevier.

Hajra, A. and Mondal, N. K. (2017) 'Utilization of aquatic fern Azolla pinnata as a green reducing agent for the synthesis of silver nanoparticles', *Indian Science Cruiser*, 31(2), pp. 10–16.

Hamida, R. S., et al. (2020) 'Synthesis of silver nanoparticles using a novel cyanobacteria Desertifilum sp. extract: Their antibacterial and cytotoxicity effects', *International Journal of Nanomedicine*, pp. 49–63.

Hamidpour, R. et al. (2015) 'Medicinal property of gotu kola (Centella asiatica) from the selection of traditional applications to the novel phytotherapy', *Archives in Cancer Research*, 3(4), p. 4.

Hamouda, R. A. et al. (2019) 'Synthesis and biological characterization of silver nanoparticles derived from the cyanobacterium Oscillatoria limnetica', *Scientific Reports*, 9(1), p. 13071.

Huang, S.-S. et al. (2013) 'Antioxidant and anti-inflammatory activities of aqueous extract of Centipeda minima', *Journal of Ethnopharmacology*, 147(2), pp. 395–405.

Hu, S. et al. (2017) 'Aquatic plant genomics: Advances, applications, and prospects', *International Journal of Genomics and Proteomics*, 2017, p. 6347874.

Husen, A., Rahman, Q. I., Iqbal, M., Yassin, M. O. and Bachheti, R. K. (2019) 'Plant-mediated fabrication of gold nanoparticles and their applications', *Nanomaterials and Plant Potential*, Springer, Cham, pp. 71–110. https://doi.org/10.1007/978-3-030-05569-1_3

Ijaz, M., Zafar, M. and Iqbal, T. (2021) 'Green synthesis of silver nanoparticles by using various extracts: A review', *Inorganic and Nano-Metal Chemistry*, pp. 744–755. DOI: 10.1080/24701556.2020.1808680.

Ismail, M. M., Alotaibi, B. S. and El-Sheekh, M. M. (2020) 'Therapeutic uses of red macroalgae', *Molecules*, 25(19). DOI: 10.3390/molecules25194411.

Janthima, R. and Siri, S. (2021) 'Cellular biogenesis of metal nanoparticles by water velvet (Azolla pinnata): Different fates of the uptake Fe3+ and Ni2+ to transform into nanoparticles', *Artificial Cells, Nanomedicine, and Biotechnology*, 49(1), pp. 471–482.

Jogee, P. and Rai, M. (2020) 'Application of nanoparticles in inhibition of mycotoxin-producing fungi', *Nanomycotoxicology*, pp. 239–250. DOI: 10.1016/b978-0-12-817998-7.00010-0.

Kalaba, M. H. et al. (2021) 'Green synthesized ZnO nanoparticles mediated by: Characterizations, antimicrobial and nematicidal activities and cytogenetic effects', *Plants*, 10(9). DOI: 10.3390/plants10091760.

Kamal, S. et al. (2019) 'Phytochemical and pharmacological potential of Enhydra fluctuans available in Bangladesh', *Journal of Pharmaceutical Research International*, pp. 1–11.

Kamath, K. A., Nasim, I. and Rajeshkumar, S. (2020) 'Evaluation of the re-mineralization capacity of a gold nanoparticle-based dental varnish: An study', *Journal of Conservative Dentistry: JCD*, 23(4), pp. 390–394.

Keskin, S. et al. (2016) 'Green synthesis of silver nanoparticles using cyanobacteria and evaluation of their photocatalytic and antimicrobial activity', *Journal of Nano Research, Trans Tech Publ*, pp. 120–127.

Khan, M. A. et al. (2014) 'Pistia stratiotes L.(Araceae): Phytochemistry, use in medicines, phytoremediation, biogas and management options', *Pakistan Journal of Botany*, 46(3), pp. 851–860.

Kim, S. K. and Wijesekara, I. (2017) 'Role of marine nutraceuticals in cardiovascular health', in: *Sustained Energy for Enhanced Human Functions and Activity* (pp. 273–279). Academic Press.

Korbekandi, H. et al. (2014) 'Green biosynthesis of silver nanoparticles using Azolla pinnata whole plant hydroalcoholic extract', *Green Processing and Synthesis*. DOI: 10.1515/GPS-2014-0042.

Kumar, V. et al. (2021) *Nanotoxicology and Nanoecotoxicology Vol. 1*. Springer.

Li, P.-H. et al. (2016) 'Biofunctional activities of Equisetum ramosissimum extract: Protective effects against oxidation, melanoma, and melanogenesis', *Oxidative Medicine and Cellular Longevity*, 2016, p. 2853543.

Linh, N. T. T. et al. (2021) 'Medicinal plant Centipeda minima: A resource of bioactive compounds', *Mini-Reviews in Medicinal Chemistry*, 21(3), pp. 273–287.

Malalavidhane, T. S., Wickramasinghe, S. M. and Jansz, E. R. (2000) 'Oral hypoglycaemic activity of Ipomoea aquatica', *Journal of Ethnopharmacology*, 72(1–2), pp. 293–298.

Malam, Y., Lim, E. J. and Seifalian, A. M. (2011) 'Current trends in the application of nanoparticles in drug delivery', *Current Medicinal Chemistry*, pp. 1067–1078. DOI: 10.2174/092986711794940860.

Mani, M., Pavithra, S., Babujanarthanam, R., Saradha Devi, N., Kumaresan, S. and Selvaraj, S. (2019) 'Rapid synthesis, characterization and antibacterial activities of Ag-NPs derived from Marsilea quadrifolia', *Int J Anal Exp Modal Anal*, 11(11), pp. 1620–1623.

Mansoori, G. A. and Soelaiman, T. A. F. (2005) 'Nanotechnology—An introduction for the standards community', *Journal of ASTM International*, 2(6), pp. 1–22.

Mody, V. V., Singh, A. and Wesley, B. (2013) 'Basics of magnetic nanoparticles for their application in the field of magnetic fluid hyperthermia', *European Journal of Nanomedicine*, 5(1), pp. 11–21.

Nagumo, Y. and Yao, H. (2021) 'Magnetic circular dichroism responses with high sensitivity and enhanced spectral resolution in multipolar plasmonic modes of silver nanoparticles with dimensions between 90 and 200 nm', *Journal of Physical Chemistry Letters*, 12(38), pp. 9377–9383.

Naik, B. S. and Shankar Naik, B. (2020) 'Biosynthesis of silver nanoparticles from endophytic fungi and their role in plant disease management', *Microbial Endophytes*, pp. 307–321. DOI: 10.1016/b978-0-12-819654-0.00012-0.

Namvar, F. et al. (2015) 'Green synthesis and characterization of gold nanoparticles using the marine macroalgae Sargassum muticum', *Research on Chemical Intermediates*, 41(8), pp. 5723–5730.

Nangare, S. N. and Patil, P. O. (2020) 'Green synthesis of silver nanoparticles: An eco-friendly approach', Nano Biomedicine and Engineering. DOI: 10.5101/nbe.v12i4.p281-296.

National Research Council et al. (2010) *Spectrum Management for Science in the 21st Century*. National Academies Press.

Nouailhat, A. (2010) *An Introduction to Nanoscience and Nanotechnology*. John Wiley & Sons.

Nunes, D. et al. (2018) *Metal Oxide Nanostructures: Synthesis, Properties and Applications*. Elsevier.

Omar, N. et al. (2019) 'The effects of Centella asiatica (L.) Urban on neural differentiation of human mesenchymal stem cells in vitro', *BMC Complementary and Alternative Medicine*, 19(1), p. 167.

Orhan, I. E. (2012) 'Centella asiatica (L.) urban: From traditional medicine to modern medicine with neuroprotective potential', *Evidence-Based Complementary and Alternative Medicine: eCAM*, 2012, p. 946259.

Pachurekar, P. and Dixit, A. K. (2017) 'A review on pharmacognostical phytochemical and ethnomedicinal properties of Hedychium coronarium J. Koenig an endangered medicine', *International Journal of Chinese Medicine*, 1(2), pp. 49–61.

Painuli, S. et al. (2020) 'Nanomaterials from nonwood forest products and their applications', in: Husen A., Jawaid M. (eds) *Nanomaterials for Agriculture and Forestry Applications*. Elsevier, pp. 15–40.

Pal, S. K. (2012) 'Study of solvent extracts of some selected ferns for antimicrobial activity'. Available at: http://inet.vidyasagar.ac.in:8080/jspui/handle/123456789/1136 (Accessed: 11 October 2021).

Parimala, M. and Shoba, F. G. (2013) 'Phytochemical analysis and in vitro antioxidant acitivity of hydroalcoholic seed extract of Nymphaea nouchali Burm. f', *Asian Pacific Journal of Tropical Biomedicine*, 3(11), pp. 887–895.

Pascu, B. et al. (2021) 'A green, simple and facile way to synthesize silver nanoparticles using soluble starch. pH studies and antimicrobial applications', *Materials*, 14(16). DOI: 10.3390/ma14164765.

Patil, M., Mehta, D. S. and Guvva, S. (2008) 'Future impact of nanotechnology on medicine and dentistry', *Journal of Indian Society of Periodontology*, 12(2), pp. 34–40.

Patra, J. K. et al. (2020) *Green Nanoparticles: Synthesis and Biomedical Applications*. Springer Nature.

Paudel, K. R. and Panth, N. (2015) 'Phytochemical profile and biological activity of Nelumbo nucifera', *Evidence-Based Complementary and Alternative Medicine: eCAM*, 2015, p. 789124.

Prabaharan, M. and Mano, J. F. (2004) 'Chitosan-based particles as controlled drug delivery systems', *Drug Delivery*, pp. 41–57. DOI: 10.1080/10717540590889781.

Prakash, V., Jaiswal, N. and Srivastava, M. (2017) 'A review on medicinal properties of Centella asiatica', *Asian Journal of Pharmaceutical and Clinical Research*, 10(10), p. 69.

Pratas, J. et al. (2014) 'Potential of aquatic plants for phytofiltration of uranium-contaminated waters in laboratory conditions', *Ecological Engineering*, 69, pp. 170–176.

Rahman, M. A. et al. (2011) 'Evaluation of antinociceptive and antidiarrhoeal properties of Pistia stratiotes (Araceae) leaves', *Journal of Pharmacology and Toxicology*, pp. 596–601. DOI: 10.3923/jpt.2011.596.601.

Rai, S. et al. (2006) 'Antioxidant activity of Nelumbo nucifera (sacred lotus) seeds', *Journal of Ethnopharmacology*, 104(3), pp. 322–327.

Rajagopal, G. et al. (2021) 'Phytofabrication of selenium nanoparticles using Azolla pinnata: Evaluation of catalytic properties in oxidation, antioxidant and antimicrobial activities', *Journal of Environmental Chemical Engineering*, 9(4), p. 105483.

Raja, M. K. M. M., Sethiya, N. K. and Mishra, S. H. (2010) 'A comprehensive review on Nymphaea stellata: A traditionally used bitter', *Journal of Advanced Pharmaceutical Technology and Research*, 1(3), pp. 311–319.

Ramya Juliet, M. et al. (2020) 'Biogenic synthesis of Copper nanoparticles using aquatic pteridophyte Marsilea quadrifolia Linn. rhizome and its antibacterial activity', *International Journal of Nano Dimension*, 11(4), pp. 337–345.

Ranu, B. C. and Chattopadhyay, K. (2009) 'Chapter 5. Green procedures for the synthesis of useful molecules avoiding hazardous solvents and toxic catalysts', in: *Green Chemistry Series*, pp. 186–219. DOI: 10.1039/9781847559760-00186.

Ravi, R. et al. (2019) 'Larvicidal effects of nano-synthesized silver particles from Azolla pinnata extract against Aedes aegypti (Diptera: Culicidae)', *Int J Innovat Technol Explore Eng*, 8, pp. 753–757.

Ravi, R. et al. (2020). 'AgNPs-azolla pinnata extract as larvicidal against aedes aegypti (diptera: culicidae)', *IOP Conference Series: Earth and Environmental Science*, 596(1), p. 012065.

Safaat, M. et al. (2021) 'Nanoparticles green synthesis macroalgae-based and its application and distribution in Indonesia – An overview', *IOP Conference Series*. https://iopscience.iop.org/article/10.1088/1755-1315/744/1/012067/meta.

Saha, P. S. et al. (2020) 'In vitro propagation, phytochemical and neuropharmacological profiles of Bacopa monnieri (L.) Wettst.: A review', *Plants*, 9(4). DOI: 10.3390/plants9040411.

Sajini, R. J., Prema, S. and Chitra, K. (2019) 'Phytoconstituents, pharmacological activities of Marsilea minuta L.(Marsileaceae) – an overview', *International Journal of Pharmaceutical Sciences and Research*, 10(4), pp. 1582–1587.

Samuelsson, G. et al. (1992) 'Inventory of plants used in traditional medicine in Somalia. II. Plants of the families Combretaceae to Labiatae', *Journal of Ethnopharmacology*, 37(1), pp. 47–70.

Sanjay, S. S. (2019) 'Safe Nano is green Nano', in: *Green Synthesis, Characterization and Applications of Nanoparticles*, pp. 27–36. DOI: 10.1016/b978-0-08-102579-6.00002-2.

Sapkal, M. and Sapkal, R. (2017) *Biosynthesis of Noble Metal Nanoparticles Using Actinomycetes: Green Synthesis of Nanoparticles*.

Scroccarello, A. et al. (2021) 'Metal nanoparticles based lab-on-paper for phenolic compounds evaluation with no sample pretreatment. Application to extra virgin olive oil samples', *Analytica Chimica Acta*, 1183, p. 338971.

Seng, R. X. et al. (2020) 'Nitrogen-doped carbon quantum dots-decorated 2D graphitic carbon nitride as a promising photocatalyst for environmental remediation: A study on the importance of hybridization approach', *Journal of Environmental Management*, 255, p. 109936.

Shanmugam, L., Ahire, M. and Nikam, T. (2020) 'Bacopa monnieri (L.) Pennell, a potential plant species for degradation of textile azo dyes', *Environmental Science and Pollution Research International*, 27(9), pp. 9349–9363.

Shukla, A. K. and Iravani, S. (2018) *Green Synthesis, Characterization and Applications of Nanoparticles*. Elsevier.

Simonsen, R. (1968) 'Sculthorpe, C. D.: The biology of aquatic vascular plants. 610 S. London: Edward Arnold Ltd. 1967, £ 66 s. net', *Internationale Revue der Gesamten Hydrobiologie und Hydrographie*, pp. 353–354. DOI: 10.1002/iroh.19680530207.

Singh, Y., Kaushal, S. and Sodhi, R. S. (2020) 'Biogenic synthesis of silver nanoparticles using cyanobacterium Leptolyngbya sp. WUC 59 cell-free extract and their effects on bacterial growth and seed germination', *Nanoscale Advances*, 2(9), pp. 3972–3982.

Soleimani, M. and Habibi-Pirkoohi, M. (2017) 'Biosynthesis of silver nanoparticles using Chlorella vulgaris and evaluation of the antibacterial efficacy against Staphylococcus aureus', *Avicenna Journal of Medical Biotechnology*, 9(3), pp. 120–125.

Sukhen, D. et al. (2015) 'Marsilea minuta plant extract mediated synthesis of gold nanoparticle for catalytic and antimicrobial applications', *International Journal of Pharmacy*, 5, pp. 600–609.

Swetha, K. G. et al. (2020) 'Characterization, in vitro cytotoxic and antibacterial exploitation of green synthesized freshwater cyanobacterial silver nanoparticles', *Journal of Applied Pharmaceutical Science*. DOI: 10.7324/japs.2020.10911.

Taniguchi, N. (1974) 'On the basic concept of nanotechnology', *Proceeding of the ICPE*. Available at: https://ci.nii.ac.jp/naid/10008480916/ (Accessed: 20 September 2021).

Tayebee, R. et al. (2018) 'Equisetum arvense as an abundant source of silica nanoparticles. SiO2 /H3 PW12 O40 nanohybrid material as an efficient and environmental benign catalyst in the synthesis of 2-amino-4H-chromenes under solvent-free conditions', *Applied Organometallic Chemistry*, 32(1), p. e3924. DOI: 10.1002/aoc.3924.

Thota, S. and Crans, D. C. (2018) *Metal Nanoparticles: Synthesis and Applications in Pharmaceutical Sciences*. John Wiley & Sons.

Torabfam, M. and Yüce, M. (2020) 'Microwave-assisted green synthesis of silver nanoparticles using dried extracts of Chlorella vulgaris and antibacterial activity studies', *Green Processing and Synthesis*, 9(1), pp. 283–293.

Tungmunnithum, D., Pinthong, D. and Hano, C. (2018) 'Flavonoids from Nelumbo nucifera Gaertn., a medicinal plant: Uses in traditional medicine, phytochemistry and pharmacological activities', *Medicines (Basel, Switzerland)*, 5(4). DOI: 10.3390/medicines5040127.

Venkatesan, J., Anil, S. and Kim, S.-K. (2017) *Seaweed Polysaccharides: Isolation, Biological and Biomedical Applications*. Elsevier.

Vent, W. (1987) 'Duke, J. A. & Ayensu, E. S., *Medicinal Plants of China*. 2 Vols. 705 S., 1300 Strichzeichnungen. Reference Publ., Inc. Algonac. Michigan, 1985. ISBN 0-917266-20-4. Preis: geb. m. Schutzumschlag $94,95', Feddes Repertorium, pp. 398–398. DOI: 10.1002/fedr.4910980707.

Vimala, A. et al. (2017) 'Moss (bryophyte) mediated synthesis and characterization of silver nanoparticles from Campylopus flexuosus (Hedw.) Bird', *Journal of Pharmaceutical Sciences and Research; Cuddalore*, 9(3), pp. 292–297.

Voliani, V. (2020) *Gold Nanoparticles: An Introduction to Synthesis, Properties and Applications*. Walter de Gruyter GmbH & Co KG.

Vysakh, A. et al. (2016) 'Traditional and therapeutic importance of Rotula aquatica Lour.: An overview', *IJPPR Hum*, 7, pp. 97–107.

Walter, T. M., Merish, S. and Tamizhamuthu, M. (2014) 'Review of Alternanthera sessilis with reference to traditional Siddha medicine', *International Journal of Pharmacognosy and Phytochemical Research*, 6(2), pp. 249–254.

Yuan, L. et al. (2018) 'Stress responses of aquatic plants to silver nanoparticles', *Environmental Science and Technology*, 52(5), pp. 2558–2565.

Zhang, C. et al. (2010) 'Recent development and application of magnetic nanoparticles for cell labeling and imaging', *Mini-Reviews in Medicinal Chemistry*, pp. 194–203. DOI: 10.2174/138955710791185073.

14 Green Synthesis of Nanoparticles from Medicinally Important Desert Plants and Their Applications

Chetan Shrivastava, Kundan Kumar Chaubey, and Shivani Tyagi

CONTENTS

14.1 INTRODUCTION

The interdisciplinary fields of biomedicine, pharmaceuticals, diagnostics, and materials science are seeing a fresh, intriguing, and quickly expanding revolution thanks to nanotechnology. By changing their molecular structure, materials can be created and manipulated using nanotechnology (Rajasekharreddy and Rani 2014). Through the use of diverse physical, chemical, and biological processes (Figure 14.1), many metals including magnesium, gold, iron, copper, zinc, silver, and titanium have all undergone nano formulation processes to be used for a variety of beneficial purposes (Painuli et al. 2018; Ponarulselvam et al. 2012). Green synthesis is a different way to create nanoparticles (NPs) by using reducing agents derived from natural assets such as bacteria, microbes, and medicinal plants. The green technique creates NPs that are extremely stable,

DOI: 10.1201/9781003213727-14

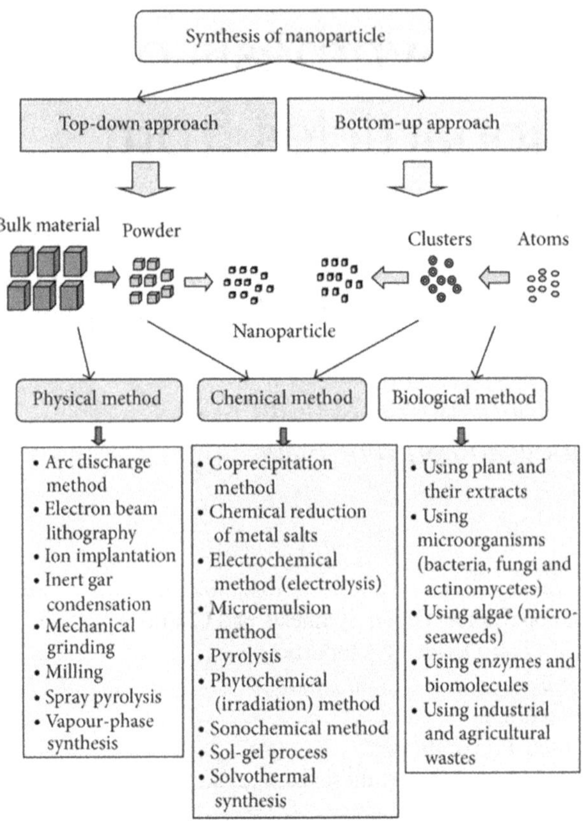

FIGURE 14.1 The approaches used to create metal nanoparticles. The various approaches are divided into top-down and bottom-up categories (*Source:* https://www.hindawi.com/journals/jnm/2014/417305/).

economical, ecologically welcoming, less toxic, biocompatible, and suitable for therapeutic and diagnostic uses (Painuli et al. 2020; Bachheti et al. 2020a, 2020b, 2021).

The fields of nanomedicine and nanotechnology benefit greatly from the use of silver-based NPs (Logeswari et al. 2015). Silver nanoparticles (AgNPs) have been shown in numerous investigations to have antibacterial action against a variety of infectious and drug-resistant microbes (Singh et al. 2016; Gonfa et al. 2022). In the realm of nanotechnology, the synthesis of NPs utilizing plant-segregated compounds may be an appealing technique due to its several benefits, including low cost, environmental friendliness, quick synthesis, and high yield. Here, a green method for the production of AgNPs using the roots of *R. Stricta* is disclosed due to the significant therapeutic characteristics of plants. It is a little, glabrous, upright shrub that has phenolic, flavonoid, and alkaloid components in it (Bukhari et al. 2017). *R. stricta*, popularly acknowledged as harmal, is a common species in South Asia and Saudi Arabia (Obaid et al. 2017; Bukhari et al. 2017). Harmal seed oil is believed to be a potentially luxuriant origin of d-tocopherol, a vital form of vitamin E (Nehdi et al. 2016). Its main use is to treat a variety of illnesses including cancer, obesity, sore throats, diabetes mellitus, metabolic, cardiac, and neurological diseases (Alagrafi et al. 2017). It has anti-inflammatory, antipyretic, antifungal, anticarcinogenic, antioxidant, antimicrobial, antidramatic, antihypertensive (Van Beek et al. 1985), herbicidal, and antianxiety properties (Van Beek et al. 1985; Marwat et al. 2012; Baeshen et al. 2010; Lantero 2014; Ali et al. 2000; Alagrafi et al. 2017). This chapter discusses the production, characterization, and applications of green NP synthesis, how to create AgNPs using the xylitol and root extract of the *R. stricta* plant as a dipping agent.

14.2 PLANT-DERIVED NANOPARTICLES: GREEN SYNTHESIS AND CHARACTERIZATION

The idea of 'green harmony' for 'sustainable growth' has been extensively researched during the past ten years (Clark et al. 2008). Meeting present requirements while simultaneously ensuring that future generations will be able to fulfil their own needs is known as sustainable development (Robert et al. 2005). Sustainable growth is crucial for some chemistry-based industries since it is concerned with the pollution evidence and the indiscriminatory use of natural assets (Omer 2008). The three most crucial prerequisites for the green synthesis of NPs are the selection of a green or ecologically friendly solvent (the most frequently used being ethanol, water, and their combinations), a conventional non-toxic reducing agent, and a safe chemical for stabilization. In fact, a wide variety of synthetic processes have been employed to create NPs, with the most common ones being physical, chemical, and biosynthetic. Chemical treatments typically cost too much money and require the use of dangerous and poisonous compounds, which pose several environmental problems (Nath and Banerjee 2013). Green synthesis, in contrast, is a secure environmentally and biocompatible sound technique of creating NPs for a variety of functions, including biomedical ones (Razavi et al. 2015). This 'green synthesis' has been accomplished using, plants, bacteria, fungi, and algae. But it has frequently been used to create different NPs from plant materials such as seeds, fruits, leaves, stems, and roots (Narayanan and Sakthivel 2011; Beshah et al. 2020; Bachheti et al. 2020, 2020a, 2022).

It is widely acknowledged that plant-derived NPs have a lower risk of having negative side effects on people than chemically produced NPs do. They also have a high biological potential and can be used in a variety of fields, including bioengineering human health protection, agriculture, nanomedicine or cosmetics, food science, and technology. To guarantee reproducibility in their safety, biological activity, and manufacture, these NPs must be accurately and fully defined dynamic light scattering, transmission electron microscopy (TEM), scanning electron microscopy (SEM), atomic force microscopy, Raman spectroscopy, attenuated total reflection, ultraviolet-visible diffuse reflectance spectroscopy, and photo-luminescence analysis are just a few of the physicochemical techniques employed to characterize the synthesized NPs very precisely for this purpose.

14.3 DIFFERENT TYPES OF PLANT-DERIVED NANOPARTICLES

The synthesis, z, and uses of various plant-derived NP types are briefly explained below:

Plant-based AgNPs are among the easiest kinds of AgNPs to create (Alshehri and Malik 2020; Khan et al. 2020; Silva Viana et al. 2020; Wahid et al. 2020; Singh et al. 2021; Mickymaray 2019). A reducing biological agent and a silver metal ion solution are required for the environmentally friendly synthesis of AgNPs. Reducing and stabilizing silver ions with a combination of macromolecule, such as alkaloids, proteins, vitamins, saponins, polysaccharides, terpenes, or phenolics, is the easiest and best cheapest way to make AgNPs (Tolaymat 2010). It is possible to make AgNPs from practically any plant (Figure 14.2).

Gold nanoparticles (AuNPs) have received a lot of consideration due to their straightforward production, surface functionalization, and distinctive features including their low noxiousness (Jeong et al. 2011; Husen et al. 2019), high latent for use in drugs (Jain et al. 2006), and extremely biocompatible environment (Sperling et al. 2008). Numerous chemical substances that are present in biological complexes act as reducing agents, reducing gold metal ions, and producing NPs as a result. Experience has shown that macromolecules like phenols, proteins, flavonoids, and others significantly affect how metal ions are reduced and how AuNPs are produced in plants.

Zinc oxide nanoparticles (ZnONPs) have in recent years seen a tremendous increase in popularity due to their numerous potential uses in optics, electronics, biomedicine, and aesthetics. Several plant parts, including flowers, roots, seeds, and leaves, can be used to make ZnONPs. Due to their large band gap of 3.37 eV and their substantial 60 meV exciton binding energy, these NPs are

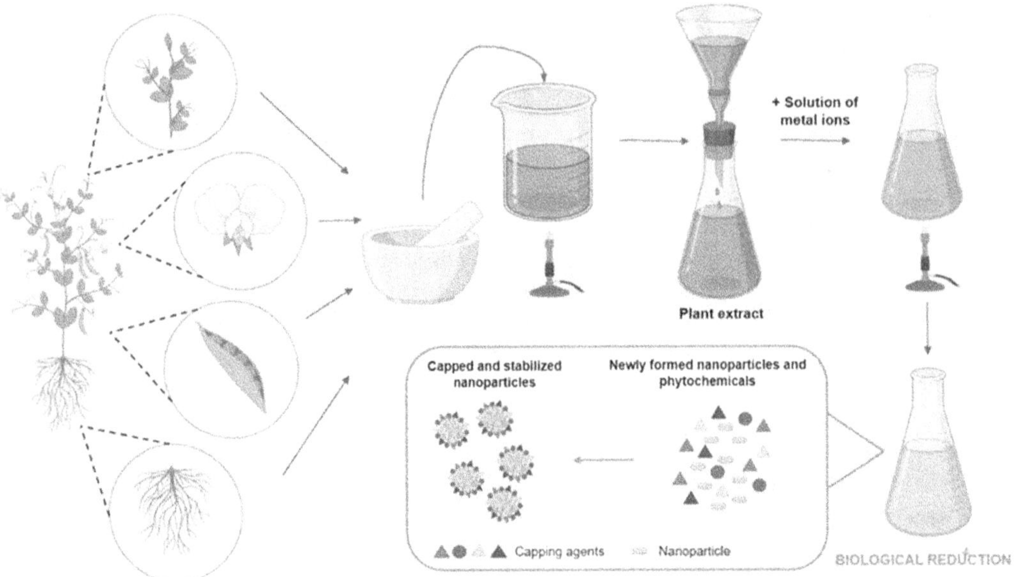

FIGURE 14.2 The general procedures for employing plant extracts in the green production of inorganic nanoparticles (*Source:* Rónavári et al.2021).

noteworthy in that they display a wide range of semiconducting characteristics (Zaeem et al. 2020; Hossain et al. 2019; Ahmad et al. 2020).

14.4 PROCEDURE

14.4.1 Requirements

In this process, 99.0% pure xylitol and silver nitrate ($AgNO_3$) are employed. Deionized water was utilized for the reaction, and 125 mm Whatman filter sheets were employed for purification and filtering.

14.4.2 Root Extracts Preparation

Rhazya stricta was harvested, and the plant's roots were cleaned three times with deionized water before being air dried. The extracts were produced by soaking 147 g of dehydrated roots within 80% methanol for three days at 25°. The extracts were then clarified using Whatman filter paper (125 mm). A rotary evaporator was used to evaporate the solvent. After drying, the extract produced 13 g.

14.4.3 Silver Nanoparticles Biosynthesis

A measure of 10 g of dehydrated extract was mixed in 18 mL of methanol and kept at 4°C to create AgNPs. $AgNO_3$ was dissolved in deionized water to a concentration of 1 mM. Drop by drop, 6 mL of plant roots extracted with methanol were added to 100 mL of 1 mM $AgNO_3$ while being continuously shaken (Figure 14.3). An equal quantity of extract, 100 mL of $AgNO_3$, and 6 mL, of xylitol were used in another reaction (10–2 M). Then, using a magnetic stirrer, both processes were heated independently for 2 h at 60°C. A shift in colour indicated that NPs had been synthesized. The material was centrifuged at 12,000 rpm for 20 min to characterize the AgNPs. To eliminate any disorganized solid, the supernatant was discarded. After that, three washes with distilled water

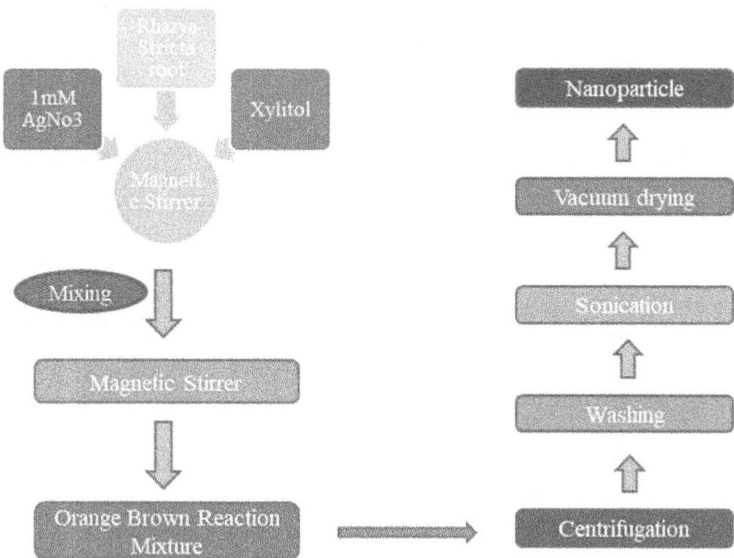

FIGURE 14.3 Methodology for synthesis of silver nanoparticles. Schematic presentation of methodology for synthesis of silver nanoparticles with root extract of *Rhazya stricta*.

were performed on the sample to remove any extra enzymes or proteins. The material was then vacuum dehydrated overnight at 50°C following sonication (Figure 14.3).

14.5 OUTCOME

14.5.1 Ultraviolet-Visible Spectroscopy

The root extract of *R. stricta* was introduced to an aqueous solution of $AgNO_3$ for the production of NPs. Two hours of heating (60°C) on a magnetic stirrer resulted in a colour change from pale yellow to orange-brown, which suggests the creation of reduced silver.

14.6 APPLICATIONS OF PLANT-DERIVED NANOPARTICLES

14.6.1 A General Idea of the Nanoparticles Derived from Plants and Their Potential Applications

Due to their numerous uses in the electronic engineering, environment, and energy industries, and more specifically in the biomedical disciplines, NPs are currently in high demand commercially. NPs, like the most recognized AgNPs and AuNPs, have been extensively researched in this field and hold great promise for biological applications. Green NPs made from plants often have fewer harmful side effects for people than those made through chemical synthesis.

- Human health safety and nanomedicine (pro- or antioxidative, antiparasitic, pro-apoptotic, pro-apoptotic, antiproliferative and antimicrobial reliant on the framework, anti-inflammatory activities, etc.) (Saleem et al. 2019; Gul et al. 2021; Anjum et al. 2021; Nadeem et al. 2020).
- Agriculture (target-specific delivery of biomolecules with controlled pesticide release, precision farming with improved nutrient uptake, detection, and management of plant diseases, etc.) (Anjum et al. 2021; Anjum et al. 2019).

- Food science and engineering (technologies for preparing, storing, and wrapping food), in bioengineering (biosensors, photocatalyst, biocatalysts, etc.) (Shafiq et al. 2020).
- Cosmetics (hair growth, nano-emulsion, bioactive compounds delivery, antiageing, sunscreen, etc.) (Abbasi et al. 2020).

Algae are undoubtedly the greatest candidates for the green synthesis of NPs because they are rich in secondary metabolites that serve as reducing and capping agents. Numerous possible uses have previously been mentioned, such as their anticancer or antibacterial properties as well as their use as biosensing agents, antifouling, or bioremediation.

Copper is a reasonably inexpensive metal that, for instance, is more economical than silver and gold. Copper nanoparticles (CuNPs) have been made by reducing aqueous copper ions with diverse plant extracts. The review by Letchumanan et al. provides a very thorough overview and most recent apprise on plant-mediated Cu/CuO (copper oxide) NPs, comprising information on their production, therapeutic applications, and procedures (Letchumanan et al. 2021). Cu/CuO NPs offer a lot of therapeutic benefits, but they also have the potential to cause catastrophic harm to healthy human cells and organs. Therefore, a thorough investigation of this potential hazardous concern is required before these NPs can be used in medicine. There is potential for making plant-based Cu/CuO NPs to treat several ailments in the future (including wounding, inflammation, cancer, or microbial infection) (Letchumanan et al. 2021)

14.6.1.1 Anticancer Potential

Nanomedicine refers to the application of nanotechnology to the prevention, detection, and cure of a wide range of illnesses, including cancer (Anjum et al. 2021; Khan et al. 2021). It also includes thorough procedures and efficient methods for combating cancer through cancer diagnosis, prevention, and treatment, as well as potential for tailored therapy (Anjum et al. 2021). Many plant-based NPs have some ability to inhibit cancer cells. In particular, ZnONPs produced from an extract of the leaves of *Cassia auriculata* displayed tumouricidal efficacy contrary to MCF-7 breast cancer cells, whereas it had no opposing effects on healthy MCF-12A human breast cells (Prasad et al. 2020). Similar to this, the apoptosis mediated by reactive oxygen species (ROS) was linked to the concentration-dependent inhibition of HepG2 cancer cell lines' proliferation by green AuNPs generated from *Trachyspermum ammi* seed extract (Perveen et al. 2021). Recent research has shown that this method may be linked to mitochondrial action via ROS-induced expression of the Caspase-3 gene and enzyme activity, which is triggered by the disturbance of the mitochondrial membrane potential brought on by plant-based NPs (Khan et al. 2021; Anjum et al. 2021).

14.6.1.2 Antimicrobial Potential

One of the most urgent challenges of recent years is antibiotic resistance, and it will only get worse. The fast development of the bacterial genome has resulted in bacteria developing resistance to antimicrobial substances. In the fight against such resistant pathophysiology in the hunt for a novel treatment, biogenic NPs have revealed promising outcomes in the treatment of multi-drug-resistant microorganisms (Nadeem et al. 2018). To increase the antimicrobial impact, several organic and inorganic compounds have been combined with NPs and other conjugates. Silver has long been recognized for its ability to fight off numerous bacterial types with its antibacterial characteristics. A variety of human pathogenic microorganisms were effectively suppressed by green AgNPs made from a leaf extract of *Carissa carandas*, with Gram-negative bacteria, particularly *Shigella flexineri*, which causes shigellosis, being more susceptible to this effect (Singh et al. 2021). *Pseudomonas aeruginosa* and *Acinetobacter baumanii*, two multidrug-resistant bacteria that cause ventilator-associated pneumonia, were successfully controlled by using AgNPs made from the Arabian desert plant *Sisymbrium irio* (Mickymaray 2019). A variety of human pathogenic pathogens were successfully suppressed by AgNPs made from the leaf extract of *Clerodendrum inerme*.

ZnONPs have also shown potential antibacterial activity, as shown by their ability to constrain the development of *E. coli* and *Staphylococcus aureus* when generated from *Cinnamomum verum* bark extract (Andleeb et al. 2021). Similar to this, ZnONPs made from a leaf extract of *C. auriculata* displayed antibacterial action because direct cell contact compromised bacterial cell integrity (Prasad et al. 2020). Other metallic NPs can demonstrate strong antimicrobial activity, including antibiofilm capabilities, such as CuONPs made from *Cymbopogon citratus* (Cherian et al. 2020). While nickel oxide nanoparticles (NiONPs) derived from stevia leaf extract were more efficient against Gram-negative bacteria, manganese oxide nanoparticles (MnONPs) developed from an *Abutilon indicum* leaf extract displayed significant antibacterial action against both Gram-positive and Gram-negative bacteria (Srihasam et al. 2020; Khan et al. 2020). This shows that the plant extract used to generate NPs controls both the sorts of NPs that are formed and the confirmation of the covered phytochemicals on their exteriors. There have been a number of hypothesized mechanisms for NPs' antibacterial activity, including DNA fragmentation, significant disruption in electron transport, loss of cellular fluids, disruption of cell walls and membranes, intense unrestricted radical fabrication, targeted or precise actions against proteins, and inhibition of crucial enzymes (Nadeem et al. 2018). Although our knowledge of the antibacterial effectiveness of plant-based NPs has increased, much is still unclear about their exact mode of action, noxiousness, and potential environmental impacts.

14.6.1.3 Effects of Antioxidants

A range of cell macromolecules, such as DNA, proteins, and membrane lipids, can experience oxidative damage from extreme oxidative pressure brought on by the act of mitochondria and other external or internal causes. This damage can result in degenerative diseases and ageing (Hano and Tungmunnithum 2020). Antioxidants may be used to treat disorders linked to ageing and old age as well as to stop this damaging process. AgNPs made from *C. carandas* leaf extract (Singh et al. 2021), green plant-derived NPs that have been demonstrated to have antioxidant activity include AgNPs and AuNPs created from *C. inerme* leaf extract (Khan et al. 2020), or NiONPs produced from stevia leaf extract (Srihasam et al. 2020). Phytochemicals that have been formed on the surface of NPs unquestionably contribute significantly to the claimed antioxidant activity.

14.6.1.4 Agricultural Applications

When agricultural pathogens are the focus, ZnONPs in particular have demonstrated their broad agricultural utility by acting as antiphytopathogenic agents in contrast to both bacteria and fungi. Examples of this include ZnONPs produced from lemon fruit in contrast to the *Dickeya dadantii* (soft rot bacteria pathogen) and the bactericidal action of ZnONPs made by consuming a eucalyptus globule extract in contrast to the main pathogens of an apple orchard (Hossain et al. 2019). AgNPs synthesized from a wheat extract considerably helped to mitigate the negative effects of salinity stress in wheat by altering the concentration of ion homeostasis, abscisic acid, and protection mechanisms made up of both enzymatic and non-enzymatic antioxidants (Wahid et al. 2020). ZnONPs showed low toxicity and the ability to boost the antioxidant response in flax seedlings, which is an interesting observation (Zaeem et al. 2020).

Metal NP synthesis and applications of some different plants are detailed in Table 14.1.

14.7 FUTURE PROSPECTS OF *RHAZYA* MOLECULES

A *Rhazya stricta* chloroform extract has been shown by Faisal et al. to have potential in treating neurological illnesses. At the beginning, it aids in neural differentiation. Pleuripotency markers (Oct4, Sox2, TRA-1-60, and Klf4) demonstrated a decrease in expression when exposed to an alkaloid extract (Shah et al. 2020). In this regard, nanoformulation is also a crucial step to control harmful bacteria. Ahmed et al. showed that *Leishmania tropica* and *Staphylococcus aureus* could be effectively combated by applying AuNPs to *Rhazya stricta* Decne leaves (Clark 2008). Diverse

TABLE 14.1
List of Medicinal plants Utilized for Metal Nanoparticle Synthesis and Their Biological Activity

Plant species	Indigenous application	Major phytochemicals	MNPs	MNP ,io-Activity
***Cyclopia* intermediates**	Regulate blood pressure as well as cough, eczema, epilepsy, and constipation (Mahomoodally 2013)	Hesperidin, isosakuranetin, mangiferin (MGF), and isomangiferin (Journal et al. 2018; Ajuwon et al. 2018)	AuNPs	Anticancer (Aboyewa et al. 2021)
Sutherlandia fructecens	Treat wounds, cancer, diabetes, rheumatoid arthritis, gonorrhoea, urinary tract infections, fever, and kidney and liver issues (Dube et al. 2020)	Cannavanine, flavonol glycosides, triterpenoids, cycloartane glycosides, flavonoids, pinitols, and aminobutyric acid	ZnONPs AgNPs	Antimicrobial, antibacterial, and anticancer (Dube et al. 2020)
Hypoxisheme-rocallidea	Purgative, laxative, and immune-stimulating tonic, treating wounds, cancer, diabetes, UTI, infertility, and TB	Sterols, norlignane, daucosterols, stanols, hypoxide, sitosterol, and sterolins	AuNPs	Antibacterial and anti-inflammatory (Elbagory et al. 2017; Elbagory et al. 2019)
Eucomis autumnalis	Treat lower backaches, syphilis, stomach, urinary tract, and other illnesses; also used occasionally to induce labour (Oguntibeju et al. 2016; Alaribe et al. 2018)	Terpenoids, diben-pyrones, and homoisoflavanones (Masondo et al. 2015)	AgNPs	Antimicrobial (Masondo et al. 2015)
Plumbago auriculata	Treat a fracture, skin infection, wounds, warts, and headaches (Lediga et al. 2018; Karishma et al. 2018)	Alkaloids, saponins, flavonoids, amyrin, capensisone, phenols, tannins, and plumbagin	AgNPs	Antimicrobial (Singh et al. 2018)
Catharanthus roseus	Treat diabetes, rheumatoid arthritis, high blood pressure, skin infections, and venereal disorders (Sumsakul et al. 2014; Oguntibeju et al. 2019)	Vinblastine, deoxyvinblastin, vinblastine, vincoline, cathananthine, rosicine, leurosine, vindoline, and vincristine (Gajalakshmi et al. 2013)	AgNPs	Antimicrobial and wound healing, larvicidal (Rajagopal et al. 2015) Antimicrobial
Aspalathus linearis	Improve appetite while treating allergies, digestive issues, stomach pains, and sleeplessness (Mordeniz et al. 2019)	Spalathin, orientin, isoquercitrin, luteolin hyperoside (Goboza et al. 2019; Joubert and de Beer 2012)	AuNPs RhNPs	Antimicrobial (Joubert et al. 2008)
Indigofera tinctoria	Treat asthma, bronchitis, stomach ache, epilepsy, and a few skin conditions (Thipe et al. 2015)	Alkaloids, flavonoids, saponins, and phenolic substances (Li et al. 2019)	AuNPs	Antibacterial, antifungal, and anticancer (Bai et al. 2014)
Artemisia herba-alba	Treat dyspepsia, anorexia, and other digestive issues (Vijayan et al. 2018; Boudjelal et al. 2015)	1,8-Cineole, alpha, and Beta-Thujone, davanone, chrysanthenone, cis-chrysanthenol (Segal et al. 1987; Thabiani et al. 2018)	AgNPs	Antibacterial and mosquito repellent (Orhan 2012)
Centella asiatica	Treat fever, syphilis, leprosy, TB, epilepsy, mental illness, and minor wounds. Both a vegetable and a spice to be eaten (Rout et al. 2013)	Triterpenes, centellose, rutin, medacassoside, triaponosides, patuletin, flavonoids quercetin, kaemferol, apigenin, polyacetylenes, phenolic acids and sterols	AgNPs	Antimicrobial (Ng'uni et al. 2018)
Galenia africana	Treat venereal sores, eye infections, coughs, wounds, skin, asthma, infections, TB, and toothaches (Mativandlela et al. 2009)	Dihydroxychalcone, trihydroxy-3-methoxychalcone, trihydroxychalcone, and trihydroxyflavanone (Mocheki et al. 2018)	AuNPs	Antibacterial (Elbagory et al. 2017)
Sclerocarya birrea	Treat diarrhoea, malaria, rheumatism, and dysentery (Virginie et al. 2016)	Alkaloids, phenols, resins, calcium, phosphorus, flavonoids, fatty acidssteroids, and glycosides (Virginie et al. 2016; Joubert et al. 2011)	AgNPs	Antimicrobial (Masondo et al. 2015)

cellular and humoral antibacterial innate immunity systems have a role in how well humans and other mammalian species can defend themselves against a variety of potentially invasive bacterial species (Khan et al. 2020).

The harmal extract, rhazimanine, inhibits NFkB and activator protein-1, which inhibits the Bcl-2 gene in vitro in breast and colon cancer, preventing metastasis and promoting apoptosis. Additionally, *R. stricta* extract exhibits colony-formation inhibition, antioxidant, and cell cycle control. Several studies employ these *R. stricta* chemicals as innovative latent antitumour agents in various cancer managements (Shah et al. 2020; Robert and Parris 2005; Omer 2008).

14.8 CONCLUSIONS AND FUTURE DIRECTIONS

The demand for green chemistry and nanotechnology has fuelled the development of green synthetic methods for producing nanomaterials using plants, microbes, and other natural resources. Plant extract-based remedies have been the subject of extensive research because they are affordable, safe, accessible, and ecologically friendly.

Rhazya stricta is a common medicinal herb with a variety of pharmacological implications. People used herbs' leaves, roots, and stems directly in the past. Due to its mono dispersion, stability, and reduced aggregation, AgNPs made utilizing xylitol and *R. stricta* root extract have shown latent antibacterial action. AgNPs were created using the plant as a reducing mediator, and xylitol improved NP dispersion. Currently, the chemical compounds of herbs derived by the separation and extraction of alkaloid and non-alkaloid exhibit a variety of biological functions (hypertension, anti-inflammatory, antifungal, antioxidant, metabolic, antimicrobial, and central nervous system (CNS)).

A fascinating and novel application of nanotechnology that has a significant impact on the environment and promotes long-term advancement in the field of nanoscience is the use of plants to create green NPs. Several possible uses of these green plant-based NPs include sensing, imaging, textile engineering, biotechnology, electronics, optics, water treatment, agriculture, food processing, dye degradation, and other biological industries. These NPs may serve as the biomedical field's future driving force in the development of drug delivery systems. These environmentally friendly NPs could potentially be used in a number of other applications, such as water decontamination for ecological clean-up or the treatment of phytopathogens in agriculture. However, future research must focus on the accumulation of these NPs in the environment and their influence, as well as their long-term effects on people and animals. It is expected that in the future, interest in this environmentally friendly approach of NP synthesis will increase significantly.

REFERENCES

Abbasi, B.H., H. Fazal, N. Ahmad, M. Ali, N. Giglioli-Guivarch, and C. Hano. 2020. *Nanomaterials for Cosmeceuticals: Nanomaterials-Induced Advancement in Cosmetics, Challenges, and Opportunities.* Amsterdam, The Netherlands: Elsevier. ISBN 9780128222867.

Aboyewa, J.A., N.R.S. Sibuyi, M. Meyer, and O.O. Oguntibeju. 2021. Gold nanoparticles synthesized using extracts of Cyclopia intermedia, commonly known as honeybush, amplify the cytotoxic effects of doxorubicin. *Nanomaterials* 11:132.

Ahmad, H., K. Venugopal, K. Rajagopal, S. De Britto, B. Nandini, H.G. Pushpalatha, N. Konappa, A.C. Udayashankar, N. Geetha, and S. Jogaiah. 2020. Green synthesis and characterization of zinc oxide nanoparticles using eucalyptus globules and their fungicidal ability against pathogenic fungi of apple orchards. *Biomolecules* 10:425.

Ajuwon, O.R., A.O. Ayeleso, and G.A. Adefolaju. 2018. The potential of South African herbal tisanes, rooibos and honeybush in the management of type 2 diabetes mellitus. *Molecules* 23:3207.

Alagrafi, F.S., A.O. Alawad, N.M. Abutaha, F.A. Nasr, O.A. Alhazzaa, S.N. Alharbi, M.N. Alkhrayef, M. Hammad, Z.A. Alhamdan, A.D. Alenazi, and M.A. Wadaan. 2017. In vitro induction of human embryonal carcinoma differentiation by a crude extract of Rhazya stricta. *BMC Complementary and Alternative Medicine* 17(1):342.

Alaribe, F.N., M.J. Maepa, N. Mkhumbeni, and S.C.K.M. Motaung. 2018. Possible roles of Eucomisautumnalis in bone and cartilage regeneration: A review. *Tropical Journal of Pharmaceutical Research* 17:741–749.

Ali, B.H., A.A. Al-Qarawi, A.K. Bashir, and M.O. Tanira. 2000. Phytochemistry, pharmacology and toxicity of Rhazya stricta Decne: A review. *Phytotherapy Research* 14(4):229–234.

Alshehri, A.A., and M.A. Malik. 2020. Phytomediated photo-induced green synthesis of silver nanoparticles using Matricaria chamomilla L. and its catalytic activity against Rhodamine, B. *Biomolecules* 10:1604.

Andleeb, A., A. Andleeb, S. Asghar, G. Zaman, M. Tariq, A. Mehmood, M. Nadeem, C. Hano, J.M. Lorenzo, and B.H. Abbasi. 2021. A systematic review of biosynthesized metallic nanoparticles as a promising anti-cancer strategy. *Cancers* 13:2818.

Anjum, S., A. Komal, B.H. Abbasi, and C. Hano. 2021. Nanoparticles as elicitors of biologically active ingredients in plants. In *Nanotechnology in Plant Growth Promotion and Protection: Recent Advances and Impacts*. Hoboken, NJ, USA: JohnWiley & Sons,pp. 170–202.

Anjum, S., A.K. Khan, A. Qamar, N. Fatima, S. Drouet, S. Renouard, J.P. Blondeau, B.H. Abbasi, and C. Hano. 2021. Light tailoring: Impact of UV-C irradiation on biosynthesis, physiognomies, and clinical activities of morusmacroura-mediated monometallic (Ag and ZnO) and bimetallic (Ag–ZnO) nanoparticles. *International Journal of Molecular Sciences* 22:11294.

Anjum, S., I. Anjum, C. Hano, and S. Kousar. 2019. Advances in nanomaterials as novel elicitors of pharmacologically active plant specialized metabolites: Current status and future outlooks. *RSC Advances* 9:40404–40423.

Anjum, S., M. Hashim, S.A. Malik, M. Khan, J.M. Lorenzo, B.H. Abbasi, and C. Hano. 2021. Recent advances in zinc oxide nanoparticles (ZnO NPs) for cancer diagnosis, target drug delivery, and treatment. *Cancers* 13:4570.

Anjum, S., S. Ishaque, H. Fatima, W. Farooq, C. Hano, B.H. Abbasi, and I. Anjum. 2021. Emerging applications of nanotechnology in healthcare systems: Grand challenges and perspectives. *Pharmaceuticals* 14:707.

Bachheti, A., R.K. Bachheti, L. Abate, and Azamal Husen. 2022. Current status of aloe-based nanoparticle fabrication, characterization and their application in some cutting-edge areas. *South African Journal of Botany* 147:1058–1069.

Bachheti, R.K., A. Fikadu, Archana Bachheti, and Azamal Husen. 2020. Biogenic fabrication of nanomaterials from flower-based chemical compounds, characterization and their various applications: A review. *Saudi Journal of Biological Sciences* 27(10):2551–2562.

Bachheti, R.K., A. Sharma, A. Bachheti, A. Husen, G.M. Shanka, and D.P. Pandey. 2020b. Nanomaterials from various forest tree species and their biomedical applications. In Husen A., Jawaid M. (eds) *Nanomaterials for Agriculture and Forestry Applications*. Elsevier, pp. 81–106.

Bachheti, R.K., L. Abate, A. Bachheti, A. Madhusudhan, and A. Husen. 2021. Algae-, fungi-, and yeast-mediated biological synthesis of nanoparticles and their various biomedical applications. In *Handbook of Greener Synthesis of Nanomaterials and Compounds*. Elsevier, pp. 701–734.

Bachheti, R.K., Y. Godebo, A. Bachheti, M.O. Yassin, and Azamal Husen. 2020a. Root-based fabrication of metal/metal-oxide nanomaterials and their various applications. In Husen A., Jawaid M. (eds) *Nanomaterials for Agriculture and Forestry Applications*. Elsevier, pp. 135–166.

Baeshen, N.A., S.A. Lari, H.A.R. Al Doghaither, and H.A.I. Ramadan. 2010. Effect of Rhazya stricta extract on rat adiponectin gene and insulin resistance. *Journal of American Science* 6(12):1237–1245.

Bai, R.R., M. Boothapandi, and A. Madhavarani. 2014. Preliminary phytochemical screening and in vitro antioxidant activities of aqueous extract of *Indigofera tinctoria* and *Indigofera astragalina*. *International Journal of Drug Research and Technology* 4:46–54.

Beshah, F., Y. Hunde, M. Getachew, R.K. Bachheti, A. Husen, and A. Bachheti. 2020. Ethnopharmacological, phytochemistry and other potential applications of *Dodonaea* genus: A comprehensive review. *Current Research in Biotechnology*, 2:103–119.

Boudjelal, A., L. Siracusa, C. Henchiri, M. Sarri, B. Abderrahim, F. Baali, and G. Ruberto. 2015. Antidiabetic effects of aqueous infusions of artemisia herba-alba and ajugaiva in alloxan-induced diabetic rats. *Planta Medica* 81:696–704.

Bukhari, N.A., R.A. Al-Otaibi, and M.M. Ibhrahim. 2017. Phytochemical and taxonomic evaluation of *Rhazya stricta* in Saudi Arabia. *Saudi Journal of Biological Sciences* 24(7):1513–1521.

Cherian, T.K. Ali, Q. Saquib, M. Faisal, R. Wahab, and J. Musarrat. 2020. *Cymbopogon citratus* functionalized green synthesis of cuo-nanoparticles: Novel prospects as antibacterial and antibiofilm agents. *Biomolecules* 10:169.

Clark, J.H., and D.J. Macquarrie. 2008. *Handbook of Green Chemistry and Technology*. Hoboken, NJ, USA: John Wiley & Sons.

Dube, P., S. Meyer, A. Madiehe, and M. Meyer. 2020. Antibacterial activity of biogenic silver and gold nanoparticles synthesized from Salvia africana-lutea and Sutherlandiafrutescens. *Nanotechnology* 31:505607.

Elbagory, A.M., M. Meyer, C.N. Cupido, and A.A. Hussein. 2017. Inhibition of bacteria associated withwound infection by biocompatible inhibition of bacteria associated with wound infection by biocompatible green synthesized gold nanoparticles from south african plant extracts. *Nanomaterials* 7:417.

Elbagory, M.A., A.A. Hussein, and M. Meyer. 2019. The in vitro immunomodulatory effects of gold nanoparticles synthesized from hypoxishemerocallidea aqueous extract and hypoxoside on macrophage and natural killer cells. *International Journal of Nanomedicine* 14:9007–9018.

Gajalakshmi, S., S. Vijayalakshmi, and V. Rajeswari. 2013. Pharmacological activities of Catharanthus roseus: A perspective review. *International Journal of Pharma and Bio Sciences* 4:431–439.

Goboza, M., Y.G. Aboua, N. Chegou, and O.O. Oguntibeju. 2019. Biomedicine & pharmacotherapy vindoline effectively ameliorated diabetes-induced hepatotoxicity by docking oxidative stress, inflammation and hypertriglyceridemia in type 2 diabetes-induced male Wistar rats. *Biomedicine & Pharmacotherapy* 112:108638.

Gul, R., H. Jan, G. Lalay, A. Andleeb, H. Usman, R. Zainab, Z. Qamar, C. Hano, and B.H. Abbasi. 2021. Medicinal plants and biogenic metal oxide nanoparticles: A paradigm shift to treat Alzheimer's disease. *Coatings* 11:717.

Hano, C., and D. Tungmunnithum. 2020. Plant polyphenols, more than just simple natural antioxidants: Oxidative stress, aging and age-related diseases. *Medicines* 7:26.

Hossain, A., Y. Abdallah, M.A. Ali, M.M.I. Masum, B. Li, G. Sun, Y. Meng, Y. Wang, and Q. An. 2019. Lemon-fruit-based green synthesis of zinc oxide nanoparticles and titanium dioxide nanoparticles against soft rot bacterial pathogen Dickeyadadantii. *Biomolecules* 9:863.

Husen, A., Q.I. Rahman, M. Iqbal, M.O. Yassin, and R.K. Bachheti. 2019. Plant-mediated fabrication of gold nanoparticles and their applications. In *Nanomaterials and Plant Potential.* Cham: Springer, pp. 71–110.

Jain, P.K., K.S. Lee, I.H. El-Sayed, and M.A. El-Sayed. 2006. Calculated absorption and scattering properties of gold nanoparticles of different size, shape, and composition: Applications in biological imaging and biomedicine. *The Journal of Physical Chemistry B* 110:7238–7248.

Jeong, S., S.Y. Choi, J. Park, J.H. Seo, J. Park, K. Cho, and S.Y. Lee. 2011. Low-toxicity chitosan gold nanoparticles for small hairpin RNA delivery in human lung adenocarcinoma cells. *Journal of Materials Chemistry A* 21:13853–13859.

Joubert, E., and D. de Beer. 2012. Phenolic content and antioxidant activity of rooibos food ingredient extracts. *Journal of Food Composition and Analysis* 27:45–51.

Joubert, E., W.C.A. Gelderblom, A. Louw, and D. de Beer. 2008. South African herbal teas: Aspalathus linearis, Cyclopia spp. and Athrixiaphylicoides-A review. *Journal of Ethnopharmacology* 119:376–412.

Joubert, E., M.E. Joubert, C. Bester, D. de Beer, and J.H. De Lange. 2011. Honeybush (Cyclopia spp.): From local cottage industry to global markets—The catalytic and supporting role of research. *South African Journal of Botany* 77:887–907.

Journal, A.I., K.C. Hembram, R. Kumar, L. Kandha, P.K. Parhi, C.N. Kundu, and B.K. Bindhani. 2018. Therapeutic prospective of plant-induced silver nanoparticles: Application as antimicrobial and anticancer agent. *Artificial Cells, Nanomedicine, and Biotechnology* 46: S38–S51.

Karishma, S., N. Yougasphree, and H. Baijnath. 2018. A comprehensive review on the genus plumbago with focus on plumbago (Plumbaginaceae). *African Journal of Traditional, Complementary and Alternative Medicines* 15:199–215.

Khan, A.K., S. Renouard, S. Drouet, J.-P. Blondeau, I. Anjum, C. Hano, B.H. Abbasi, and S. Anjum. 2021. Effect of UV irradiation (A and C) on Casuarinaequisetifolia-mediated biosynthesis and characterization of antimicrobial and anticancer activity of biocompatible zinc oxide nanoparticles. *Pharmaceutics* 13:1977.

Khan, S.A., S. Shahid, and C.-S. Lee. 2020. Green synthesis of gold and silver nanoparticles using leaf extract of Clerodendrum inerme; characterization, antimicrobial, and antioxidant activities. *Biomolecules* 10:835.

Khan, S.A., S. Shahid, B. Shahid, U. Fatima, and S.A. Abbasi. 2020. Green synthesis of MnO nanoparticles using Abutilon indicum leaf extract for biological, photocatalytic, and adsorption activities. *Biomolecules* 10:785.

Lantero, A. 2014. Sewarine, an indole alkaloid from Rhazya stricta and a k opioid receptor antagonist, induces apoptosis via caspase activation in various cancer cell lines, and inhibits NF-kB activation. *Intrinsic Activity* 2 (Suppl. 1):A1.20.

Lediga, M.E., T.S. Malatjie, D.K. Olivier, D.T. Ndinteh, and S.F. Vuuren. 2018. Biosynthesis and characterisation of antimicrobial silver nanoparticles from a selection of fever-reducing medicinal plants of South Africa. *South African Journal of Botany* 119:172–180.

Letchumanan, D., S.P.M. Sok, S. Ibrahim, N.H. Nagoor, and N.M. Arshad. 2021. Plant-based biosynthesis of copper/copper oxide nanoparticles: An update on their applications in biomedicine, mechanisms, and toxicity. *Biomolecules* 11:564.

Li, S., A.B. Cunningham, R. Fan, and Y. Wang. 2019. Identity blues: The ethnobotany of the indigo dyeing by Landian Yao (Iu Mien) in Yunnan, Southwest China. *Journal of Ethnobiology and Ethnomedicine* 15:13.

Logeswari, P., S. Silambarasan, and J. Abraham. 2015. Synthesis of silver nanoparticles using plants extract and analysis of their antimicrobial property. *Journal of Saudi Chemical Society* 19(3):311–317.

Mahomoodally, M.F. 2013. Traditional medicines in Africa: An appraisal of ten potent African medicinal plants. *Evidence-Based Complementary and Alternative Medicine* 2013:617459.

Marwat, S.K., F. Rehman, K. Usman, S. Syed Shah, and N. Anwar. 2012. A review of phytochemistry, bioactivities and ethno medicinal uses of Rhazya stricta Decsne (Apocynaceae). *African Journal of Microbiology Research* 6(8):1629–1641.

Masondo, N.A., A.O. Aremu, J.F. Finnie, and J Van Staden. 2015. Growth and phytochemical levels in micropropagated Eucomisautumnalis subspecies autumnalis using different gelling agents, explant source, and plant growth regulators. *In Vitro Cellular & Developmental Biology Plant* 51:102–110.

Mativandlela, S.P.N., T. Muthivhi, H. Kikuchi, Y. Oshima, C. Hamilton, A.A. Hussein, M.L. Van Der Walt, P.J. Houghton, and N. Lall. 2009. Antimycobacterial flavonoids from the leaf extract of Galenia africana. *Journal of Natural Products* 72:2169–2171.

Mickymaray, S. 2019. One-step synthesis of silver nanoparticles using Saudi Arabian desert seasonal plant Sisymbrium irio and antibacterial activity against multidrug-resistant bacterial strains. *Biomolecules* 9:662.

Mocheki, T.A., M.H. Ligavha-Mbelengwa, M.P. Tshisikhawe, N. Swelankomo, T.R. Tshivhandekano, M.G. Mokganya, L.I. Ramovha, and N.A. Masevhe. 2018. Comparative population ecology of Sclerocarya birrea (A. rich.) hochst. subspeciescaffra (sond) in two rural villages of limpopo province. *South African Journal of Botany* 50:2339–2345.

Mordeniz, C. 2019. Introductory chapter: Traditional and complementary medicine. *Cengiz Mordeniz* 395:116–124.

Nadeem, M., R. Khan, K. Afridi, A. Nadhman, S. Ullah, S. Faisal, Z.U.I. Mabood, C. Hano, and B.H. Abbasi. 2020. Green synthesis of cerium oxide nanoparticles (CeO_2 NPs) and their antimicrobial applications: A review. *International Journal of Nanomedicine* 15:5951.

Nadeem, M., D. Tungmunnithum, C. Hano, B.H. Abbasi, S.S. Hashmi, W. Ahmad, and A. Zahir. 2018. The current trends in the green syntheses of titanium oxide nanoparticles and their applications. *Green Chemistry Letters and Reviews* 11:492–502.

Narayanan, K.B., and N. Sakthivel. 2011. Green synthesis of biogenic metal nanoparticles by terrestrial and aquatic phototrophic and heterotrophic eukaryotes and biocompatible agents. *Advances in Colloid and Interface Science* 169:59–79.

Nath, D., and P. Banerjee. 2013. Green nanotechnology–A new hope for medical biology. *Environmental Toxicology and Pharmacology* 36:997–1014.

Nehdi I.A., H.M. Sbihi, C.P. Tan, and S.I. Al-Resayes. 2016. Seed oil from Harmal (Rhazya stricta Decne) grown in Riyadh (Saudi Arabia): A potential source of d-tocopherol. *Journal of Saudi Chemical Society* 20(1):107–113.

Ng'uni, T., J.A. Klaasen, and B.C. Fielding. 2018. Acute toxicity studies of the South African medicinal plant *Galenia africana. Toxicology Reports* 5:813–818.

Obaid, A.Y., S. Voleti, R.S. Bora, N.H. Hajrah, A.M.S. Omer, J.S.M. Sabir, and K.S. Saini. 2017. Cheminformatics studies to analyze the therapeutic potential of phytochemicals from Rhazya stricta. *Chemistry Central Journal* 11(1):1–21.

Oguntibeju, O.O., Y. Aboua, and M. Goboza. 2019. Vindoline–A natural product from Catharanthus roseus reduces hyperlipidemia and renal pathophysiology in experimental Type 2 Diabetes. *Biomedicines* 7:59.

Oguntibeju, O.O., S. Meyer, Y.G. Aboua, and M. Goboza. 2016. Hypoxishemerocallidea significantly reduced hyperglycaemia and hyperglycaemic-induced oxidative stress in the liver and kidney tissues of streptozotocin-induced diabetic male Wistar rats. *Evidence-Based Complementary and Alternative Medicine* 2016:8934362.

Omer, A.M. 2008. Energy, environment and sustainable development. *Renewable and Sustainable Energy Reviews* 12:2265–2300.

Orhan, I.E. 2012. Centella asiatica (L.) urban: From traditional medicine to modern medicine with neuroprotective potential. *Evidence-Based Complementary and Alternative Medicine* 2012:946259.

Painuli, R., P. Joshi, and D. Kumar. 2018. Cost-effective synthesis of bifunctional silver nanoparticles for simultaneous colorimetric detection of Al(III) and disinfection. *Sensors and Actuators B: Chemical* 272:79–90.

Painuli, S., P. Semwal, A. Bacheti, R.K. Bachheti, and A. Husen. 2020. Nanomaterials from nonwood forest products and their applications. In Husen A., Jawaid M. (eds) *Nanomaterials for Agriculture and Forestry Applications*. Elsevier, pp. 15–40.

Perveen, K., F.M. Husain, F.A. Qais, A. Khan, S. Razak, T. Afsar, P. Alam, A.M. Almajwal, and M.M.A. Abulmeaty. 2021. Microwave- assisted rapid green synthesis of gold nanoparticles using seed extract of *Trachyspermum ammi*: ROS mediated biofilm inhibition and anticancer activity. *Biomolecules* 11:197.

Ponarulselvam, S., C. Panneerselvam, K. Murugan, N. Aarthi, K. Kalimuthu, and S. Thangamani. 2012. Synthesis of silver nanoparticles using leaves of *Catharanthus roseus* Linn. G. Don and their antiplasmodial activities. *Asian Pacific Journal of Tropical Biomedicine* 2(7):574–580.

Prasad, K.S., S.K. Prasad, M.A. Ansari, M.A. Alzohairy, M.N. Alomary, S. AlYahya, C. Srinivasa, M. Murali, V.M. Ankegowda, and C. Shivamallu. 2020. Tumoricidal and bactericidal properties of ZnONPs synthesized using Cassia auriculata leaf extract. *Biomolecules* 10:982.

Rajagopal, T., P. Ponmanickam, and M. Ayyanar. 2015. Synthesis of silver nanoparticles using *Catharanthus roseus* root extract and its larvicidal effects. *Journal of Environmental Biology* 36:1283–1289.

Rajasekharreddy, P., and P.U. Rani. 2014. Biofabrication of Ag nanoparticles using *Sterculia foetida* L. seed extract and their toxic potential against mosquito vectors and HeLa cancer cells. *Materials Science and Engineering C* 39(1):203–212.

Razavi, M., E. Salahinejad, M. Fahmy, M. Yazdimamaghani, D. Vashaee, and L. Tayebi. 2015. Green chemical and biological synthesis of nanoparticles and their biomedical applications. In *Green Processes for Nanotechnology*. Berlin/Heidelberg, Germany: Springer, pp. 207–235.

Robert, K.W., T.M. Parris, and A.A. Leiserowitz. 2005. What is sustainable development? Goals, indicators, values, and practice. *Environment: Science and Policy for Sustainable Development* 47:8–21.

Rónavári, A., N. Igaz, D.I. Adamecz, B. Szerencsés, C. Molnar, Z. Kónya, and M. Kiricsi. 2021. Green silver and gold nanoparticles: Biological synthesis approaches and potentials for biomedical applications. *Molecules* 26(4):844.

Rout, A., P.K. Jena, U.K. Parida, and B.K. Bindhani. 2013. Green synthesis of silver nanoparticles using leaves extract of Centella asiatica L. for studies against human pathogens. *International Journal of Pharma and Bio Sciences* 4:661–674.

Saleem, K., Z. Khursheed, C. Hano, I. Anjum, and S. Anjum. 2019. Applications of nanomaterials in Leishmaniasis: A focus on recent advances and challenges. *Nanomaterials* 9:1749.

Segal, R., I. Feuerstein, and A. Danin. 1987. Chemotypes of Artemisia herba-alba in Israel based on their sesquiterpene lactone and essential oil constitution. *Biochemical Systematics and Ecology* 15:411–416.

Shafiq, M., S. Anjum, C. Hano, I. Anjum, and B.H. Abbasi. 2020. An overview of the applications of nanomaterials and nanodevices in the food industry. *Foods* 9:148.

Shah, M., S. Nawaz, H. Jan, N. Uddin, A. Ali, S. Anjum, N. Giglioli-Guivarc'H, C. Hano, and B.H. Abbasi. 2020. Synthesis of biomediated silver nanoparticles from Silybummarianum and their biological and clinical activities. *Materials Science and Engineering C* 112:110889.

Silva Viana, R.L., G. Pereira Fidelis, M. Jane Campos Medeiros, M. Antonio Morgano, M. Gabriela Chagas Faustino Alves, L.F. Domingues Passero, D. Lima Pontes, R. Cordeiro Theodoro, T. Domingos Arantes, and D. AraujoSabry, et al. 2020. Green synthesis of antileishmanial and antifungal silver nanoparticles using corn cob xylan as a reducing and stabilizing agent. *Biomolecules* 10:1235.

Singh, P., Y.J. Kim, C. Wang, R. Mathiyalagan, and D.C. Yang. 2016. The development of a green approach for the biosynthesis of silver and gold nanoparticles by using *Panax ginseng* root extract, and their biological applications. *Artificial Cells, Nanomedicine, and Biotechnology* 44(4):1150–1157.

Singh, K., Y. Naidoo, C. Mocktar, and H. Baijnath. 2018. Biosynthesis of silver nanoparticles using Plumbago auriculata leaf and calyx extracts and evaluation of their antimicrobial activities. *Advances in Natural Sciences: Nanoscience and Nanotechnology* 9:035004.

Singh, R., C. Hano, G. Nath, and B. Sharma. 2021. Green biosynthesis of silver nanoparticles using leaf extract of *Carissa carandas* L. and their antioxidant and antimicrobial activity against human pathogenic bacteria. *Biomolecules* 11:299.

Sperling, R.A., P.R. Gil, F. Zhang, M. Zanella, and W.J. Parak. 2008. Biological applications of gold nanoparticles. *Chemical Society Reviews* 37:1896–1908.

Srihasam, S., K. Thyagarajan, M. Korivi, V.R. Lebaka, and S.P.R. Mallem. 2020. Phytogenic generation of nio nanoparticles using Stevia leaf extract and evaluation of their in-vitro antioxidant and antimicrobial properties. *Biomolecules* 10:89.

Sumsakul, W., T. Plengsuriyakarn, W. Chaijaroenkul, V. Viyanant, J. Karbwang, and K. Na Bangchang. 2014. Antimalarial activity of plumbagin in vitro and in animal models. *BMC Complementary Medicine and Therapies* 14:15.

Thabiani, A., M. Ali, C. Panneerselvam, K. Murugan, S. Trivedi, J.A. Mahyoub, M. Hassan, F. Maggi, S. Sut, S. Dall, et al. 2018. The desert wormwood Artemisia herba alba: From folk medicine to a source of green and effective nanoinsecticides against mosquito vectors. *Journal of Photochemistry and Photobiology B: Biology* 180:225–234.

Thipe, V.C., P.B. Njobeh, and S.D. Mhlanga. 2015. Optimization of commercial antibiotic agents using gold nanoparticles against toxigenic Aspergillus spp. *Materials Today: Proceedings* 2:4136–4148.

Tolaymat, T.M., A.M. El Badawy, A. Genaidy, K.G. Scheckel, T.P. Luxton, and M. Suidan. 2010. An evidence-based environmental perspective of manufactured silver nanoparticle in syntheses and applications: A systematic review and critical appraisal of peer-reviewed scientific papers. *Science of the Total Environment* 408:999–1006.

Van Beek, T.A., R. Verpoorte, A.B. Svendsen, and R. Fokkens. 1985. Antimicrobially active alkaloids from *Tabernaemontana chippii*. *Journal of Natural Products* 48(3):400–423.

Vijayan, R., S. Joseph, and B. Mathew. 2018. *Indigofera tinctoria* leaf extract mediated green synthesis of silver and gold nanoparticles and assessment of their anticancer, antimicrobial, antioxidant and catalytic properties catalytic properties. *Artificial Cells, Nanomedicine, and Biotechnology* 46:861–871.

Virginie, A., K. Dago Pierre, M.G. Francois, and A.M. Franck. 2016. Hytochemical Screening of Sclerocarya birrea (Anacardiaceae) and *Khayasenegalensis* (Meliaceae), antidiabetic plants. *International Journal of Pharmaceutical Chemistry* 2:1–5.

Wahid, I., S. Kumari, R. Ahmad, S.J. Hussain, S. Alamri, M.H. Siddiqui, and M.I.R. Khan. 2020. Silver nanoparticle regulates salt tolerance in wheat through changes in ABA concentration, ion homeostasis, and defense systems. *Biomolecules* 10:1506.

Zaeem, A., S. Drouet, S. Anjum, R. Khurshid, M. Younas, J.P. Blondeau, D. Tungmunnithum, N. Giglioli-Guivarc'h, C. Hano, and B.H. Abbasi. 2020. Effects of biogenic zinc oxide nanoparticles on growth and oxidative stress response in flax seedlings vs. in vitro cultures: A comparative analysis. *Biomolecules* 10:918.

15 Green and Cost-Effective Nanoparticles Synthesis from Some Frequently Used Medicinal Plants and Their Various Applications

Meseret Zebeaman, Rakesh Kumar Bachheti, Archana Bachheti, D.P. Pandey, Deepti, and Azamal Husen

CONTENTS

15.1 INTRODUCTION

Medicinal plants have recently been used for preparation of nanoparticles (NPs) as they contain secondary metabolites which can reduce the metal ion to free metal. For instance, the medicinal plant *Aloe vera* is recorded as a most bioactive plant. Anthraquinones, polysaccharides, naphthalenones, proteins, enzymes, and organic acid are the main phytochemicals present in *Aloe vera* leaves and other *Aloe* spp. Some of the important phytochemicals present in *Aloe* spp. are aloin, chrysophanol,

DOI: 10.1201/9781003213727-15

emodin, aloe-emodine, helminthosporin. A possible mechanism of the formation of silver nanoparticles (AgNPs) from *Aloe vera* and the chemistry involved is presented in Figure 15.1. There was the presence of a hydroxyl (–OH) group (as in aloin) in most phytochemicals obtained from *Aloe* spp. and this –OH served as a reducing agent, converting metal ions into metal/metal oxide NPs. Also, carbonyl functional groups present in the phytochemicals of *Aloe* spp. play a significant role in NP fabrication (Logaranjan et al. 2016; Bachheti et al. 2021).

The use of medicinal plant extract to synthesize NPs is the outcome of nanotechnology advancement. It categorizes the above synthesis as green nanotechnology. Nanotechnology studies properties and their multiple uses in nano scale. NPs are solid particles that have a size of less than 100 nm. The unique properties of NPs are that they have a large surface area. This property enables NPs to have higher reactivity. Due to these properties NPs are now widely used in various industries (Drummer et al. 2021).

Surprisingly, nowadays there are over 720 products on the global market which contain nanomaterials (Shand et al. 2006). Still more study is underway to bring more advanced nanotechnology that will transform human life in every aspect. Even in 2000 it was suggested that soon nanotechnology will control every industry.

The conventional techniques used for nanomaterial production also follow either top-down, bottom-up, and/or green methods (Figure 15.2). The two main reasons the green method advanced more than the other two ways are due to cost effectiveness. The reasons are the other methods do not use eco-friendly chemicals and consume too much energy. The second reason is the processes are difficult to scale up to industry (Drummer et al. 2021).

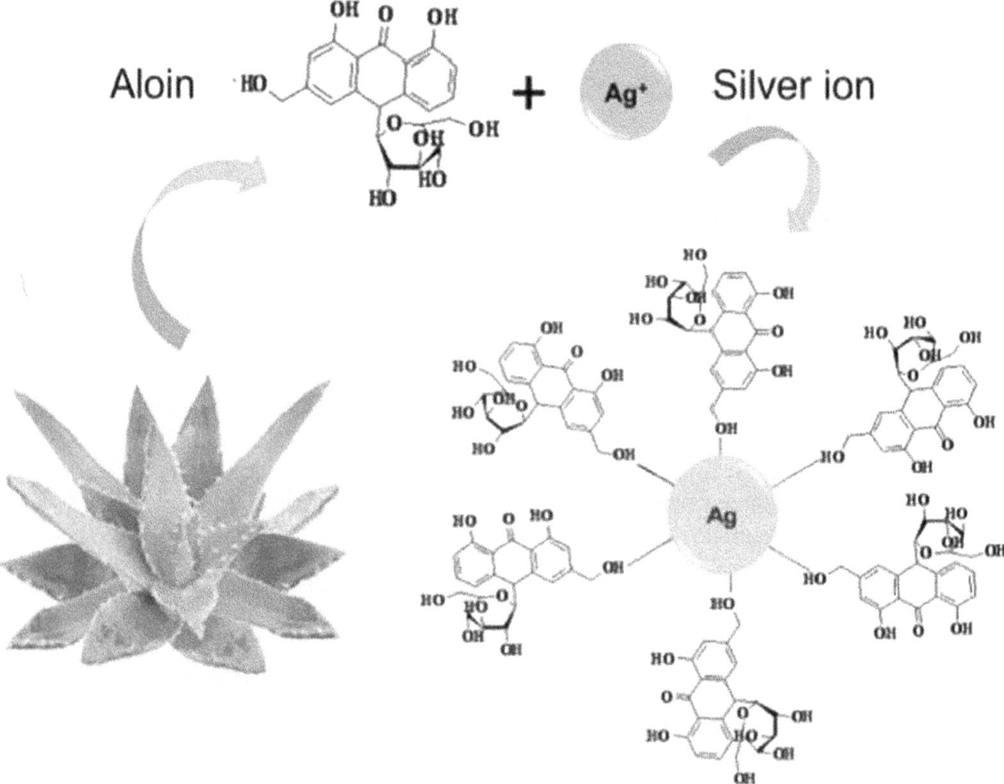

FIGURE 15.1 Mechanism of silver nanoparticles synthesis from *Aloe vera* plant extract (Adopted from Logaranjan et al. (2016).

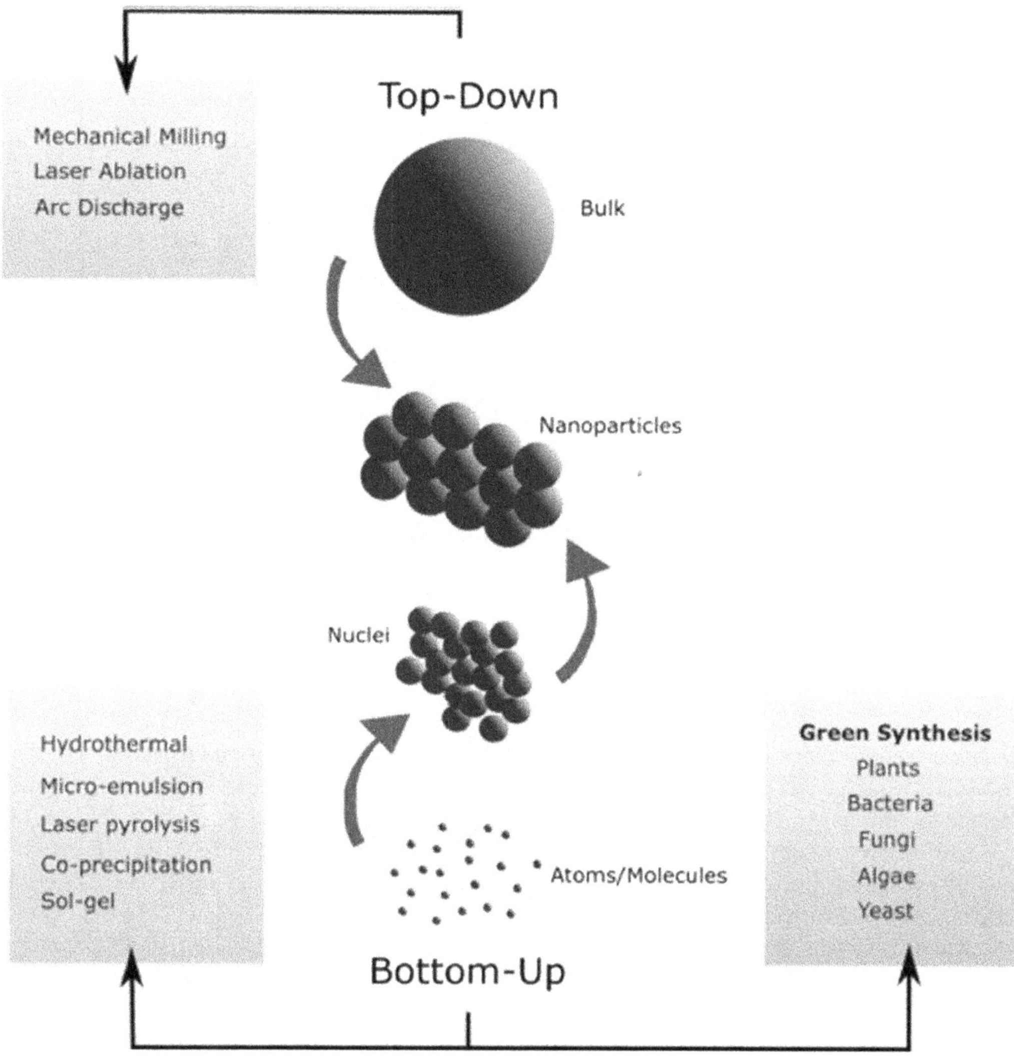

FIGURE 15.2 The three approaches used for nanoparticles synthesis.

The green method uses metabolites of microorganisms and plant extract which are used to reduce and stabilize the product. The plants are more suitable for synthesis. NPs synthesized by the green method are more effective than others (Gunalan et al. 2012). The other importance of the green method is that it is easy to scale up to industry since the inputs are easily available and have low energy consumption (Drummer et al. 2021).

The transition metal oxide NPs from the main element groups are mostly applied in nanotechnology (Sivakami et al. 2020). With this advantage there is also one main drawback of this method to scale up to industry. The problem is it is difficult to control the nature of the NP during synthesis (Olawale et al. 2021).

Besides the available domestic and industry appliances containing nanomaterials on the market, recent studies show NP applications in medicine and chemistry, such as cancer treatment (Chen et al. 2021), antibacterial activity (Ramesh et al. 2021), drug delivery systems (Yew et al. 2020), antioxidant activity (Khalil et al. 2020), catalytic activity (Boudiaf et al. 2021), and the removal of toxic pollutants from wastewater (Sarwar et al. 2021).

Herein, the aim of this chapter is to point out the advantage of the green method of NPs synthesis using the ten most frequently used medicinal plants. The ten medicinal plants are *Ocimum tenuiflorum*, *Aloe vera*, *Tanacetum parthenium*, *Allium sativum*, *Zingiber officinale*, *Curcuma longa*, *Cinnamomum verum*, *Mentha × piperita*, *Azadirachta indica*, and *Tagetes*. Similarly, applications of the synthesized NPs are also discussed. The applications discussed include antibacterial, antioxidant, catalytic, anticancer, and miscellaneous

15.2 NANOPARTICLES FABRICATION FROM FREQUENTLY USED MEDICINAL PLANTS AND THEIR CHARACTERIZATION

Before organic antibiotics emerged in the mid-20th century, metals like silver, copper, brass, and gold were used to fight microbial infection. However due to the advancement of nanotechnology and organic antibiotics, resistance by bacteria to the use of metal as antibiotics is growing. This is because nanometals target multiple cellular processes and it leads to pleiotropic effects (Kamala et al. 2020).

The ten most frequently used medicinal plants which are used for metal nanoparticle synthesis are shown in Table 15.1 and are discussed as follows by reviewing recent publications. To alleviate confusion in NP synthesis the green method has different names which include biogenic, biochemical, eco-friendly, in situ generated, and biosynthesis. Among the transition metals, silver, iron, copper, and zinc and their oxides are mostly studied and some recent publications are presented here.

15.2.1 *Ocimum tenuiflorum*

Ocimum tenuiflorum is also known as holy basil or tulsi. A recent study by Olawal et al. (2021) shows silver (Ag) core – selenium (Se) shell NP (Ag@SeNP) synthesis was possible using a green method of this plant flower. These bimetallic nanoparticles (BNPs) were characterized using UV-visible spectroscopy (UV-Vis), Fourier transform infrared spectroscopy (FT-IR), NP tracking analysis, scanning electron microscopy (SEM), and energy dispersive x-ray analysis (EDX). The shapes are spherical, pentagonal, and hexagonal with the size of Ag 23.5–24 and Se 3.6–4.1nm. The size of BNPs, as SEM measurement indicates, is 33.1 ± 2.7 nm. While Figure 15.3 shows the UV-Vis and FT-IR results of Ag@SeNPs, Figure 15.4 shows the SEM, EDX, and zeta potential of Ag@SeNPs.

Similarly, Pandey et al. (2021) synthesized AgNPs using light induced from this plant. The synthesized AgNPs were analyzed using SEM, DLS, and zeta potential measurement techniques. In 2021 Sharma et al. (2021) also synthesized copper oxide nanoparticles (CuONPs) from the root of the plant by using a green method. To characterize the NP, they used UV-Vis and XRD. The shape and size of the CuONPs are elliptical/spherical/rice-like and 6–18 nm, respectively. In 2015 Raut et al. (2015) synthesized zinc oxide nanoparticles (ZnONPs) from the leaf of the holy basil using the green method. The characterization of the NPs is done by XRD, SEM, and FT-IR. The shape and size of the NP is hexagonal and 13.8 nm, respectively. Similarly, in 2020 Kaura et al (2020) synthesized iron nanoparticles (FeNPs) from the leaf of the plant. Table 15.1 shows the synthesis, characterization, and applications of NPs synthesized from some frequently used medicinal plants

15.2.2 *Aloe vera*

Aloe vera is a succulent shrub with medicinal properties due to its richness in biomolecules. *Aloe vera* gel contains pectin, cellulose, hemicelluloses, flavonoids, polyphenols, anthraquinones, ascorbic acid, citric acid, acetic acid, and acetylated galactoglucomannan called acemannan. Due to the richness of the biomolecules in *Aloe vera* it has nutritional, medicinal, and cosmetic values (Kamala et al. 2020).

TABLE 15.1
Synthesis, Characterization, and Applications of Nanoparticles Synthesized from Some Frequently Used Medicinal Plants

Plant botanical name	Nanoparticles	Plant part used	Size (nm)	Shape	Characterization techniques	Application	Synthesis Method	Key reference
Aloe vera	AgNPs, AuNPs	Gel	12–14, and less than 15	Poly-dispersed spherical	UV-Vis, FT-IR, TEM, and XRD	Antibacterial fabric	Green	Nalini et al. (2020)
	AgNPs	Gel	66.6	Spherical	UV-Vis, FT-IR, and TEM,	Antioxidant activity	Biochemical	Sohal et al. (2019)
	AgNPs	Leaf	46	Dispersed spherical	UV-Vis, FT-IR, TEM, and DLS	Antibacterial activity	Microwave	Ahmadi et al. (2018)
	AgNPs	Leaf	5–50	Octahedron	UV-Vis, SEM, TEM, FT-IR, XRD, and SERS	Antimicrobial activity	Biogenic	Logaranjan et al. (2016)
	AgNPs, FeNPs	Leaf	36.61, 34.93	Spherical	UV-Vis, FT-IR, and TEM	Antimicrobial activity	Green	Yadav et al. (2016)
	AgNPs	Leaf	70–120	Crystalline	XRD and SEM	Antibacterial	Hydrothermal	Tippayawat et al. (2016)
	AgNPs	Leaf	70	Rectangular, triangular, and spherical	UV-Vis SEM, and FT-IR	Antifungal activity	Green	Medda et al. (2015)
	ZnONPs	Leaf	50–220	Hexagonal	UV-Vis, FT-IR, SEM, TEM, and XRD	Antibacterial, and antifungal activities	Biosynthesis	Chaudhary et al. (2019)
	ZnONPs	Gel/leaf	–	Hexagonal rod	FT-IR, UV-Vis, XRD, SEM, and EDX	Antibiotic, and antioxidant activities	Biosynthesis	Mahendiran et al. (2017)
	ZnONPs	Leaf	15		UV-Vis, FT-IR SEM, EDX, TEM, and XRD	Antibacterial activity	Biogenic	Ali et al. (2016)
	CuNPs	Flower			UV-Vis, FESEM, and FT-IR		Green	Karimi et al. (2015)
***Ocimum tenuiflorum* L.**	Ag Core Se Shell (Se-capped Ag NPs) (Ag@Se NPs)	Flower	33.1±2.7 (Ag 23.5–24, Se 3.6–4.1)	Near spherical, pentagonal, and hexagonal	UV-Vis, FT-IR, nanoparticle tracking analysis, SEM, and EDX	Antioxidant, cytotoxicity, and allium cepa assay activities	Biogenic	Olawale et al. (2021)

(Continued)

TABLE 15.1 (CONTINUED)
Synthesis, Characterization, and Applications of Nanoparticles Synthesized from Some Frequently Used Medicinal Plants

Plant botanical name	Nanoparticles	Plant part used	Size (nm)	Shape	Characterization techniques	Application	Synthesis Method	Key reference
	AgNPs	Leaf	85±37	Spherical AgNPs without any agglomeration	UV-Vis, SEM, DLS, and zeta potential	Antibacterial activity	Light induced	Pandey et al. (2021)
	AgNPs	Leaf	28	Irregular	UV-Vis, XRD, AFM, and SEM	Antibacterial activity	–	Logeswari et al. (2015)
	AgNPs	Leaf	20	Circular with rough edges	UV-vis, TEM, XRD, and FT-IR	Toxic activity	Green	Daniel et al. (2011)
	AgNPs	Endophyte Fungal	-	-	TEM, SEM, DLS, XRD, and FT-IR,	Antibiotic, and antiproliferative activities	Biogenically	Bagur et al. (2020)
	AgNPs		5–40	Spherical shape	FT-IR, AFM, and SERS	Drug delivery		Jha et al. (2017)
	AgNPs	Leaf	-	-	FT-IR, XRD, TGA, SEM, and TEM	Antibacterial activity	*In situ*-generated	Sadanand et al. (2016)
	AgNPs	Leaf	25–40	-	UV-Vis, TEM, DLS, SAED, and zeta potential	Antibacterial activity	Bioinspired	Patil et al. (2012)
	Capped AuNPs (C-AuNPs)	Flower and leaf	20–30	Spherical	UV-Vis, FT-IR, ATM, and TEM	Antibacterial activity	Green	Rao et al. (2017)
	CuNPs	Leaf	22.09	Spherical	FT-IR, XRD, PSA, and TEM	-		Ramadhan et al. (2019)
	CuONPs	Green route	6–18	Elliptical/ spherical/rice	UV-Vis, and XRD,	Photocatalytic and antibacterial activities	Eco-friendly, green	Sharma et al. (2021)
	ZnONPs	Leaf	13.86	Hexagonal	XRD, SEM, and FT-IR		Green	Sagar Raut et al. (2015)
	FeNPs	Leaf				Anti-inflammatory activity		Kaur (2020)

(Continued)

TABLE 15.1 (CONTINUED)
Synthesis, Characterization, and Applications of Nanoparticles Synthesized from Some Frequently Used Medicinal Plants

Plant botanical name	Nanoparticles	Plant part used	Size (nm)	Shape	Characterization techniques	Application	Synthesis Method	Key reference
Tanacetum parthenium	Au and Au/ZnO NPs	Herb (leaf)	5 and 7	Spherical	UV-Vis, TEM, AFM, and FT-IR	Phytotoxicity	Biologically	Szymanski et al. (2021)
	CuONPs	Leaf and flower	16	Globular	UV-Vis, XRD, FT-IR, FESEM, and EDX	Antimicrobial, cytotoxic, and catalytic activities	Green	Ranjbar et al. (2021)
Allium sativum	AgNPs	Clove	7.3 ± 4.4	Spherical	UV-Vis, TEM, GA-XRD (glancing angle-XRD), and FT-IR	Antibacterial activity	Sunlight assisted, green	Rastogi et al. (2011)
	AgNP-plant extract	–	26±7	Spherical	UV-Vis, FT-IR, TEM, EDX, and XPS	Antimycotic activity	Chemical (Lee–Meisel seed synthesis)	Robles-Martínez et al. (2019)
	AgNPs	Garlic	5.28	Spherical	UV-Vis, FT-IR, XRD, TEM, and SEM,	Antibacterial activity	Synthesis	Otunola et al. (2017)
	ZnONPs	Garlic	27	Hexagonal wurtiz	UV-Vis, FT-IR, XRD, and TEM	Antibacterial and antioxidant activities	Chemical, biological	Stan et al. (2016)
	CuONPs	Raw garlic	20–40	Spherical	UV-Vis, FT-IR, XRD, SEM, HRTEM, and EDX	Antimicrobial, antioxidant, and antilarvicidal activities	Green	Velsankar et al. (2020)
Zingiber officinale	AgNPS	Leaf	18.93	Spherical	UV-Vis, FT-IR, FESEM, and TEM	Anti-human pancreatic cancer activity	Green	Wang et al. (2021)
	AgNPs	Ginger powder	11–24	Semi-spherical	UV-Vis, FT-IR, XRD, SEM, and EDX	Catalytic activity	Clean, green	Eisa et al. (2019)

(Continued)

TABLE 15.1 (CONTINUED)
Synthesis, Characterization, and Applications of Nanoparticles Synthesized from Some Frequently Used Medicinal Plants

Plant botanical name	Nanoparticles	Plant part used	Size (nm)	Shape	Characterization techniques	Application	Synthesis Method	Key reference
	AgNPs	Ginger	12.97	Spherical	UV-Vis, FT-IR XRD, TEM, and SEM	Antimicrobial activity	Synthesis	Otunola et al. (2017)
	ZnONPs	Root	17–40	Spherical	UV–Vis, FESEM, and ED	Biosensor	Green	Dönmez (2020)
Curcuma longa	AgNPs	Leaf	15–40	Spherical	UV-Vis, FT-IR HRTEM, SEM, and EDX	Antibacterial Fabric	Green	Maghimaa et al. (2020)
	ZnONPs	Rhizome	25	Spherical	XRD and TEM	Antibacterial and antioxidant activities	Eco-friendly	Venoy (2019)
Cinnamomum verum	AgNPs	Bark	60–80	Spherical	SEM and AFM	Subacute toxicity activity	Green	Alwan et al. (2021)
	ZnONPs	Bark	40–48	Hexagonal wurtiz	UV-Vis, FT-IR, TEM, and SEM		Green	Ansari et al. (2020)
Mentha × *piperita*	AgNPs	Leaves	448–800	Spherical	UV-Vis, FT-IR, XRD, TEM, SEM, and EDX	Antibacterial activity	Green	Mojally et al. (2022)
	ZincNPs	Leaves	75	Spherical	XRD, FT-IR, SEM, and zeta potential	Antibacterial activity	Eco-friendly	Hasnain et al. (2020)
Azadirachta indica	AgNPs	Leaves	33	Spherical	UV-Vis, EDX, SEM, and TEM	Antimicrobial activity	Synthesis	Chinnasamy et al. (2021)
	ZnONPs	Leaf	20–40	Spherical	UV-Vis, FT-IR XRD, FESEM and EDAX	DNA-interaction / stabilization	Green	Singh et al. (2019)
Tagetes	AgNPs	Flower	24–49	Spherical	UV-Vis, FT-IR, XRD, SEM, and EDXA	Plasmonic devices	Green	Katta et al. (2021)
	ZnONPs	Flower	30–50	Spherical	UV-Vis, FT-IR, XRD, and SEM	Cytotoxicity activity	Green	Ilangovan et al. (2021)

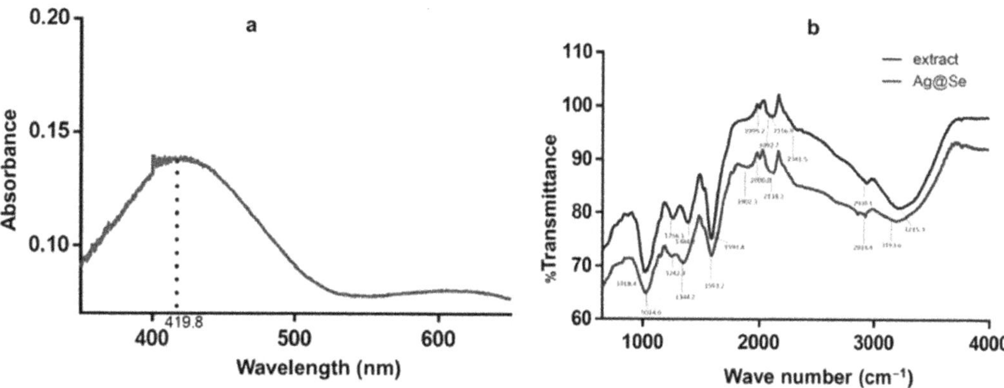

FIGURE 15.3 (a) Ultraviolet-visible spectroscopy and (b) Fourier transform infrared spectroscopy spectrum of silver core selenium shell nanoparticles (Adopted from Olawale et al. 2021).

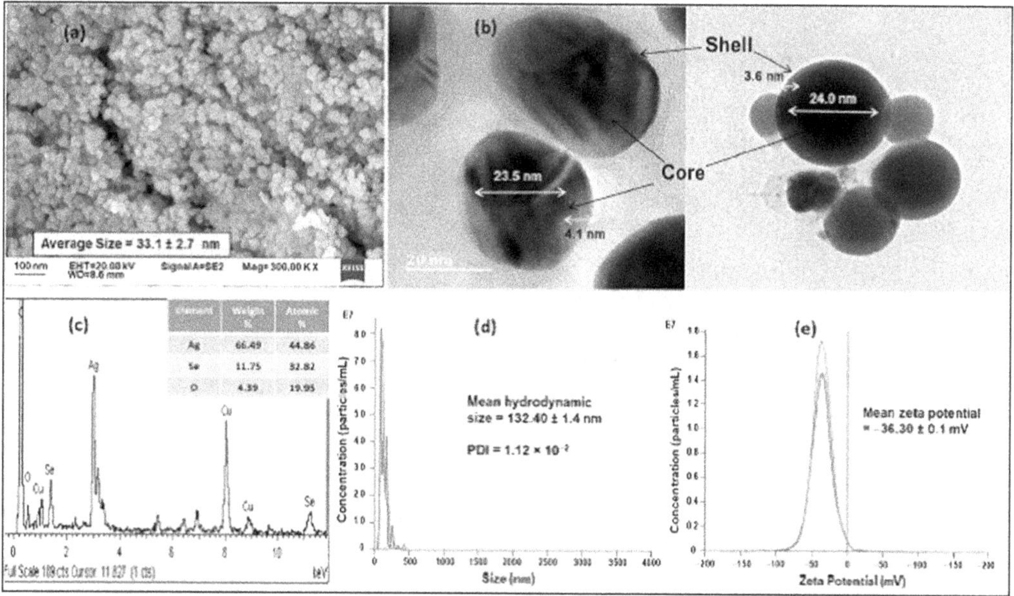

FIGURE 15.4 (a) Scanning electron microscopy, (b) high resolution transmission electron microscopy, (c) energy dispersive x-ray analysis, (d) nanoparticle tracking analysis (NTA) and, (e) zeta potential image of silver core selenium shell nanoparticles (Adopted from Olawale et al. 2021).

In 2020 Kamala et al. (2020) synthesized AgNPs and gold nanoparticles (AuNPs) 10–40 nm in size and of polydispersed spherical shape. AgNPs and AuNPs were synthesized by the green method. The NPs are characterized by UV-Vis, FT-IR, TEM, and XRD. Figures 15.5 and 15.6 show AgNPs and AuNPs UV-Vis and TEM images, respectively.

Sohal and his co-worker synthesized AgNPs from plant gel. It is found that the NPs are 66.6 nm in size and spherical in shape as TEM analysis confirmed. The difference between the two studies is the size of the NPs (10–40 vs 66.6 nm) they synthesized while the shape was similar, spherical. The other difference is the biological activity they measure. The first one measured antibacterial activity while the second one measured antioxidant activity in the NPs. In 2017 a microwave-assisted synthesis of AgNPs from the leaf of *Aloe vera* was done by Ahmadi et al. (2018). The size and shape of the NPs are 46 nm and dispersed spherical, respectively. The characterization is done via UV-Vis,

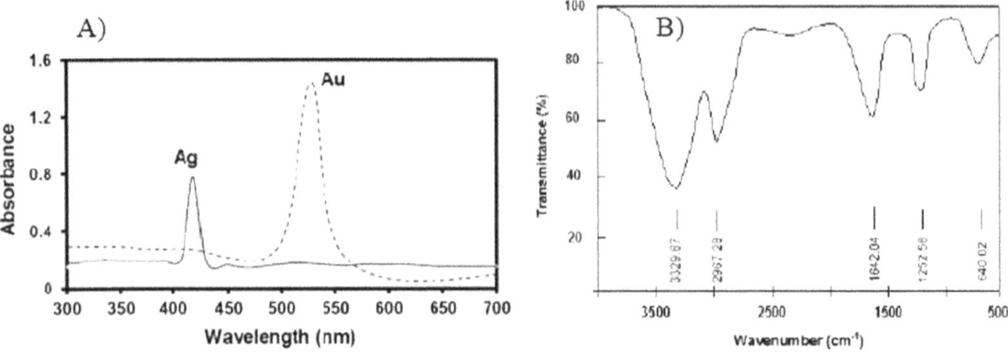

FIGURE 15.5 (A) UV-visible spectroscopy and (B) Fourier transform infrared spectroscopy of silver and gold nanoparticles (Adopted from Nalini et al. 2020).

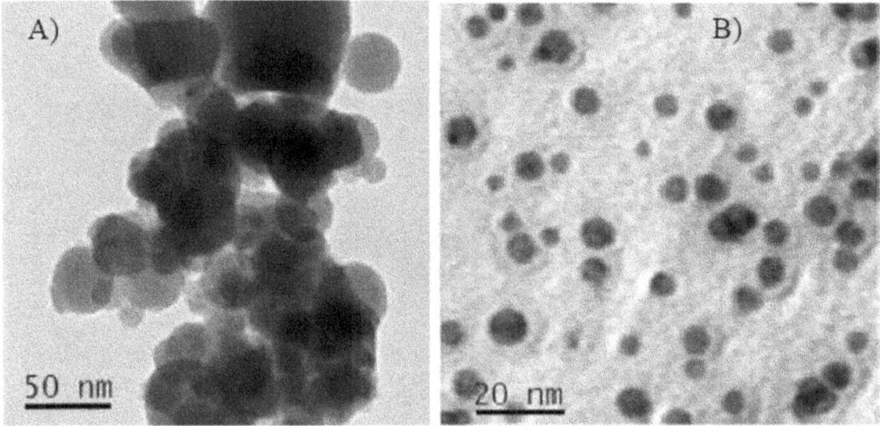

FIGURE 15.6 Transmission electron microscopy images of (A) silver and (B) gold nanoparticles (Adopted from Nalini et al. 2020).

FT-IR, TEM, and DLS. The other transition metals studied are ZnONPs (Chaudhary et al. 2019), CuNPs (Karimi et al. 2015), and FeNPs (Yadav et al. 2016). They are tabulated Table 15.1.

15.2.3 *Tanacetum parthenium*

Tanacetum parthenium is called the medieval aspirin. While Ranjbar et al. (2021) synthesized CuONPs by the reducing agent of feverfew leaf extract, Szymanski et al. (2021) synthesized AuNPs. The characterization technique they used is quite different, the first used UV-Vis, TEM, atomic force microscopy (AFM), and FT-IR and the second used UV-Vis, XRD, FT-IR, field emission SEM (FESEM), and EDX. The authors of this chapter suggest that the second research is good at characterization because it used EDX to characterize the presence of the transition metals. For more information like size and shape as well as plant part used and biological activity test type measured, refer to Table 15.1.

15.2.4 *Allium sativum*

Allium sativum belongs to the Amaryllidaceae family. Velsankar et al. (2020) synthesized CuONPs by using green method. They characterized it by using UV-Vis, FT-IR, XRD, SEM, high resolution

TEM (HRTEM), and EDX. The size and shape are 17–40 nm and spherical, respectively. Similarly, Stan et al. (2016) synthesized ZnONPs by using both biological and chemical methods whereas Otunola et al. (2017) synthesized AgNPs by using a green method (Cf. Table 15.1).

15.2.5 *Zingiber officinale*

Zingiber officinale is widely used as a spice and in traditional medicine. Recently Wang et al. (2021) synthesized AgNPs from this plant using the green method. Similarly, Dönmez (2020) synthesized ZnONPs. The characterization technique, shape and size of NPs, and biological activity they measured are tabulated in Table 15.1. A detailed discussion is presented in the next section of this chapter.

15.2.6 *Curcuma longa*

Curcuma longa is known as turmeric in English and mostly used in the home as an additive. As much research shows it is one of the best known medicinal plants with many uses to treat different illnesses. Recently Maghimaa et al. (2020) synthesized AgNPs from the leaves of turmeric and adhered them on cloth to evaluate their antibiotic activity for wound healing. The NP structure they synthesized is spherical and characterized by UV-Vis, FT-IR, HRTEM, SEM, and EDX. Similarly, Venoy (2019) synthesized ZnONPs from the leaves of turmeric using the eco-friendly method. He evaluated its antibacterial and antioxidant activity.

15.2.7 *Cinnamomum verum*

Cinnamomum verum is called the true cinnamon tree or Ceylon cinnamon tree; it is a small evergreen tree belonging to the Lauraceae family. It is also mostly used as a food additive. Recently, Alwan et al. (2021) synthesized AgNPs from the bark of the plant and evaluated their sub-acute toxicity. They found spherical and 60 nm-sized AgNPs by characterizing it with SEM and AFM. Similarly, Ansari et al. (2020) synthesized ZnONPs from the bark of the plant using the green method. They found a hexagonal wurtiz structure and 40 nm-sized ZnONPs as they characterized it with UV-Vis, FT-IR, XRD, TEM, and SEM.

15.2.8 *Mentha piperita*

Mentha piperita is a known flavouring additive with many medicinal applications. It is a hybrid mint, a cross between watermint and spearmint. Recently Mojally et al. (2022) synthesized AgNPs from the leaves of the plant using the green method and evaluated their antibacterial activity. They found AgNPs 480 nm in size and spherical in shape as they characterized it with UV-Vis, FT-IR, XRD, TEM, SEM, and EDX. Hasnain et al. (2020) also synthesized ZnNPs from the leaves and evaluated their antibacterial activity. They were 75 nm in size and spherical in shape as they characterized the NPs with XRD, FT-IR, SEM, and zeta potential.

15.2.9 *Azadirachta indica*

Azadirachta indica is known as an insecticidal medicinal plant in addition to many known traditional medicinal applications. Its oil is marketed globally. Recently, Chinnasamy et al. (2021) synthesized AgNPs from the leaves of the plant and evaluated their antibacterial application *in vitro*. They were 33 nm and spherical AgNPs by characterizing them with UV-Vis, EDX, SEM, and TEM. Similarly, Singh et al. (2019) synthesized ZnONPs from the leaves of the plant using the green method and evaluated their DNA interaction and stability. They found 20 nm and spherical ZnONPs as indicated by the following characterization techniques: UV-Vis, FT-IR, XRD, FESEM, and EDAX.

FIGURE 15.7 The ten most widely studied medicinal plants used for metal nanoparticle synthesis. (A) *Ocimum tenuiflorum* L., (B) *Aloe vera*, (C) *Tanacetum parthenium*, (D) *Allium sativum*, (E) *Zingiber officinale*, (F) *Curcuma longa*, (G) *Cinnamomum verum*, (H) *Mentha piperita*, (I) *Azadirachta indica*, and (J) *Tagetes*.

15.2.10 *Tagetes*

Tagetes, or marigold flower, is a genus in the sunflower or aster family that includes 49 species of flowers. They are used for digestive tract problems including poor appetite, gas, stomach pain, colic, intestinal worms, and dysentery. It is also used for coughs, colds, mumps, fluid retention, and sore eyes; and they cause sweating. Recently, Ilangovan et al. (2021) synthesized ZnONPs from the flower by using the green method and evaluated their cytotoxicity. As they characterized the particle with UV-Vis, FT-IR, XRD, an SEM, they found 30 nm, spherical ZnONPs. Similarly, Katta et al. (2021) synthesized AgNPs from the flower and found 24 nm in size and a spherical shape. To characterize the particles, they used the following characterization techniques: UV-Vis, FT-IR, XRD, SEM, and EDAX. Figure 15.7 shows the ten most widely studied medicinal plants used for transition metal nanoparticle synthesis.

15.3 APPLICATIONS OF NANOPARTICLES

As mentioned above, NPs have various applications. Some biological and chemical applications are discussed here as follows.

15.3.1 Antimicrobial Activities

The most studied antimicrobial NP activity is antibacterial and antifungal properties. AgNPs and AuNPs prepared with *Aloe vera* extract in 2020 inhibit *Escherichia coli* and *Staphylococcus aureus* growth at 30 μl/ml. Similarly, these NPs attached to cloth inhibited the pathogens *Escherichia coli* and *Staphylococcus aureus* at a dosage of 10 μl/ml using a streak plate (Kamala et al. 2020). Chaudhary et al. (2019) also synthesized ZnONPs from *Aloe vera* and found the effective against

Escherichia coli bacteria and *Aspergillus niger* fungi. The synergetic or the combined NPs have antibiotic as well as antifungal effects, however, it was low as compared to individual NPs. Therefore, more work is needed to study ZnONPs as antibiotics. In 2021, Pandy and his colleagues synthesized AgNPs from *Ocimum tenuiflorum* L. These AgNPs inhibit both Gram-positive and -negative bacteria. Surprisingly, an indirect relationship was observed between the antimicrobial efficacy and the size of AgNPs (Pandy et al. 2021). The other finding, as the Medda et al. (2015) study reveals, AgNPs synthesized from *Aloe vera* showed selective inhibition of some fungi. They found that AgNPs inhibit *Aspergillus* sp. growth more than *Rhizopus* spp. Similarly, and recently Maghimaa et al. (2020) synthesized AgNPs from the leaves of *Curcuma longa* and coated a cloth with them to evaluate their antibiotic activity for wound healing. The crude extract, as well as formulated AgNPs, also exhibited a noticeable antimicrobial potency against *S. aureus*, *P. aeruginosa*, *S. pyogenes*, and *C. albicans*.

At the same time Chinnasamy et al. (2021) synthesized AgNPs from the leaves of the *Azadirachta indica* and evaluated their antibacterial application *in vitro* using the disc diffusion method. The AgNPs inhibited the bacterial growth at 1000 mg/l with a zone of inhibition against *Bacillus cereus* (17.7 mm), *Escherichia coli* (18.7 mm), *Pseudomonas aeruginosa* (10.3 mm), and *Staphylococcus aureus* (17.7 mm).

15.3.2 Antioxidant Activities

Like secondary metabolites, NPs' antioxidant properties are mostly studied using 2,2-diphenyl-1-picrylhydrazyl (DPPH) assays. The bimetallic nanoparticle, Ag@Se, synthesized from *Ocimum tenuiflorum* showed that as concentration increased antioxidant activity also increased (Olawal et al. 2021). This study showed that these NPs scavenge fewer DPPH radicals than ferric ions. Surprisingly, the plant extract showed more activity than both the NPs and standard ascorbic acid (Olawal et al. 2021). Similarly, CuONPs synthesized with the help of *Allium sativum* are also antioxidant (Velsankar et al. 2020). All the above results show NPs are good antioxidants. In addition, AgNPs synthesized by Wang et al. (2021) from *Zingiber officinale* scavenge the DPPH molecules at 172 mg/ml by half. Similarly, Venoy (2019) synthesized ZnONPs from the leaves of *Curcuma longa* using the eco-friendly method. He evaluated its antioxidant activity using DPPH and hydrogen peroxide assay. They found ZnONPs to be a good scavenger of superoxide radical, nitric oxide, and hydrogen peroxide and to have reducing power, which is greater than ascorbic acid at a higher concentration.

15.3.3 Anticancer Activities

Different cancers are now treated by anticancer AgNP drugs. AgNPs synthesized by Wang et al. (2021) from *Zingiber officinale* showed anticancer activity from its (3-(4, 5-dimethylthiazolyl-2)-2, 5-diphenyltetrazolium bromide (MTT) assay confirmed. The study concludes that a diagnosis can be pancreatic cancer. Olawal et al. (2021) synthesized Ag@SeNPs using an *Ocimum tenuiflorum* extract. The bimetallic Ag@Se NPs anticancer or cytotoxic activity is evaluated using MTT assay. Ranjbar et al. (2021) synthesized CuONPs from the feverfew plant and found that CuONPs showed good anticancer activity.

15.3.4 Catalytic Activities

When NPs come to catalytic use, CuONPs are currently good catalysts. Besides, as NPs have a high surface area this is good for catalytic activity. Hence, Ranjbar et al. (2021) synthesized CuONPs from the *Tanacetum parthenium* plant using the green method. CuONPs showed good catalytic reduction on industrial dyes. Similarly, Eisa et al. (2019) synthesized AgNPs from *Zingiber officinale* and evaluated their catalytic activity. The catalytic activity is measured by using dyes. They also found a way to recover the catalyst.

15.3.5 Miscellaneous Applications

Other applications for NPs include drug delivery, anti-inflammatory, biosensor, DNA-interaction/stabilization, and plasmonic devices (photocatalyst). Jha et al. (2017) synthesized AgNPs using *Ocimum tenuiflorum* and used them in drug delivery by encapsulating them in carbon nanotubes. In 2020 it was also possible to make biosensors from ZnONPs synthesized by using *Zingiber officinale* and used them as glucose detectors (Dönmez 2020). The other study which shows that NPs can stabilize DNA could be a good indication that NPs can serve as biosensing and biomedical applications (Singh et al. 2019). According to the study done by Kaur (2020), FeNPs are good anti-inflammatory candidates.

15.4 CONCLUSION AND FUTURE PROSPECTS

NPs synthesized by the green method using medicinal plants can be used to make antibiotic dressings for wound-healing purposes. Similarly, the other biological applications, of antioxidant and anti-inflammatory use of NPs, are also confirmed by various research. For this purpose, mostly AgNPs and ZnONPs are used and studied The catalytic activities of NPs are studied by dye removal because NPs have a high surface area. CuONPs and AgNPs are mostly used for catalytic activity. The other biological application of NPs is in anticancer use. For this purpose, metallic like AgNPs, bimetallic NPs like Ag@Se, and also metallic oxides like CuONPs are showing good results from the MTT assay. The other application is drug delivery and as a biosensor. Therefore, with all the above advantages of NPs and the available few NP products on the market more work is indeed in the future. The other limitation is more *in vivo* work, and the biological mechanism of action is not studied.

REFERENCES

Ahmadi, O., Jafarizadeh-Malmiri, H., & Jodeiri, N. 2018. Eco-friendly microwave-enhanced green synthesis of silver nanoparticles using Aloe vera leaf extract and their physico-chemical and antibacterial studies. *Green Processing and Synthesis*, *7*(3), 231–240.

Ali, K., Dwivedi, S., Azam, A., Saquib, Q., Al-Said, M. S., Alkhedhairy, A. A., & Musarrat, J. 2016. Aloe vera extract functionalized zinc oxide nanoparticles as nanoantibiotics against multi-drug resistant clinical bacterial isolates. *Journal of Colloid and Interface Science*, *472*, 145–156.

Alwan, S., Al-Saeed, M., & Abid, H. 2021. Safety assessment and biochemical evaluation of biogenic silver nanoparticles (using bark extract of C. zeylanicum) in Rattus norvegicus rats: Safety of biofabricated AgNPs (using Cinnamomum zeylanicum extract). *Baghdad Journal of Biochemistry and Applied Biological Sciences*, *2*(3), 138–150.

Ansari, M. A., Murali, M., Prasad, D., Alzohairy, M. A., Almatroudi, A., Alomary, M. N., ... Niranjana, S. R. 2020. Cinnamomum verum bark extract mediated green synthesis of ZnO nanoparticles and their antibacterial potentiality. *Biomolecules*, *10*(2), 336.

Bachheti, A., Bachheti, R. K., Abate, L., & Husen, A. 2021. Current status of Aloe-based nanoparticle fabrication, characterization and their application in some cutting-edge areas. *South African Journal of Botany*, *147*, 1058–1069.

Bagur, H., Poojari, C. C., Melappa, G., Rangappa, R., Chandrasekhar, N., & Somu, P. 2020. Biogenically synthesized silver nanoparticles using endophyte fungal extract of Ocimum tenuiflorum and evaluation of biomedical properties. *Journal of Cluster Science*, *31*(6), 1241–1255.

Boudiaf, M., Messai, Y., Bentouhami, E., Schmutz, M., Blanck, C., Ruhlmann, L., ... Mekki, D. E. 2021. Green synthesis of NiO nanoparticles using Nigella sativa extract and their enhanced electro-catalytic activity for the 4-nitrophenol degradation. *Journal of Physics and Chemistry of Solids*, *153*, 110020.

Chaudhary, A., Kumar, N., Kumar, R., & Salar, R. K. 2019. Antimicrobial activity of zinc oxide nanoparticles synthesized from Aloe vera peel extract. *SN Applied Sciences*, *1*(1), 1–9.

Chen, J., Li, Y., Fang, G., Cao, Z., Shang, Y., Alfarraj, S., ... Duan, X. 2021. Green synthesis, characterization, cytotoxicity, antioxidant, and anti-human ovarian cancer activities of Curcumae kwangsiensis leaf aqueous extract green-synthesized gold nanoparticles. *Arabian Journal of Chemistry*, *14*(3), 103000.

Chinnasamy, G., Chandrasekharan, S., Koh, T. W., & Bhatnagar, S. 2021. Synthesis, characterization, antibacterial and wound healing efficacy of silver nanoparticles from Azadirachta indica. *Frontiers in Microbiology, 12*, 204.

Daniel, S. C. G. K., Kumar, R., Sathish, V., Sivakumar, M., Sunitha, S., & Sironmani, T. A. 2011. Green synthesis (Ocimum tenuiflorum) of silver nanoparticles and toxicity studies in zebra fish (Danio rerio) model. *International Journal of Nanoscience and Nanotechnology, 2*, 103–117.

Dönmez, S. 2020. Green synthesis of zinc oxide nanoparticles using Zingiber officinale root extract and their applications in glucose biosensor. *El-Cezeri Journal of Science and Engineering, 7*(3), 1191–1200.

Drummer, S., Madzimbamuto, T., & Chowdhury, M. 2021. Green synthesis of transition-metal nanoparticles and their oxides: A review. *Materials, 14*(11), 2700.

Eisa, W. H., Zayed, M. F., Anis, B., Abbas, L. M., Ali, S. S., & Mostafa, A. M. 2019. Clean production of powdery silver nanoparticles using Zingiber officinale: The structural and catalytic properties. *Journal of Cleaner Production, 241*, 118398.

Gunalan, S., Sivaraj, R., & Rajendran, V. 2012. Green synthesized ZnO nanoparticles against bacterial and fungal pathogens. *Progress in Natural Science: Materials International, 22*(6), 693–700.

Hasnain, Z., Zafar, S., Shafqat, U., Perveen, S., Iqbal, N., Qaisrani, S. A., ... Mumtaz, S. 2020. Antibacterial activity of eco-friendly zinc nanoparticles prepared from leaf extract of Mentha piperita L. *Pakistan Journal of Pharmaceutical Sciences, 33*(5 (Special)), 2413–2416.

Ilangovan, A., Venkatramanan, A., Thangarajan, P., Saravanan, A., Rajendran, S., & Kaveri, K. 2021. Green synthesis of zinc oxide nanoparticles (ZnO NPs) using aqueous extract of tagetes erecta flower and evaluation of its antioxidant, antimicrobial, and cytotoxic activities on HeLa cell line. *Current Biotechnology, 10*(1), 61–76.

Jha, P. K., Jha, R. K., Rout, D., Gnanasekar, S., Rana, S. V., & Hossain, M. 2017. Potential targetability of multi-walled carbon nanotube loaded with silver nanoparticles photosynthesized from Ocimum tenuiflorum (tulsi extract) in fertility diagnosis. *Journal of Drug Targeting, 25*(7), 616–625.

Kamala Nalini, S. P., & Vijayaraghavan, K. 2020. Green synthesis of silver and gold nanoparticles using Aloe vera gel and determining its antimicrobial properties on nanoparticle impregnated cotton fabric. *Journal of Nanotechnology Research, 2*(3), 42–50.

Karimi, J., & Mohsenzadeh, S. 2015. Rapid, green, and eco-friendly biosynthesis of copper nanoparticles using flower extract of Aloe vera. *Synthesis and Reactivity in Inorganic, Metal-Organic, and Nano-Metal Chemistry, 45*(6), 895–898.

Katta, V. K. M., & Dubey, R. S. 2021. Green synthesis of silver nanoparticles using Tagetes erecta plant and investigation of their structural, optical, chemical and morphological properties. *Materials Today: Proceedings, 45*, 794–798.

Kaur, M. 2020. Impact of response surface methodology–optimized synthesis parameters on in vitro anti-inflammatory activity of iron nanoparticles synthesized using Ocimum tenuiflorum Linn. *BioNanoScience, 10*(1), 1–10.

Khalil, A. T., Ovais, M., Ullah, I., Ali, M., Shinwari, Z. K., & Maaza, M. 2020. Physical properties, biological applications and biocompatibility studies on biosynthesized single phase cobalt oxide (Co3O4) nanoparticles via Sageretia thea (Osbeck.). *Arabian Journal of Chemistry, 13*(1), 606–619.

Logaranjan, K., Raiza, A. J., Gopinath, S. C., Chen, Y., & Pandian, K. 2016. Shape-and size-controlled synthesis of silver nanoparticles using Aloe vera plant extract and their antimicrobial activity. *Nanoscale Research Letters, 11*(1), 1–9.

Logaranjan, K., Raiza, A. J., Gopinath, S. C., Chen, Y., & Pandian, K. 2016. Shape-and sizecontrolled synthesis of silver nanoparticles using Aloe vera plant extract and their antimicrobial activity. *Nanoscale Research Letters, 11*(1), 1–9.

Logeswari, P., Silambarasan, S., & Abraham, J. 2015. Synthesis of silver nanoparticles using plants extract and analysis of their antimicrobial property. *Journal of Saudi Chemical Society, 19*(3), 311–317.

Maghimaa, M., & Alharbi, S. A. 2020. Green synthesis of silver nanoparticles from Curcuma longa L. and coating on the cotton fabrics for antimicrobial applications and wound healing activity. *Journal of Photochemistry and Photobiology B: Biology, 204*, 111806.

Mahendiran, D., Subash, G., Arumai Selvan, D., Rehana, D., Senthil Kumar, R., & Kalilur Rahiman, A. 2017. Biosynthesis of zinc oxide nanoparticles using plant extracts of Aloe vera and Hibiscus sabdariffa: Phytochemical, antibacterial, antioxidant and anti-proliferative studies. *BioNanoScience, 7*(3), 530–545.

Medda, S., Hajra, A., Dey, U., Bose, P., & Mondal, N. K. 2015. Biosynthesis of silver nanoparticles from Aloe vera leaf extract and antifungal activity against Rhizopus sp. and Aspergillus sp. *Applied Nanoscience, 5*(7), 875–880.

Mojally, M., Sharmin, E., Obaid, N. A., Alhindi, Y., & Abdalla, A. N. 2022. Polyvinyl alcohol/corn starch/castor oil hydrogel films, loaded with silver nanoparticles biosynthesized in Mentha piperita leaves' extract. *Journal of King Saud University – Science, 34*(4), 101879.

Olawale, F., Ariatti, M., & Singh, M. 2021. Biogenic synthesis of silver-core selenium-shell nanoparticles using Ocimum tenuiflorum L.: Response surface methodology-based optimization and biological activity. *Nanomaterials, 11*(10), 2516.

Otunola, G. A., Afolayan, A. J., Ajayi, E. O., & Odeyemi, S. W. 2017. Characterization, antibacterial and antioxidant properties of silver nanoparticles synthesized from aqueous extracts of Allium sativum, Zingiber officinale, and Capsicum frutescens. *Pharmacognosy Magazine, 13*(Suppl 2), S201.

Pandey, V. K., Upadhyay, S. N., & Mishra, P. K. 2021. Light-induced synthesis of silver nanoparticles using Ocimum tenuiflorum extract: Characterisation and application. *Journal of Chemical Research, 45*(1–2), 179–186.

Patil, R. S., Kokate, M. R., & Kolekar, S. S. 2012. Bioinspired synthesis of highly stabilized silver nanoparticles using Ocimum tenuiflorum leaf extract and their antibacterial activity. *Spectrochimica Acta Part A: Molecular and Biomolecular Spectroscopy, 91*, 234–238.

Ramadhan, V. B., Ni'mah, Y. L., Yanuar, E., & Suprapto, S. 2019. Synthesis of copper nanoparticles using Ocimum tenuiflorum leaf extract as capping Agent. In *AIP Conference Proceedings* (Vol. 2202, No. 1, p. 020067). AIP Publishing LLC.

Ramesh, P., Saravanan, K., Manogar, P., Johnson, J., Vinoth, E., & Mayakannan, M. 2021. Green synthesis and characterization of biocompatible zinc oxide nanoparticles and evaluation of its antibacterial potential. *Sensing and Bio-Sensing Research, 31*, 100399.

Ranjbar, M., Dastani, M., & Khakdan, F. (2021). *Evaluation of Antimicrobial, Cytotoxicity and Catalytic Activities of CuO-NPs Synthesized by Tanacetum parthenium Extract.*

Rao, Y., Inwati, G. K., & Singh, M. 2017. Green synthesis of capped gold nanoparticles and their effect on Gram-positive and Gram-negative bacteria. *Future Science OA, 3*(4), FSO239.

Rastogi, L., & Arunachalam, J. 2011. Sunlight based irradiation strategy for rapid green synthesis of highly stable silver nanoparticles using aqueous garlic (Allium sativum) extract and their antibacterial potential. *Materials Chemistry and Physics, 129*(1–2), 558–563.

Raut, S., Thorat, P., & Thakre, R. 2015. Green synthesis of zinc oxide (ZnO) nanoparticles using Ocimumtenuiflorum leaves. *International Journal of Science and Research, 4*(5), 1225–1228.

Robles-Martínez, M., González, J. F. C., Pérez-Vázquez, F. J., Montejano-Carrizales, J. M., Pérez, E., & Patiño-Herrera, R. 2019. Antimycotic activity potentiation of Allium sativum extract and silver nanoparticles against Trichophyton rubrum. *Chemistry and Biodiversity, 16*(4), e1800525.

Sadanand, V., Rajini, N., Satyanarayana, B., & Rajulu, A. V. 2016. Preparation and properties of cellulose/silver nanoparticle composites with in situ-generated silver nanoparticles using Ocimum sanctum leaf extract. *International Journal of Polymer Analysis and Characterization, 21*(5), 408–416.

Sarwar, N., Humayoun, U. B., Kumar, M., Zaidi, S. F. A., Yoo, J. H., Ali, N., ... Yoon, D. H. 2021. Citric acid mediated green synthesis of copper nanoparticles using cinnamon bark extract and its multifaceted applications. *Journal of Cleaner Production, 292*, 125974.

Shand, H., & Wetter, K. 2006Shrinking science: An introduction to nanotechnology. Chapter 5. *In State of the World 2006: Special Focus: China and India.* The Worldwatch Institute. New York, USA: WW Norton & Company, 83.

Sharma, S., Kumar, K., Thakur, N., Chauhan, S., & Chauhan, M. S. 2021. Eco-friendly Ocimum tenuiflorum green route synthesis of CuO nanoparticles: Characterizations on photocatalytic and antibacterial activities. *Journal of Environmental Chemical Engineering, 9*(4), 105395.

Singh, A., & Kaushik, M. 2019. Physicochemical investigations of zinc oxide nanoparticles synthesized from Azadirachta indica (Neem) leaf extract and their interaction with Calf-Thymus DNA. *Results in Physics, 13*, 102168.

Sivakami, M., Renuka, R., & Thilagavathi, T. 2020. Green synthesis of magnetic nanoparticles via Cinnamomum verum bark extract for biological application. *Journal of Environmental Chemical Engineering, 8*(5), 104420.

Sohal, J. K., Saraf, A., Shukla, K., & Shrivastava, M. 2019. Determination of antioxidant potential of biochemically synthesized silver nanoparticles using Aloe vera gel extract. *Plant Science Today, 6*(2), 208–217.

Stan, M., Popa, A., Toloman, D., Silipas, T. D., & Vodnar, D. C. 2016. Antibacterial and antioxidant activities of ZnO nanoparticles synthesized using extracts of Allium sativum, Rosmarinus officinalis and Ocimum basilicum. *Acta Metallurgica Sinica (English Letters), 29*(3), 228–236.

Szymanski, M., & Dobrucka, R. 2021. Application of phytotests to study of environmental safety of biologicaly synthetised au and Au/ZnO nanoparticles using Tanacetum parthenium extract. *Journal of Inorganic and Organometallic Polymers and Materials, 32*, 1–16.

Tippayawat, P., Phromviyo, N., Boueroy, P., & Chompoosor, A. 2016. Green synthesis of silver nanoparticles in aloe vera plant extract prepared by a hydrothermal method and their synergistic antibacterial activity. *PeerJ*, *4*, e2589.

Velsankar, K., Ashwin Kumar, R.M., Preethi, R., Muthulakshmi, V., & Sudhahar, S. 2020. Green synthesis of CuO nanoparticles via Allium sativum extract and its characterizations on antimicrobial, antioxidant, antilarvicidal activities. *Journal of Environmental Chemical Engineering*, *8*(5), 104123.

Vinoy Jacob, R. P. 2019. In vitro analysis: The antimicrobial and antioxidant activity of zinc oxide nanoparticles from Curcuma longa. *In Vitro*, *12*(1), 200–204.

Wang, Y., Chinnathambi, A., Nasif, O., & Alharbi, S. A. 2021. Green synthesis and chemical characterization of a novel anti-human pancreatic cancer supplement by silver nanoparticles containing Zingiber officinale leaf aqueous extract. *Arabian Journal of Chemistry*, *14*(4), 103081.

Yadav, J. P., Kumar, S., Budhwar, L., Yadav, A., & Yadav, M. 2016. Characterization and antibacterial activity of synthesized silver and iron nanoparticles using Aloe vera. *Journal of Nanomedicine and Nanotechnology*, *7*(384), 2.

Yew, Y. P., Shameli, K., Miyake, M., Khairudin, N. B. B. A., Mohamad, S. E. B., Naiki, T., & Lee, K. X. 2020. Green biosynthesis of superparamagnetic magnetite Fe3O4 nanoparticles and biomedical applications in targeted anticancer drug delivery system: A review. *Arabian Journal of Chemistry*, *13*(1), 2287–2308.

16 Aromatic Oils from Medicinal Plants and Their Role in Nanoparticles Synthesis, Characterization, and Applications

Sandeep Singh, Neetu Panwar, Smita S. Kumar, Rajesh Singh, Gagan Anand, Amit Kumar, and Sandeep K. Malyan

CONTENTS

ABBREVIATIONS

MAPs Medicinal and aromatic plants
MB Methylene
MOP Morphine
MTT assay, 2,5-diphenyl-2H-tetrazolium bromide

16.1 INTRODUCTION

Aromatic and essential oils are obtained from several medicinal plant parts such as roots, leaves, flower, barks, fruits, etc. These essential and aromatic oils are generally extracted through important techniques, namely extraction with solvent and azeotropic (steam, distillation, hydro-distillation, and hydro-diffusion) (Akhtar et al. 2019; Elyemni et al. 2019). Medicinal and aromatic plants (MAPs) have high economic value due to their role as drugs, cosmetics, and food (Kralova and Jampilek 2021; Kumari et al. 2019; Nair et al. 2022; Raut and Karuppayil 2014). Population explosions simulate the disease and infection rate which simultaneously enhances the demand for and production of chemical and MOP drugs. Synthetic drugs often have side-effects/toxicity on human health while on the other hand side-effects due to the use of plant-based drugs are negligible. Recently, Reddy (2019) reported that there are about 100 herbs from which essential oil can be extracted and used for

DOI: 10.1201/9781003213727-16

different purposes beneficial to mankind. MOP is also used for fragrance and some of the important plants are listed in Table 16.1. The plant-based treatment of several diseases has been documented in ancient Indian and global literature. Kumari et al. (2019) recently quoted that about 80% of total global diseases are directly or indirectly cured by traditional MAPs. The phyto molecules present in MAPs play a significant role and therefore are considered vital constituents of MAPs. Aromatic oils have low molecular weight and phenylpropanoids, terpenoids, and short chain hydrocarbons are major components. There are several approaches for the application of aromatic oils. Generally, the extracted pure oil is diluted with solvent and then consumed later. Direct application of pure aromatic oils on the skin or other parts of the body may cause skin irritation, burning, etc. (Bilia et al. 2021; Orchard and Van Vuuren 2017; Sindle and Martin 2021).

Green nanoparticles (NPs) can be synthesized from the extract of MOP and their efficiency in drugs and industrial wastewater treatment is significant. In this chapter, we compile the previous studies describing the role of different green NP synthesis in the fields of drug, textile, wastewater treatment, and other sectors.

TABLE 16.1
Examples of a Few Important Medicinal and Aromatic Plants

Common name of plant	Scientific name	Aromatic oil potential	Reference
Basil	*Ocimum basilicum* L.	Basil oil	Kiec-Swierczynska et al. (2010); Martínez-González et al. (2007); Pinto et al. (2021)
Lavender	*Lavandula angustifolia*	Lavender oil	De Groot and Schmidt (2016); Pasias et al. (2021); Pinto et al. (2021); Samadi et al. (2021)
Bergamot orange	*Citrus bergamia*	Bergamot oil	Kaddu et al. (2001); Navarra et al. (2015); Salvino et al. (2022); Sarkic and Stappen (2018); Sindle and Martin (2021); Valussi et al. (2021a)
Rose	*Rosa rubiginosa*	Rose Oil	Boutekedjiret et al. (2003); Sindle and Martin (2021)
Cumin	*Cuminum cyminum*	Cumin oil	Karimirad et al. (2018); Kumari et al. (2019)
Common jasmine	*Jasminum officinale*	Jasmine oil	Al-snafi (2015); Al-Snafi (2020); Engelberg et al. (2021); Lo et al. (2020); Lubbe and Verpoorte (2011)
Peppermint	*Mentha piperita*	Mint Oil	Lubbe and Verpoorte (2011); Pinto et al. (2021)
Patchouli	*Pogostemon cablin*	Patchouli oil	Kusuma and Mahfud (2017); Lubbe and Verpoorte (2011); Shah et al. (2017)
Damask rose	*Rosa damascena*	Rose oil	Babu et al. (2002); Lubbe and Verpoorte (2011); Manouchehri et al. (2018); Younis et al. (2007)
Vetiver	*Chrysopogon zizanioides*	Vetiver oil	Danh et al. (2010); Jindapunnapat et al. (2018); Kusuma et al. (2017); Lubbe and Verpoorte (2011)
Rosemary	*Rosmarinus officinalis*	Rosemary oil	González-Mahave et al. (2006); Lešnik et al. (2021); Lubbe and Verpoorte (2011); Oualdi et al. (2021); Pieracci et al. (2021); Pinto et al. (2021); Sharma et al. (2020)
Indian sandalwood	*Santalum album*	Sandalwood oil	Kusuma and Mahfud (2018); Lubbe and Verpoorte (2011); Nautiyal (2014); Xin-Hua et al. (2012); Zhang et al. (2012)
Java citronella	*Cymbopogon winterianus*	Citronella oil	Lubbe and Verpoorte (2011); Saha et al. (2021); Wany et al. (2014); Wu et al. (2019)
Bitter orange	*Citrus aurantium*	Bergamot oil	Boussaada and Chemli (2007); Figoli et al. (2006); Lubbe and Verpoorte (2011); Valussi et al. (2021b); Valussi et al. (2021a)
Clove	*Salvia multicaulis*	Clove oil	Guan et al. (2007); Ratri et al. (2020); Sinico et al. (2005)

16.2 METHOD OF AROMATIC OIL EXTRACTION FROM HERBAL PLANTS

Aromatic oil is extracted from herbal plants and used for several purposes. The demand for aromatic oil is increasing with human population growth. The selection of appropriate technology is one of the critical factors to improve the economical yield and quality of aromatic oil. Steam distillation, solvent extraction, maceration, enfleurage, cold-press extraction, water and steam distillation, CO_2 extraction, and water distillation are the most commonly and widely used extraction technologies (Figure 16.1) (Zhang et al. 2018).

16.3 AROMATIC OIL-BASED NANOPARTICLES AND THEIR APPLICATIONS

Many chemical compounds have adverse effects on humans and the environment; this has led to increased demand for green or natural innovations. NPs have a wide range of uses which can be applied for the remediation of environmental pollution, enhancing drug efficiency, etc. The chemical-treated approach for the remediation of dyes and nitrophenol from wastewater and contaminated soil has limitations. The nature-based innovations for environmental pollution remediation and drug efficiency are gaining global interest (Bharti et al. 2022; Hosny et al. 2021; Kaplan 2022; Kumari and Meena 2020; Malyan et al. 2020; Nair et al. 2022, Malyan et al. 2021). The role of aromatic oil-based NPs in the research area of wastewater and drugs is of great significance.

16.4 NANOPARTICLES AND WASTEWATER TREATMENT

Pollutants such as dyes, pharmaceutical compounds, metals etc., can be efficiently removed from aqueous solution through modified NPs. Methylene blue (MB) dye removal from an aqueous solution using NPs extracted from *Ricinus communis* seed was reported by Abdelfatah et al. (2021b). Biomolecules extracted from seed enhance reactivity by reducing aggregation of nZVI which results in the removal of up to 96.8% of the dye (Abdelfatah et al. 2021b). Green platinum NPs synthesized from *Atriplex halimus* leaves have antimicrobial and antioxidant properties which enhance the removal rate of MB dye from aqueous solution (Eltaweil et al. 2022). In a few seconds the addition of platinum NPs completely degraded the MB dye up to 60 ppm. Degradation of 100 ppm MB dye, platnium NPs required 5 min of duration (Eltaweil et al. 2022). The platinum NPs act as strong catalytic agents which degrade the dimethylamino groups of MB dye and result in 100% removal of MB dye. The addition of platinum NPs-degraded MB dye converts MB dye to leuco dye which leads to

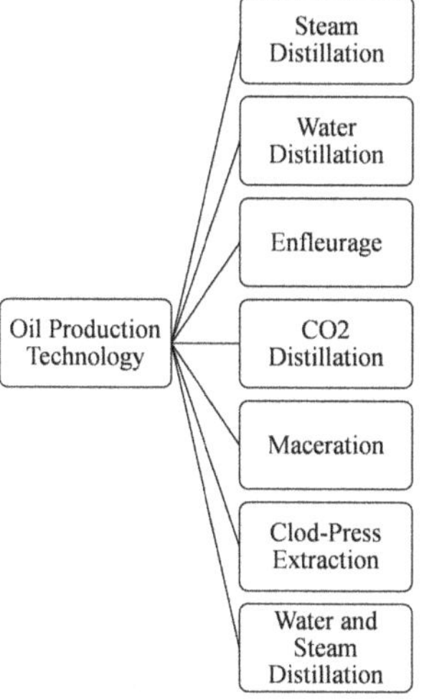

FIGURE 16.1 Widely used technologies for the extraction of aromatic oil from medicinal and other plants.

complete removal of MB dye in just a few seconds (Eltaweil et al. 2022; Sallam et al. 2018). Similar findings were also reported by Hosny et al. (2022b) for MB dye degradation. Hosny et al. (2022) found that the platinum NPs synthesized from *Polygonum salicifolium* leaves have high degradation efficiency and they degrade up to 100 ppb of MB dye immediately. Platinum NPs synthesized from *Polygonum salicifolium* leaves have high anti-Gram-negative bacterial properties which give them potential as antidiabetic agents (Hosny et al. 2022b). Recently, Hosny et al. (2022a) observed that gold NPs synthesized from *Ziziphus spina-christi* leaf extract removed 81.14% of malachite green after 100 min under ultraviolet light (UV). The removal of malachite green is significantly affected by pH, and it was observed that at very high and low pH the rate of dye degradation was low. The application of gold NPs enhanced the rate of malachite green by controlling aqueous pH (Hosny et al. 2022a). The removal of 98% of the tetracycline from the aqueous solution through iron NPs was also reported by Abdelfatah et al. (2021a). Textile wastewater dyes such as MB, rhodamine, and nigrosine photo-degradation under green synthesis zinc oxide NPs were recently observed by (Raghavendra et al. 2022) which indicated that green NPs can be used for the removal of multiple wastewater pollutants.

The removal of copper ions (Cu^{2+}) from an aqueous solution through silver NPs has been reported by Al-Senani and Al-Kadhi (2020). Al-Senani and Al-Kadhi (2020) observed that the silver NPs synthesized from the leaf extract of the *Convolvulus arvensis* plant removed 98.99% of 10 ppm of Cu^{2+} ions through a dose of 0.2 g silver NPs. Degradation of crystal violet dye up to 80% through green iron NPs (synthesis from *Ruellia tuberosa* leaf extract) was reported by Vasantharaj et al. (2019). On the basis of the above discussion, it can be concluded that different aromatic plant-based green NPs can be used for the removal of dyes, metals, antibacterial activity etc., in the field of wastewater treatment and dependence on chemical-based technology can be reduced (Table 16.2).

16.5 NANOPARTICLES AND DRUGS

The biomedical potential of several green NPs for medicinal activities is reported globally (Ahmed and Mustafa 2020; Alwhibi et al. 2022; Hosny et al. 2022a; Hosny and Fawzy 2021; Kocak et al. 2022; Kumari and Meena 2020; Mohanasundari et al. 2022; Rehman et al. 2022). The antibacterial, antifungal, anticancer, and antioxidant activity of different green nanomaterials is widely reported (Table 16.3). Hosny et al. (2022a) observed that gold NPs synthesized from *Ziziphus spina-christi* leaves with a particle size of 1–10 nm have anticancer properties. Particle size plays a significant role in biological activity and particle size is inversely proportional (Ahmed and Mustafa 2020). Nanomaterials smaller than 15 nm are highly suitable for medicinal applications (Ahmed and Mustafa 2020; Zulfiqar et al. 2019). Hosny et al. (2021) synthesized gold NPs from *Chenopodiun amperosides* and *Atriplex halimus* and investigated the MTT assay for both and found that the *Atriplex halimus*-based gold NPs have higher anticancer potential (Table 16.3). Silver NPs synthesized from *Ferula pseudalliacea* have inhibiting potential for both Gram-negative and Gram-positive bacteria along with antifungal properties (Kocak et al. 2022). Kumari and Meena (2020) synthesized green NPs from the leaf extract of *Lawsonia inermis* and found that the gold NPs significantly degraded dyes and nitrophenol without having any adverse effect on human lung cells. The synthesis of green silver NPs with particle sizes 41.43–60.51 nm with a spherical shape from the extract of *Aristolochia bracteolate* shoots shows excellent antioxidant potential (Thanh et al. 2022) (Table 16.3).

16.6 CONCLUSIONS

The synthesis of NPs from medicinal and other plants has green potential in the fields of food, medicine, wastewater treatment, and cosmetics. The size of the NPs depends on the extraction and manufacturing technique. This review describes the potential and diversity of aromatic oils from

TABLE 16.2
Different Aromatic Nanoparticles in Wastewater Treatment

Nanoparticle (NP)	Size (nm)	Aromatic plants	Remarks	Reference
Platinum NP	1–3	*Atriplex halimus*	Antimicrobial, and antioxidant, used for removal of MB dye from aqueous solution.	Eltaweil et al. (2022)
Platinum NP	4.40	*Hibiscus sabdariffa*	Photodegradation of MB dye with high antibacterial activity for the bacteria (Gram-positive).	Seckin et al. (2022)
Zinc oxide NP	–	*Areca catechu*	Antibacterial and photo-degradation for different textile dyes	Raghavendra et al. (2022)
Gold NP	1–10	*Ziziphus spina-christi*	Reported for the removal of malachite green dye removal.	Hosny et al. (2022a)
Platinum NP	1–3	*Polygonum salicifolium*	Immediately removed 100% of 100 ppm concentration MB dye.	Hosny et al. (2022b)
Zinc oxide NP	55–70	*Crataegus monogyna*	Have high antibacterial activity.	Fouladi-Fard et al. (2022)
Zinc oxide NP	40–50	*Ruellia tuberosa*	Degraded 92% and 94% of malachite green and methylene blue respectively from textile industrial wastewater.	Vasantharaj et al. (2021a)
Copper oxide NP	10–15	*Justicia gendarussa*	Has antibacterial and dye degradation potential in wastewater.	Vasantharaj et al. (2021b)
Iron NP	–	*Ricinus communis*	96.8% of MB dye removal is achieved on 25°C and 6 pH.	Abdelfatah et al. (2021b)
Iron NP	–	*Ricinus communis*	Significantly removed tetracycline from wastewater.	Abdelfatah et al. (2021a)
Gold NP	18	*Phragmites australis*	Removed MB dye in just 1 min.	El-Borady et al. (2021)
Silver NP	–	*Convolvulus arvensis*	Used for the removal of copper ions removal	Al-Senani and Al-Kadhi (2020)
Iron NP	52.78	*Ruellia tuberosa*	This iron NP has antibacterial and violet dye degradation potential.	Vasantharaj et al. (2019)
Silver NP	15–40	*Phoenix dactylifera*	100% potency for *C. albicans* and *E. coli* at 40 µg/l of sliver NP.	Oves et al. (2018)

medicinal plants and their role in NP synthesis, characterization, and applications. Aromatic oil-mediated synthesis of platinum, gold, iron, and silver NPs showed antibacterial, antifungal, and anticancer activity, and use in wastewater treatment. Dyes from industrial wastewater from fibres such as MB, malachite green, and violet dyes can be completely (100%) decomposed using NPs formed by aromatic oils from medicinal plants. Based on scanning electron microscope (SEM) results, the size of the synthesized NPs is typically in the range of 1–70 nm. Aromatic plants are widespread all over the world, but the synthesis of NPs from aromatic oils is an unexplored region. No detailed work has been done in this area. Therefore, it is necessary to study aromatic medicinal plants for synthesizing NPs for various purposes.

ACKNOWLEDGEMENTS

The authors (Sandeep Singh, Rajesh Singh, and Sandeep K. Malyan) gratefully acknowledge the National Institute of Hydrology, Roorkee, India, for all the necessary support.

TABLE 16.3
Different Green Nanoparticles with Medicinal Activity

Nanoparticle (NP)	Particles size (nm)	Aromatic plants	Application	Reference
Platinum NP	1–3	*Atriplex halimus*	Antimicrobial and antioxidant	Eltaweil et al. (2022)
Gold NP	1–10	*Ziziphus spina-christi*	Anticancer properties	Hosny et al. (2022a)
Zinc oxide NP	8–10	*Morinda tinctoria*	Anticancer activity	Anitha et al. (2022)
Sliver NP	41.43–60.51	*Aristolochia bracteolata*	Antioxidant potential	Thanh et al. (2022)
Silver NP	–	*Ferula pseudalliacea*	Antibacterial Gram-negative and -positive) and antifungal properties	Kocak et al. (2022)
Gold NP	2–10 40	*Atriplex halimus* *Chenopodium amperosidies*	Anticancer properties	Hosny et al. (2021)
Gold NP	18	*Phragmites australis*	Anticancer	El-Borady et al. (2021)
Gold NP	5.23	*Persicaria salicifolia*	Anticancer potential	Hosny and Fawzy (2021)
Gold NP	5–23	*Persicaria salicifolia*	Anticancer properties	Hosny and Fawzy (2021)
Sliver NP	25–35	*Ziziphus jujuba*	Antibacterial and anticancer properties	Naghizadeh et al. (2021)
Gold NP	20	*Lawsonia inermis*	Degradation of dyes and nitrophenol	Kumari and Meena (2020)

REFERENCES

Abdelfatah, A.M., Fawzy, M., El-Khouly, M.E., Eltaweil, A.S., 2021a. Efficient adsorptive removal of tetracycline from aqueous solution using phytosynthesized nano-zero valent iron. *J. Saudi Chem. Soc.* 25(12), 101365. https://doi.org/10.1016/j.jscs.2021.101365

Abdelfatah, A.M., Fawzy, M., Eltaweil, A.S., El-Khouly, M.E., 2021b. Green synthesis of nano-zero-valent iron using Ricinus communis Seeds extract: Characterization and application in the treatment of methylene blue-polluted water. *ACS Omega* 6(39), 25397–25411. https://doi.org/10.1021/acsomega.1c03355

Ahmed, R.H., Mustafa, D.E., 2020. Green synthesis of silver nanoparticles mediated by traditionally used medicinal plants in Sudan. *Int. Nano Lett.* 10(1), 1–14. https://doi.org/10.1007/s40089-019-00291-9

Akhtar, M.S., Swamy, M.K., Sinniah, U.R., 2019. *Natural Bio-active Compounds, Natural Bio-active Compounds: Volume 1: Production and Applications.* Springer Singapore, Singapore. https://doi.org/10.1007/978-981-13-7154-7

Al-Senani, G.M., Al-Kadhi, N., 2020. The synthesis and effect of silver nanoparticles on the adsorption of Cu2+ from aqueous solutions. *Appl. Sci.* 10(14). https://doi.org/10.3390/app10144840

Al-Snafi, A.E., 2015. Pharmacology and medicinal properties of Caesalpinia crista, an overview. *Int. J. Pharm.* 5, 71–83. https://doi.org/10.5281/zenodo.1214994

Al-Snafi, A.E., 2020. Clinically tested medicinal plants view project medical education view project phenolics and flavonoids contents of medicinal plants, as natural ingredients for many therapeutic purposes-A review. *IOSR J Pharm.* 10, 42–81.

Alwhibi, M.S., Ortashi, K.M.O., Hendi, A.A., Awad, M.A., Soliman, D.A., El-Zaidy, M., 2022. Green synthesis, characterization and biomedical potential of Ag@Au core–shell noble metal nanoparticles. *J. King Saud Univ. Sci.* 34(4), 102000. https://doi.org/10.1016/j.jksus.2022.102000

Anitha, J., Selvakumar, R., Hema, S., Murugan, K., Premkumar, T., 2022. Facile green synthesis of nano-sized ZnO using leaf extract of Morinda tinctoria: MCF-7 cell cycle arrest, antiproliferation, and apoptosis studies. *J. Ind. Eng. Chem.* 105, 520–529. https://doi.org/10.1016/j.jiec.2021.10.008

Babu, K.G.D., Singh, B., Joshi, V.P., Singh, V., 2002. Essential oil composition of damask rose (Rosa damascena Mill.) distilled under different pressures and temperatures. *Flavour Fragr. J.* 17(2), 136–140. https://doi.org/10.1002/ffj.1052

Bharti, Jangwan J.S., Kumar, S.S., Kumar, V., Kumar, A., Kumar, D., 2022. A review on the capability of zinc oxide and iron oxides nanomaterials, as a water decontaminating agent: adsorption and photocatalysis. *Appl. Water Sci.* 12(3), 1–14. https://doi.org/10.1007/s13201-021-01566-3

Bilia, A.R., Guccione, C., Isacchi, B., Righeschi, C., Firenzuoli, F., Bergonzi, M.C., 2021. Retraction: Essential oils loaded in nanosystems: A developing strategy for a successful therapeutic approach (Evidence-Based Complementary and Alternative Medicine (2014) 2014 (651593) DOI: 10.1155/2014/651593). *Evid. Based Complement. Alternat. Med.* 2021, 2088–2097. https://doi.org/10.1155/2021/7259208

Boussaada, O., Chemli, R., 2007. Seasonal variation of essential oil composition of Citrus Aurantium L. var. amara. *J. Essent. Oil Bear. Plants* 10(2), 109–120. https://doi.org/10.1080/0972060X.2007.10643528

Boutekedjiret, C., Bentahar, F., Belabbes, R., Bessiere, J.M., 2003. Extraction of rosemary essential oil by steam distillation and hydrodistillation. *Flavour Fragr. J.* 18(6), 481–484. https://doi.org/10.1002/ffj.1226

Danh, L.T., Truong, P., Mammucari, R., Foster, N., 2010. Extraction of vetiver essential oil by ethanol-modified supercritical carbon dioxide. *Chem. Eng. J.* 165(1), 26–34. https://doi.org/10.1016/j.cej.2010.08.048

De Groot, A., Schmidt, E., 2016. Essential oils, part V: Peppermint oil, lavender oil, and lemongrass oil. *Dermatitis* 27(6), 325–332. https://doi.org/10.1097/DER.0000000000000218

El-Borady, O.M., Fawzy, M., Hosny, M., 2021. Antioxidant, anticancer and enhanced photocatalytic potentials of gold nanoparticles biosynthesized by common reed leaf extract. *Appl. Nanosci.* https://doi.org/10.1007/s13204-021-01776-w

Eltaweil, A.S., Fawzy, M., Hosny, M., Abd El-Monaem, E.M., Tamer, T.M., Omer, A.M., 2022. Green synthesis of platinum nanoparticles using Atriplex halimus leaves for potential antimicrobial, antioxidant, and catalytic applications. *Arab. J. Chem.* 15(1), 103517. https://doi.org/10.1016/j.arabjc.2021.103517

Elyemni, M., Louaste, B., Nechad, I., Elkamli, T., Bouia, A., Taleb, M., Chaouch, M., Eloutassi, N., 2019. Extraction of essential oils of Rosmarinus officinalis L. by two different methods: Hydrodistillation and microwave assisted hydrodistillation. *Sci. World J.* 2019. https://doi.org/10.1155/2019/3659432

Engelberg, S., Lin, Y., Assaraf, Y.G., Livney, Y.D., 2021. Targeted nanoparticles harboring jasmine-oil-entrapped paclitaxel for elimination of lung cancer cells. *Int. J. Mol. Sci.* 22(3), 1–13. https://doi.org/10.3390/ijms22031019

Figoli, A., Donato, L., Carnevale, R., Tundis, R., Statti, G.A., Menichini, F., Drioli, E., 2006. Bergamot essential oil extraction by pervaporation. *Desalination* 193(1–3), 160–165. https://doi.org/10.1016/j.desal.2005.06.060

Fouladi-Fard, R., Aali, R., Mohammadi-Aghdam, S., Mortazavi-derazkola, S., 2022. The surface modification of spherical ZnO with Ag nanoparticles: A novel agent, biogenic synthesis, catalytic and antibacterial activities. *Arab. J. Chem.* 15(3), 103658. https://doi.org/10.1016/j.arabjc.2021.103658

González-Mahave, I., Lobesa, T., Del Pozo, M.D., Blasco, A., Venturini, M., 2006. Rosemary contact dermatitis and cross-reactivity with other labiate plants. *Contact Dermatitis* 54(4), 210–212. https://doi.org/10.1111/j.0105-1873.2006.00794.x

Guan, W., Li, S., Yan, R., Tang, S., Quan, C., 2007. Comparison of essential oils of clove buds extracted with supercritical carbon dioxide and other three traditional extraction methods. *Food Chem.* 101(4), 1558–1564. https://doi.org/10.1016/j.foodchem.2006.04.009

Hosny, M., Eltaweil, A.S., Mostafa, M., El-Badry, Y.A., Hussein, E.E., Omer, A.M., Fawzy, M., 2022a. Facile synthesis of gold nanoparticles for anticancer, antioxidant applications, and photocatalytic degradation of toxic organic pollutants. *ACS Omega* 7(3), 3121–3133. https://doi.org/10.1021/acsomega.1c06714

Hosny, M., Fawzy, M., 2021. Instantaneous phytosynthesis of gold nanoparticles via Persicaria salicifolia leaf extract, and their medical applications. *Adv. Powder Technol.* 32(8), 2891–2904. https://doi.org/10.1016/j.apt.2021.06.004

Hosny, M., Fawzy, M., Abdelfatah, A.M., Fawzy, E.E., Eltaweil, A.S., 2021. Comparative study on the potentialities of two halophytic species in the green synthesis of gold nanoparticles and their anticancer, antioxidant and catalytic efficiencies. *Adv. Powder Technol.* 32(9), 3220–3233. https://doi.org/10.1016/j.apt.2021.07.008

Hosny, M., Fawzy, M., El-Fakharany, E.M., Omer, A.M., El-Monaem, E.M.A., Khalifa, R.E., Eltaweil, A.S., 2022b. Biogenic synthesis, characterization, antimicrobial, antioxidant, antidiabetic, and catalytic applications of platinum nanoparticles synthesized from Polygonum salicifolium leaves. *J. Environ. Chem. Eng.* 10(1), 106806. https://doi.org/10.1016/j.jece.2021.106806

Jindapunnapat, K., Reetz, N.D., MacDonald, M.H., Bhagavathy, G., Chinnasri, B., Soonthornchareonnon, N., Sasnarukkit, A., Chauhan, K.R., Chitwood, D.J., Meyer, S.L.F., 2018. Activity of vetiver extracts and essential oil against Meloidogyne incognita. *J. Nematol.* 50(2), 147–162. https://doi.org/10.21307/jofnem-2018-008

Kaddu, S., Kerl, H., Wolf, P., 2001. Accidental bullous phototoxic reactions to bergamot aromatherapy oil. *J. Am. Acad. Dermatol.* 45(3), 458–461. https://doi.org/10.1067/mjd.2001.116226

Kaplan, A., 2022. The nanocomposites designs of phytomolecules from medicinal and aromatic plants: Promising anticancer-antiviral applications. *Beni Suef Univ. J. Basic Appl. Sci.* 11(1), 17. https://doi.org/10.1186/s43088-022-00198-z

Karimirad, R., Behnamian, M., Dezhsetan, S., 2018. Development and characterization of Nano biopolymer containing cumin oil as a new approach to enhance antioxidant properties of button mushroom. *Int. J. Biol. Macromol.* 113, 662–668. https://doi.org/10.1016/j.ijbiomac.2018.02.043

Kiec-Swierczynska, M., Krecisz, B., Chomiczewska, D., Swierczynska-Machura, D., Palczynski, C., 2010. Occupational allergic contact dermatitis caused by basil (Ocimum basilicum). *Contact Dermatitis* 63(6), 365–367. https://doi.org/10.1111/j.1600-0536.2010.01821.x

Kocak, Y., Oto, G., Meydan, I., Seckin, H., Gur, T., Aygun, A., Sen, F., 2022. Assessment of therapeutic potential of silver nanoparticles synthesized by ferula Pseudalliacea Rech. F. Plant. *Inorg. Chem. Commun.* 140, 109417. https://doi.org/10.1016/j.inoche.2022.109417

Kralova, K., Jampilek, J., 2021. Responses of medicinal and aromatic plants to engineered nanoparticles. *Appl. Sci.* 11(4), 1813. https://doi.org/10.3390/app11041813

Kumari, P., Luqman, S., Meena, A., 2019. Application of the combinatorial approaches of medicinal and aromatic plants with nanotechnology and its impacts on healthcare. *DARU, J. Pharm. Sci.* 27(1), 475–489. https://doi.org/10.1007/s40199-019-00271-6

Kumari, P., Meena, A., 2020. Green synthesis of gold nanoparticles from Lawsoniainermis and its catalytic activities following the Langmuir-Hinshelwood mechanism. *Colloids Surf. A Physicochem. Eng. Asp.* 606, 125447. https://doi.org/10.1016/j.colsurfa.2020.125447

Kusuma, H.S., Altway, A., Mahfud, M., 2017. Alternative to conventional extraction of vetiver oil: Microwave hydrodistillation of essential oil from vetiver roots (Vetiveria zizanioides). *IOP Conf. Ser. Earth Environ. Sci.* 101, 012015. https://doi.org/10.1088/1755-1315/101/1/012015

Kusuma, H.S., Mahfud, M., 2018. Kinetic studies on extraction of essential oil from sandalwood (Santalum album) by microwave air-hydrodistillation method. *Alex. Eng. J.* 57(2), 1163–1172. https://doi.org/10.1016/j.aej.2017.02.007

Kusuma, H.S., Mahfud, M., 2017. The extraction of essential oils from patchouli leaves (Pogostemon cablin benth) using a microwave air-hydrodistillation method as a new green technique. *RSC Adv.* 7(3), 1336–1347. https://doi.org/10.1039/c6ra25894h

Lešnik, S., Furlan, V., Bren, U., 2021. Rosemary (Rosmarinus officinalis L.): Extraction techniques, analytical methods and health-promoting biological effects. *Phytochem. Rev.*. https://doi.org/10.1007/s11101-021-09745-5

Lo, C.M., Han, J., Wong, E.S.W., 2020. Chemistry in aromatherapy – Extraction and analysis of essential oils from plants of Chamomilla recutita, Cymbopogon nardus, Jasminum officinale and Pelargonium graveolens. *Biomed. Pharmacol. J.* 13(3), 1339–1350. https://doi.org/10.13005/bpj/2003

Lubbe, A., Verpoorte, R., 2011. Cultivation of medicinal and aromatic plants for specialty industrial materials. *Ind. Crops Prod.* 34(1), 785–801. https://doi.org/10.1016/j.indcrop.2011.01.019

Malyan, S.K., Singh, S., Bachheti, A., Chahar, M., Sah, M.K., Narender, Y., Kumar,A., Yadav, A.N., Kumar, S.S., 2020. Cyanobacteria: A perspective paradigm for agriculture and environment. In Rastegari, A.A., Yadav, A.N., Yadav, N., Awasthi, A.K. (Eds.), *New and Future Developments in Microbial Biotechnology and Bioengineering*. Elsevier, pp. 215–224. https://doi.org/10.1016/b978-0-12-820526-6.00014-2

Malyan, S.K., Yadav, S., Sonkar, V., Goyal, V.C., Singh, O., Singh, R., 2021. Mechanistic understanding of the pollutant removal and transformation processes in the constructed wetland system. *Water Environ. Res.* 93(10), 1882–1909. https://doi.org/10.1002/wer.1599

Manouchehri, R., Saharkhiz, M.J., Karami, A., Niakousari, M., 2018. Extraction of essential oils from damask rose using green and conventional techniques: Microwave and ohmic assisted hydrodistillation versus hydrodistillation. *Sustain. Chem. Pharm.* 8, 76–81. https://doi.org/10.1016/j.scp.2018.03.002

Martínez-González, M.C., Goday Buján, J.J., Martínez Gómez, W., Fonseca Capdevila, E., 2007. Concomitant allergic contact dermatitis due to Rosmarinus officinalis (rosemary) and Thymus vulgaris (thyme). *Contact Dermatitis* 56(1), 49–50. https://doi.org/10.1111/j.1600-0536.2007.00951.x

Mohanasundari, C., Anbalagan, S., Srinivasan, K., Narayanan, M., Saravanan, M., Alharbi, S.A., Salmen, S.H., Nhung, T.C., Pugazhendhi, A., 2022. Antibacterial activity potential of leaf extracts of Blepharis maderaspatensis and Ziziphus oenoplia against antibiotics resistant Pseudomonas strains isolated from pus specimens. *Process Biochem.* 249, 105596. https://doi.org/10.1016/j.procbio.2022.04.008

Naghizadeh, A., Mizwari, Z.M., Ghoreishi, S.M., Lashgari, S., Mortazavi-Derazkola, S., Rezaie, B., 2021. Biogenic and eco-benign synthesis of silver nanoparticles using jujube core extract and its performance in catalytic and pharmaceutical applications: Removal of industrial contaminants and in-vitro antibacterial and anticancer activities. *Environ. Technol. Innov.* 23, 101560. https://doi.org/10.1016/j.eti.2021.101560

Nair, A., Mallya, R., Suvarna, V., Khan, T.A., Momin, M., Omri, A., 2022. Nanoparticles—attractive carriers of antimicrobial essential oils. *Antibiotics* 11(1), 108. https://doi.org/10.3390/antibiotics11010108

Nautiyal, O.H., 2014. Process optimization of sandalwood (Santalum album) oil extraction by subcritical carbon dioxide and conventional techniques. *Indian J. Chem. Technol.* 21, 290–297.

Navarra, M., Mannucci, C., Delbò, M., Calapai, G., 2015. Citrus bergamia essential oil: from basic research to clinical application. *Front. Pharmacol.* 6. https://doi.org/10.3389/fphar.2015.00036

Orchard, A., Van Vuuren, S., 2017. Commercial essential oils as potential antimicrobials to treat skin diseases. *Evid. Based Complement. Alternat. Med.* 2017. https://doi.org/10.1155/2017/4517971

Oualdi, I., Brahmi, F., Mokhtari, O., Abdellaoui, S., Tahani, A., Oussaid, A., 2021. Rosmarinus officinalis from Morocco, Italy and France: Insight into chemical compositions and biological properties. *Mater. Today Proc.* 45, 7706–7710. https://doi.org/10.1016/j.matpr.2021.03.333

Oves, M., Aslam, M., Rauf, M.A., Qayyum, S., Qari, H.A., Khan, M.S., Alam, M.Z., Tabrez, S., Pugazhendhi, A., Ismail, I.M.I., 2018. Antimicrobial and anticancer activities of silver nanoparticles synthesized from the root hair extract of Phoenix dactylifera. *Mater. Sci. Eng. C* 89, 429–443. https://doi.org/10.1016/j.msec.2018.03.035

Pasias, I.N., Ntakoulas, D.D., Raptopoulou, K., Gardeli, C., Proestos, C., 2021. Chemical composition of essential oils of aromatic and medicinal herbs cultivated in Greece—Benefits and drawbacks. *Foods* 10(10), 2354. https://doi.org/10.3390/foods10102354

Pieracci, Y., Ciccarelli, D., Giovanelli, S., Pistelli, L., Flamini, G., Cervelli, C., Mancianti, F., Nardoni, S., Bertelloni, F., Ebani, V.V., 2021. Antimicrobial activity and composition of five Rosmarinus (Now Salvia spp. and varieties) essential oils. *Antibiotics* 10(9). https://doi.org/10.3390/antibiotics10091090

Pinto, T., Aires, A., Cosme, F., Bacelar, E., Morais, M.C., Oliveira, I., Ferreira-Cardoso, J., Anjos, R., Vilela, A., Gonçalves, B., 2021. Bioactive (poly)phenols, volatile compounds from vegetables, medicinal and aromatic plants. *Foods* 10(1), 106. https://doi.org/10.3390/foods10010106

Raghavendra, V.B., Shankar, S., Govindappa, M., Pugazhendhi, A., Sharma, M., Nayaka, S.C., 2022. Green synthesis of zinc oxide nanoparticles (ZnO NPs) for effective degradation of dye, polyethylene and antibacterial performance in waste water treatment. *J. Inorg. Organomet. Polym. Mater.* 32(2), 614–630. https://doi.org/10.1007/s10904-021-02142-7

Ratri, P.J., Ayurini, M., Khumaini, K., Rohbiya, A., 2020. Clove oil extraction by steam distillation and utilization of clove buds waste as potential candidate for eco-friendly packaging. *J. Bahan Alam Terbarukan* 9(1), 47–54. https://doi.org/10.15294/jbat.v9i1.24935

Raut, J.S., Karuppayil, S.M., 2014. A status review on the medicinal properties of essential oils. *Ind. Crops Prod.* 62, 250–264. https://doi.org/10.1016/j.indcrop.2014.05.055

Reddy, D.N., 2019. Essential oils extracted from medicinal plants and their applications. *Natural Bio-active Compounds: Volume 1: Production and Applications*, 237–283.

Rehman, K. ur, Khan, S.U., Tahir, K., Zaman, U., Khan, D., Nazir, S., Khan, W.U., Khan, M.I., Ullah, K., Anjum, S.I., Bibi, R., 2022. Sustainable and green synthesis of novel acid phosphatase mediated platinum nanoparticles (ACP-PtNPs) and investigation of its in vitro antibacterial, antioxidant, hemolysis and photocatalytic activities. *J. Environ. Chem. Eng.* 107623. https://doi.org/10.1016/j.jece.2022.107623

Saha, A., Basak, B.B., Manivel, P., Kumar, J., 2021. Valorization of Java citronella (Cymbopogon winterianus Jowitt) distillation waste as a potential source of phenolics/antioxidant: Influence of extraction solvents. *J. Food Sci. Technol.* 58(1), 255–266. https://doi.org/10.1007/s13197-020-04538-8

Sallam, S.A., El-Subruiti, G.M., Eltaweil, A.S., 2018. Facile synthesis of Ag–γ-Fe2O3 superior nanocomposite for catalytic reduction of nitroaromatic compounds and catalytic degradation of methyl orange. *Catal. Lett.* 148(12), 3701–3714. https://doi.org/10.1007/s10562-018-2569-z

Salvino, R.A., Aroulanda, C., De Filpo, G., Celebre, G., De Luca, G., 2022. Metabolic composition and authenticity evaluation of bergamot essential oil assessed by nuclear magnetic resonance spectroscopy. *Anal. Bioanal. Chem.* 414(6), 2297–2313. https://doi.org/10.1007/s00216-021-03869-5

Samadi, Z., Jannati, Y., Hamidia, A., Mohammadpour, R.A., Hesamzadeh, A., 2021. The effect of aromatherapy with lavender essential oil on sleep quality in patients with major depression. *J. Nurs. Midwif. Sci.* 8(2), 67–73. https://doi.org/10.4103/JNMS.JNMS_26_20

Sarkic, A., Stappen, I., 2018. Essential oils and their single compounds in cosmetics-a critical review. *Cosmetics* 5(1), 1–21. https://doi.org/10.3390/cosmetics5010011

Seckin, H., Tiri, R.N.E., Meydan, I., Aygun, A., Gunduz, M.K., Sen, F., 2022. An environmental approach for the photodegradation of toxic pollutants from wastewater using Pt–Pd nanoparticles: Antioxidant, antibacterial and lipid peroxidation inhibition applications. *Environ. Res.* 208, 112708. https://doi.org/10.1016/j.envres.2022.112708

Shah, K.A., Bhatt, D.R., Desai, M.A., Jadeja, G.C., Parikh, J.K., 2017. Extraction of essential oil from patchouli leaves using hydrodistillation: Parametric studies and optimization. *Indian J. Chem. Technol.* 24, 405–410.

Sharma, Y., Schaefer, J., Streicher, C., Stimson, J., Fagan, J., 2020. Qualitative analysis of essential oil from French and Italian varieties of rosemary (Rosmarinus officinalis L.) grown in the Midwestern United States. *Anal. Chem. Lett.* 10(1), 104–112. https://doi.org/10.1080/22297928.2020.1720805

Sindle, A., Martin, K., 2021. Art of prevention: Essential oils - Natural products not necessarily safe. *Int. J. Womens Dermatol.* 7(3), 304–308. https://doi.org/10.1016/j.ijwd.2020.10.013

Sinico, C., De Logu, A., Lai, F., Valenti, D., Manconi, M., Loy, G., Bonsignore, L., Fadda, A.M., 2005. Liposomal incorporation of Artemisia arborescens L. essential oil and in vitro antiviral activity. *Eur. J. Pharm. Biopharm.* 59(1), 161–168. https://doi.org/10.1016/j.ejpb.2004.06.005

Thanh, N.C., Pugazhendhi, A., Chinnathambi, A., Alharbi, S.A., Subramani, B., Brindhadevi, K., Whangchai, N., Pikulkaew, S., 2022. Silver nanoparticles (AgNPs) fabricating potential of aqueous shoot extract of Aristolochia bracteolata and assessed their antioxidant efficiency. *Environ. Res.* 208, 112683. https://doi.org/10.1016/j.envres.2022.112683

Valussi, M., Donelli, D., Firenzuoli, F., Antonelli, M., 2021a. Bergamot oil: Botany, production, pharmacology. *Encyclopedia* 1(1), 152–176. https://doi.org/10.3390/encyclopedia1010016

Valussi, M., Donelli, D., Firenzuoli, F., Antonelli, M., 2021b. Bergamot oil: Botany, production, pharmacology. *Encyclopedia* 1(1), 152–176. https://doi.org/10.3390/encyclopedia1010016

Vasantharaj, S., Sathiyavimal, S., Senthilkumar, P., Kalpana, V.N., Rajalakshmi, G., Alsehli, M., Elfasakhany, A., Pugazhendhi, A., 2021a. Enhanced photocatalytic degradation of water pollutants using bio-green synthesis of zinc oxide nanoparticles (ZnO NPs). *J. Environ. Chem. Eng.* 9(4), 105772. https://doi.org/10.1016/j.jece.2021.105772

Vasantharaj, S., Sathiyavimal, S., Senthilkumar, P., LewisOscar, F., Pugazhendhi, A., 2019. Biosynthesis of iron oxide nanoparticles using leaf extract of Ruellia tuberosa: Antimicrobial properties and their applications in photocatalytic degradation. *J. Photochem. Photobiol. B Biol.* 192, 74–82. https://doi.org/10.1016/j.jphotobiol.2018.12.025

Vasantharaj, S., Shivakumar, P., Sathiyavimal, S., Senthilkumar, P., Vijayaram, S., Shanmugavel, M., Pugazhendhi, A., 2021b. Antibacterial activity and photocatalytic dye degradation of copper oxide nanoparticles (CuONPs) using Justicia Gendarussa. *Appl. Nanosci.* https://doi.org/10.1007/s13204-021-01939-9

Wany, A., Kumar, A., Nallapeta, S., Jha, S., Nigam, V.K., Pandey, D.M., 2014. Extraction and characterization of essential oil components based on geraniol and citronellol from Java citronella (Cymbopogon winterianus Jowitt). *Plant Growth Regul.* 73(2), 133–145. https://doi.org/10.1007/s10725-013-9875-7

Wu, H., Li, J., Jia, Y., Xiao, Z., Li, P., Xie, Y., Zhang, A., Liu, R., Ren, Z., Zhao, M., Zeng, C., Li, C., 2019. Essential oil extracted from Cymbopogon citronella leaves by supercritical carbon dioxide: Antioxidant and antimicrobial activities. *J. Anal. Methods Chem.* 2019, 1–10. https://doi.org/10.1155/2019/8192439

Xin-Hua, Z., Silva, J.A.T. da, Yong-Xia, J., Jian, Y., Guo-Hua, M., 2012. Essential oils composition from roots of Santalum album l. *J. Essent. Oil Bear. Plants* 15(1), 1–6. https://doi.org/10.1080/0972060X.2012.10644011

Younis, A., Khan, M., Khan, A., Riaz, A., Pervez, M., 2007. Effect of different extraction methods on yield and quality of essential oil from four Rosa species. *Floric Ornam Biotechnol.* 1, 73–76.

Zhang, Q., Li-Gen, L., Wen-Cai, Y., 2018. Techniques for extraction and isolation of natural products: A comprehensive review. *Chin. Med.* 13(1), 1–26.

Zhang, X.H., Teixeira Da Silva, J.A., Jia, Y.X., Zhao, J.T., Ma, G.H., 2012. Chemical composition of volatile oils from the pericarps of Indian sandalwood (Santalum album) by different extraction methods. *Nat. Prod. Commun.* 7(1), 93–96. https://doi.org/10.1177/1934578x1200700132

Zulfiqar, H., Zafar, A., Rasheed, M.N., Ali, Z., Mehmood, K., Mazher, A., Hasan, M., Mahmood, N., 2019. Synthesis of silver nanoparticles using: Fagonia cretica and their antimicrobial activities. *Nanoscale Adv.* https://doi.org/10.1039/c8na00343b

Index

G

H

I

J

K

L

M

N

O

P

Q

R

S

T

U

V

W

X

Z

For Product Safety Concerns and Information please contact our EU representative GPSR@taylorandfrancis.com
Taylor & Francis Verlag GmbH, Kaufingerstraße 24, 80331 München, Germany

www.ingramcontent.com/pod-product-compliance
Lightning Source LLC
Chambersburg PA
CBHW080922220326
41598CB00034B/5649

* 9 7 8 1 0 3 2 1 0 0 9 9 9 *